An Introduction to Applied Multivariate Statistics

An Introduction to Applied Multivariate Statistics

M. S. Srivastava
University of Toronto

and

E. M. Carter
University of Guelph

North-Holland
New York · Amsterdam · Oxford

Elsevier Science Publishing Co., Inc.
52 Vanderbilt Avenue, New York, NY 10017

Distributors outside the United States and Canada:

Elsevier Science Publishers B. V.
Post Office Box 211, 1000 AE Amsterdam, The Netherlands

Library of Congress Cataloging in Publication Data

Srivastava, M. S.
An introduction to applied multivariate statistics.

Bibliography: p.
Includes index.
1. Multivariate analysis. I. Carter, E. M. (Edward M.)
II. Title.
QA278.S687 519.5′35 82-2385
ISBN 0-444-00621-4 AACR2

Manufactured in the United States of America

Contents

Preface

This book is based on lectures given by the authors to first-year graduate students at the University of Toronto and the University of Guelph. Our primary aim is to teach those multivariate techniques applicable to the data available in such varied disciplines as forestry, biology, medicine, and education. Theoretical details have accordingly been kept to a bare minimum. A knowledge of matrix notation and manipulations will be helpful, and Chapter 1 should assist readers deficient in this area. The reader is, however, expected to have sufficient knowledge of elementary univariate theory.

Our emphasis is on methods in current use in multivariate statistics. For each new topic we present not only a description of the problem and its solution, but also several worked examples, chosen from many different fields. Each chapter ends with a discussion of the computer packages available and additional worked examples.

The likelihood ratio approach has been adopted for tests of the significance of a given hypothesis, and Roy's union–intersection principle and Bonferoni's inequalities for confidence intervals are introduced. For each test statistic, formulas for the observed significance levels are given. The percentage points needed to calculate confidence intervals appear as Appendixes.

Chapters 1 and 2 provide a review of the necessary matrix theory and statistical theory; Chapter 1 also contains a discussion of the SAS matrix procedures for calculating eigenvalues, eigenvectors, and the generalized inverse of a matrix. Chapters 3–7 are multivariate generalizations of univariate procedures for t-tests, analysis of variance, and multiple regression.

Chapter 12 gives tests of the assumptions required for these multivariate procedures to be valid, including tests for equality of covariance and independence. It is Chapters 8–11 that contain strictly multivariate procedures, beginning with discriminant analysis. Methods for finding functions that will discriminate among populations or groups are discussed, and procedures to reduce the number of characteristics necessary for discrimination are given in a section on stepwise discriminant analysis. Chapters 9–11 then introduce dimension-reducing procedures, including canonical correlation, which investigates the correlation between linear combinations of variables, and principal component analysis, which reduces the set of measured characteristics to fewer components. For example, 20 measurements on bulls might be reduced to two or three components for size and shape. Factor analysis performs the same reduction but assuming that the observations have an underlying structure; this method is often used to group responses to questionnaires and psychological tests.

Advanced topics are given toward the end of a chapter. For example, Chapter 3 covers the problem of incomplete data and tests for shift in the mean. The core material may be used in a one-semester course in applied multivariate statistics at the senior or first-year graduate level. Two semesters are required to cover the text in its entirety.

This book is dedicated to Jagdish Bahadur Srivastava and Ivy May Carter.

An Introduction to Applied Multivariate Statistics

Chapter 1
Some Resul

1.1 Notation and Definitions

Suppose that $a_{11}, a_{12}, \ldots, a_{pq}$ is a collection of pq real numbers. The rectangular array of these elements

$$\begin{pmatrix} a_{11} & a_{12} & \cdots & a_{1q} \\ a_{21} & a_{22} & \cdots & a_{2q} \\ \vdots & \vdots & & \vdots \\ a_{p1} & a_{p2} & \cdots & a_{pq} \end{pmatrix},$$

consisting of p rows and q columns, is called a $p \times q$ matrix. We write $A = (a_{ij})$: $p \times q$, where a_{ij} is the element in the ith row and jth column. For example, the matrix

$$A = \begin{pmatrix} 6 & 8 & 9 \\ 1 & 3 & 5 \end{pmatrix}$$

is a 2×3 matrix with $a_{11} = 6$, $a_{12} = 8$, $a_{13} = 9$, $a_{21} = 1$, $a_{22} = 3$, $a_{23} = 5$.

We shall now define some special matrices that will be used later.

Null Matrix If all the elements of A are zero, then A is said to be a zero or null matrix, denoted 0_{pq} or, if there is no confusion, simply 0.

That is,

$$0_{32} = \begin{pmatrix} 0 & 0 \\ 0 & 0 \\ 0 & 0 \end{pmatrix}.$$

Square Matrix If $p = q$, then A is said to be a square matrix of order p.

Column Vector If $q = 1$, then A is said to be a p-column vector, or simply a p-vector, and the vector will be written

$$\mathbf{a} = \begin{pmatrix} a_1 \\ \vdots \\ a_p \end{pmatrix}.$$

For example, the 3-vector with entries 6, 8, and 9 is written

$$\mathbf{a} = \begin{pmatrix} 6 \\ 8 \\ 9 \end{pmatrix}.$$

Row Vector If $p = 1$, then A is said to be a q-row vector. A will be written $\mathbf{a}' = (a_1, \ldots, a_q)$.

For example, the 4-row vector $\mathbf{a}'$ with entries 2, 4, 6, and 8 is written

$$\mathbf{a}' = (2, 4, 6, 8).$$

Lower Triangular Matrix A square matrix with all elements above the main diagonal equal to zero is called a lower triangular matrix.

Examples are

$$\begin{pmatrix} 2 & 0 & 0 \\ 8 & 4 & 0 \\ 10 & 9 & 6 \end{pmatrix}, \quad \begin{pmatrix} 1 & 0 \\ 2 & 3 \end{pmatrix}, \quad \begin{pmatrix} 2 & 0 & 0 \\ 0 & 4 & 0 \\ 0 & 5 & 0 \end{pmatrix}.$$

Upper Triangular Matrix A square matrix with all elements below the main diagonal equal to zero is called an upper triangular matrix.

Diagonal Matrix A square matrix A with the off-diagonal elements equal to zero is called a diagonal matrix. If the diagonal entries are $a_1, \ldots, a_p$, then A is sometimes written D_a or $\operatorname{diag}(a_1, \ldots, a_p)$.

For example, if $a_1 = 1$, $a_2 = 2$, and $a_3 = 4$, then

$$A = \begin{pmatrix} 1 & 0 & 0 \\ 0 & 2 & 0 \\ 0 & 0 & 4 \end{pmatrix}.$$

Identity Matrix If A is a diagonal $p \times p$ matrix with all p diagonal entries equal to 1, then A is called an identity matrix, denoted I_p.

For example,

$$I_1=[1], \quad I_2=\begin{pmatrix}1 & 0\\ 0 & 1\end{pmatrix}, \quad I_3=\begin{pmatrix}1 & 0 & 0\\ 0 & 1 & 0\\ 0 & 0 & 1\end{pmatrix}.$$

Transpose Matrix If the rows and columns of a matrix are interchanged, the resulting matrix is called the transpose of A, denoted A'. Thus if $A=(a_{ij})\colon p\times q$, then $A'=(a_{ji})\colon q\times p$.

For example, if

$$A=\begin{pmatrix}1 & 2 & 3\\ 4 & 5 & 6\end{pmatrix}$$

then

$$A'=\begin{pmatrix}1 & 4\\ 2 & 5\\ 3 & 6\end{pmatrix}.$$

Symmetric Matrix A square matrix is said to be symmetric if $A=A'$.

Examples of symmetric matrices are

$$\begin{pmatrix}1 & 3\\ 3 & 2\end{pmatrix}, \quad \begin{pmatrix}1 & 0\\ 0 & 2\end{pmatrix}, \quad \begin{pmatrix}1 & 2 & 3\\ 2 & 4 & 5\\ 3 & 5 & 6\end{pmatrix}.$$

Skew Symmetric Matrix A matrix A is said to be skew symmetric if $A=-A'$. If A is skew symmetric, then all the diagonal entries are zero.

Examples are

$$\begin{pmatrix}0 & 1\\ -1 & 0\end{pmatrix}, \quad \begin{pmatrix}0 & 0 & 1\\ 0 & 0 & 2\\ -1 & -2 & 0\end{pmatrix}, \quad \begin{pmatrix}0 & -1 & -2\\ 1 & 0 & 3\\ 2 & -3 & 0\end{pmatrix}.$$

1.2 Matrix Operations

Sometimes, it is convenient to represent a matrix A as

$$A=\left(\begin{array}{c|c}A_{11} & A_{12}\\ \hline A_{21} & A_{22}\end{array}\right),$$

where A_{ij} is an $m_i\times m_j$ submatrix of A $(i, j=1,2)$. A submatrix of A is a matrix obtained from A by deleting certain rows and columns. The above representation of A is called a partitioned matrix. If there are more than two partitions of rows (or columns), we can write $A=(A_{ij})$, where A_{ij} is the submatrix of A in the ith row and jth column partition. For example, if

$$A_{11}=\begin{pmatrix}1&2\\3&4\end{pmatrix},\qquad A_{12}=\begin{pmatrix}2\\3\end{pmatrix},\qquad A_{21}=(6,7),\qquad A_{22}=(8),$$

then

$$A=\left(\begin{array}{cc|c}1&2&2\\3&4&3\\\hline 6&7&8\end{array}\right).$$

Another type of partitioned matrix can be written

$$A=\left(\begin{array}{c|c}A_{11}&\\\cline{1-1}A_{21}&A_2\end{array}\right).$$

For example, if

$$A_{11}=\begin{pmatrix}1&2\\3&4\end{pmatrix},\qquad A_{21}=(6,7),\qquad A_2=\begin{pmatrix}2\\3\\8\end{pmatrix},$$

then

$$A=\left(\begin{array}{cc|c}1&2&2\\3&4&3\\\cline{1-2}6&7&8\end{array}\right).$$

Let $A:m\times n$ and $B:p\times q$ be two matrices. The usual addition of two matrices is defined only if the matrices are of the same order. Thus if $A=(a_{ij}):p\times q$ and $B=(b_{ij}):p\times q$, then their *sum* is defined by

$$A+B=(a_{ij}+b_{ij}):p\times q.$$

The usual multiplication of A by B is defined only if the number of columns of A is equal to the number of rows of B. Thus if $A=(a_{ij}):m\times n$ and $B=(b_{ij}):n\times q$, then the product of A and B (denoted AB) is defined as an $m\times q$ matrix $AB=C=(c_{ij})$, where

$$c_{ij}=\sum_{k=1}^{n}a_{ik}b_{kj},\qquad i=1,2,\ldots,m,\quad j=1,2,\ldots,q.$$

The following results can be obtained from the above definitions:

i. $(A+B)'=A'+B'$, $(A+B+C)'=A'+B'+C'$;
ii. $(AB)'=B'A'$, $(ABC)'=C'B'A'$;
iii. $A(B_1+B_2)=AB_1+AB_2$; and
iv. $\Sigma_{\alpha=1}^{k}AB_\alpha=A(\Sigma_{\alpha=1}^{k}B_\alpha)$.

Examples Let

$$A=\begin{pmatrix}1&2&3\\4&5&6\end{pmatrix},\qquad B=\begin{pmatrix}7&8&9\\10&11&12\end{pmatrix},\qquad C=\begin{pmatrix}3&1\\1&2\end{pmatrix}.$$

Then

1. $A' = \begin{pmatrix} 1 & 4 \\ 2 & 5 \\ 3 & 6 \end{pmatrix}$, $B' = \begin{pmatrix} 7 & 10 \\ 8 & 11 \\ 9 & 12 \end{pmatrix}$;
2. $A+B = \begin{pmatrix} 8 & 10 & 12 \\ 14 & 16 & 18 \end{pmatrix}$;
3. $(A+B)' = A' + B' = \begin{pmatrix} 8 & 14 \\ 10 & 16 \\ 12 & 18 \end{pmatrix}$;
4. $CA = \begin{pmatrix} 3 & 1 \\ 1 & 2 \end{pmatrix}\begin{pmatrix} 1 & 2 & 3 \\ 4 & 5 & 6 \end{pmatrix} = \begin{pmatrix} 7 & 11 & 15 \\ 9 & 12 & 15 \end{pmatrix}$.

Note that AC is not defined.

We now define a few more matrices.

Semiorthogonal Matrix A matrix $A: p \times q$ is said to be semiorthogonal if $AA' = I_p$ $(q \geq p)$.

Examples are

$$(0.5, 0.5, 0.5, 0.5), \qquad \begin{pmatrix} 1 & 0 & 0 \\ 0 & 1 & 0 \end{pmatrix}, \qquad \begin{pmatrix} 0 & 0 & -1 \\ 1 & 0 & 0 \end{pmatrix}.$$

Orthogonal Matrix A square matrix A is said to be orthogonal if $AA' = I_p$.

Idempotent Matrix A square matrix A is said to be idempotent if $A = A^2$.

Examples are

$$\begin{pmatrix} 1 & 0 \\ 0 & 1 \end{pmatrix}, \qquad \begin{pmatrix} 1 & 0 \\ 0 & 0 \end{pmatrix}, \qquad \begin{pmatrix} \frac{1}{2} & \frac{1}{2} \\ \frac{1}{2} & \frac{1}{2} \end{pmatrix}.$$

Kronecker Product Let $A: m \times n$ and $B: p \times q$ be two matrices. Then the Kronecker product (or direct product) of A and B is defined as the $mp \times nq$ matrix $A \otimes B = (a_{ij}B)$, where $A = (a_{ij})$.

For example, for

$$A = \begin{pmatrix} 4 & 2 \\ 3 & 1 \end{pmatrix} \qquad \text{and} \qquad B = \begin{pmatrix} 1 & 0 \\ 0 & 1 \end{pmatrix}$$

then

$$A \otimes B = \left(\begin{array}{cc|cc} 4 & 0 & 2 & 0 \\ 0 & 4 & 0 & 2 \\ \hline 3 & 0 & 1 & 0 \\ 0 & 3 & 0 & 1 \end{array}\right).$$

1.3 Determinants

The determinant of a square matrix $A=(a_{ij}): n\times n$ is defined as

$$|A| \equiv \sum_{\boldsymbol{\alpha}} (-1)^{N(\boldsymbol{\alpha})} \prod_{j=1}^{n} a_{\alpha_j, j},$$

where $\Sigma_{\boldsymbol{\alpha}}$ denotes the summation over the distinct permutations $\boldsymbol{\alpha}$ of the numbers $1,2,\ldots,n$ and $N(\boldsymbol{\alpha})$ is the total number of inversions of a permutation. An inversion of a permutation $\boldsymbol{\alpha}=(\alpha_1,\alpha_2,\ldots,\alpha_n)$ is an arrangement of two indices such that the larger index comes after the smaller index. For example, $N(2,1,4,3)=1+N(1,2,4,3)=2+N(1,2,3,4)=2$ because $N(1,2,3,4)=0$. Similarly, $N(4,3,1,2)=1+N(3,4,1,2)=3+N(3,1,2,4)=5$. The above expression of a determinant is denoted $|A|$ or $\det A$. If $|A|$ is real, then $|A|_+$ denotes the positive value of $|A|$. Note that

$$|A'| = \sum_{\boldsymbol{\alpha}} (-1)^{N(\boldsymbol{\alpha})} \prod a_{j,\alpha_j} = \sum_{\boldsymbol{\alpha}} (-1)^{N(\boldsymbol{\alpha})} \prod a_{\alpha_j, j} = |A|.$$

Examples

1. $\begin{vmatrix} 2 & 1 \\ 1 & 5 \end{vmatrix} = 2\times 5 - 1\times 1 = 9,$
2. $\begin{vmatrix} 2 & 1 \\ 1 & 2 \end{vmatrix} = 4-1=3.$

The following are immediate consequences of the definition of $|A|$.

i. If the ith row (or column) is multiplied by a constant c, the value of the determinant is multiplied by c. Hence $|cA|=c^n|A|$ if A is an $n\times n$ matrix.
ii. If any two rows (or columns) of a matrix are interchanged, the sign of the determinant is changed. Hence if two rows (or columns) of a matrix are identical, the value of the determinant is zero.
iii. The value of the determinant is unchanged if in the ith row (or column) a cth multiple of the jth row (or column) is added. Hence the value of the determinant is zero if a row (or column) is a linear combination of other rows (or columns).
iv. $|I_n|=1$, $|D_a|=\prod a_i$.
v. $|AB|=|A||B|$ if $A: p\times p$ and $B: p\times p$,
vi. $|AA'|\geq 0$.
vii.

$$\begin{vmatrix} I & 0 \\ C & A \end{vmatrix} = |A|, \qquad \begin{vmatrix} A & C \\ 0 & I \end{vmatrix} = |A|, \qquad \begin{vmatrix} A & C \\ 0 & B \end{vmatrix} = \begin{vmatrix} A & 0 \\ C & B \end{vmatrix} = |A||B|,$$

where A and B are square matrices.
viii. $|I_p+AB|=|I_q+BA|$, where A and B are $p\times q$ and $q\times p$ matrices.

1.3.1 Cofactors of a Square Matrix

Let A_j^i be the submatrix of $A: n \times n$ obtained by deleting the ith row and the jth column of A. Let $m_{ij} = |A_j^i|$. Then $c_{ij}(A) = (-1)^{i+j}|A_j^i| = (-1)^{i+j}m_{ij}$ is known as the *cofactor* of a_{ij}. Note that

$$|A| = \sum_{j=1}^{n} c_{ij}(A)a_{ij} = \sum_{i=1}^{n} c_{ij}(A)a_{ij},$$

$$0 = \sum_{j=1}^{n} c_{ij}a_{kj}, \quad k \neq i, \qquad 0 = \sum_{i=1}^{n} c_{ij}a_{ik}, \quad j \neq k.$$

Example Let

$$A = \begin{pmatrix} 1 & 6 & 5 & 7 \\ 6 & 9 & 10 & 12 \\ 3 & 7 & 8 & 10 \\ 2 & 5 & 9 & 11 \end{pmatrix},$$

and let $i = 2$ and $j = 3$. Then

$$A_3^2 = \begin{pmatrix} 1 & 6 & 7 \\ 3 & 7 & 10 \\ 2 & 5 & 11 \end{pmatrix},$$

and $c_{23}(A) = (-1)^{2+3}|A_3^2|$.

1.3.2 Minor, Principal Minor, and Trace of a Matrix

Let $A_{(j_1,\ldots,j_t)}^{(i_1,\ldots,i_t)}$ be the submatrix of $A: m \times n$ obtained by taking the $i_1, i_2, \ldots, i_t$ rows and $j_1, j_2, \ldots, j_t$ columns; note that it is a square submatrix of A. Then $|A_{(j_1,\ldots,j_t)}^{(i_1,\ldots,i_t)}|$ is know as a *minor* of order t. If $i_1 = j_1, i_2 = j_2, \ldots, i_t = j_t$, then it is known as a *principal minor* of order t.

For any square matrix, the sum of all principal minors of order t is called the tth trace of A. Symbolically, $\text{tr}_t(A) = \Sigma_\alpha |A_{(i_1,\ldots,i_t)}^{(i_1,\ldots,i_t)}|$.
Thus, when $t = 1$, we have $\text{tr}_1 A \equiv \text{tr} A = \Sigma_{i=1}^n a_{ii}$.

Example Let A be the 4×4 matrix given in Section 1.3.1. Let $i_1 = 2$, $i_2 = 3$, $j_1 = 2$, and $j_2 = 3$. Then

$$A_{(2,3)}^{(2,3)} = \begin{pmatrix} 9 & 10 \\ 7 & 8 \end{pmatrix}, \qquad A_{(1)}^{(1)} = 1, \quad A_{(2)}^{(2)} = 9$$

and $\begin{vmatrix} 9 & 10 \\ 7 & 8 \end{vmatrix}$ is a principal minor of order 2. The sum of all principal minors of order 2,

$$\begin{vmatrix} 9 & 10 \\ 7 & 8 \end{vmatrix} + \begin{vmatrix} 1 & 6 \\ 6 & 9 \end{vmatrix} + \begin{vmatrix} 1 & 5 \\ 3 & 8 \end{vmatrix} + \begin{vmatrix} 9 & 12 \\ 5 & 11 \end{vmatrix} + \begin{vmatrix} 1 & 7 \\ 2 & 11 \end{vmatrix} + \begin{vmatrix} 8 & 10 \\ 9 & 11 \end{vmatrix} = 4 = \text{tr}_2(A),$$

and $\text{tr}_1 A = 1 + 9 + 8 + 11$.

From the definition of a determinant, we get the following property of a determinant.

Let A be a square matrix of order n. Then the determinant $|A+\lambda I_n|$ is a polynomial of degree n in λ and can be written

$$|A+\lambda I_n| = \sum_{i=1}^{n} \lambda^i \mathrm{tr}_{n-i} A, \quad \text{where} \quad \mathrm{tr}_0 A = 1.$$

If $\lambda_1, \lambda_2, \dots, \lambda_n$ denote the roots of this polynomial, then

$$\mathrm{tr}_1(A) = \sum_{i=1}^{n} \lambda_i,$$

$$\mathrm{tr}_2(A) = \lambda_1\lambda_2 + \lambda_1\lambda_3 + \cdots + \lambda_{n-1}\lambda_n,$$

$$\vdots$$

$$\mathrm{tr}_n(A) = \lambda_1\lambda_2 \cdots \lambda_n = |A|.$$

We shall write $\mathrm{tr}\,A$ for $\mathrm{tr}_1 A$. It can easily be verified that

i. $\mathrm{tr}\,AB = \mathrm{tr}\,BA$,
ii. $\mathrm{tr}\,ABC = \mathrm{tr}\,BCA = \mathrm{tr}\,CAB$,
iii. $\mathrm{tr}\,A = \mathrm{tr}\,A'$,
iv. $\mathrm{tr}(A+B) = \mathrm{tr}\,A + \mathrm{tr}\,B$,
v. $\mathrm{tr}(A+B+C) = \mathrm{tr}\,A + \mathrm{tr}\,B + \mathrm{tr}\,C$,
vi. $\mathrm{tr}(\Sigma_{\alpha=1}^{k} A_\alpha) = \Sigma_{\alpha=1}^{k}(\mathrm{tr}\,A_\alpha)$, and
vii. $\mathrm{tr}\,c = c$, where c is a scalar.

1.4 Rank of a Matrix

An $m \times n$ matrix A is said to be of rank r, denoted $\rho(A) = r$, if and only if (iff) there is at least one nonzero minor of order r from A and all of the minors of order $r+1$ are zero. It is easy to establish the following:

i. $\rho(A) = 0$ iff $A = 0$.
ii. If A is an $m \times n$ matrix and $A \neq 0$, then $1 \leq \rho(A) \leq \min(m, n)$. Further, $\rho(A) = \rho(A')$.
iii. $\max(\rho(A), \rho(B)) \leq \rho(A \,\vdots\, B) \leq \min(n, \rho(A) + \rho(B))$, where n is the number of rows in A.
iv.
$$\rho\begin{pmatrix} 0 & Q \\ R & 0 \end{pmatrix} = \rho\begin{pmatrix} Q & 0 \\ 0 & R \end{pmatrix} = \rho(R) + \rho(Q).$$
v. $\rho(AB) \leq \min(\rho(A), \rho(B))$.
vi. $\rho(AB) = \rho(A)$ if $\rho(B) = p$, where $B: p \times q$ and $p \leq q$.

Note that $\rho(A)$ need not be equal to $\rho(A^2)$. For example, if

$$A = \begin{pmatrix} 1 & -1 \\ 1 & -1 \end{pmatrix},$$

then, as shown in Example 1.4.1, $\rho(A)=1$, but $A^2=0$; hence the rank of A^2 is zero.

Example 1.4.1 Let

$$A=\begin{pmatrix} 1 & -1 \\ 1 & -1 \end{pmatrix}.$$

The minors of order 1 are 1 and -1 (each repeated twice). The minor of order 2 is $|A|=-1+1=0$. Hence the matrix A is of rank 1.

Example 1.4.2 Let

$$A=\begin{pmatrix} 6 & 2 \\ 2 & 3 \end{pmatrix}.$$

Then since $|A|=18-4=14\neq 0$ the rank of A is 2.

Example 1.4.3 Let

$$A=\begin{pmatrix} 0 & 1 \\ -1 & 0 \end{pmatrix}.$$

Then since $|A|=0+1=1\neq 0$ the rank of A is 2.

An alternative and at times easier way is to check the linear independency of the row or column vectors, where the vectors $\mathbf{a}_1,\ldots,\mathbf{a}_p$ are said to be linearly independent if there does not exist a nonnull vector $\mathbf{c}$ such that $\Sigma_{i=1}^{p} c_i\mathbf{a}_i=\mathbf{0}$.

Theorem 1.4.1 *Let $A: m\times n$ be a matrix of rank r. Then there exist two matrices $F: m\times r$ and $G: r\times n$ of rank r such that $A=FG$.*

Corollary 1.4.1 *If $A^2=A$ (idempotent), then $\operatorname{tr} A=\rho(A)\equiv r$ (say). The proof follows by taking $A=FG$ and $GF=I_r$ applying the assumption $A^2=A$.*

Corollary 1.4.2 *Let A be an $m\times n$ matrix of rank m $(\leq n)$. Then there exists a matrix $G:(n-m)\times n$ of rank $n-m$ such that $AG'=0$ and $(A':G')$ is nonsingular.*

PROOF: Let $B=I_n-A'(AA')^{-1}A$. Note that $AB=0$, $B^2=B$, and $\rho(B)=\operatorname{tr} B=n-\operatorname{tr} A'(AA')^{-1}A=n-m$. Then by Theorem 1.4.1 we can write $B=G'G_1$, where $G:(n-m)\times n$ and $G_1:(n-m)\times n$ are each matrices of rank $n-m$. Thus $AG'G_1=0$ gives $AG'=0$, and (A',G') is a nonsingular matrix. □

Example 1.4.4 Let $A:1\times 4$; that is, let A be a row vector of ones, $A\equiv \mathbf{e}'=(1,1,1,1)$. Then G is the 3×4 matrix given by

$$G=\begin{pmatrix} 1 & -1 & 0 & 0 \\ 0 & 1 & -1 & 0 \\ 0 & 0 & 1 & -1 \end{pmatrix}, \qquad \mathbf{e}'G'=\mathbf{0}'.$$

Another G could be taken as

$$G=\begin{pmatrix} 1 & -1 & 0 & 0 \\ 1 & 0 & -1 & 0 \\ 1 & 0 & 0 & -1 \end{pmatrix}.$$

There will be several Gs such that $G\mathbf{e}=\mathbf{0}$. G is a matrix of contrasts.

Example 1.4.5 Let A be the 3×4 matrix given by

$$\begin{pmatrix} 1 & 1 & 1 & 1 \\ 0 & 1 & 2 & 3 \\ 0 & 1 & 4 & 9 \end{pmatrix}.$$

Then G is a 1×4 matrix; that is, a row vector $\mathbf{b}'$. One such $\mathbf{b}'$ is given by $(1,-3,3,-1)$.

1.5 Inverse of a Nonsingular Matrix

A square matrix A is said to be *nonsingular* if $|A|\neq 0$. For a nonsingular matrix A, let $B'=(c_{ij}(A))/|A|$, where $c_{ij}(A)$ is the cofactor of a_{ij} in A. Then $I=B'A'=AB$. Such a matrix B will be said to be an inverse of A and will be denoted $B=A^{-1}$. Notice that $|B|=1/|A|\neq 0$ and B is a nonsingular matrix. Hence there exists a matrix C such that $BC=I_n$. Then $I=AB\rightarrow C=ABC$; i.e., $A=C$. This shows that

$$(A^{-1})^{-1}=A, \qquad AA^{-1}=A^{-1}A=I,$$

and A^{-1} is unique. Now we shall establish the following results.

Theorem 1.5.1 *Let P and Q be nonsingular matrices with proper orders. Then*
(a) $(PQ)^{-1}=Q^{-1}P^{-1}$;
(b) *if* $Q=P+UV$,

$$Q^{-1}=P^{-1}-P^{-1}U(I+VP^{-1}U)^{-1}VP^{-1}.$$

PROOF: (a) follows from

$$(Q^{-1}P^{-1})(PQ)=I.$$

For (b), we note that $Q^{-1}P+Q^{-1}UV=I$ or $Q^{-1}+Q^{-1}UVP^{-1}=P^{-1}$. Hence $Q^{-1}U(I+VP^{-1}U)=P^{-1}U$, or $Q^{-1}U=P^{-1}U(I+VP^{-1}U)^{-1}$. Using this in $Q^{-1}=P^{-1}-Q^{-1}UVP^{-1}$, we get the required result. □

Corollary 1.5.1 *Let P and Q be nonsingular square matrices. Then*

$$\begin{pmatrix} P & 0 \\ R & Q \end{pmatrix}^{-1} = \begin{pmatrix} P^{-1} & 0 \\ -Q^{-1}RP^{-1} & Q^{-1} \end{pmatrix},$$

$$\begin{pmatrix} P & S \\ 0 & Q \end{pmatrix}^{-1} = \begin{pmatrix} P^{-1} & -P^{-1}SQ^{-1} \\ 0 & Q^{-1} \end{pmatrix}.$$

Corollary 1.5.2 *Let P and Q be nonsingular matrices. Then*

$$\begin{pmatrix} P & S \\ R & Q \end{pmatrix}^{-1} = \begin{pmatrix} (P-SQ^{-1}R)^{-1} & -P^{-1}S(Q-RP^{-1}S)^{-1} \\ -(Q-RP^{-1}S)^{-1}RP^{-1} & (Q-RP^{-1}S)^{-1} \end{pmatrix}.$$

Example 1.5.1 Let

$$Q = \begin{pmatrix} 2 & 1 & \cdots & 1 \\ 1 & 2 & \cdots & 1 \\ \vdots & \vdots & & \vdots \\ 1 & 1 & \cdots & 2 \end{pmatrix} : p \times p,$$

that is, a matrix with diagonal elements of 2 and the off-diagonal elements of 1. Thus if $\mathbf{e}' = (1, 1, \ldots, 1)$,

$$Q = I + \mathbf{e}\mathbf{e}'.$$

Hence

$$Q^{-1} = I - (p+1)^{-1}\mathbf{e}\mathbf{e}'.$$

Example 1.5.2 Let

$$A = \begin{pmatrix} 9 & 0 \\ 5 & 3 \end{pmatrix}.$$

Then

$$A^{-1} = \begin{pmatrix} \frac{1}{9} & 0 \\ -\frac{5}{27} & \frac{1}{3} \end{pmatrix}.$$

1.6 Generalized Inverse of a Matrix

A matrix $A^- : n \times m$ is said to be a generalized inverse of $A : m \times n$ if it satisfies

$$AA^-A = A.$$

This generalized inverse A^- of A is not unique. However, there is a unique way to define the generalized inverse, which is called the Moore–Penrose generalized inverse of A, denoted A^+. A^+ is uniquely defined by the following four properties:

i. $AA^+A = A$,
ii. $A^+AA^+ = A^+$,
iii. $(A^+A)' = A^+A$, and
iv. $(AA^+)' = AA^+$.

For further details see Rao and Mitra (1971) or Srivastava and Khatri (1979).

1.7 Idempotent Matrices

A square matrix A will be called an idempotent matrix iff $A^2 = A$. Two of the properties of an idempotent matrix are stated below:

i. If A is idempotent, then $\operatorname{tr} A = \rho(A)$.
ii. $A : n \times n$ is idempotent iff $\rho(A) + \rho(I - A) = n$.

Theorem 1.7.1 *If $A : p \times p$ is symmetric and idempotent of rank r, then there exists a matrix $G : p \times r$ of rank r such that $A = G(G'G)^{-1}G'$.*

PROOF: By Theorem 1.4.1, we can find two matrices, $G : p \times r$ and $F : r \times p$, of rank r such that $A = GF$. Since A is symmetric and idempotent, we get $GF = F'G'$ and $FG = I_r$. Hence $G = GFG = F'G'G$, and $F' = G(G'G)^{-1}$, or $F = (G'G)^{-1}G'$. □

1.8 Characteristic Roots and Characteristic Vectors

Let A be an $n \times n$ square matrix, and let $\mathbf{x}$ be an $n \times 1$ nonnull vector such that $A\mathbf{x} = \lambda\mathbf{x}$. Then λ is called a characteristic (ch) root or an eigenvalue of A, and $\mathbf{x}$ is called a ch vector or an eigenvector corresponding to ch root λ. The ch roots are the solutions of

$$|A - \lambda I| = 0.$$

Hence if A is a $p \times p$ matrix, then there are p ch roots of A. All the ch roots of an identity matrix are ones, and those of a diagonal matrix are the diagonal elements.

Example 1.8.1 Let

$$A = \begin{pmatrix} 1 & -1 \\ 1 & -1 \end{pmatrix}.$$

Then

$$|A - \lambda I| = \lambda^2,$$

and hence the ch roots of A are zero. Note that the matrix A is of rank 1.

Thus the number of nonzero roots need not equal the rank of the matrix (except for symmetric matrices; see Srivastava and Khatri, 1979).

Example 1.8.2 Let

$$A = \begin{pmatrix} 0 & 1 \\ -1 & 0 \end{pmatrix}.$$

Then

$$|A - \lambda I| = \lambda^2 + 1.$$

Hence there are no real ch roots of A.

Example 1.8.3 Let

$$A = \begin{pmatrix} 6 & 2 \\ 2 & 3 \end{pmatrix}.$$

Then

$$|A - \lambda I| = (6-\lambda)(3-\lambda) - 4 = 14 - 9\lambda + \lambda^2 = (\lambda - 7)(\lambda - 2).$$

Hence the ch roots of A are 2 and 7.

Theorem 1.8.1 *For any real symmetric matrix A, there exists an orthogonal matrix with real elements such that UAU' is a diagonal matrix. Note that if the roots are distinct, positive, and ordered, then we can choose U uniquely by taking all the diagonal elements of U (or the first row of U) as positive.*

1.9 Positive Definite and Positive Semidefinite Matrices

Let A be an $n \times n$ symmetric matrix and $\mathbf{x}$ be any $n \times 1$ vector. Then $\mathbf{x}'A\mathbf{x}$ will be called a quadratic form of A. Notice that $\mathbf{x}'A\mathbf{x}$ is a second-degree polynomial in the elements of $\mathbf{x}$.

A will be said to be positive (semi-)definite iff $\mathbf{x}'A\mathbf{x} > 0$ (≥ 0) for all nonnull vectors $\mathbf{x}$ and to be negative (semi-)definite iff $\mathbf{x}'A\mathbf{x} < 0$ (≤ 0) for all nonnull vectors $\mathbf{x}$. In this book the positive (semi-)definite matrices will be assumed to be symmetric.

Theorem 1.9.1 *If A is positive (semi-)definite (pd or psd), then all the ch roots are positive (nonnegative).*

PROOF: We can write $A = U'D_\lambda U$, where U is an orthogonal matrix and $D_\lambda = \text{diag}(\lambda_1, \ldots, \lambda_n)$, the λ_is being the ch roots of A. Then

$$\mathbf{x}'A\mathbf{x} = \mathbf{x}'U'D_\lambda U\mathbf{x} = \mathbf{y}'D_\lambda \mathbf{y} > 0 \qquad (\geq 0)$$

for all nonnull vectors $\mathbf{x}$ and $\mathbf{y} = U\mathbf{x}$. By choosing $\mathbf{y}' = (0, \ldots, 0, 1, 0, \ldots, 0)$, where 1 is in the ith place, we find that $\lambda_i > 0$ (or ≥ 0). □

Corollary 1.9.1 *A matrix A is positive semidefinite iff A can be written as XX'. X can be taken as a nonsingular matrix if A is positive definite.*

Corollary 1.9.2 *Let $S: p\times p$ be pd, and let $B: p\times m$ with rank of $B=m\leq p$ and $B_1: p\times(p-m)$ be two matrices such that $\rho(B_1)=p-m$ and $B'B_1=0$. Then $S^{-1}-S^{-1}B(B'S^{-1}B)^{-1}B'S^{-1}=B_1(B_1'SB_1)^{-1}B_1'$ is a psd matrix of rank $p-m$.*

1.10 Some Inequalities

Any ch root of $A: n\times n$ will be denoted $\text{ch}(A)$. If A is a symmetric matrix, then the ith largest ch root of A will be denoted $\text{ch}_i(A)$, $i=1,2,\ldots,n$; i.e., $\text{ch}_1(A)\geq \text{ch}_2(A)\geq\cdots\geq \text{ch}_n(A)$.

Theorem 1.10.1 *Let A be any $n\times n$ symmetric matrix and $\mathbf{x}$ be any $n\times 1$ vector. Then*

$$\text{ch}_n(A)\leq \mathbf{x}'A\mathbf{x}/\mathbf{x}'\mathbf{x}\leq \text{ch}_1(A).$$

The equality on the right side holds iff $\mathbf{x}$ is the ch vector corresponding to $\text{ch}_1(A)$, and the equality on the left side holds iff $\mathbf{x}$ is the ch vector corresponding to $\text{ch}_n(A)$.

PROOF: Since A is symmetric, by Theorem 1.8.1 we can write $A=UD_\lambda U'$, where $\lambda_i=\text{ch}_i(A)$, $i=1,2,\ldots,n$, and U is orthogonal. Let $\mathbf{y}=U'\mathbf{x}/\sqrt{\mathbf{x}'\mathbf{x}}$. Then $\mathbf{y}'\mathbf{y}=1$, and

$$\mathbf{x}'A\mathbf{x}/\mathbf{x}'\mathbf{x}=\mathbf{y}'D_\lambda\mathbf{y}=\sum_{i=1}^{n}\lambda_i y_i^2,$$

which is the weighted mean of $\lambda_1,\lambda_2,\ldots,\lambda_n$ with weights $y_1^2, y_2^2,\ldots,y_n^2$ such that $\Sigma_{i=1}^n y_i^2=1$. Hence

$$\lambda_n\leq \sum_{i=1}^{n}\lambda_i y_i^2\leq\lambda_1.$$

The right-hand equality holds iff $y_1^2=1$, $y_j=0$ $(j\neq 1)$, and the left-hand equality holds iff $y_n^2=1$ and $y_j=0$ $(j=1,2,\ldots,n-1)$. This proves the required result. □

Corollary 1.10.1 *Let $A: n\times n$ be symmetric, and let $B: n\times n$ be pd. Then*

$$\text{ch}_n(B^{-1}A)\leq \mathbf{x}'A\mathbf{x}/\mathbf{x}'B\mathbf{x}\leq \text{ch}_1(B^{-1}A).$$

The equality on the right side holds iff $\mathbf{x}$ is the ch vector corresponding to $\text{ch}_1(B^{-1}A)$, and the equality on the left side holds iff $\mathbf{x}$ is the ch vector corresponding to $\text{ch}_n(B^{-1}A)$.

The proof follows from Theorem 1.10.1 by noting that

$$\mathbf{x}'A\mathbf{x}/\mathbf{x}'B\mathbf{x} = \mathbf{y}'A_1\mathbf{y}/\mathbf{y}'\mathbf{y},$$

where $B = PP'$, $\mathbf{y} = P\mathbf{x}$, and $A_1 = P'^{-1}AP^{-1}$.

Corollary 1.10.2 $(\mathbf{x}'\mathbf{y})^2 \leq (\mathbf{x}'B\mathbf{x})(\mathbf{y}'B^{-1}\mathbf{y})$, *where* B *is any pd matrix. The equality holds iff* $\mathbf{x} = \lambda B^{-1}\mathbf{y}$, *where* λ *is an arbitrary positive quantity.*

PROOF: Notice that

$$|(\mathbf{x}'\mathbf{y})|^2/(\mathbf{x}'B\mathbf{x}) = \mathbf{x}'(\mathbf{y}\mathbf{y}')\mathbf{x}/\mathbf{x}'B\mathbf{x}.$$

There is only one nonzero ch root of $B^{-1}\mathbf{y}\mathbf{y}'$, namely $\mathbf{y}'B^{-1}\mathbf{y}$, and its ch vector will be $\lambda B^{-1}\mathbf{y}$, where λ is a constant. Then by Corollary 1.10.1 we get the required result. □

Corollary 1.10.3 (Cauchy–Schwarz Inequality)

$$(\mathbf{x}'\mathbf{y})^2 \leq (\mathbf{x}'\mathbf{x})(\mathbf{y}'\mathbf{y}).$$

PROOF: $(\mathbf{x}'\mathbf{y})^2 = \mathbf{x}'\mathbf{y}\mathbf{y}'\mathbf{x}$. There is only one nonzero ch root of $\mathbf{y}\mathbf{y}'$, namely $\mathbf{y}'\mathbf{y}$, and the result follows from Theorem 1.10.1. □

1.11 Computational Procedures

Computations involving matrices are best handled by high-speed computer programs. The SAS matrix procedure is one set of programs that is quite useful and handles many of the computations covered in this text. We now give some examples using this package. For readers without access to high-speed computers, we give numerical procedures at the end of this section for use with a calculator.

The eigenvalues and eigenvectors as well as the generalized inverse of the following matrix was calculated with SAS.

$$X = \begin{pmatrix} 1 & 2 & 3 & 4 & 5 \\ 2 & 4 & 7 & 8 & 9 \\ 3 & 7 & 10 & 15 & 20 \\ 4 & 8 & 15 & 30 & 20 \\ 5 & 9 & 20 & 20 & 40 \end{pmatrix}.$$

The following file was created:

```
PROC MATRIX;
X=1 2  3  4  5/
   2 4  7  8  9/
   3 7 10 15 20/
   4 8 15 30 20/
   5 9 20 20 40;
G=GINV (X);
E=EIGVAL (X);
D=EIGVEC (X);
PRINT G E D;
```

Table 1.1 Statistical Analysis System

G	Col. 1	Col. 2	Col. 3	Col. 4	Col. 5
Row 1	9.78873	−2.1831	−1.8592	0.112656	0.140845
Row 2	−2.1831	0.774648	0.788732	−0.16901	−0.21127
Row 3	−1.8592	0.788732	0.039437	−0.00845	0.039437
Row 4	0.112676	−0.16901	−0.00845	0.073239	−0.00845
Row 5	0.140845	−0.21127	0.039437	−0.00845	0.039437

E	Col. 1
Row 1	70.3349
Row 2	14.4402
Row 3	1.99761
Row 4	0.093745
Row 5	−1.8665

D	Col. 1	Col. 2	Col. 3	Col. 4	Col. 5
Row 1	0.105139	−0.00733	−0.26674	−0.95627	−0.05731
Row 2	0.205967	−0.0555	−0.82859	0.226294	0.46554
Row 3	0.397077	0.025855	−0.32383	0.184029	−0.83841
Row 4	0.546217	0.78569	0.258517	−0.01391	0.131559
Row 5	0.700358	0.615535	0.265699	−0.01648	0.244436

The resulting output is displayed in Table 1.1, where G is the G inverse of X. E contains the five eigenvalues, the columns of D being the corresponding eigenvectors. It is left to the reader to check that G is the Moore–Penrose inverse of X. To calculate the Moore–Penrose inverse G of the matrix

$$Y = \begin{pmatrix} 1 & 2 \\ 3 & 4 \\ 5 & 6 \end{pmatrix}',$$

the following file was created:

```
PROC MATRIX;
Y=1 2/
   3 4/
   5 6;
G=GINV (Y);
PRINT G Y;
```

The output produced is given in Table 1.2.

Finally, to obtain the singular value decomposition of a matrix Z into $U(\text{diag } Q)V'$ (where U and V are semiorthogonal matrices and diag Q is a

Table 1.2

G	Col. 1	Col. 2	Col. 3
Row 1	−0.61111	−0.38889	0.555556
Row 2	0.574074	0.425926	−0.37037
Y	**Col. 1**	**Col. 2**	
Row 1	1	2	
Row 2	3	4	
Row 3	5	5	

diagonal matrix of the singular values of Z), we can use the following file:

```
PROC MATRIX;
Z=1.0676 0.1848/
   0.1848 1.130;
SVD U Q V Z;
PRINT Z;
PRINT U Q V;
```

The matrix to be decomposed in this case was

$$Z=\begin{pmatrix} 1.0676 & 0.1848 \\ 0.1848 & 1.130 \end{pmatrix}.$$

The printout of Table 1.3 ensued.

The matrix Z may be written

$$\begin{pmatrix} 1.0676 & 0.1848 \\ 0.1848 & 1.130 \end{pmatrix}=\begin{pmatrix} 0.7637 & 0.64557 \\ -0.64557 & 0.7637 \end{pmatrix}\begin{pmatrix} 0.911385 & 0 \\ 0 & 1.28622 \end{pmatrix}$$
$$\times\begin{pmatrix} 0.7637 & -0.64557 \\ 0.64557 & 0.7637 \end{pmatrix}.$$

Note that as Z is symmetric, $U=V$ contains the eigenvectors of Z, and Q contain the eigenvalues of Z. For other procedures, the reader is referred to the SAS manual. A symmetric square root of Z can be obtained by taking

$$\begin{pmatrix} 1.0676 & 0.1848 \\ 0.1848 & 1.130 \end{pmatrix}^{1/2}=\begin{pmatrix} 0.7637 & 0.64557 \\ -0.64557 & 0.7637 \end{pmatrix}\begin{pmatrix} 0.911385 & 0 \\ 0 & 1.28622 \end{pmatrix}^{1/2}$$
$$\times\begin{pmatrix} 0.7637 & -0.64557 \\ 0.64557 & 0.7637 \end{pmatrix}=\begin{pmatrix} 1.029 & 0.0885 \\ 0.0885 & 1.059 \end{pmatrix}.$$

For readers without access to other programs, we now give some numerical procedures for the calculation of determinants, inverses, and ranks of matrices. From the definition of the determinants in Section 1.3 it follows that for

$$A=\begin{pmatrix} a_{11} & a_{12} \\ a_{21} & a_{22} \end{pmatrix}, \qquad \det(A)=a_{11}a_{22}-a_{12}a_{21},$$

Table 1.3

Z	Col. 1	Col. 2
Row 1	1.0676	0.1848
Row 2	0.1848	1.13
U	Col. 1	Col. 2
Row 1	0.7637	0.64557
Row 2	−0.64557	0.7637
Q	Col. 1	
Row 1	0.911385	
Row 2	1.28622	
V	Col. 1	Col. 2
Row 1	0.7637	0.64557
Row 2	−0.64557	0.7637

and for

$$A=\begin{pmatrix} a_{11} & a_{12} & a_{13} \\ a_{21} & a_{22} & a_{23} \\ a_{31} & a_{32} & a_{33} \end{pmatrix},$$

$$\det(A)=a_{11}a_{22}a_{33}+a_{12}a_{23}a_{31}+a_{21}a_{32}a_{13} \\ -a_{12}a_{21}a_{33}-a_{11}a_{23}a_{32}-a_{13}a_{31}a_{22}.$$

For higher-order matrices the calculations involved in the definition of a determinant become too time consuming. One thus uses the property that for a $p\times p$ nonsingular matrix H_1 and a $p\times p$ matrix A,

$$\det(H_1A)=\det(A)\det(H_1).$$

Hence if we multiply row i of A by a constant and add it to row j, the determinant of A is unchanged. If two rows of A are interchanged, then only the sign of the determinant is changed, and if a row is multiplied by a constant, so is the determinant. Therefore, given

$$A=\begin{pmatrix} a_{11} & \cdots & a_{1p} \\ \vdots & & \vdots \\ a_{p1} & \cdots & a_{pp} \end{pmatrix},$$

if $a_{11}\neq 0$, then we multiply row 1 by a_{i1}/a_{11} and subtract it from row i.

If $a_{11}=0$, then we interchange two rows to obtain a nonzero a_{11} entry. We obtain a matrix B with $|A|=|B|(-1)^j$, where j equals the number of rows interchanged and B is of the form

$$B=\begin{pmatrix} b_{11} & b_{12} & \cdots & b_{1p} \\ 0 & b_{22} & \cdots & b_{2p} \\ \vdots & \vdots & & \vdots \\ 0 & b_{p2} & \cdots & b_{pp} \end{pmatrix}.$$

The above procedure is repeated on the submatrix

$$\begin{pmatrix} b_{2i} & \cdots & b_{ip} \\ \vdots & & \vdots \\ b_{pi} & \cdots & b_{pp} \end{pmatrix}.$$

Continuing in this manner, we obtain $|A|=|T|(-1)^j$, where j is the number of rows interchanged and T is an upper triangular matrix of the form

$$T=\begin{pmatrix} t_{11} & t_{12} & \cdots & t_{1p} \\ 0 & & & \vdots \\ \vdots & & & \\ 0 & \cdots & 0 & t_{pp} \end{pmatrix}$$

and $|T|=\prod_{i=1}^{p} t_{ii}$.

Example 1.11.1

$$A=\begin{pmatrix} 1 & 2 \\ 3 & 4 \end{pmatrix}, \qquad \det(A)=4-6=-2.$$

Example 1.11.2

$$A=\begin{pmatrix} 1 & -5 & 6 \\ 2 & 3 & 4 \\ 0 & 1 & 3 \end{pmatrix}, \qquad \det(A)=9+0+12-4+30-0=47.$$

Example 1.11.3

$$A=\begin{pmatrix} 1 & 2 & 0 & 1 \\ 6 & 14 & 1 & 0 \\ 4 & 2 & 2 & 1 \\ 1 & 0 & 1 & 1 \end{pmatrix}.$$

We subtract six times row 1 from row 2, four times row 1 from row 3, and row 1 from row 4 to obtain the matrix

$$\begin{pmatrix} 1 & 2 & 0 & 1 \\ 0 & 2 & 1 & -6 \\ 0 & -6 & 2 & -3 \\ 0 & -2 & 1 & 0 \end{pmatrix}.$$

Adding three times row 2 to row 3 and adding row 2 to row 3, we obtain

$$\begin{pmatrix} 1 & 2 & 0 & 1 \\ 0 & 2 & 1 & -6 \\ 0 & 0 & 5 & -21 \\ 0 & 0 & 2 & -6 \end{pmatrix}.$$

Finally, subtracting $\frac{2}{5}$ times row 3 from row 4, we obtain the upper triangular matrix

$$T=\begin{pmatrix} 1 & 2 & 0 & 1 \\ 0 & 2 & 1 & -6 \\ 0 & 0 & 5 & -21 \\ 0 & 0 & 0 & 2.4 \end{pmatrix}.$$

Hence

$$\det(A)=\det(T)=\prod_{i=1}^{4} t_{ii}=1\times2\times5\times2.4=24.$$

The reduction of the matrix A by elementary row operations to the upper triangular matrix T is equivalent to the multiplication of A by a nonsingular H matrix such that $HA=T$. If we look at the matrix $A_1=(A\,\vdots\,I_p)$ and perform the same elementary row operations on A_1, we obtain the new matrix $HA_1=(HA\,\vdots\,H)=(T\,\vdots\,H)$. Hence H can be evaluated by simply looking at $(A\,\vdots\,I_p)$ instead of A.

For Example 1.11.3,

$$\left(\begin{array}{cccc|cccc} 1 & 2 & 0 & 1 & 1 & 0 & 0 & 0 \\ 6 & 14 & 1 & 0 & 0 & 1 & 0 & 0 \\ 4 & 2 & 2 & 1 & 0 & 0 & 1 & 0 \\ 1 & 0 & 1 & 1 & 0 & 0 & 0 & 1 \end{array}\right)$$

reduces to

$$\left(\begin{array}{cccc|cccc} 1 & 3 & 5 & 1 & 1 & 0 & 0 & 0 \\ 0 & 2 & 1 & -6 & -6 & 1 & 0 & 0 \\ 0 & 0 & 5 & -21 & -22 & 3 & 1 & 0 \\ 0 & 0 & 0 & 2.4 & 1.8 & -0.2 & -0.4 & 1 \end{array}\right)=\left(T\,\vdots\,H\right).$$

Note that $HA=T$.

We now have that given any $p\times p$ matrix A, there exists a nonsingular matrix H_1 such that $H_1A=T$, where T is an upper triangular matrix. By repeating the same procedure to the rows of T we obtain a nonsingular matrix H_2 such that $TH_2=D$, where D is a diagonal matrix. Hence we have that $H_1AH_2=D$, and therefore $D^{-1}=(H_1AH_2)^{-1}=H_2^{-1}A^{-1}H_1^{-1}$. The inverse of A can be calculated from $A^{-1}=H_1D^{-1}H_2$.

Suppose now that $A_{p\times q}$ is of rank r. We can still proceed as above to produce matrices $H_1: p\times p$ and $H_2: q\times q$ such that

$$H_1AH_2=D \qquad \text{where} \quad D=\begin{pmatrix} D_1 & 0\\ 0 & 0\end{pmatrix}$$

and D_1 is an $r\times r$ diagonal matrix. Hence the rank of A is the number of nonzero entries of D. A generalized inverse of D is given by

$$\begin{pmatrix} D_1^{-1} & 0\\ 0 & 0\end{pmatrix},$$

as

$$\begin{pmatrix} D_1 & 0\\ 0 & 0\end{pmatrix}\begin{pmatrix} D_1^{-1} & 0\\ 0 & 0\end{pmatrix}\begin{pmatrix} D_1 & 0\\ 0 & 0\end{pmatrix}=\begin{pmatrix} D_1 & 0\\ 0 & 0\end{pmatrix},$$

and hence we have that

$$(H_1AH_2)^{-}=H_2^{-1}A^{-}H_1^{-1}=D^{-}.$$

A generalized inverse of A is therefore given by

$$A^{-}=H_2D^{-}H_1.$$

Example 1.11.4 Consider the matrix

$$A=\begin{pmatrix} 1 & 2 & 3 & 4\\ 1 & 3 & 0 & 1\end{pmatrix}.$$

We first reduce A to an upper triangular matrix of elementary row operations. That is, for the matrix

$$\left(\begin{array}{cc|cccc} 1 & 0 & 1 & 2 & 3 & 4\\ 0 & 1 & 1 & 3 & 0 & 1\end{array}\right),$$

we subtract row 1 from row 2 to obtain

$$\left(\begin{array}{cc|cccc} 1 & 0 & 1 & 2 & 3 & 4\\ \hline -1 & 1 & 0 & 1 & -3 & -3\end{array}\right)=(H_1 \vdots T).$$

Then

$$H_1A=\begin{pmatrix} 1 & 2 & 3 & 4\\ 0 & 1 & -3 & -3\end{pmatrix}.$$

Now look at

$$\left(\begin{array}{cccc} 1 & 2 & 3 & 4\\ 0 & 1 & -3 & -3\\ \hline 1 & 0 & 0 & 0\\ 0 & 1 & 0 & 0\\ 0 & 0 & 1 & 0\\ 0 & 0 & 0 & 1\end{array}\right).$$

Subtract two times column 1 from column 2, three times column 1 from column 3, and four times column 1 from column 4 to obtain the matrix

$$\begin{pmatrix} 1 & 0 & 0 & 0 \\ 0 & 1 & -3 & -3 \\ \hline 1 & -2 & -3 & -4 \\ 0 & 1 & 0 & 0 \\ 0 & 0 & 1 & 0 \\ 0 & 0 & 0 & 1 \end{pmatrix}.$$

Now add three times column 2 to column 3 and to column 4 to obtain the matrix

$$\begin{pmatrix} 1 & 0 & 0 & 0 \\ 0 & 1 & 0 & 0 \\ \hline 1 & -2 & -9 & -10 \\ 0 & 1 & 3 & 3 \\ 0 & 0 & 1 & 0 \\ 0 & 0 & 0 & 1 \end{pmatrix} = \begin{pmatrix} 1 & 0 & 0 & 0 \\ 0 & 1 & 0 & 0 \\ \hline & & H_2 & \end{pmatrix}.$$

We can now write

$$H_1 A H_2 = \begin{pmatrix} 1 & 0 & 0 & 0 \\ 0 & 1 & 0 & 0 \end{pmatrix} = D,$$

as the reader should check. Because H_1 and H_2 are nonsingular matrices, the rank of A is the rank of D. Therefore, the rank of A is 2.

A generalized inverse of D is

$$D^- = \begin{pmatrix} 1 & 0 \\ 0 & 1 \\ 0 & 0 \\ 0 & 0 \end{pmatrix} \quad \text{as} \quad D^- D = \begin{pmatrix} 1 & 0 \\ 0 & 1 \end{pmatrix}.$$

Therefore, a generalized inverse of A is given by

$$\begin{aligned} A^- &= H_2 D^- H_1 \\ &= \begin{pmatrix} 1 & -2 & -9 & -10 \\ 0 & 1 & 3 & 3 \\ 0 & 0 & 1 & 0 \\ 0 & 0 & 0 & 1 \end{pmatrix} \begin{pmatrix} 1 & 0 \\ 0 & 1 \\ 0 & 0 \\ 0 & 0 \end{pmatrix} \begin{pmatrix} 1 & 0 \\ -1 & 1 \end{pmatrix} \\ &= \begin{pmatrix} -3 & -2 \\ -1 & 1 \\ 0 & 0 \\ 0 & 0 \end{pmatrix}. \end{aligned}$$

Note that $AA^- = \left(\begin{smallmatrix} 1 & 0 \\ 0 & 1 \end{smallmatrix}\right)$, $AA^-A = A$, and hence A^- is indeed a generalized inverse of A.

Problems

1.1 Let $A=(a_{ij}): p\times p$, where $a_{ii}=a$, $i=1,2,\ldots,p$, and $a_{ij}=b$, $i\neq j$, $i,j=1,2,\ldots,p$. Find the determinant of A.

1.2 Let A be defined as in Problem 1.1. Find A^{-1}.

1.3 Prove Corollary 1.5.2.

1.4 Let $X: p\times n$, of rank $p\leq n$. Show that the matrix $B\equiv I_n = X'(XX')^{-1}X$ is idempotent. What is the rank of B?

1.5 Let $B=I_n-n^{-1}\mathbf{ee}'$, where $\mathbf{e}'=(1,\ldots,1)$, an n-row vector of ones. Show that B is idempotent. What is the rank of B?

1.6 Find the inverse of the following matrices:

$$\textbf{(a)}\quad \begin{pmatrix} 12 & 3 \\ 21 & 10 \end{pmatrix}, \qquad \textbf{(b)}\quad \begin{pmatrix} 1 & 2 & 3 \\ 2 & 3 & 1 \\ 3 & 1 & 2 \end{pmatrix}.$$

1.7 Find the inverse of the matrix

$$\begin{pmatrix} 1 & 3 & 5 & 1 \\ 3 & 1 & 0 & 4 \\ 5 & 0 & 6 & 1 \\ 1 & 4 & 1 & 6 \end{pmatrix}.$$

1.8 Find a generalized inverse of each of the following matrices:

$$\textbf{(a)}\quad \begin{pmatrix} 1 & 1 \\ 1 & 1 \end{pmatrix}, \qquad \textbf{(b)}\quad \begin{pmatrix} 1 & 1 \\ 0 & 1 \end{pmatrix},$$

$$\textbf{(c)}\quad \begin{pmatrix} 1 & 2 & 3 \\ 3 & 4 & 5 \\ 6 & 7 & 8 \end{pmatrix}, \qquad \textbf{(d)}\quad \begin{pmatrix} 2 & 4 & 6 \\ 1 & 2 & 4 \\ 1 & 4 & 8 \end{pmatrix}.$$

1.9 For the matrices X below, calculate the eigenvalues of $X'X$.

$$\textbf{(a)}\quad X=\begin{pmatrix} 1 & 1 \\ 1 & 2 \\ 1 & 3 \end{pmatrix}, \qquad \textbf{(b)}\quad X=\begin{pmatrix} 1 & 2 & 3 \\ 4 & 5 & 6 \\ 7 & 8 & 9 \end{pmatrix}.$$

1.10 For the intraclass correlation matrix $\Sigma=\sigma^2[I(1-\rho)+\rho\mathbf{ee}']$, find the distinct eigenvalues.

1.11 Find a matrix C_2 such that $[C_1 : C_2]$ is an orthogonal matrix for the following matrices.

$$\textbf{(a)}\quad C_1=\begin{pmatrix} \frac{1}{2} & -\frac{1}{2} \\ \frac{1}{2} & -\frac{1}{2} \\ \frac{1}{2} & \frac{1}{2} \\ \frac{1}{2} & \frac{1}{2} \end{pmatrix}, \qquad \textbf{(b)}\quad C_1=\begin{pmatrix} 1/\sqrt{3} & -1/\sqrt{2} \\ 1/\sqrt{3} & 0 \\ 1/\sqrt{3} & 1/\sqrt{2} \end{pmatrix}.$$

1.12 Show that

(a) $(A\otimes B)(C\otimes D)=AC\otimes BD$;

(b) $(A\otimes B)^{-1}=A^{-1}\otimes B^{-1}$;

(c) $I\otimes\Gamma$ is orthogonal if Γ is orthogonal.

1.13 Find the eigenvalues and eigenvectors of the following matrices:

(a) $\begin{pmatrix} 2 & 0 \\ 0 & 3 \end{pmatrix}$, **(b)** $\begin{pmatrix} 2 & 1 \\ 1 & 2 \end{pmatrix}$,

(c) $\begin{pmatrix} 4 & 2 & 0 \\ 2 & 3 & 1 \\ 0 & 1 & 5 \end{pmatrix}$, **(d)** $\begin{pmatrix} 3 & 1 & 1 \\ 1 & 3 & 1 \\ 1 & 1 & 6 \end{pmatrix}$,

(e) $\begin{pmatrix} 1 & 0.1 & 0.04 & 0.1 & 0.4 \\ 0.1 & 1 & 0.9 & 0.9 & -0.1 \\ 0.04 & 0.9 & 1 & 0.8 & -0.3 \\ 0.1 & 0.9 & 0.8 & 1 & -0.1 \\ 0.4 & -0.1 & -0.3 & -0.1 & 1 \end{pmatrix}$.

Chapter 2
Multivariate Normal Distributions

2.1 Some Notation

For convenience of presentation we shall not distinguish between a random variable and its observed values. Capital letters are generally used for matrices and small letters for random variables and their observed values. When we have more than one random variable, vector notation will be used. For example, if we have p random variables $x_1,\ldots,x_p$, we shall write $\mathbf{x}'=(x_1,\ldots,x_p)$, which may be read as either a vector of p random variables $x_1,\ldots,x_p$ or a random p-vector $\mathbf{x}$. Similarly, we can have a matrix $U=(u_{ij})$ of random variables u_{ij}, $i,j=1,2,\ldots,p$. The expectation of a random variable x is denoted $E(x)$. If we have p random variables $x_1,\ldots,x_p$, their expectations are $E(x_1),\ldots,E(x_p)$. Thus $E(\mathbf{x}')=(E(x_1),\ldots,E(x_p))$. Similarly, $E(U)=(E(u_{ij}))$; $E(AU)=AE(U)$; $E(AUB)=AE(U)B$; and $E(AUB+C)=AE(U)B+C$, where A,B,C are matrices (of appropriate orders) of constants. The variance of a random variable x is defined as

$$\operatorname{var}(x)=E(x^2)-[E(x)]^2=E(x-\mu)^2 \quad \text{if} \quad \mu=E(x),$$

usually denoted σ^2. Suppose now we have two random variables x_1 and x_2. We define the covariance between x_1 and x_2 as

$$\operatorname{cov}(x_1,x_2)=E(x_1x_2)-E(x_1)E(x_2)\equiv E([x_1-E(x_1)][x_2-E(x_2)]),$$

usually denoted by σ_{12}. Denoting the variances of x_1 and x_2 by σ_{11} and σ_{22}, respectively, we define the *covariance matrix* or the *dispersion matrix* of the random vector $\mathbf{x}=(x_1,x_2)'$ as

$$\Sigma=\operatorname{cov}(\mathbf{x})=\begin{pmatrix}\sigma_{11} & \sigma_{12}\\ \sigma_{12} & \sigma_{22}\end{pmatrix}.$$

Letting

$$y_1=x_1-E(x_1) \quad \text{and} \quad y_2=x_2-E(x_2),$$

we get

$$\operatorname{var}(y_1)=\operatorname{var}(x_1), \qquad \operatorname{var}(y_2)=\operatorname{var}(x_2), \qquad \operatorname{cov}(y_1,y_2)=\operatorname{cov}(x_1,x_2).$$

Let

$$\mu_1 = E(x_1), \qquad \mu_2 = E(x_2), \qquad \text{and} \qquad \boldsymbol{\mu} = (\mu_1, \mu_2)'.$$

Then $\mathbf{y} = (\mathbf{x} - \boldsymbol{\mu})$, and

$$\Sigma = \operatorname{cov}(\mathbf{x}) = E((\mathbf{x}-\boldsymbol{\mu})(\mathbf{x}-\boldsymbol{\mu})') = \operatorname{cov}(\mathbf{y}) = E(\mathbf{y}\mathbf{y}')$$

$$= E\begin{pmatrix} y_1^2 & y_1 y_2 \\ y_1 y_2 & y_2^2 \end{pmatrix} = \begin{pmatrix} E(y_1^2) & E(y_1 y_2) \\ E(y_1 y_2) & E(y_2^2) \end{pmatrix}.$$

Since $(\mathbf{y}\mathbf{y}')' = (\mathbf{y}\mathbf{y}')$, Σ is symmetric. Also, if $\sigma_{11} > 0$, $\sigma_{22} > 0$, and $-1 < \rho_{12} < 1$, where $\rho_{12} = \sigma_{12}/(\sigma_{11}\sigma_{22})^{1/2}$, then Σ is positive definite. If $\mathbf{x}$ is a random p-vector, then $\operatorname{cov}(\mathbf{x})$ will be a $p \times p$ matrix given by

$$E((\mathbf{x}-\boldsymbol{\mu})(\mathbf{x}-\boldsymbol{\mu})') = E(\mathbf{y}\mathbf{y}')$$

$$= E\begin{pmatrix} y_1^2 & y_1 y_2 & \cdots & y_1 y_p \\ y_2 y_1 & y_2^2 & \cdots & y_2 y_p \\ \vdots & \vdots & & \vdots \\ y_p y_1 & y_p y_2 & \cdots & y_p^2 \end{pmatrix},$$

where $\mathbf{y} = \mathbf{x} - \boldsymbol{\mu}$ and $\boldsymbol{\mu} = E(\mathbf{x})$.

2.2 Multivariate Normal Distributions

We begin with the definition of a standard normal random variable u, defined by its probability density function (hereafter, pdf)

$$(2\pi)^{-1/2}\exp\left(-\tfrac{1}{2}u^2\right) \tag{2.2.1}$$

and denoted $u \sim N(0,1)$. In this case, as we know, the mean of the random variable u is 0, and the variance is 1. A random variable x has a normal distribution with mean μ and variance $\sigma^2 > 0$ if x has the same distribution as

$$\mu + \sigma u \tag{2.2.2}$$

where $u \sim N(0,1)$. In this case its pdf is given by

$$(2\pi\sigma^2)^{-1/2}\exp\left[-(1/2\sigma^2)(x-\mu)^2\right], \tag{2.2.3}$$

where $-\infty < \mu$, $x < \infty$, and $\sigma > 0$. The moment-generating function of the random variable x is given by

$$\phi(t) = E(e^{tx}) = E(x^{t(\mu+\sigma u)}) = e^{t\mu}E(e^{bu}) \qquad (b = t\sigma)$$

$$= e^{t\mu + b^2/2} = e^{t\mu + t^2\sigma^2/2}. \tag{2.2.4}$$

Let us now consider p independent standard normal random variables $u_1,\ldots,u_p$. Their joint pdf is given by

$$(2\pi)^{-p/2}\exp\left[-\tfrac{1}{2}\left(u_1^2+\cdots+u_p^2\right)\right].$$

In vector notation $u_1^2+\cdots+u_p^2=\mathbf{u}'\mathbf{u}$ and we write $\mathbf{u}=(u_1,\cdots,u_p)'$ for p random variables. Thus the pdf of $\mathbf{u}$ is given by

$$(2\pi)^{-p/2}\exp-\tfrac{1}{2}\mathbf{u}'\mathbf{u}, \qquad -\infty<u_i<\infty. \tag{2.2.5}$$

We note that $E(\mathbf{u})=E(u_1,\ldots,u_p)'=(E(u_1),\ldots,E(u_p))'=\mathbf{0}'$. Also

$$\operatorname{cov}(\mathbf{u})=E(\mathbf{u}\mathbf{u}')=E\begin{pmatrix} u_1^2 & u_1u_2 & \cdots & u_1u_p \\ u_2u_1 & u_2^2 & \cdots & u_2u_p \\ \vdots & \vdots & & \vdots \\ u_pu_1 & u_pu_2 & \cdots & u_p^2 \end{pmatrix}$$

$$=\begin{pmatrix} 1 & 0 & \cdots & 0 \\ 0 & 1 & \cdots & 0 \\ \vdots & \vdots & & \vdots \\ 0 & 0 & \cdots & 1 \end{pmatrix}=I$$

since the u_is are independent (thus giving zero covariances). Hence the random vector $\mathbf{u}$ has mean vector $\mathbf{0}$, covariance matrix identity I, and pdf (2.2.5). This we write $u\sim N_p(\mathbf{0},I)$, where $\mathbf{0}$ denotes the mean of $\mathbf{u}$, I the dispersion or covariance matrix of $\mathbf{u}$, and the subscript p the dimension of the random vector $\mathbf{u}$. In a manner analogous to the univariate case, we now give a definition of a multivariate normal distribution.

Definition 2.2.1 A p-dimensional random vector $\mathbf{x}$ is said to have a nonsingular multivariate normal distribution $N_p(\boldsymbol{\mu},\Sigma)$ if $\mathbf{x}$ has the same distribution as $\boldsymbol{\mu}+A\mathbf{u}$, where A is a $p\times p$ matrix such that $\Sigma=AA'$ and $\mathbf{u}\sim N_p(\mathbf{0},I)$. $\boldsymbol{\mu}$ is called the mean of the distribution, and Σ is the covariance matrix, also known as the dispersion matrix.

Note that if $\mathbf{x}=\boldsymbol{\mu}+A\boldsymbol{\mu}$, the Jacobian of the transformation (see the Appendix to this chapter) $J(\mathbf{u}\rightarrow\mathbf{x})=|A^{-1}|=|AA'|^{-1/2}=|\Sigma|^{-1/2}$. Hence the pdf of $\mathbf{x}$ is given by

$$(2\pi)^{-p/2}|\Sigma|^{-1/2}\exp-\tfrac{1}{2}(\mathbf{x}-\boldsymbol{\mu})'\Sigma^{-1}(\mathbf{x}-\boldsymbol{\mu}), \tag{2.2.6}$$

which is in conformity with (2.2.3).

An alternative definition is given as follows [see Srivastava and Khatri (1979) for a proof]:

Definition 2.2.2 A p-dimensional random vector $\mathbf{x}$ is said to have a multivariate normal distribution iff *every linear* combination of $\mathbf{x}$ has a univariate normal distribution.

REMARK 2.2.1: Definition 2.2.1 can be extended to a *singular* multivariate normal distribution. In this case Σ is psd, of rank $r \leq p$, and can be written as $\Sigma = AA'$, where A is a $p \times r$ matrix of rank r. Thus $\mathbf{x} = \boldsymbol{\mu} + A\mathbf{u}$, where $\mathbf{u} \sim N_r(\mathbf{0}, I)$. However, the pdf of $\mathbf{x}$ is not given by (2.2.6). In fact, there is no unique way to write it. One expression is given by Srivastava and Khatri (1979).

2.3 Some Properties of Multivariate Normal Distributions

In this section we list some properties of multivariate normal distributions.

Let $\mathbf{x} \sim N_p(\boldsymbol{\mu}, \Sigma)$. Then the moment-generating function defined as $m(\mathbf{t}) \equiv E[\exp(t_1 x_1 + t_2 x_2 + \cdots + t_p x_p)] = E[\exp \mathbf{t}'\mathbf{x}] = E[\exp(\mathbf{t}'\boldsymbol{\mu} + \mathbf{t}'A\mathbf{u})]$, where $\mathbf{u} \sim N_p(\mathbf{0}, I)$, is given by

$$
\begin{aligned}
m(\mathbf{t}) &= E\big[\exp(\mathbf{t}'\boldsymbol{\mu} + \mathbf{t}'A\mathbf{u})\big] \\
&= (\exp \mathbf{t}'\boldsymbol{\mu}) E(\exp \mathbf{b}'\mathbf{u}) \qquad (b' = \mathbf{t}'A) \\
&= (\exp \mathbf{t}'\boldsymbol{\mu}) E\big[\exp(b_1 u_1 + \cdots + b_p u_p)\big] \\
&= (\exp \mathbf{t}'\boldsymbol{\mu}) \prod_{i=1}^{p} E(e^{b_i u_i}) \\
&= (\exp \mathbf{t}'\boldsymbol{\mu}) \prod_{i=1}^{p} \left(\exp \tfrac{1}{2} b_i^2\right) \\
&= (\exp \mathbf{t}'\boldsymbol{\mu}) \left(\exp \tfrac{1}{2} \sum_{i=1}^{p} b_i^2\right) \\
&= \exp(\mathbf{t}'\boldsymbol{\mu})\left(\exp \tfrac{1}{2} \mathbf{b}'\mathbf{b}\right) \\
&= \exp(\mathbf{t}'\boldsymbol{\mu})\left(\exp \tfrac{1}{2} \mathbf{t}'AA'\mathbf{t}\right) \\
&= \exp\left(\mathbf{t}'\boldsymbol{\mu} + \tfrac{1}{2}\mathbf{t}'\Sigma\mathbf{t}\right),
\end{aligned}
$$

where we have used (2.2.4) and the independence of the u_is. We have obtained the following theorem.

Theorem 2.3.1 *Let* $\mathbf{x} \sim N_p(\boldsymbol{\mu}, \Sigma)$; *then the moment-generating function is given by* $m(t) = \exp(\mathbf{t}'\boldsymbol{\mu} + \frac{1}{2}\mathbf{t}'\Sigma\mathbf{t})$.

The uniqueness of the moment-generating function can be used to prove the following results.

Theorem 2.3.2 *Let* $\mathbf{x} \sim N_p(\boldsymbol{\mu}, \Sigma)$. *Then for any* $r \times p$ *matrix* C, $C\mathbf{x} \sim N_r(C\boldsymbol{\mu}, C\Sigma C')$.

Example 2.3.1 Suppose $\mathbf{b}: p \times 1$. Then $\mathbf{b}'\mathbf{x} \sim N(b'\boldsymbol{\mu}, \mathbf{b}'\Sigma\mathbf{b})$, the if part of Definition 2.1.2.

Example 2.3.2 Let $\mathbf{x}' = (x_1, x_2)$, with $\mathbf{x} \sim N_2(\boldsymbol{\mu}, \Sigma)$, where $\boldsymbol{\mu}' = (\mu_1, \mu_2)$ and

$$\Sigma = \begin{pmatrix} \sigma_1^2 & \rho\sigma_1\sigma_2 \\ \rho\sigma_1\sigma_2 & \sigma_2^2 \end{pmatrix}.$$

Then $x_1 - x_2$ is normally distributed with mean $\mu_1 - \mu_2$ and variance

$$\sigma_1^2 + \sigma_2^2 - 2\rho\sigma_1\sigma_2.$$

Theorem 2.3.3 *Let $\mathbf{x} \sim N_p(\boldsymbol{\mu}, \Sigma)$. Then any subvector of $\mathbf{x}$ is also normally distributed with the mean as the corresponding subvector of $\boldsymbol{\mu}$ and the dispersion matrix as the corresponding submatrix of Σ.*

Example 2.3.3 Suppose

$$\mathbf{x}' = (\mathbf{x}_1', \mathbf{x}_2'), \qquad \boldsymbol{\mu}' = (\boldsymbol{\mu}_1', \boldsymbol{\mu}_2'), \qquad \text{and} \qquad \Sigma = \begin{pmatrix} \Sigma_{11} & \Sigma_{12} \\ \Sigma_{12}' & \Sigma_{22} \end{pmatrix},$$

where $\mathbf{x}$ and $\boldsymbol{\mu}$ are p-vectors, $\mathbf{x}_1$ and $\boldsymbol{\mu}_1$ are k-vectors, and Σ_{11} is $k \times k$, $k \le p$. Then $\mathbf{x}_1 \sim N_k(\boldsymbol{\mu}_1, \Sigma_{11})$.

Example 2.3.4 Suppose $\mathbf{x} = (x_1, \ldots, x_p)'$, $\boldsymbol{\mu} = (\mu_1, \ldots, \mu_p)'$, and $\Sigma = (\sigma_{ij})$. Then the *marginal* distribution of each x_i is $N(\mu_i, \sigma_{ii})$. It should, however, be pointed out that if the marginal distribution of each x_i is normal, it does *not* necessarily imply that the joint distribution of $(x_1, \ldots, x_p)$ is also normal [see Srivastava and Khatri (1979)].

Theorem 2.3.4 *Let $\mathbf{x}_r$, $r = 1, 2, \ldots, k$, be independently distributed as $N_p(\boldsymbol{\mu}_r, \Sigma_r)$. Then for fixed matrices $A_i: m \times p$*

$$\sum_{r=1}^{k} A_r\mathbf{x}_r \sim N_m\left(\sum_{r=1}^{k} A_r\boldsymbol{\mu}_r, \sum_{r=1}^{k} (A_r\Sigma_r A_r') \right).$$

Example 2.3.5 Let $\boldsymbol{\mu}_r \equiv \boldsymbol{\mu}$ and $\Sigma_r \equiv \Sigma$, $A_r = k^{-1}I_p$, $m = p$, $r = 1, 2, \ldots, k$; that is, $\mathbf{x}_1, \ldots, \mathbf{x}_k$ are independent observations from $N_p(\boldsymbol{\mu}, \Sigma)$. Then $\bar{\mathbf{x}} = k^{-1}\sum_{r=1}^{k}\mathbf{x}_r \sim N_p(\boldsymbol{\mu}, k^{-1}\Sigma)$; that is, the sample mean also has a multivariate normal distribution with mean $\boldsymbol{\mu}$ and covariance matrix $k^{-1}\Sigma$, where k is the size of the sample.

Example 2.3.6 Let $\mathbf{x}_1$ and $\mathbf{x}_2$ be independent $N_p(\boldsymbol{\mu}_i, \Sigma_i)$, $i = 1, 2$. Then $\mathbf{x}_1 + \mathbf{x}_2 \sim N_p(\boldsymbol{\mu}_1 + \boldsymbol{\mu}_2, \Sigma_1 + \Sigma_2)$.

Actually, the converse of Example 2.3.6 is also true. That is, if $\mathbf{x}_1$ and $\mathbf{x}_2$ are independently distributed such that $\mathbf{x}_1 + \mathbf{x}_2$ is distributed normally, then

$\mathbf{x}_1$ and $\mathbf{x}_2$ are normally distributed [see Srivastava and Khatri (1979, p. 46), Cramèr (1937)].

Theorem 2.3.5 *Let* $\mathbf{x}=(\mathbf{x}_1',\mathbf{x}_2')' \sim N_p(\boldsymbol{\mu},\Sigma)$, $\Sigma>0$, *where the partitioning of* Σ *is similar to that of Example* 2.3.3. *Then* $\mathbf{x}_1$ *and* $\mathbf{x}_2$ *are independently distributed iff* $\Sigma_{12}=0$.

Note that the independence of $\mathbf{x}_1$ and $\mathbf{x}_2$, no matter what their joint distribution, implies that the covariance between $\mathbf{x}_1$ and $\mathbf{x}_2$ is zero. However, the zero covariance between $\mathbf{x}_1$ and $\mathbf{x}_2$ does not always imply the independence of $\mathbf{x}_1$ and $\mathbf{x}_2$ unless we assume that they are jointly normally distributed. So the normality assumption is crucial for this part of the theorem.

Theorem 2.3.6 *Let* $\mathbf{x}=(\mathbf{x}_1',\mathbf{x}_2')' \sim N_p(\boldsymbol{\mu},\Sigma)$, $\Sigma>0$, *where* $\mathbf{x}_1$ *and* $\mathbf{x}_2$ *are, respectively, r- and s-vectors,* $r+s=p$, *and the partitioning of* $\boldsymbol{\mu}$ *and* Σ *is similar to that of Example* 2.3.3. *Then the conditional distribution of* $\mathbf{x}_1$, *given* $\mathbf{x}_2$, *is*

$$N_r\left(\boldsymbol{\mu}_1+\Sigma_{12}\Sigma_{22}^{-1}(\mathbf{x}_2-\boldsymbol{\mu}_2),\Sigma_{1\cdot 2}\right),$$

where

$$\Sigma_{1\cdot 2}=\Sigma_{11}-\Sigma_{12}\Sigma_{22}^{-1}\Sigma_{12}'.$$

REMARK: This theorem is fundamental to the theory of linear regression. That is, we have the classical model for $r=1$,

$$E(x_1|\mathbf{x}_2)=\beta_0+\boldsymbol{\beta}'\mathbf{x}_2$$

where $\beta_0=\mu_1-\Sigma_{12}\Sigma_{22}^{-1}\boldsymbol{\mu}_2$ and $\boldsymbol{\beta}'=\Sigma_{12}\Sigma_{22}^{-1}$.

2.4 Random Sample from $N_p(\boldsymbol{\mu},\Sigma)$

Let $\mathbf{x}\sim N_p(\boldsymbol{\mu},\Sigma)$, $\Sigma>0$. Let $\mathbf{x}_1,\mathbf{x}_2,\ldots,\mathbf{x}_N$ be an independent sample of size N on $\mathbf{x}$. We shall write X for the observation matrix

$$X\equiv(\mathbf{x}_1,\ldots,\mathbf{x}_N)=\begin{pmatrix} x_{11} & x_{12} & \cdots & x_{1N}\\ x_{21} & x_{22} & \cdots & x_{2N}\\ \vdots & \vdots & & \vdots\\ x_{p1} & x_{p2} & \cdots & x_{pN}\end{pmatrix}. \tag{2.4.1}$$

Hence

$$E(X)=(\boldsymbol{\mu},\ldots,\boldsymbol{\mu})=\boldsymbol{\mu}\mathbf{e}'=\begin{pmatrix} \mu_1 & \mu_1 & \cdots & \mu_1\\ \mu_2 & \mu_2 & \cdots & \mu_2\\ \vdots & \vdots & & \vdots\\ \mu_p & \mu_p & \cdots & \mu_p\end{pmatrix}. \tag{2.4.2}$$

The joint pdf of $\mathbf{x}_1, \dots, \mathbf{x}_N$ is given by

$$p(X) = \left[(2\pi)^p|\Sigma|\right]^{-N/2}\exp -\tfrac{1}{2}\sum_{\alpha=1}^{N}(\mathbf{x}_\alpha - \boldsymbol{\mu})'\Sigma^{-1}(\mathbf{x}_\alpha - \boldsymbol{\mu}). \quad (2.4.3)$$

Using properties of the trace of a matrix, we can write

$$\begin{aligned}\sum_{\alpha=1}^{N}(\mathbf{x}_\alpha - \boldsymbol{\mu})'\Sigma^{-1}(\mathbf{x}_\alpha - \boldsymbol{\mu}) &= \operatorname{tr}\sum_{\alpha=1}^{N}(\mathbf{x}_\alpha - \boldsymbol{\mu})'\Sigma^{-1}(\mathbf{x}_\alpha - \boldsymbol{\mu})\\ &= \sum_{\alpha=1}^{N}\operatorname{tr}(\mathbf{x}_\alpha - \boldsymbol{\mu})'\Sigma^{-1}(\mathbf{x}_\alpha - \boldsymbol{\mu})\\ &= \sum_{\alpha=1}^{N}\operatorname{tr}\Sigma^{-1}(\mathbf{x}_\alpha - \boldsymbol{\mu})(\mathbf{x}_\alpha - \boldsymbol{\mu})'\\ &= \operatorname{tr}\Sigma^{-1}\sum_{\alpha=1}^{N}(\mathbf{x}_\alpha - \boldsymbol{\mu})(\mathbf{x}_\alpha - \boldsymbol{\mu})'.\end{aligned}$$

Hence (2.4.3) can be rewritten

$$p(X) = \left[(2\pi)^p|\Sigma|\right]^{-N/2}\operatorname{etr} -\tfrac{1}{2}\Sigma^{-1}\left[\sum_{\alpha=1}^{N}(\mathbf{x}_\alpha - \boldsymbol{\mu})(\mathbf{x}_\alpha - \boldsymbol{\mu})'\right], \quad (2.4.4)$$

where etr stands for the exponential of a trace of a matrix.

Alternatively, (2.4.3) can be written in another matrix notation as given below. Although this alternative expression will not be used elsewhere in this book, it may help in reading advanced books and technical papers. Let

$$Z = (\mathbf{z}_1, \mathbf{z}_2, \dots, \mathbf{z}_N) : p \times N \qquad \text{and} \qquad Y = (\mathbf{y}_1, \dots, \mathbf{y}_N) : p \times N.$$

Then

$$\begin{aligned} YZ' &= (\mathbf{y}_1, \mathbf{y}_2, \dots, \mathbf{y}_N)\begin{pmatrix}\mathbf{z}_1'\\ \mathbf{z}_2'\\ \vdots\\ \mathbf{z}_N'\end{pmatrix}\\ &= \mathbf{y}_1\mathbf{z}_1' + \mathbf{y}_2\mathbf{z}_2' + \cdots + \mathbf{y}_N\mathbf{z}_N'\\ &= \sum_{\alpha=1}^{N}\mathbf{y}_\alpha\mathbf{z}_\alpha',\end{aligned}$$

and

$$\begin{aligned}\operatorname{tr} YZ' &= \operatorname{tr}\left(\sum_{\alpha=1}^{N}\mathbf{y}_\alpha\mathbf{z}_\alpha'\right) = \sum_{\alpha=1}^{N}\operatorname{tr}\mathbf{y}_\alpha\mathbf{z}_\alpha'\\ &= \sum_{\alpha=1}^{N}\mathbf{z}_\alpha'\mathbf{y}_\alpha = \sum_{\alpha=1}^{N}\mathbf{y}_\alpha'\mathbf{z}_\alpha.\end{aligned}$$

Labeling $\mathbf{y}_\alpha = \Sigma^{-1}(\mathbf{x}_\alpha - \boldsymbol{\mu})$ and $\mathbf{z}_\alpha = (\mathbf{x}_\alpha - \boldsymbol{\mu})$, we find that

$$\begin{aligned}
&\operatorname{tr}\Sigma^{-1}(X-\boldsymbol{\mu}\mathbf{e}')(X-\boldsymbol{\mu}\mathbf{e}')' \\
&\quad = \operatorname{tr}\left[\Sigma^{-1}(\mathbf{x}_1-\boldsymbol{\mu}),\Sigma^{-1}(\mathbf{x}_2-\boldsymbol{\mu}),\ldots,\Sigma^{-1}(\mathbf{x}_N-\boldsymbol{\mu})\right](\mathbf{x}_1-\boldsymbol{\mu},\ldots,\mathbf{x}_N-\boldsymbol{\mu})' \\
&\quad = \sum_{\alpha=1}^{N}(\mathbf{x}_\alpha-\boldsymbol{\mu})'\Sigma^{-1}(\mathbf{x}_\alpha-\boldsymbol{\mu}).
\end{aligned}$$

Hence we have the following theorem.

Theorem 2.4.1 *The pdf of X is given by* (2.4.3), (2.4.4), *or equivalently*

$$p(X) = \left[(2\pi)^{p/2}|\Sigma|\right]^{-N/2}\operatorname{etr} - \tfrac{1}{2}\Sigma^{-1}(X-\boldsymbol{\mu}\mathbf{e}')(X-\boldsymbol{\mu}\mathbf{e}')'. \qquad (2.4.5)$$

As in the univariate case, we have

$$\begin{aligned}
\sum_{\alpha=1}^{N}(\mathbf{x}_\alpha-\boldsymbol{\mu})(\mathbf{x}_\alpha-\boldsymbol{\mu})' &= \sum_{\alpha=1}^{N}(\mathbf{x}_\alpha-\bar{\mathbf{x}}+\bar{\mathbf{x}}-\boldsymbol{\mu})(\mathbf{x}_\alpha-\bar{\mathbf{x}}+\bar{\mathbf{x}}-\boldsymbol{\mu})' \\
&= \sum_{\alpha=1}^{N}(\mathbf{x}_\alpha-\bar{\mathbf{x}})(\mathbf{x}_\alpha-\bar{\mathbf{x}})' \\
&\quad + N(\bar{\mathbf{x}}-\boldsymbol{\mu})(\bar{\mathbf{x}}-\boldsymbol{\mu})' + \sum_{\alpha=1}^{N}(\bar{\mathbf{x}}-\boldsymbol{\mu})(\mathbf{x}_\alpha-\bar{\mathbf{x}})' + \sum_{\alpha=1}^{N}(\mathbf{x}_\alpha-\bar{\mathbf{x}})(\bar{\mathbf{x}}-\boldsymbol{\mu})' \\
&= \sum_{\alpha=1}^{N}(\mathbf{x}_\alpha-\bar{\mathbf{x}})(\mathbf{x}_\alpha-\bar{\mathbf{x}})' + N(\bar{\mathbf{x}}-\boldsymbol{\mu})(\bar{\mathbf{x}}-\boldsymbol{\mu})',
\end{aligned}$$

where

$$\bar{\mathbf{x}} = N^{-1}\sum_{\alpha=1}^{N}\mathbf{x}_\alpha \qquad (2.4.6)$$

and

$$\sum_{\alpha=1}^{N}(\bar{\mathbf{x}}-\boldsymbol{\mu})(\mathbf{x}_\alpha-\bar{\mathbf{x}})' = (\bar{\mathbf{x}}-\boldsymbol{\mu})\sum_{\alpha=1}^{N}(\mathbf{x}_\alpha-\bar{\mathbf{x}})' = 0.$$

Hence we find from the pdf (2.4.3) that $\bar{\mathbf{x}}$ and

$$V = \sum_{\alpha=1}^{N}(\mathbf{x}_\alpha-\bar{\mathbf{x}})(\mathbf{x}_\alpha-\bar{\mathbf{x}})' = XX' - N\bar{\mathbf{x}}\bar{\mathbf{x}}' \qquad (2.4.7)$$

are sufficient for $(\boldsymbol{\mu}, \Sigma)$. We have proved the following theorem:

Theorem 2.4.2 *Let $\mathbf{x}_1,\ldots,\mathbf{x}_N$ be a random sample of size N from $N_p(\boldsymbol{\mu},\Sigma)$. Then $\bar{\mathbf{x}}$ and V, defined by* (2.4.6) *and* (2.4.7), *respectively, are sufficient statistics for $\boldsymbol{\mu}$ and Σ.*

Proceeding as in Problem 2.7, it can be shown that $\bar{x}$ and V are independently distributed. $\bar{\mathbf{x}} \sim N_p(\boldsymbol{\mu}, N^{-1}\Sigma)$ and $V = \Sigma_{\alpha=2}^{N} \mathbf{y}_\alpha \mathbf{y}_\alpha'$ where $\mathbf{y}_2, \ldots, \mathbf{y}_n$ are independent and identically distributed (iid) as $N_p(\mathbf{0}, \Sigma)$. The distribution of V is known as the Wishart distribution with $n \equiv N-1$ degrees of freedom (df), a multivariate generalization of the chi-square distribution. Its pdf is given by

$$c|\Sigma|^{-n/2}|V|^{(n-p-1)/2} \operatorname{etr} - \tfrac{1}{2}\Sigma^{-1}V, \tag{2.4.8}$$

where $V > 0$, $\Sigma > 0$, and c is a constant of integration chosen so that the total probability is 1. When $p = 1$, this distribution reduces to the chi-square distribution with n df. Hence we have the following theorem:

Theorem 2.4.3 *Let $\mathbf{x}_1, \ldots, \mathbf{x}_n$ be a random sample of size N from $N_p(\boldsymbol{\mu}, \Sigma)$. Then $\bar{\mathbf{x}}$ and V are independently distributed. The distribution of $\bar{\mathbf{x}}$ is $N_p(\boldsymbol{\mu}, N^{-1}\Sigma)$ and that of V is given by (2.4.7), known as the Wishart distribution, denoted by $W_p(\Sigma, n)$, where $n = N-1$. Also V is distributed as $\Sigma_{\alpha=2}^{N} \mathbf{y}_\alpha \mathbf{y}_\alpha'$ where the $\mathbf{y}_\alpha$s are iid $N_p(\mathbf{0}, \Sigma)$.*

Since

$$V = \sum_{\alpha=2}^{N} \mathbf{y}_\alpha \mathbf{y}_\alpha', \tag{2.4.9}$$

where the $\mathbf{y}$s are iid $N_p(\mathbf{0}, \Sigma)$, we get

$$E(V) = (N-1)\Sigma = n\Sigma. \tag{2.4.10}$$

Hence

$$S = n^{-1}V \tag{2.4.11}$$

is an unbiased estimate of Σ. Similarly, $\bar{\mathbf{x}}$ is an unbiased estimate of $\boldsymbol{\mu}$. Hence we have the following theorem:

Theorem 2.4.4 *Let $\mathbf{x}_1, \ldots, \mathbf{x}_N$ be independent $N_p(\boldsymbol{\mu}, \Sigma)$. Then $\bar{\mathbf{x}} = N^{-1}\Sigma_{\alpha=1}^{N} \mathbf{x}_\alpha$ and $S = n^{-1}V$ are unbiased estimates of $\boldsymbol{\mu}$ and Σ, respectively.*

It may be noted that the normality assumption is not required.

We shall now define the

Sample Correlation Matrix Let $S = (s_{ij})$. Here $s_{11}, \ldots, s_{pp}$ are the p diagonal elements of S. Then the sample correlation matrix is $R = D_{S^{-1/2}} S D_{S^{-1/2}}$, where $D_{S^{-1/2}} = \operatorname{diag}(s_{11}^{-1/2}, \ldots, s_{pp}^{-1/2})$, a diagonal matrix.

2.4.1 Maximum Likelihood Estimates of $\boldsymbol{\mu}$ and Σ

Let $\mathbf{x} \sim N_p(\boldsymbol{\mu}, \Sigma)$, $\Sigma > 0$. Let $\mathbf{x}_1, \ldots, \mathbf{x}_N$ be an independent sample of size N on $\mathbf{x}$. For any vector $\mathbf{a} \neq \mathbf{0}$, let $y = \mathbf{a}'\mathbf{x}$. Then $y \sim N(\theta, \sigma^2)$, where $\theta = \mathbf{a}'\boldsymbol{\mu}$ and $\sigma^2 = \mathbf{a}'\Sigma\mathbf{a}$. Hence $y_1, \ldots, y_N$ is an independent sample of size N on y, where

$y_i = \mathbf{a}'\mathbf{x}_i$, $i=1,2,\ldots,N$. From univariate theory, the maximum likelihood estimates of θ and σ^2 are $\bar{y}$ and $N^{-1}\Sigma_{\alpha=1}^{N}(y_\alpha - \bar{y})^2$, respectively. Note that

$$\bar{y} = N^{-1}\sum_{\alpha=1}^{N} y_\alpha = N^{-1}\sum_{\alpha=1}^{N} \mathbf{a}'\mathbf{x}_\alpha = N^{-1}\mathbf{a}'\sum_{\alpha=1}^{N} \mathbf{x}_\alpha = \mathbf{a}'\bar{\mathbf{x}},$$

$$\sum_{\alpha=1}^{N}(y_\alpha - \bar{y})^2 = \sum_{\alpha=1}^{N}(y_\alpha - \bar{y})(y_\alpha - \bar{y}) = \mathbf{a}'\sum_{\alpha=1}^{N}(\mathbf{x}_\alpha - \bar{\mathbf{x}})(\mathbf{x}_\alpha - \bar{\mathbf{x}})'\mathbf{a} = \mathbf{a}'V\mathbf{a}.$$

Hence $\mathbf{a}'\bar{\mathbf{x}}$ and $N^{-1}(\mathbf{a}'V\mathbf{a})$ are the maximum likelihood estimates of $\mathbf{a}'\boldsymbol{\mu}$ and $\mathbf{a}'\Sigma\mathbf{a}$, respectively. Since it is true for every $\mathbf{a} \neq \mathbf{0}$, it can be shown that $\bar{\mathbf{x}}$ and $N^{-1}V$ are the maximum likelihood estimates of $\boldsymbol{\mu}$ and Σ, respectively.

For example, choosing $\mathbf{a}' = (0,\ldots,0,1,\ldots,0)$, i.e., a 1 in the ith place, we get the maximum likelihood estimates of μ_i and σ_{ii} as $\bar{x}_i$ and $N^{-1}v_{ii}$, respectively, where $\bar{\mathbf{x}} = (\bar{x}_1,\ldots,\bar{x}_p)'$ and $V=(v_{ij})$. Similarly, by choosing $\mathbf{a}' = (1,1,0,\ldots,0)$ and using the above results, it can be shown that the maximum likelihood estimate of σ_{12} is $N^{-1}v_{12}$, and so on.

Thus we get the following theorem.

Theorem 2.4.5 *Let $\mathbf{x}_1,\ldots,\mathbf{x}_N$ be an independent sample of size N from $N_p(\boldsymbol{\mu},\Sigma)$. Then the maximum likelihood estimates of $\boldsymbol{\mu}$ and Σ are given by $\bar{\mathbf{x}}$ and $N^{-1}V$, respectively.*

Note that, as in the univariate case, $N^{-1}V$ is not an unbiased estimate of Σ.

The likelihood function is again defined by (2.4.3)–(2.4.5) except that here we consider it as a function of $\boldsymbol{\mu}$ and Σ for given observations $\mathbf{x}_1,\ldots,\mathbf{x}_N$. It can be written

$$L(\boldsymbol{\mu},\Sigma) = (2\pi)^{-Np/2}|\Sigma|^{-N/2}\operatorname{etr} - \tfrac{1}{2}\Sigma^{-1}[V + N(\bar{\mathbf{x}}-\boldsymbol{\mu})(\bar{\mathbf{x}}-\boldsymbol{\mu})'], \tag{2.4.12}$$

where

$$V = \sum_{\alpha=1}^{N}(\mathbf{x}_\alpha - \bar{\mathbf{x}})(\mathbf{x}_\alpha - \bar{\mathbf{x}})' = XX' - N\bar{\mathbf{x}}\bar{\mathbf{x}}'.$$

2.5 Some Results on Quadratic Forms

Let $\mathbf{x} \sim N_p(\mathbf{0},\sigma^2 I)$. Then $\mathbf{x}'\mathbf{x}/\sigma^2$ is distributed as a chi-square random variable with p degrees of freedom. However, if $\mathbf{x} \sim N_p(\boldsymbol{\mu},\sigma^2 I)$, then $\mathbf{x}'\mathbf{x}/\sigma^2$ is said to have a noncentral chi-square distribution with $p$df and a noncentrality parameter $\boldsymbol{\mu}'\boldsymbol{\mu}/\sigma^2$. We now list a few properties of the chi-square distribution. Without loss of generality we shall assume $\sigma^2=1$. In the following we also assume that A and the A_is are symmetric matrices.

i. Let $\mathbf{x} \sim N_p(\mathbf{0}, I)$. Then $\mathbf{x}'A\mathbf{x}$ has a chi-square distribution with r df if and only if $\rho(A)=r$ and $A^2=A$.
ii. Let $\mathbf{x} \sim N_p(\mathbf{0}, I)$. Then $\mathbf{x}'A_1\mathbf{x}$ and $\mathbf{x}'A_2\mathbf{x}$ are independently distributed iff $A_1A_2=0$.
iii. Let $\mathbf{x} \sim N_p(\boldsymbol{\theta}, I)$, and let $q_j=\mathbf{x}'A_j\mathbf{x}$, $j=1,2,\ldots,k$, where $\Sigma_{j=1}^k A_j=I$. Then $q_1,\ldots,q_k$ are independently distributed as noncentral chi-squares iff $\Sigma_{j=1}^k\rho(A_j)=p$, or, alternatively, iff $A_1,\ldots,A_k$ are idempotent matrices such that $A_iA_j=0$ for $i\neq j$.

Appendix

Lemma 2.A.1 *Let* $\mathbf{x}=A\mathbf{u}$. *Then the Jacobian of the transformation* $J(\mathbf{u}\to\mathbf{x})=|A^{-1}|$ *or* $J(\mathbf{x}\to\mathbf{u})=|A|$.

PROOF: Since $\mathbf{x}=A\mathbf{u}$, we get $\mathbf{u}=A^{-1}\mathbf{x}\equiv B\mathbf{x}$. Let

$$\mathbf{u}=(u_1,\ldots,u_p)',\qquad \mathbf{x}=(x_1,\ldots,x_p)',\qquad \text{and}\qquad B=\begin{pmatrix}\mathbf{b}_i'\\ \vdots\\ \mathbf{b}_p'\end{pmatrix}.$$

Then

$$\begin{aligned}
u_1&=\mathbf{b}_1'\mathbf{x}=b_{11}x_1+\cdots+b_{1p}x_p,\\
u_2&=\mathbf{b}_2'\mathbf{x}=b_{21}x_1+\cdots+b_{2p}x_p,\\
&\vdots\\
u_p&=\mathbf{b}_p'\mathbf{x}=b_{p1}x_1+\cdots+b_{pp}x_p.
\end{aligned}$$

	u_1	u_2	$\cdots$	u_p
x_1	$\partial u_1/\partial x_1$	$\partial u_2/\partial x_1$	$\cdots$	$\partial u_p/\partial x_1$
$\vdots$	$\vdots$	$\vdots$		$\vdots$
x_p	$\partial u_1/\partial x_p$	$\partial u_2/\partial x_p$	$\cdots$	$\partial u_p/\partial x_p$

Hence $J(\mathbf{u}\to\mathbf{x})$ is the absolute value of the above determinant and is equal to

$$|B|_+=|A^{-1}|=|AA'|^{-1/2}=|\Sigma|^{-1/2}$$

since $|A|=|A'|$. □

Problems

2.1 Let $u\sim N(0,1)$. Show that $E(\exp bu)=\exp\frac{1}{2}b^2$.

2.2 Let $\mathbf{x}\sim N_2(\boldsymbol{\mu},\Sigma)$, where $\mathbf{x}=(x_1,x_2)'$, $\boldsymbol{\mu}=(\mu_1,\mu_2)'$, and

$$\Sigma=\sigma^2\begin{pmatrix}1&\rho\\ \rho&1\end{pmatrix}.$$

Show that x_1+x_2 and x_1-x_2 are independently distributed.

2.3 Let $\mathbf{x}=(\mathbf{x}_1',\mathbf{x}_2')' \sim N_{2p}(\boldsymbol{\mu},\Sigma)$, where $\boldsymbol{\mu}'=(\boldsymbol{\mu}_1',\boldsymbol{\mu}_2')$, $\mathbf{x}_1$, $\mathbf{x}_2$, $\boldsymbol{\mu}_1$, and $\boldsymbol{\mu}_2$ are p-vectors and

$$\Sigma=\begin{pmatrix}\Sigma_1 & \Sigma_2\\ \Sigma_2 & \Sigma_1\end{pmatrix}.$$

Show that $\mathbf{x}_1+\mathbf{x}_2$ and $\mathbf{x}_1-\mathbf{x}_2$ are independently distributed.

2.4 Let $\mathbf{x}\sim N_3(\boldsymbol{\mu},\Sigma)$ and

$$A=\begin{pmatrix}\frac{1}{2} & -1 & \frac{1}{2}\\ -\frac{1}{2} & 0 & -\frac{1}{2}\end{pmatrix}.$$

Find the distribution of $\mathbf{y}=A\mathbf{x}$.

2.5 Let $\mathbf{x}\sim N_2(\mathbf{0},I)$, where $\mathbf{x}=(x_1,x_2)$. Find the conditional distribution of x_1 given x_1+x_2.
Hint: First obtain the joint distribution of x_1 and x_1+x_2 by a linear transformation of the type $\mathbf{y}=A\mathbf{x}$.

2.6 Let

$$A=\begin{pmatrix}1/\sqrt{4} & 1/\sqrt{4} & 1/\sqrt{4} & 1/\sqrt{4}\\ 1/\sqrt{2} & -1/\sqrt{2} & 0 & 0\\ 1/\sqrt{6} & 1/\sqrt{6} & -2/\sqrt{6} & 0\\ 1/\sqrt{12} & 1/\sqrt{12} & 1/\sqrt{12} & -3/\sqrt{12}\end{pmatrix}.$$

(a) Show that $AA'=I$; that is, A is an orthogonal matrix.
(b) Let $\mathbf{y}=A\mathbf{x}$, where $\mathbf{x}\sim N_4(\mu\mathbf{e},\sigma^2 I)$; $\mathbf{e}'=(1,1,1,1)$. Show that

$$y_2^2+y_3^2+y_4^2=\sum_{i=1}^{4}x_i^2-\frac{1}{4}\left(\sum_{i=1}^{4}x_i\right)^2$$

and that y_1, y_2, y_3, and y_4 are independently distributed, where $y_1\sim N(2\mu,\sigma^2)$ and $y_i\sim N(0,\sigma^2)$, $i=2,3,4$.

2.7 Let $\mathbf{x}\sim N_n(\boldsymbol{\mu},\sigma^2 I)$. Show that $\bar{x}=n^{-1}\Sigma_{i=1}^n x_i$ and $s^2=\Sigma(x_i-\bar{x})^2=\Sigma x_i^2-n\bar{x}^2=\Sigma x_i^2-n^{-1}(x_i)^2$ are independently distributed.

2.8 **(a)** Let $\mathbf{x}$ be $N_3(\mathbf{0},\Sigma)$, where

$$\Sigma=\begin{pmatrix}1 & \rho_{12} & \rho_{13}\\ \rho_{12} & 1 & \rho_{23}\\ \rho_{13} & \rho_{23} & 1\end{pmatrix}.$$

What is the distribution of (x_1,x_2), given x_3? Find the covariance between x_1 and x_2, given x_3.
(b) Repeat (a) for $\rho_{12}=0.4$, $\rho_{13}=0.7$, and $\rho_{23}=0.6$.

2.9 Let

$$\mathbf{x} = \begin{pmatrix} x_1 \\ x_2 \end{pmatrix} \sim N_2\left(\begin{pmatrix} \mu_1 \\ \mu_2 \end{pmatrix}, \begin{pmatrix} \sigma_1^2 & \rho\sigma_1\sigma_2 \\ \rho\sigma_1\sigma_2 & \sigma_2^2 \end{pmatrix} \right).$$

What is $E(x_1 \mid x_2)$?

2.10 Let $\mathbf{x} \sim N_p(\mu, \Sigma)$. Show that

(a) $E(\mathbf{x}\mathbf{x}') = \Sigma + \boldsymbol{\mu}\boldsymbol{\mu}'$.

(b) $E(\mathbf{x}'A\mathbf{x}) = E\,\mathrm{tr}(\mathbf{x}\mathbf{x}'A) = \mathrm{tr}\,\Sigma A + \boldsymbol{\mu}'A\boldsymbol{\mu}$.

(c) Suppose $\boldsymbol{\mu} = \mu\mathbf{e}$, $\Sigma = \sigma^2 I$, and $A = I - \mathbf{e}\mathbf{e}'p^{-1}$. Use the results of (a) and (b) to show that

$$\sigma^{-2}E(\mathbf{x}'A\mathbf{x}) = \mathrm{tr}\,A = p - 1.$$

Alternatively, if $x_1 \cdots x_p \sim N(\mu, \sigma^2)$, then $E(\Sigma(x_i - \bar{x})^2) = (p-1)\sigma^2$.

Chapter 3
Inference on Location—Hotelling's T^2

3.1 Introduction

In this chapter we consider several problems concerning testing and confidence intervals for the mean of a multivariate distribution. The tests are based on the likelihood ratio principle, and the confidence intervals are based on Roy's union–intersection principle and Bonferroni inequalities. Certain assumptions are necessary: For the analysis to be valid the data must follow a multivariate normal distribution and, in the two-population case, the covariances must be equal. The problem of unequal covariances, discussed in Section 3.3.8, poses no difficulty, but nonnormal data cannot be handled so easily. Tests for multivariate normality are not very powerful. Even if a null hypothesis of normality is not rejected, the data might not be normal, and if the null hypothesis is rejected, one still needs to analyze the data. We suggest making power and modulus transformations and looking at univariate normal probability plots, as discussed in Section 3.7.

We begin with test statistics and confidence intervals for the mean of a multivariate normal distribution. Since most multivariate procedures are generalizations of univariate procedures, we shall review them briefly.

3.2 Univariate Testing Problems

3.2.1 One-Sample Student's t-Test

Example 3.2.1 Eight men received 2.0 mg/kg of a certain drug. Their change in blood sugar level was recorded:

Subject	1	2	3	4	5	6	7	8
Blood sugar change	30	90	−10	10	30	60	0	40

With the assumption that the observations $x_1,\ldots,x_N$, $N=8$, are iid $N(\mu,\sigma^2)$, the test of hypothesis that the drug produces a change of μ_0 in the blood sugar level is given by H: $\mu = \mu_0$, $\sigma^2 > 0$, vs A: $\mu \neq \mu_0$, $\sigma^2 > 0$. H is rejected if

$$|N^{1/2}(\bar{x} - \mu_0)|/s \geq t_{N-1,\alpha/2}, \tag{3.2.1}$$

where $\bar{x} = (\Sigma x_i)/N$, $s^2 = [1/(N-1)]\Sigma(x_i - \bar{x})^2$, and $t_{\alpha/2}$ is chosen such that the size of the test is α.

Since under H, $N^{1/2}(\bar{x} - \mu_0)/s$ has a Student's t-distribution with $n \equiv N - 1$ degrees of freedom, $t_{N-1,\alpha/2}$ is the upper $(\alpha/2)100\%$ point of the t-distribution with $N-1$ df. Alternatively, since under H, $N(\bar{x} - \mu_0)^2/s^2 \sim F_{1,n}$, where $F_{1,n}$ denotes the F-distribution with 1 and n degrees of freedom, we reject H if

$$N(\bar{x} - \mu_0)^2/s^2 > F_{1,n,\alpha} \equiv t^2_{N-1,\alpha/2}, \tag{3.2.2}$$

where $F_{1,n,\alpha}$ is the upper $\alpha 100\%$ point of the F-distribution with 1 and n degrees of freedom. Under the alternative A, the distribution of the statistic in (3.2.1) is a noncentral t-distribution with $n \equiv N-1$ df and that of (3.2.2) is a noncentral F-distribution with $(1, n)$ df. The noncentrality parameter in both the cases is

$$\delta^2 = N(\mu - \mu_0)^2/\sigma^2. \tag{3.2.3}$$

A $(1-\alpha)100\%$ confidence interval for μ is given by

$$\bar{x} - st_{n,\alpha/2}N^{-1/2} \leq \mu \leq \bar{x} + st_{n,\alpha/2}N^{-1/2}. \tag{3.2.4}$$

In Example 3.2.1, $\bar{x} = 31.25$, $s^2 = 10.70$, and $N = 8$. Hence the test of the hypothesis that there was a change in blood sugar level is as follows: Reject H: $\mu = 0$ vs A: $\mu \neq 0$ if

$$t^2 = N(\bar{x} - 0)^2/s^2 \geq F_{1,7,0.05}.$$

The value of t^2 is 7.1 with $F_{1,7,0.05} = 5.59$. Hence we reject H: $\mu = 0$.

The above Student's t-test is the likelihood ratio test. It has the property that if the scale of the observations is changed, the test remains unchanged. That is, if $\bar{x} \to c\bar{x}$ and $s^2 \to c^2 s^2$, $c > 0$, then the value of t remains unchanged. Also, among all unbiased tests (tests whose power for all parameter values is greater than or equal to significance level), the t-test has uniformly maximum power.

3.2.2 *Two-Sample Student's t-Test*

Let $x_1, x_2, \ldots, x_{N_1}, y_1, \ldots, y_{N_2}$ be independently distributed, where the first N_1 observations are from $N(\mu_1, \sigma^2)$ and the last N_2 observations are from $N(\mu_2, \sigma^2)$. The set of sufficient statistics is

$$\bar{x} = N_1^{-1} \sum_{i=1}^{N_1} x_i, \qquad \bar{y} = N_2^{-1} \sum_{i=1}^{N_2} y_i;$$

$$s_1^2 = n_1^{-1} \sum (x_i - \bar{x})^2, \qquad \text{and} \qquad s_2^2 = n_2^{-1} \sum (y_i - \bar{y})^2,$$

where $n_1 = N_1 - 1$ and $n_2 = N_2 - 1$. The pooled estimate of σ^2 is given by

$$s_p^2 = f^{-1}(n_1 s_1^2 + n_2 s_2^2),$$

where $f = n_1 + n_2$. Suppose we wish to test the hypothesis H: $\mu_1 = \mu_2$ vs A: $\mu_1 \neq \mu_2$. Then the likelihood ratio test is as follows: Reject H if

$$\left(\frac{N_1 N_2}{N_1 + N_2}\right)^{1/2} \left|\frac{\bar{x} - \bar{y}}{s_p}\right| > t_{f,\alpha/2},$$

where $t_{f,\alpha/2}$ is the upper $(\alpha/2)100\%$ point of the Student's t-distribution with $f = n_1 + n_2 = N_1 + N_2 - 2$ df. Alternatively, reject H if

$$\left[\left(\frac{N_1 N_2}{N_1 + N_2}\right) \frac{(\bar{x} - \bar{y})^2}{s_p^2}\right] > t_{f,\alpha/2}^2 \equiv F_{1,f,\alpha}.$$

3.3 Multivariate Generalization

In Example 3.2.1, eight men had only their blood sugar levels measured. If more characteristics are to be measured, then the data must be analyzed by multivariate techniques. Hence we have the following data set, where blood pressure levels were also measured.

Example 3.3.1 Eight men each received a certain drug. The recorded changes in blood sugar and blood pressure (systolic and diastolic) are listed in Table 3.1. The null hypothesis to be tested is that the change in characteristics due to the drug is a specified amount (zero).

In general, let $\mathbf{x}_1, \ldots, \mathbf{x}_N$ be independently normally distributed with mean $\boldsymbol{\mu}$ and covariance Σ, written $N_p(\boldsymbol{\mu}, \Sigma)$, where p is the dimension of $\boldsymbol{\mu}$. Then estimates of $\boldsymbol{\mu}$ and Σ are given by

$$\bar{\mathbf{x}} = N^{-1} \sum_{i=1}^{N} \mathbf{x}_i \qquad \text{and} \qquad S = n^{-1} \sum_{i=1}^{N} (\mathbf{x}_i - \bar{\mathbf{x}})(\mathbf{x}_i - \bar{\mathbf{x}})',$$

where $n = N - 1$. Note that in contrast to the univariate case we take

Table 3.1

	Subject							
Characteristic	1	2	3	4	5	6	7	8
Blood sugar	30	90	−10	10	30	60	0	40
Blood pressure (systolic)	−8	7	−2	0	−2	0	−2	1
Blood pressure (diastolic)	−1	6	4	2	5	3	4	2

S—and not S^2—as an estimator of Σ. For testing the hypothesis

$$H: \boldsymbol{\mu} = \boldsymbol{\mu}_0 \qquad \text{vs} \qquad A: \boldsymbol{\mu} \neq \boldsymbol{\mu}_0$$

we reject H if

$$T^2 = N(\bar{\mathbf{x}} - \boldsymbol{\mu}_0)' S^{-1} (\bar{\mathbf{x}} - \boldsymbol{\mu}_0) \geq T_\alpha^2, \tag{3.3.1}$$

where T_α^2 is chosen such that the error of the first kind is at a specified level α. The statistic in (3.3.1) is called Hotelling's T^2-statistic, a natural generalization of Student's t-statistic. It is shown in Section 3.3.4 that it is the likelihood ratio test. In Section 3.3.5 another heuristic derivation is given; this derivation (Roy, 1957) is known as Roy's union–intersection test. Under the hypothesis H, the statistic

$$\frac{n-p+1}{pn} T^2, \tag{3.3.2}$$

where $n = N - 1$, has an F-distribution with $(p, n-p+1)$ degrees of freedom. Thus, equivalently, the hypothesis H is rejected if

$$\frac{(n-p+1)N}{pn} (\bar{\mathbf{x}} - \boldsymbol{\mu}_0)' S^{-1} (\bar{\mathbf{x}} - \boldsymbol{\mu}_0) \geq F_{p, n-p+1, \alpha} \tag{3.3.3}$$

where $F_{p, n-p+1, \alpha}$ is the upper $\alpha 100\%$ point of the F-distribution with $(p, n-p+1)$ degrees of freedom. Under the alternative A, the distribution of the statistic in (3.3.2) has a noncentral $F_{p, n-p+1}$ distribution. The noncentrality parameter δ^2 is given by

$$\delta^2 = N(\boldsymbol{\mu} - \boldsymbol{\mu}_0)' \Sigma^{-1} (\boldsymbol{\mu} - \boldsymbol{\mu}_0).$$

It can be seen that when $p = 1$ (3.3.3) coincides with (3.2.2). Since there are p unknown μs, $\mu_1, \ldots, \mu_p$, we expect to lose p degrees of freedom, rather than just one as in the univariate case.

The above test has many optimum properties. For example, the Hotelling's T^2-test is uniformly most powerful among all tests whose power depends on δ^2. Also, the statistic (and hence the distribution) T^2 remains unchanged under any nonsingular linear transformation of the original variables $\mathbf{x} \to C\mathbf{x}$ for C nonsingular. For the data of Example 3.3.2, we have

$$\bar{\mathbf{x}} = \begin{pmatrix} 31.25 \\ -0.75 \\ 3.125 \end{pmatrix}, \qquad S = \begin{pmatrix} 1069.64 & 82.5 & 16.964 \\ 82.5 & 17.357 & 6.393 \\ 16.964 & 6.393 & 4.696 \end{pmatrix}, \qquad N = 8.$$

The null hypothesis is

$$H: \boldsymbol{\mu} = \mathbf{0} \qquad \text{vs} \qquad A: \boldsymbol{\mu} \neq 0,$$

and the T^2 value is

$$T^2 = N\bar{\mathbf{x}}' S^{-1} \bar{\mathbf{x}} = 79.064.$$

From (3.3.2), the value of the F-statistic is

$$\frac{n-p+1}{np}T^2=\frac{5}{7\times 3}\times 79.064=18.825>F_{3,5,0.05}=5.41.$$

Hence we may claim at the 5% level that there is a difference between the mean responses before and after the drug was administered. Note that this change could be due to the administering of a drug and not to the drug itself. Sometimes there will be a significant change in characteristics when a placebo is given. An experimenter should use a control group to test for any change with a placebo. If there is a significant change with the placebo, the experimenter can then test for a difference between the effects of the drug and the effects of the placebo with the two sample T^2-tests given in Sections 3.3.4 and 3.3.8. For simplicity, the data here are artificial. For subsequent data sets, taken from actual experiments, we give a more thorough discussion of the conclusions and assumptions.

3.3.1 Confidence Regions

In the univariate case the confidence regions are intervals. However, in the multivariate case we get an ellipsoidal region. This $(1-\alpha)100\%$ confidence region for $\boldsymbol{\mu}$ is given by

$$\left\{\boldsymbol{\mu}: N(\bar{\mathbf{x}}-\boldsymbol{\mu})'S^{-1}(\bar{\mathbf{x}}-\boldsymbol{\mu})\leq[np/(n-p+1)]F_{p,n-p+1,\alpha}\right\}. \quad (3.3.4)$$

In the multivariate case, however, we may also be interested in the simultaneous confidence intervals on all linear functions of $\boldsymbol{\mu}$. By some results in matrix theory (see Section 3.3.5) it can be shown that (3.3.4) is equivalent to

$$\mathbf{a}'\bar{\mathbf{x}}-T_\alpha N^{-1/2}(\mathbf{a}'S\mathbf{a})^{1/2}\leq\mathbf{a}'\boldsymbol{\mu}\leq\mathbf{a}'\bar{\mathbf{x}}+T_\alpha N^{-1/2}(\mathbf{a}'S\mathbf{a})^{1/2} \quad (3.3.5)$$

for all $\mathbf{a}\neq 0$, where

$$T_\alpha^2=[pn/(n-p+1)]F_{p,n-p+1,\alpha}. \quad (3.3.6)$$

Thus, with a confidence coefficient greater than $1-\alpha$, (3.3.5) gives confidence intervals for any linear function $\mathbf{a}'\boldsymbol{\mu}$. Equation (3.3.4) can also be used to obtain confidence intervals for $\boldsymbol{\mu}'\boldsymbol{\mu}$. If $\mathrm{ch}_i(A)$ denotes the ith largest root of a matrix A, $i=1,2,\ldots,p$ $(\mathrm{ch}_1(A)\geq\mathrm{ch}_2(A)\geq\cdots\geq\mathrm{ch}_p(A))$, then a $(1-\alpha)100\%$ confidence interval for $(\boldsymbol{\mu}'\boldsymbol{\mu})^{1/2}$ is given by

$$\left[\mathrm{ch}_p(S)\right]^{1/2}(T-T_\alpha)\leq(\boldsymbol{\mu}'\boldsymbol{\mu})^{1/2}\leq[\mathrm{ch}_1(S)]^{1/2}(T+T_\alpha), \quad (3.3.7)$$

where

$$T^2=N\bar{\mathbf{x}}'S^{-1}\bar{\mathbf{x}} \quad \text{and} \quad T_\alpha^2=[pn/(n-p+1)]F_{p,n-p+1,\alpha}.$$

For the univariate case, procedures to find confidence intervals for all linear combinations of parameters, such as Scheffé's procedure in ANOVA, are too conservative if we only wish to look at a few combinations. This problem carries over to the multivariate situation. For example, if we wish to obtain $(1-\alpha)100\%$ simultaneous confidence intervals for $\mathbf{a}_i'\mu$, $i=1,\ldots,k$, and if k is small then the intervals in (3.3.5) will be too wide. *Bonferroni inequalities* tend to give smaller intervals if the number k of linear combinations of interest is small. The confidence intervals are given by

$$\mathbf{a}_i'\bar{\mathbf{x}}-t_{N-1,\alpha/2k}N^{-1/2}(\mathbf{a}_i'S\mathbf{a}_i)^{1/2}\leq\mathbf{a}_i'\boldsymbol{\mu}\leq\mathbf{a}_i'\bar{\mathbf{x}}+t_{N-1,\alpha/2k}N^{-1/2}(\mathbf{a}_i'S\mathbf{a}_i)^{1/2},$$
$$i=1,\ldots,k. \quad (3.3.8)$$

The choice of the type of confidence interval depends on whether $t_{N-1,\alpha/2k} \leq T_\alpha$.

We justify the use of $t_{N-1,\alpha/2k}$ with a probability argument. If $E_1,\ldots,E_k$ are k events then

$$1-P(E_1\cap\cdots\cap E_k)\leq\sum_{i=1}^{k}P(E_i^c),$$

where E_i^c is the complement of E_i for $i=1,\ldots,k$. For our purposes $E_1^c,\ldots,E_k^c$ are rejection regions. If we perform k tests at levels $\alpha_1,\ldots,\alpha_k$, then the probability of accepting all hypotheses when they are true is

$$P(E_1\cap\cdots\cap E_k)\geq 1-\sum_{i=1}^{k}P(E_i^c)=1-\sum_{i=1}^{k}\alpha_i.$$

Hence we can obtain an overall significance level of no more than α by choosing $\Sigma_{i=1}^{k}\alpha_i=\alpha$. In the absence of any preference for choosing $\alpha_1,\ldots,\alpha_k$ we choose, by symmetry, $\alpha_1=\cdots=\alpha_k=\alpha/k$. Simultaneous $(1-\alpha)100\%$ confidence intervals for the parameters consist of all values of the parameters that would be accepted by the test procedure.

Consider Example 3.3.1 for the case of systolic and diastolic blood pressure changes only. That is, for $N=8$ observations we have

$$\begin{pmatrix}\bar{x}_1\\ \bar{x}_2\end{pmatrix}=\begin{pmatrix}-0.75\\ 3.125\end{pmatrix}\quad\text{and}\quad S=\begin{pmatrix}17.357 & 6.393\\ 6.393 & 4.696\end{pmatrix}.$$

Simultaneous ellipsoidal 95% confidence regions for the population means μ_1 and μ_2 are given by the set of all (μ_1,μ_2) that satisfy the relationship

$$[(-0.75-\mu_1),(3.125-\mu_2)]\begin{pmatrix}0.116 & -0.157\\ -0.157 & 0.427\end{pmatrix}\begin{pmatrix}-0.75-\mu_1\\ 3.125-\mu_2\end{pmatrix}$$
$$\leq[(7\times2)/(6\times8)]F_{2,6,0.05}=0.292\times5.1433=1.50.$$

Equivalently,

$$0.116(\mu_1+0.75)^2-0.314(\mu_1+0.75)(\mu_2-3.125)+0.427(\mu_2-3.125)^2$$
$$\leq1.50.$$

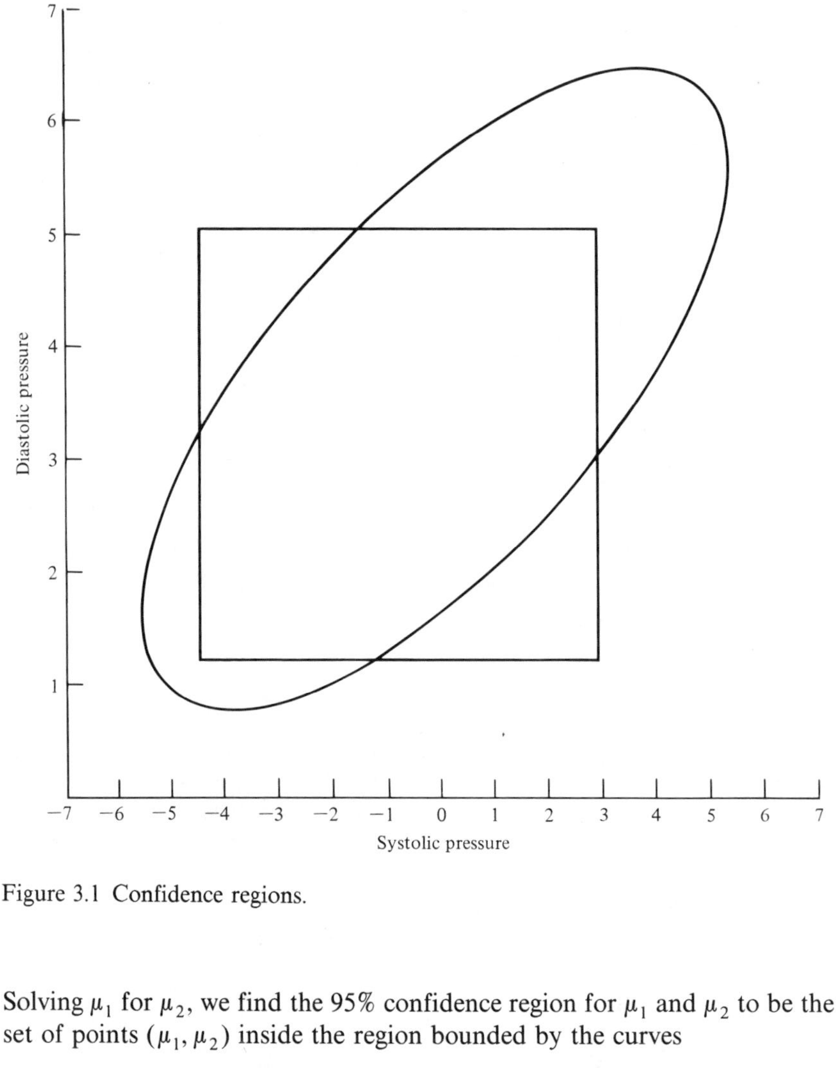

Figure 3.1 Confidence regions.

Solving μ_1 for μ_2, we find the 95% confidence region for μ_1 and μ_2 to be the set of points (μ_1, μ_2) inside the region bounded by the curves

$$\mu_1 = (1.352\mu_2 - 4.978) \pm (9.71\mu_2 - 5.12 - 1.849\mu_2^2)^{1/2}.$$

The ellipsoidal confidence region is given in Fig. 3.1. Since, for $k=2$, t_{n}, $\alpha/2k \le T_\alpha$, we may use Bonferroni's bounds. We obtain 95% simultaneous intervals for μ_1 and μ_2 as

$$\bar{x}_1 - t_{7,0.0125}\sqrt{17.357/8} \le \mu_1 \le \bar{x}_1 + t_{7,0.0125}\sqrt{17.357/8}$$

and

$$\bar{x}_2 - t_{7,0.0125}\sqrt{4.696/8} \le \mu_2 \le \bar{x}_2 + t_{7,0.0125}\sqrt{4.696/8}$$

or

$$\begin{pmatrix} -5.18 \leq \mu_1 \leq 3.68 \\ 0.825 \leq \mu_2 \leq 5.425 \end{pmatrix}.$$

In the case of Bonferroni's inequalities, the confidence regions are rectangular. The two types of regions are given in Fig. 3.1. The Bonferroni-type regions give narrower confidence intervals for μ_1 or μ_2 but wider intervals for a combination of the two.

3.3.2 *Fixed-Width Confidence Regions*

If we wish to fix the width of our confidence region, we must let the sample size vary. In the univariate situation $x_1,\ldots,x_N$ are independently distributed as $N(\mu,\sigma^2)$. The sample mean is $\bar{x}$. If σ^2 is known then $P(N(\bar{x}-\mu)^2 \leq \sigma^2 z^2_{\alpha/2}) = 1-\alpha$ where z_α is such that for $z \sim N(0,1)$ $P(z \geq z_\alpha) = \alpha$. A $(1-\alpha)100\%$ confidence interval for μ is given by $(\bar{x} - \sigma N^{1/2} z_{\alpha/2}, \bar{x} + \sigma N^{-1/2} z_{\alpha/2})$. We wish to fix the width of the interval as no more than $2d$. For some fixed value d, then, we must have

$$\sigma N^{-1/2} z_{\alpha/2} < d.$$

Equivalently,

$$N > \sigma^2 z^2_{\alpha/2} / d^2.$$

Suppose now that we have multivariate observations $\mathbf{x}_1,\ldots,\mathbf{x}_N \sim N_p(\boldsymbol{\mu},\Sigma)$ with Σ known. We may wish k confidence intervals for parameters of the form $\mathbf{a}'_j\boldsymbol{\mu}$ for some vector $\mathbf{a}_j$, $j=1,\ldots,k$. If $\bar{\mathbf{x}}$ is the sample mean, then k simultaneous $(1-\alpha)100\%$ confidence intervals for $\mathbf{a}'_j\boldsymbol{\mu}, j=1,\ldots,k$, are given by $\mathbf{a}'_j\bar{\mathbf{x}} \pm (\chi^2_{p,\alpha}\mathbf{a}'_j\Sigma\mathbf{a}_j/N)^{1/2}$ where $P(\chi^2_p > \chi^2_{p,\alpha}) = \alpha$; if $z_{\alpha/2k} < \chi_{p,\alpha}$, then we use $\mathbf{a}'_j\bar{\mathbf{x}} \pm z_{\alpha/2k}(\mathbf{a}'_j\Sigma\mathbf{a}_j/N)^{1/2}$. In most cases unless k is large we have $z_{\alpha/2k} < \chi_{p,\alpha}$. In this case, if we wish the width of the jth interval to be no more than $2d_j$, we choose N such that $N > z^2_{\alpha/2k} a'_j\Sigma a_j/d^2_j$ for all $j = 1,\ldots,k$. If we use $\chi^2_{p,\alpha}$ instead of $z^2_{\alpha/2k}$, then $N > \chi^2_{p,\alpha}(a'_j\Sigma a_j)/d^2_j$ for all $j=1,\ldots,k$.

If Σ is unknown, we may estimate it with data from previous experiments or surveys.

Example 3.3.2 Bernard (1981) gives the means and covariance of three measurements for male perch off the Columbia River. We let x_1 be the asymptotic length (in millimeters), x_2 the coefficient of growth, and x_3 the time (in years) at which length is zero. The covariance matrix and means on

75 observations were

$$S=\begin{pmatrix} 294.70 & -0.60 & -32.57 \\ -0.60 & 0.0073 & -0.730 \\ -32.57 & 0.0730 & 4.23 \end{pmatrix}$$

and $\bar{\mathbf{x}}=(441.16,\ 0.13,\ -3.36)'$.

Suppose we wish to estimate the measurements on ocean perch in a follow-up study a year later. What sample size would be necessary to estimate with 95% confidence the mean of x_1 to within 10 mm, the mean of x_2 to within 0.01, and the mean of x_3 to within 1 year? With 75 past observations we may consider the covariance matrix known. We want three intervals with $\mathbf{a}_1=(1,0,0)'$, $\mathbf{a}_2=(0,1,0)'$, and $\mathbf{a}_3=(0,0,1)'$. Because $z^2_{0.0083}=5.61<\chi^2_{3,0.05}=7.81$, we use $z^2_{0.0083}$ in our calculations. N must satisfy

$$N>(5.61)(294.7)/100=16.5,$$

$$N>(5.61)(0.0013)/(0.01)^2=72.9,$$

$$N>(5.61)(4.23)/1=23.7;$$

hence we should take at least 73 observations.

It was shown by Dantzig (1940) that, if there is no prior information on $\sigma^2(\Sigma)$, then no fixed-sample-size procedure exists to provide a fixed-width confidence region. Stein (1945) proposed a two-stage procedure, whereas Anscombe (1953) and Chow and Robbins (1965) gave sequential procedures. The multivariate generalization has been made by Srivastava (1967) and Srivastava and Bhargava (1979).

3.3.3 Some Remarks

In the above tests and confidence intervals a sample of N observations is taken, and naturally the estimates of $\boldsymbol{\mu}$ and Σ are based on these N observations. However, in many practical situations N may be small, and the estimate of Σ may not be very reliable. If it so happens that we have a sample of M observations from another population having the same covariance as the one from which N observations have been taken but different mean, we can combine (pool) the estimates of Σ from both the samples:

$$S_p=[(N-1)S_1+(M-1)S_2]/f, \qquad f=N+M-2. \qquad (3.3.9)$$

This will improve the estimate of Σ. Thus we can have an improved estimate of Σ based on many samples. Naturally, the tests and confidence intervals should use this improved estimate (the pooled estimate) of Σ. Hence, if we have a pooled estimate S_p of Σ on f df (3.3.3) and (3.3.5) change to

$$[(f-p+1)N/fp](\bar{\mathbf{x}}-\boldsymbol{\mu}_0)'S_p^{-1}(\bar{\mathbf{x}}-\boldsymbol{\mu}_0)>F_{p,f-p+1,\alpha} \qquad (3.3.10)$$

and

$$\mathbf{a}'\bar{\mathbf{x}} - T_\alpha N^{-1/2}(\mathbf{a}'S_p\mathbf{a})^{1/2} \leq \mathbf{a}'\boldsymbol{\mu} \leq \mathbf{a}'\bar{\mathbf{x}} + T_\alpha N^{-1/2}(\mathbf{a}'S_p\mathbf{a})^{1/2}, \quad (3.3.11)$$

respectively, where

$$T_\alpha^2 = [pf/(f-p+1)]F_{p,f-p+1,\alpha}. \quad (3.3.12)$$

3.3.4 *Two-Sample Problems*

In the preceding sections hypotheses concerning the mean of a single multivariate population were considered. The procedure given can be generalized to the comparison of means from two or more populations. The case of two populations will be considered in this section, and the case of three or more populations will be considered in Chapter 4 in the multivariate analysis of variance (MANOVA).

For the problem of comparing two population means let $\mathbf{x} \sim N_p(\boldsymbol{\mu}_1, \Sigma)$ and $\mathbf{y} \sim N_p(\boldsymbol{\mu}_2, \Sigma)$ represent the distribution of the two populations. On the basis of independent samples $\mathbf{x}_1, \ldots, \mathbf{x}_{N_1}$ on $\mathbf{x}$ and $\mathbf{y}_1, \ldots, \mathbf{y}_{N_2}$ on $\mathbf{y}$ we wish to test the hypothesis

$$H: \boldsymbol{\mu}_1 = \boldsymbol{\mu}_2 \qquad \text{vs} \qquad A: \boldsymbol{\mu}_1 \neq \boldsymbol{\mu}_2.$$

Note that both samples have to be independent among themselves. In other words, the observations $\mathbf{y}_1, \ldots, \mathbf{y}_{N_2}$ cannot be another set of observations taken on the same subjects from which the observations $\mathbf{x}_1, \ldots, \mathbf{x}_{N_1}$ have been obtained.

To test the hypothesis we need to calculate only the estimates of $\boldsymbol{\mu}_1$, $\boldsymbol{\mu}_2$, and Σ, which are

$$\bar{\mathbf{x}} = N_1^{-1} \sum_{i=1}^{N_1} \mathbf{x}_i, \qquad \bar{\mathbf{y}} = N_2^{-1} \sum_{i=1}^{N_2} \mathbf{y}_i$$

and

$$S_p = [(N_1 - 1)S_1 + (N_2 - 1)S_2]/f, \qquad f = N_1 + N_2 - 2,$$

where

$$(N_1 - 1)S_1 = \sum_{i=1}^{N_1} (\mathbf{x}_i - \bar{\mathbf{x}})(\mathbf{x}_i - \bar{\mathbf{x}})'$$

and

$$(N_2 - 1)S_2 = \sum_{i=1}^{N_2} (\mathbf{y}_i - \bar{\mathbf{y}})(\mathbf{y}_i - \bar{\mathbf{y}})'.$$

As a generalization of Student's t-test statistic in the univariate case, we get Hotelling's T^2-test statistic, given by

$$T^2 = \left(\frac{1}{N_1}+\frac{1}{N_2}\right)^{-1}(\bar{\mathbf{x}}-\bar{\mathbf{y}})'S_p^{-1}(\bar{\mathbf{x}}-\bar{\mathbf{y}}) = \frac{N_1N_2}{N_1+N_2}(\bar{\mathbf{x}}-\bar{\mathbf{y}})'S_p^{-1}(\bar{\mathbf{x}}-\bar{\mathbf{y}}), \tag{3.3.13}$$

where S_p is based on f degrees of freedom. As in Section 3.3.3, S_p could be a pooled estimate from many samples; we still write the total degrees of freedom of S_p as f. Hence, as in (3.3.10), the hypothesis H is rejected if

$$[f-p+1/fp]T^2 \geq F_{p,f-p+1,\alpha} \tag{3.3.14}$$

where T^2 is as defined above in (3.3.13). A $(1-\alpha)100\%$ confidence interval for $\mathbf{a}'(\boldsymbol{\mu}_1-\boldsymbol{\mu}_2)$ for any $\mathbf{a}\neq 0$ is given by

$$\mathbf{a}'(\bar{\mathbf{x}}-\bar{\mathbf{y}}) - \left[T_\alpha^2\frac{N_1+N_2}{N_1N_2}\mathbf{a}'S_p\mathbf{a}\right]^{1/2}$$
$$\leq a'(\mu_1-\mu_2) \leq \mathbf{a}'(\bar{\mathbf{x}}-\bar{\mathbf{y}}) + \left[T_\alpha^2\frac{N_1+N_2}{N_1N_2}\mathbf{a}'S_p\mathbf{a}\right]^{1/2}, \tag{3.3.15}$$

where as in (3.3.12) T_α^2 is given by

$$T_\alpha^2 = [pf/(f-p+1)]F_{p,f-p+1,\alpha}.$$

Example 3.3.3 Two measurements, stride and strain, were taken on eight dogs for two treatments (see Table 3.2). The dogs were randomly assigned, four to each treatment. Treatment 1 was the control, and treatment 2 was a metal plate placed on the leg of the dogs. The means for the control group and the treatment group are

$$\bar{\mathbf{x}} = \begin{pmatrix}141.875\\21.75\end{pmatrix} \quad\text{and}\quad \bar{\mathbf{y}} = \begin{pmatrix}65.125\\63.375\end{pmatrix}.$$

The pooled estimate of the covariance matrix Σ is given by

$$S = \begin{pmatrix}309.90 & 86.36\\86.36 & 124.99\end{pmatrix}.$$

Table 3.2

	Control				Plate			
Dog	1	2	3	4	5	6	7	8
Stride	131.5	145	141	150	40.5	80	50	90
Strain	9	12	30	36	54	74.5	64.5	60.5

Now

$$T^2 = \frac{N_1 N_2}{N_1 + N_2} (\bar{\mathbf{x}} - \bar{\mathbf{y}})' S^{-1} (\bar{\mathbf{x}} - \bar{\mathbf{y}}) = 116.7,$$

$$T_\alpha^2 = \frac{(N_1 + N_2 - 2)p}{N_1 + N_2 - p - 1} F_{p, N_1 + N_2 - p - 1} = \frac{6 \times 2}{5} \times 5.79 = 13.9.$$

We therefore reject the hypothesis that the two treatments are the same.

With Bonferroni's bounds, simultaneous 95% confidence limits for the difference in strain and stride for the two treatments are given by

$$\bar{x}_1 - \bar{y}_1 \pm 3.03(0.5 \times 309.9)^{1/2} = 76.75 \pm 37.7$$

and

$$\bar{x}_2 - \bar{y}_2 \pm 3.03(0.5 \times 124.99)^{1/2} = -41.625 \pm 24.0.$$

If $T_{0.05}$ is used instead of $t_{0.0125,7} = 3.03$, the confidence intervals are

$$\bar{x}_1 - \bar{y}_1 \pm (13.9 \times 0.5 \times 309.9)^{1/2} = 76.75 \pm 46.41$$

and

$$\bar{x}_2 - \bar{y}_2 \pm (13.9 \times 0.5 \times 124.99)^{1/2} = -41.625 \pm 29.47.$$

We therefore use the shorter intervals given by the Bonferroni inequality. Because the confidence intervals for the difference in mean values of stride and strain between the two groups do not contain zero, we claim at the 5% level that the treatment affects both the strain on the leg of the dog and the stride of the dog. We point out, however, that checks on the assumption of equality of covariances and normality are difficult with only eight observations.

3.3.5 Likelihood Ratio Test of $\boldsymbol{\mu} = \boldsymbol{\mu}_0$

Let us consider the univariate case. Here $x \sim N(\mu, \sigma^2)$, and an independent sample $x_1, \ldots, x_N$ of size N is given. Hence the sufficient statistics are

$$\bar{x} = N^{-1} \sum_{i=1}^{N} x_i \qquad \text{and} \qquad v = \sum_{i=1}^{N} (x_i - \bar{x})^2,$$

and the likelihood function is given by

$$L(\mu, \sigma^2) \equiv \text{const} \times (\sigma^2)^{-N/2} \exp -\tfrac{1}{2}\sigma^{-2}\left[v + N(\bar{\mathbf{x}} - \mu)^2\right].$$

Suppose we wish to test the hypothesis H: $\mu = \mu_0$ vs A: $\mu \neq \mu_0$, μ_0 specified. Then the likelihood ratio test rejects H if

$$\max_A L(\mu, \sigma^2) / \max_H L(\mu, \sigma^2) = \left[(\hat{\sigma}_H^2) / (\hat{\sigma}_A^2) \right]^{N/2} > c_\alpha. \quad (3.3.16)$$

Here c_α is chosen so that the level of significance is α and $\hat{\sigma}_A^2$ and $\hat{\sigma}_H^2$ are the maximum likelihood estimates of σ^2 under the alternative A and the hypothesis H, respectively, given by

$$\hat{\sigma}_A^2 = N^{-1}v \qquad \text{and} \qquad \hat{\sigma}_H^2 = N^{-1}\sum_{i=1}^{N}(x_i-\mu_0)^2.$$

Since

$$\sum_{i=1}^{N}(x_i-\mu_0)^2 = \sum_{i=1}^{N}(x_i-\bar{x})^2 + N(\bar{x}-\mu_0)^2 \equiv v + N(\bar{x}-\mu_0)^2,$$

the test is based on the statistic

$$\left[N^{-1}v+(\bar{x}-\mu_0)^2\right]/N^{-1}v = 1+N(\bar{x}-\mu_0)^2/v, \tag{3.3.17}$$

the ratio of the two estimates of the variances. Hence the hypothesis H is rejected if

$$t^2 = N(\bar{x}-\mu_0)^2/s^2 \geq t_{n,\alpha}^2, \qquad s^2 = n^{-1}v, \quad n = N-1,$$

where $t_{n-1,\alpha}^2$ is the upper $\alpha 100\%$ of the t^2-distribution with n df, or the upper $\alpha 100\%$ point of the F-distribution with 1 and n df.

Considering the multivariate case, we now have the likelihood function given by [see (2.4.1)]

$$L(\boldsymbol{\mu},\Sigma) = \text{const} \times |\Sigma|^{-N/2}\operatorname{etr} -\tfrac{1}{2}\Sigma^{-1}\left[V+N(\bar{\mathbf{x}}-\boldsymbol{\mu})(\bar{\mathbf{x}}-\boldsymbol{\mu})'\right],$$

where $V = \Sigma_{\alpha=1}^{N}(\mathbf{x}_\alpha-\bar{\mathbf{x}})(\mathbf{x}_\alpha-\bar{\mathbf{x}})'$ and $\bar{\mathbf{x}} = N^{-1}\Sigma_{\alpha=1}^{N}\mathbf{x}_\alpha$. Hence the likelihood ratio test rejects H if

$$\left[|\hat{\Sigma}_H|/|\hat{\Sigma}_A|\right]^{N/2} \geq c_\alpha, \tag{3.3.18}$$

where c_α is chosen so that the size of the test is α and where $\hat{\Sigma}_H$ and $\hat{\Sigma}_A$ denote the maximum likelihood estimates of Σ under H and A, respectively. These are given by

$$\begin{aligned}N\hat{\Sigma}_H &= \sum_{\alpha=1}^{N}(\mathbf{x}_\alpha-\boldsymbol{\mu}_0)(\mathbf{x}_\alpha-\boldsymbol{\mu}_0)' = \sum_{\alpha=1}^{N}(\mathbf{x}_\alpha-\bar{\mathbf{x}})(\mathbf{x}_\alpha-\bar{\mathbf{x}})' + N(\bar{\mathbf{x}}-\boldsymbol{\mu}_0)(\bar{\mathbf{x}}-\boldsymbol{\mu}_0)'\\ &= V+N(\bar{\mathbf{x}}-\boldsymbol{\mu}_0)(\bar{\mathbf{x}}-\boldsymbol{\mu}_0)'\end{aligned}$$

and

$$N\hat{\Sigma}_A \equiv V.$$

Hence the likelihood ratio test rejects H if the ratio of the determinants of the two estimates of covariances

$$\begin{aligned}\left|N^{-1}V+(\bar{\mathbf{x}}+\boldsymbol{\mu}_0)(\bar{\mathbf{x}}-\boldsymbol{\mu}_0)'\right|/|N^{-1}V| &= \left|I+NV^{-1}(\bar{\mathbf{x}}-\boldsymbol{\mu}_0)(\bar{\mathbf{x}}-\boldsymbol{\mu}_0)'\right|\\ &= 1+N(\bar{\mathbf{x}}-\boldsymbol{\mu}_0)'V^{-1}(\bar{\mathbf{x}}-\boldsymbol{\mu}_0) \geq d_\alpha\end{aligned} \tag{3.3.19}$$

(since $|I_p + AB| = |I_q + BA|$, $A: p \times q$, $B: q \times p$), where d_α is determined from the size of the test. Equivalently, the likelihood ratio test rejects H if

$$\frac{n-p+1}{p} N(\bar{\mathbf{x}}-\boldsymbol{\mu}_0)'V^{-1}(\bar{\mathbf{x}}-\boldsymbol{\mu}_0) = \frac{n-p+1}{p(N-1)} N(\bar{\mathbf{x}}-\boldsymbol{\mu}_0)'S^{-1}(\bar{\mathbf{x}}-\boldsymbol{\mu}_0)$$
$$\geq F_{p,n-p+1,\alpha},$$

where $F_{p,n-p+1,\alpha}$ is the upper $\alpha 100\%$ point of the F-distribution with p and $n-p+1$ degrees of freedom, $S = n^{-1}V$, and $n = N-1$. Thus (3.3.3) gives the likelihood ratio test.

3.3.6 *Heuristic Method (Roy's Union–Intersection Method)*

Let $\mathbf{x} \sim N_p(\boldsymbol{\mu}, \Sigma)$, and let $\mathbf{x}_1, \ldots, \mathbf{x}_N$ be an independent sample of size N on $\mathbf{x}$. The sufficient statistics are $\bar{\mathbf{x}}$ and $V = \Sigma(\mathbf{x}_\alpha - \bar{\mathbf{x}})(\mathbf{x}_\alpha - \bar{\mathbf{x}})'$. We shall reduce the problem of testing H: $\boldsymbol{\mu} = \boldsymbol{\mu}_0$ vs A: $\boldsymbol{\mu} \neq \boldsymbol{\mu}_0$ to a univariate problem. Let $y = \mathbf{a}'\mathbf{x}$; then $y \sim N(\theta, \sigma^2)$, where $\theta = \mathbf{a}'\boldsymbol{\mu}$ and $\sigma^2 = \mathbf{a}'\Sigma\mathbf{a}$. The sufficient statistics for θ and σ^2 are

$$\bar{y} = N^{-1} \sum_{i=1}^{N} y_i, \qquad y_i = \mathbf{a}'\mathbf{x}_i, \quad i = 1, 2, \ldots, N$$

and

$$v = \sum_{i=1}^{N} (y_i - \bar{y})^2.$$

In terms of the vector variables $\bar{\mathbf{x}}$ and V, we get

$$\bar{y} = \mathbf{a}'\bar{\mathbf{x}} \quad \text{and} \quad v = \mathbf{a}'V\mathbf{a}. \tag{3.3.20}$$

Let $H_\mathbf{a}$: $\mathbf{a}'\boldsymbol{\mu} = \mathbf{a}'\boldsymbol{\mu}_0$ and $A_\mathbf{a}$: $\mathbf{a}'\boldsymbol{\mu} \neq \mathbf{a}'\boldsymbol{\mu}_0$. The hypothesis H is accepted if and only if $H_\mathbf{a}$ is accepted for all $\mathbf{a} \neq \mathbf{0}$. The statistic for testing $H_\mathbf{a}$ vs $A_\mathbf{a}$ is based on

$$N[\mathbf{a}'(\bar{\mathbf{x}}-\boldsymbol{\mu}_0)]^2/(\mathbf{a}'S\mathbf{a}) \tag{3.3.21}$$

where $S = n^{-1}V$ and $n = N-1$. Hence H is accepted iff

$$N[\mathbf{a}'(\bar{\mathbf{x}}-\boldsymbol{\mu}_0)]^2/(\mathbf{a}'S\mathbf{a}) < T_\alpha^2 \qquad \text{for every} \quad \mathbf{a} \neq 0, \tag{3.3.22}$$

i.e., iff

$$\sup_{\mathbf{a}} \left\{ N[\mathbf{a}'(\bar{\mathbf{x}}-\boldsymbol{\mu}_0)]^2/(\mathbf{a}'S\mathbf{a}) \right\} < T_\alpha^2, \tag{3.3.23}$$

where T_α^2 is chosen so that the error of the first kind is at a specified level α. Since

$$[\mathbf{a}'(\bar{\mathbf{x}}-\boldsymbol{\mu}_0)]^2 = [\mathbf{a}'S^{1/2}S^{-1/2}(\bar{\mathbf{x}}-\boldsymbol{\mu}_0)]^2 \leq (\mathbf{a}'S\mathbf{a})(\bar{\mathbf{x}}-\boldsymbol{\mu}_0)'S^{-1}(\bar{\mathbf{x}}-\boldsymbol{\mu}_0) \tag{3.3.24}$$

from the Cauchy–Schwarz inequality, with equality holding for at least one **a**, the hypothesis H is rejected if

$$N(\bar{\mathbf{x}}-\boldsymbol{\mu}_0)'S^{-1}(\bar{\mathbf{x}}-\boldsymbol{\mu}_0)\geq T_\alpha^2. \tag{3.3.25}$$

The confidence interval given in (3.3.5) is obtainable from (3.3.22). Note that the hypothesis is accepted iff the specified value of $\boldsymbol{\mu}$ is such that

$$N[\mathbf{a}'(\bar{\mathbf{x}}-\boldsymbol{\mu})]^2/(\mathbf{a}'S\mathbf{a})\leq T_\alpha^2, \tag{3.3.26}$$

that is, iff for every $\mathbf{a}\neq\mathbf{0}$,

$$N^{1/2}|\mathbf{a}'(\bar{\mathbf{x}}-\boldsymbol{\mu})|\leq T_\alpha(\mathbf{a}'S\mathbf{a})^{1/2}. \tag{3.3.27}$$

Hence simultaneous confidence intervals for linear combinations $\mathbf{a}'\boldsymbol{\mu}$ of $\boldsymbol{\mu}$ are given by

$$\mathbf{a}'\bar{\mathbf{x}}-N^{-1/2}T_\alpha(\mathbf{a}'S\mathbf{a})^{1/2}\leq\mathbf{a}'\boldsymbol{\mu}\leq\mathbf{a}'\bar{\mathbf{x}}+N^{-1/2}T_\alpha(\mathbf{a}'S\mathbf{a})^{1/2}. \tag{3.3.28}$$

3.3.7 Paired T^2-Test

In Section 3.3.6 the problem of comparing the means of two independent populations was considered; that is, the observations $\mathbf{x}_1,\ldots,\mathbf{x}_{N_1}$ and $\mathbf{y}_1,\ldots,\mathbf{y}_{N_2}$ were assumed to be independently distributed. However, in many practical situations the two populations may not be independently distributed. For example, the observations $\mathbf{x}_1,\ldots,\mathbf{x}_N$ could be measurements on N individuals before breakfast, and $\mathbf{y}_1,\ldots,\mathbf{y}_N$ could be the measurements on the same N individuals 2 hours after breakfast; the p characteristics could be sugar level, cholesterol level, and glycerol level in blood. We may then want to know if there has been a significant change in the mean levels before and after breakfast. In this illustration we have chosen $N_1=N_2=N$, a situation we shall describe in this section. The case when $N_1\neq N_2$ can be handled on the lines of results in Section 3.8.3 on missing observations.

Let

$$\mathbf{d}_i=\mathbf{x}_i-\mathbf{y}_i,\qquad i=1,2,\ldots,N.$$

Then, for $i=1,\ldots,N$, $\mathbf{d}_i$ is independently distributed with mean $\boldsymbol{\delta}=\boldsymbol{\mu}_1-\boldsymbol{\mu}_2$ and some unknown covariance matrix Σ, where $\boldsymbol{\mu}_1$ and $\boldsymbol{\mu}_2$ are the population means of the two populations. We may wish to test the hypothesis H: $\boldsymbol{\delta}=\mathbf{0}$ against the alternative A: $\boldsymbol{\delta}\neq\mathbf{0}$. The problem is now reduced to the one-population case. Thus provided $\mathbf{d}_i\sim N_p(\delta,\Sigma)$, $i=1,\ldots,N$, the hypothesis H is rejected if

$$\frac{N(n-p+1)}{p(N-1)}\bar{\mathbf{d}}'S_d^{-1}\bar{\mathbf{d}}\geq F_{p,n-p+1,\alpha}, \tag{3.3.29}$$

where $F_{p,n-p+1,\alpha}$ is the upper $\alpha 100\%$ point of the F-distribution with p and $n-p+1$ degrees of freedom, and

$$\bar{\mathbf{d}}=\bar{\mathbf{x}}-\bar{\mathbf{y}}\qquad\text{and}\qquad(N-1)S_d=\sum_{i=1}^{N}(\mathbf{d}_i-\bar{\mathbf{d}})(\mathbf{d}_i-\bar{\mathbf{d}})'. \tag{3.3.30}$$

3.3.8 *Behrens – Fisher Problem*

Let $\mathbf{x}_1,\ldots,\mathbf{x}_{N_1}$ and $\mathbf{y}_1,\ldots,\mathbf{y}_{N_2}$ be independently distributed such that $\mathbf{x}_i \sim N_p(\boldsymbol{\mu}_1, \Sigma_1)$, $i=1,2,\ldots,N_1$, and $\mathbf{y}_i \sim N_p(\boldsymbol{\mu}_2, \Sigma_2)$, $i=1,2,\ldots,N_2$. Suppose we wish to test the hypothesis

$$H: \boldsymbol{\mu}_1 = \boldsymbol{\mu}_2, \quad \Sigma_1 > 0, \quad \Sigma_2 > 0, \qquad \text{vs} \qquad A: \boldsymbol{\mu}_1 \neq \boldsymbol{\mu}_2, \quad \Sigma_1 > 0, \quad \Sigma_2 > 0.$$

That is, we wish to test the equality of the means of two multivariate normal populations with different covariance matrices. This is known in the literature as the Behrens–Fisher problem. A solution first given by Bennett (1951), is a multivariate generalization of Scheffé's (1943) univariate solution. However, it does not make use of all the information; for details see Srivastava and Khatri (1979). It should be mentioned that we recommend using the two-sample Hotelling's T^2-test if the inequality between the two covariances is slight (Carter et al., 1979). However, where the two covariances appear to differ significantly Yao (1965) suggested an approximate method which is a generalization of Welch's (1937, 1947) solution for the univariate case. This can be described as follows. Let

$$T^2 = (\bar{\mathbf{x}} - \bar{\mathbf{y}})'(S_1/N_1 + S_2/N_2)^{-1}(\bar{\mathbf{x}} - \bar{\mathbf{y}}), \tag{3.3.31}$$

where

$$\bar{\mathbf{x}} = N_1^{-1} \sum_{i=1}^{N_1} \mathbf{x}_i, \qquad \bar{\mathbf{y}} = N_2^{-1} \sum_{i=1}^{N_2} \mathbf{y}_i,$$

$$(N_1 - 1)S_1 = \sum_{i=1}^{N_1} (\mathbf{x}_i - \bar{\mathbf{x}})(\mathbf{x}_i - \bar{\mathbf{x}})',$$

$$(N_2 - 1)S_2 = \sum_{i=1}^{N_2} (\mathbf{y}_i - \bar{\mathbf{y}})(\mathbf{y}_i - \bar{\mathbf{y}})',$$

and

$$f^{-1} = \sum_{i=1}^{2} (N_i^3 - N_i^2)^{-1} \left[(\bar{\mathbf{x}} - \bar{\mathbf{y}})' \left(\frac{S_1}{N_1} + \frac{S_2}{N_2} \right)^{-1} S_i \left(\frac{S_1}{N_1} + \frac{S_2}{N_2} \right)^{-1} (\bar{\mathbf{x}} - \bar{\mathbf{y}}) \right]^2 T^{-4}. \tag{3.3.32}$$

Then $[(f-p+1)/fp]T^2$ has approximately an F-distribution with p and $f-p+1$ degrees of freedom. Thus H is rejected if

$$[(f-p+1)/fp]T^2 > F_{p, f-p+1, \alpha},$$

where $F_{p, f-p+1, \alpha}$ is the upper $\alpha 100\%$ of the F-distribution with p and $f-p+1$ degrees of freedom. It should be mentioned that if N_1 and N_2 are large (that is, $\min(N_1, N_2) \to \infty$), T^2 has a chi-squared distribution with p df.

Example 3.3.4 Consider the measurements in Table 3.3 on the flea beetle *Halticus* for two species *Haltica oleracea* and *Haltica carduorum*. These are the thorax length x_1 (in microns) and elytra length x_2 (in 0.01 mm).

Table 3.3[a]

	Haltica oleracea ($N_1=10$)			*Haltica carduorum* ($N_2=13$)	
	x_1	x_2		x_1	x_2
	180	245		181	305
	192	260		158	237
	217	276		184	300
	221	299		171	273
	171	239		181	297
	192	262		181	308
	213	278		177	301
	192	255		198	308
	170	244		180	286
	201	276		177	299
mean	194.9	263.4		176	317
				192	312
				164	265
			mean	178.5	292.9

[a]Data abridged from Lubischew (1962).

We have the covariances

$$S_1=\begin{pmatrix}330.32 & 325.27\\ 325.27 & 354.71\end{pmatrix},\qquad S_2=\begin{pmatrix}109.27 & 189.79\\ 189.79 & 505.41\end{pmatrix}.$$

If these are not too unequal, then the T^2-procedure should be used [see Carter et al. (1979)]. The significance level of the t-test and Hotelling's T^2-procedures is affected only slightly if the variances of the characteristics of one species are not more than twice those of the other species and if the sample sizes are roughly the same. If the sample sizes are not equal, however, it is better that more observations be taken from the species with the larger variances. In this example, the variance of the thorax length for *oleracea* is three times that of *carduorum*, and the variance of the elytra length of *carduorum* is twice that of *oleracea*. Hence rather than Hotelling's T^2-procedure we should use the T^2-test given by (3.3.31).

Note, however, that a test for the equality of the two covariances may not be useful. If the sample sizes are large, then so is f; hence any error in estimating f will have almost no effect on the critical value of the t-test, and (3.3.31) should be used. If the sample sizes are small, then not rejecting the null hypothesis of equal covariances is insufficient to ensure that equality. Suppose we do not assume equal covariances but apply Yao's approximation. We then obtain a T^2 value of $T_f^2=118.5$, where f is calculated to be 16. Hence H is rejected if

$$[(f-p+1)/(fp)]T_f^2=[15/(2\times 16)]\times 118.5=55.5\geq F_{2,16,0.05}\simeq 3.63.$$

Hence the null hypothesis is again rejected, and we claim that there is a difference in the means of the two species.

If we wish to find the linear combination of characteristics that differentiates between the two species, then we need to perform a discriminant analysis (discussed in Chapter 8). If, however, we are interested only in the confidence intervals for the difference in thorax and elytra length between the two species, then we use Bonferroni bounds. The variances of the two species are not the same, so the approximate degrees of freedom method is applied separately to each characteristic. That is, if we are interested in obtaining k confidence intervals of the form $\mathbf{a}_i'(\boldsymbol{\mu}_1-\boldsymbol{\mu}_2)$, $i=1,\ldots,k$, then we define

$$c_i=(\mathbf{a}_i'S_1\mathbf{a}_i/N_i)/(\mathbf{a}_i'S_1\mathbf{a}_i/N_1+\mathbf{a}_i'S_2\mathbf{a}_i/N_2), \qquad i=1,\ldots,k.$$

The k confidence intervals are then given by

$$\mathbf{a}_i'(\mathbf{x}-\mathbf{y})\pm t_{f_i,\alpha/2k}(\mathbf{a}_i'S_1\mathbf{a}_i/N_1+\mathbf{a}_i'S_2\mathbf{a}_i/N_2)^{1/2}, \qquad i=2,\ldots,k,$$

where $f_i^{-1}=c_i^2/(N_1-1)+(1-c_i)^2/(N_2-1)$. For thorax length, $c_1=0.80$ and $f_1=13.4\simeq 13$. For elytra length, $c_2=0.477$ and $f_2=20.8\simeq 21$. The 95% confidence intervals for the difference in the thorax and elytra lengths, respectively, between the two species are given by

$$(194.9-178.5)\pm t_{13,0.0125}(330.32/10+109.27/13)^{1/2}=16.4\pm 2.47\times 6.44$$
$$=(0.5,32.3),$$

and

$$(263.4-292.9)\pm t_{21,0.0125}(354.71/10+505.41/13)^{1/2}$$
$$=-29.5\pm 2.35\times 8.62$$
$$=(-49.8,-9.2).$$

Hence the lengths of both the thorax and the elytra differ from one species to the other.

3.4 Comparing Means of Correlated Random Variables

Let $\mathbf{x}\sim N_p(\boldsymbol{\mu},\Sigma)$, and let $\mathbf{x}_1,\ldots,\mathbf{x}_N$ be an independent sample of size N on $\mathbf{x}$. In Section 3.3 we considered the usual problem of testing the hypothesis H: $\boldsymbol{\mu}=\boldsymbol{\mu}_0$ vs A: $\boldsymbol{\mu}\neq\boldsymbol{\mu}_0$. However, in multivariate situations the vector $\boldsymbol{\mu}$ is the mean of p correlated random variables, and there are situations where we may like to compare these means among themselves. An interesting example is given by Rao (1948). Suppose we are interested in finding whether cork trees have the same amount of cork on each side of the tree. A boring can be carried out on four sides of the tree: north, south, east, and west. Let N represent the amount of cork in a boring from the north, S from the south, E from the east, and W from the west. We would like to test whether the means of these correlated variables are the same; that is, we wish to test the hypothesis H: $\mu_1=\mu_2=\mu_3=\mu_4$ vs A: $\mu_i\neq\mu_j$ for at least one pair $(i, j), i \neq j, i, j=1,2,3,4$. Since the four random variables N, S, E, W are correlated, the usual univariate analysis of variance (one-way classification) is not applicable. In another example, the p correlated variables could be the response of an individual under p conditions—repeated observations on the same individual—and we may like to know whether the response of the individual changes over p "conditions." Again the usual univariate analysis of variance is not applicable, although it may be assumed that the covariance matrix is of the intraclass correlation form. These problems will be discussed in Chapters 6 and 7, respectively.

3.5 A Test for a Subvector

In some situations partial information concerning the population means may be available to an experimenter. For example, it may be inexpensive to obtain a large number of observations on a subvector so that the corresponding population values may be assumed to be known. In a two-sample problem, rats may have their weights recorded each week under two diets. Assuming that the rats were randomly assigned to the two diets, the expected values of their initial weight should be the same for both groups. In growth curve analysis we may have the information that cubic and higher order terms are zero according to the model. We now discuss the one- and two-sample cases.

Let $\mathbf{x} \sim N_p(\boldsymbol{\mu}, \Sigma)$, $\boldsymbol{\mu}'=(\boldsymbol{\mu}_1', \boldsymbol{\mu}_2')$, and $\mathbf{x}'=(\mathbf{x}_1', \mathbf{x}_2')$, where $\boldsymbol{\mu}_1$ and $\mathbf{x}_1$ are s-vectors and $\boldsymbol{\mu}_2$ and $\mathbf{x}_2$ are t-vectors, $s+t=p$. Suppose N observations are taken and the sample mean and covariance are given by $\bar{\mathbf{x}}$ and S. Then we partition $\bar{x}$ and S as

$$\bar{x} = \begin{pmatrix} \bar{\mathbf{x}}_1 \\ \bar{\mathbf{x}}_2 \end{pmatrix} \quad \text{and} \quad S = \begin{pmatrix} S_{11} & S_{12} \\ S_{21} & S_{22} \end{pmatrix},$$

where $\bar{\mathbf{x}}_1$ is an s-vector and S_{11} is an $s \times s$ symmetric matrix.

We wish to test the hypothesis

$$H: \boldsymbol{\mu}_2 = \boldsymbol{\mu}_{20}, \quad \text{given} \quad \boldsymbol{\mu}_1 = \boldsymbol{\mu}_{10} \qquad \text{vs} \qquad A: \boldsymbol{\mu}_2 \neq \boldsymbol{\mu}_{20}, \quad \text{given} \quad \boldsymbol{\mu}_1 = \boldsymbol{\mu}_{10}. \tag{3.5.1}$$

We first define T_p^2 to be the usual test for H: $\boldsymbol{\mu} = \boldsymbol{\mu}_0$ vs A: $\boldsymbol{\mu} \neq \boldsymbol{\mu}_0$, where $\boldsymbol{\mu}_0' = (\boldsymbol{\mu}_{10}', \boldsymbol{\mu}_{20}')$. That is,

$$T_p^2 = N(\bar{\mathbf{x}} - \boldsymbol{\mu}_0)' S^{-1} (\bar{\mathbf{x}} - \boldsymbol{\mu}_0). \tag{3.5.2}$$

We also define T_s^2 to be the T^2-statistic for testing H: $\boldsymbol{\mu}_1 = \boldsymbol{\mu}_{10}$ vs A: $\boldsymbol{\mu}_1 \neq \boldsymbol{\mu}_{10}$. That is,

$$T_s^2 = N(\bar{\mathbf{x}}_1 - \boldsymbol{\mu}_{10})' S_{11}^{-1} (\bar{\mathbf{x}}_1 - \boldsymbol{\mu}_{10}). \tag{3.5.3}$$

Then we reject H: $\boldsymbol{\mu}_2 = \boldsymbol{\mu}_{20}$ given $\boldsymbol{\mu}_1 = \boldsymbol{\mu}_{10}$ if

$$\frac{n-p+1}{t}\,\frac{T_p^2 - T_s^2}{n+T_s^2} > F_{t,n-p+1,\alpha}, \tag{3.5.4}$$

where $n = N-1$. This is the likelihood ratio test, and the same test is obtained even if Σ_{11} is known. This problem has been considered by Rao (1949), Olkin and Shrikhande (1953), and Giri (1964).

In the two-sample problem we obtain the two sample means $\bar{\mathbf{x}}$ and $\bar{\mathbf{y}}$, based on N_1 and N_2 observations. That is, $\bar{\mathbf{x}} \sim N_p(\boldsymbol{\mu}, N_1^{-1}\Sigma)$ and $\bar{\mathbf{y}} \sim N_p(\boldsymbol{\gamma}, N_2^{-1}\Sigma)$. We also obtain S, the pooled estimate of Σ, on $f = N_1 + N_2 - 2$ degrees of freedom. We partition $\bar{\mathbf{x}}$, $\bar{\mathbf{y}}$, $\boldsymbol{\mu}$, $\boldsymbol{\gamma}$, and S as

$$\bar{\mathbf{x}} = \begin{pmatrix} \bar{\mathbf{x}}_1 \\ \bar{\mathbf{x}}_2 \end{pmatrix}, \qquad \boldsymbol{\mu} = \begin{pmatrix} \boldsymbol{\mu}_1 \\ \boldsymbol{\mu}_2 \end{pmatrix}, \qquad \boldsymbol{\gamma} = \begin{pmatrix} \gamma_1 \\ \gamma_2 \end{pmatrix},$$

$$\mathbf{y} = \begin{pmatrix} \bar{\mathbf{y}}_1 \\ \bar{\mathbf{y}}_2 \end{pmatrix}, \qquad \text{and} \qquad S = \begin{pmatrix} S_{11} & S_{12} \\ S_{21} & S_{22} \end{pmatrix},$$

where $\bar{\mathbf{x}}_1$, $\boldsymbol{\mu}_1$, γ_1, and $\bar{\mathbf{y}}_1$ are s-vectors and S_{11} is an $s \times s$ matrix. Then the hypothesis

$$H: \boldsymbol{\mu}_2 = \gamma_2 \quad \text{given} \quad \boldsymbol{\mu}_1 = \gamma_1 \qquad \text{vs} \qquad A: \boldsymbol{\mu}_2 \neq \gamma_2 \quad \text{given} \quad \boldsymbol{\mu}_1 = \gamma_1 \tag{3.5.5}$$

is rejected if

$$\frac{f-p+1}{t}\,\frac{T_p^2 - T_s^2}{f+T_s^2} \geq F_{t,f-p+1,\alpha} \tag{3.5.6}$$

where

$$T_p^2 = \frac{N_1N_2}{N_1+N_2}(\bar{\mathbf{x}} - \bar{\mathbf{y}})'S^{-1}(\bar{\mathbf{x}} - \bar{\mathbf{y}}),$$

$$T_s^2 = \frac{N_1N_2}{N_1+N_2}(\bar{\mathbf{x}}_1 - \bar{\mathbf{y}}_1)'S_{11}^{-1}(\bar{\mathbf{x}}_1 - \bar{\mathbf{y}}_1),$$

and $f = N_1 + N_2 - 2$.

3.5.1 Confidence Intervals

To find a confidence interval for $\mathbf{a}'\boldsymbol{\mu}_2$ let

$$f = N-1, \qquad \mathbf{z} = \bar{\mathbf{x}}_1 - \boldsymbol{\mu}_{10} \qquad c_\alpha^2 = tb^{-1}(f-p+1)^{-1}F_{t,f-p+1,\alpha},$$

$$S_{2.1} = S_{22} - S_{12}'S_{11}^{-1}S_{12}, \qquad b = N.$$

Then a $(1-\alpha)100\%$ confidence interval is given by

$$\mathbf{a}'\left(\bar{\mathbf{x}}_2 - S_{21}S_{11}^{-1}\mathbf{z}\right) \pm \left[f + N\mathbf{z}'S_{11}^{-1}\mathbf{z}\right]^{1/2} c_\alpha(\mathbf{a}'S_{2.1}\mathbf{a})^{1/2}.$$

In the case of two-sample problems, a $(1-\alpha)100\%$ confidence interval for $\mathbf{a}'(\boldsymbol{\mu}_2-\boldsymbol{\gamma}_2)$ is given by

$$\mathbf{a}'[\bar{\mathbf{x}}_2-\bar{\mathbf{y}}_2-S_{21}S_{11}^{-1}(\bar{\mathbf{x}}_1-\bar{\mathbf{y}}_1)]\pm[f+b(\bar{\mathbf{x}}_1-\bar{\mathbf{y}}_1)'S_{11}^{-1}(\bar{\mathbf{x}}_1-\bar{\mathbf{y}}_1)]^{1/2}c_\alpha(\mathbf{a}'S_{2.1}\mathbf{a})^{1/2},$$

where

$$f=N_1+N_2-2, \qquad b=N_1N_2/(N_1+N_2),$$
$$c_\alpha^2=tb^{-1}(f-p+1)^{-1}F_{t,f-p+1,\alpha}$$

For k confidence intervals, Bonferroni's inequality enables us to replace c_α^2 in both cases with $b^{-1}(f-p+1)^{-1}t_{f,\alpha/2k}^2$ if the latter is smaller.

Example 3.5.1 The following artificial example gives the scores of ten randomly selected students on an achievement test. It is known that the mean for all students writing the test was 50. The subjects underwent training for a similar test given nationally. The data of Table 3.4 resulted. The sample covariance and correlation matrices are

$$S=\begin{pmatrix}250 & 159\\ 159 & 148\end{pmatrix} \qquad \text{and} \qquad R=\begin{pmatrix}1 & 0.83\\ 0.83 & 1\end{pmatrix},$$

respectively. Note that there is a high correlation (0.83) between the two test scores.

To test whether training affects average score, we test the hypothesis

$$H\colon \mu_1=50, \quad \mu_2=50 \qquad \text{vs} \qquad A_1\colon \mu_1=50, \quad \mu_2\neq 50.$$

The chosen value $\mu_2=50$ was the mean nationally for all students taking the test. Calculations yield $T_2^2=14.1$, $f=10-1=9$, $s=1$, $t=1$, $p=2$,

Table 3.4

	Before training	After training
	70	75
	60	58
	65	70
	50	55
	43	48
	40	41
	80	78
	45	65
	30	55
	40	50
sample mean	52.3	59.5
population mean	μ_1	μ_2

$T_1^2 = 0.21$, and

$$\frac{f-p+1}{t}\,\frac{T_p^2 - T_s^2}{f+T_s^2} = \frac{8}{1}\times\frac{14.1-0.21}{9.21}$$
$$= 10.39 \geq F_{1,8,0.05} = 5.32.$$

We conclude that the training has an effect on the scores of the students. A 95% confidence interval for μ_2 is given by

$$\mathbf{a}'(\bar{\mathbf{x}}_2 - S_{21}S_{11}^{-1}\mathbf{z}) \pm (f + N\mathbf{z}'S_{11}^{-1}\mathbf{z})^{1/2} c_\alpha(\mathbf{a}'S_{2.1}\mathbf{a})^{1/2}$$

with $\mathbf{a}=1$, $c_\alpha^2 = (0.1)(8)^{-1}5.32 = 0.0666$, and $S_{2.1} = 148-(159)^2/250 = 46.876$. The interval is then calculated to be

$$59.5 - 148(250)^{-1}2.3 \pm \left[9+10(2.3)^2/250\right]^{1/2} 0.258[46.876]^{1/2}$$
$$= 58.14 \pm 5.34 = (52.80, 63.48).$$

Example 3.5.2 Table 3.5 shows gains in fish length over a two-week period under a standard and a test diet. As the fish were randomly assigned to the diets, the expected initial lengths of the two groups are the same. We now wish to test for the difference in gain in length for the two groups. That is, we test

$$H: \begin{pmatrix}\mu_1\\ \mu_2\\ \mu_3\end{pmatrix} = \begin{pmatrix}\gamma_1\\ \gamma_2\\ \gamma_3\end{pmatrix} \quad \text{vs} \quad A: \mu_1 = \gamma_1,\ \mu_i \neq \gamma_i \quad i = 2,3.$$

Table 3.5

	Standard diet			Test diet		
	Initial length	Week 1	Week 2	Initial length	Week 1	Week 2
	12.3	2.5	2.9	12.0	2.3	2.7
	12.1	2.2	2.5	11.8	2.0	2.4
	12.8	2.9	3.0	12.7	3.1	3.6
	12.0	2.1	2.2	12.4	2.8	3.2
	12.1	2.2	2.4	12.1	2.5	2.8
	11.8	1.9	2.0	12.0	2.2	2.7
	12.7	2.9	3.3	11.7	2.0	2.4
	12.5	2.7	3.0	12.2	2.5	3.0
sample means	12.2875	2.425	2.6625	12.1125	2.425	2.85
population means	μ_1	μ_2	μ_3	γ_1	γ_2	γ_3

The pooled covariance matrix is given by

$$S=\begin{pmatrix}0.114 & 0.128 & 0.140\\ 0.128 & 0.146 & 0.161\\ 0.140 & 0.161 & 0.186\end{pmatrix}.$$

Calculations yield $T_1^2=1.22$, $f=7+7=14$, $T_3^2=22.155$, $s=1$, $t=2$, $p=3$, and

$$F_{2,12}=\frac{12}{2}\times\frac{22.155-1.22}{14+1.22}=8.25$$
$$\geq F_{2,12,0.05}=3.89.$$

Hence we reject H at $\alpha=0.05$ and claim that the test diet differs from the standard diet with respect to the effect on gain in length.

Simultaneous 95% confidence intervals for each week for the difference in gain in length for the two diets (standard − test) are given by Bonferroni bounds:

$$\mathbf{a}'\left\{\begin{pmatrix}2.425-2.425\\ 2.6625-2.800\end{pmatrix}-\begin{pmatrix}0.128\\ 0.140\end{pmatrix}(0.114)^{-1}(12.2875-12.1125)\right\}$$
$$\pm\left[14+4(12.2875-12.1125)^2/0.114\right]^{1/2}2.68(24)^{-1/2}$$
$$\times\left\{\mathbf{a}'\left[\begin{pmatrix}0.146 & 0.161\\ 0.161 & 0.186\end{pmatrix}-(0.114)^{-1}\begin{pmatrix}0.128\\ 0.140\end{pmatrix}(0.128,0.140)\right]\mathbf{a}\right\}^{1/2},$$

for $\mathbf{a}=(1,0)$ and $(0,1)$, where 2.68 corresponds to $t_{14,0.0125}$. Calculations yield the following 95% simultaneous confidence intervals for the difference in gain in length between the two diets: For week 1, -0.196 ± 0.092; for week 2, -0.402 ± 0.244. Hence the test diet and the standard diet produced different length gains both weeks, and because both intervals are below zero, we can state that the gain in length of a fish is lower with the standard diet than with the new test diet.

The variable initial length is called a *concomitant variable* or *covariable*. It should be included in the model only if it is correlated with the response variables of interest. If the initial length of the fish did not affect the gain in length of the fish then the addition of the covariable would only add to the variation in the experiment and decrease the power of the tests.

3.6 Tests for Detecting a Change in Mean

A problem of considerable practical interest is the following: Given one observation from each of N random variables $x_1,\ldots,x_N$, how can we decide whether the means of the x_i can be considered to be the same or whether one needs to consider two models of the form

$$\begin{aligned} x_i&=\mu+\varepsilon_i \qquad (1\leq i\leq r),\\ x_i&=\mu^*+\varepsilon_i \qquad (r+1\leq i\leq N),\end{aligned}\tag{3.6.1}$$

Table 3.6

	1962	1963	1964	1965	1966	1967	1968	1969	1970	1971
Deaths ($\times 100$)	18.90	20.28	22.07	22.56	25.22	24.93	24.99	25.33	23.46	24.00
Injuries ($\times 1000$)	112.31	119.89	134.16	145.54	149.14	149.51	148.73	157.45	159.88	148.83
Accidents ($\times 1000$)	252.02	258.68	281.16	324.07	329.42	337.56	351.07	405.51	409.17	393.57
Deaths per 10^8 vehicle miles	4.9	5.1	5.2	5.1	5.3	5.1	4.9	4.7	4.2	4.2

where the ε_is are independent error terms and r is unknown? Apart from its obvious applications to the detection of shifts in production processes, the problem is also important for the study of the impact of treatments since the point when the treatment (e.g., a drug or an advertising campaign) might take effect is usually unknown. For univariate results, see Sen and Srivastava (1975a–c) and the references therein.

To fix our ideas, consider the annual data from 1962 to 1971 on hundreds of traffic deaths, thousands of traffic injuries, thousands of accidents, and the number of deaths per hundred million vehicle miles in the state of Illinois (Table 3.6).

It was felt that the increase in traffic fatalities and accidents each year would tend to be constant. A *change* in the mean increase each year could then be attributed to external influences, such as new regulations and safety standards. For example, variables of interest would be the increase in traffic deaths or injuries from 1962 to 1963. Our interest in the increase in traffic deaths and injuries means that we must look at the new set of variables formed by taking differences.

Sen and Srivastava (1975c) considered these four types of traffic data separately and applied univariate results. Here we apply multivariate techniques. Let

$$x_i^{(j)} = y_{i+1}^{(j)} - y_i^{(j)} = \mu_i^{(j)} + \varepsilon_i^{(j)}, \qquad i=1,2,\ldots,N, \quad j=1,2,3,4, \quad (3.6.2)$$

where $y_i^{(j)}$ is the jth type of traffic data for the ith year. Let

$$\mathbf{x}_i = \begin{pmatrix} x_i^{(1)} \\ x_i^{(2)} \\ x_i^{(3)} \\ x_i^{(4)} \end{pmatrix}, \qquad \boldsymbol{\mu}_i = \begin{pmatrix} \mu_i^{(1)} \\ \mu_i^{(2)} \\ \mu_i^{(3)} \\ \mu_i^{(4)} \end{pmatrix}, \qquad \text{and} \qquad \boldsymbol{\varepsilon}_i = \begin{pmatrix} \varepsilon_i^{(1)} \\ \varepsilon_i^{(2)} \\ \varepsilon_i^{(3)} \\ \varepsilon_i^{(4)} \end{pmatrix}. \quad (3.6.3)$$

Then the model becomes

$$\mathbf{x}_i = \boldsymbol{\mu}_i + \boldsymbol{\varepsilon}_i, \qquad i=1,2,\ldots,N. \quad (3.6.4)$$

We assume that the $\boldsymbol{\varepsilon}_i$s are independently normally distributed with mean vector zero and covariance matrix Σ. When Σ is known or $\Sigma = \sigma^2 I$, σ^2 unknown, Sen and Srivastava (1973) proposed tests for the hypothesis

$$H: \boldsymbol{\mu}_1 = \cdots = \boldsymbol{\mu}_N \tag{3.6.5}$$

vs

$$A: \boldsymbol{\mu}_1 = \cdots = \boldsymbol{\mu}_r \neq \boldsymbol{\mu}_{r+1} = \cdots = \boldsymbol{\mu}_N, \tag{3.6.6}$$

where the point of change r is unknown. The following sections consider the case of Σ unknown.

3.6.1 An Estimate of the Change Point

Let $\mathbf{x}_1, \mathbf{x}_2, \ldots, \mathbf{x}_N$ be independently distributed random p-vectors obeying the model (3.6.4). Let

$$\bar{\mathbf{x}}_r = r^{-1} \sum_{i=1}^{r} \mathbf{x}_i, \qquad \bar{\mathbf{x}}_{N-r} = (N-r)^{-1} \sum_{i=r+1}^{N} \mathbf{x}_i \tag{3.6.7}$$

and

$$W_r = \sum_{i=1}^{r} (\mathbf{x}_i - \bar{\mathbf{x}}_r)(\mathbf{x}_i - \bar{\mathbf{x}}_r)' + \sum_{i=r+1}^{N} (\mathbf{x}_i - \bar{\mathbf{x}}_{N-r})(\mathbf{x}_i - \bar{\mathbf{x}}_{N-r})'. \tag{3.6.8}$$

Let

$$T_r^2 = N^{-1} r(N-r)(\bar{\mathbf{x}}_r - \bar{\mathbf{x}}_{N-R})' W_r^{-1} (\bar{\mathbf{x}}_r - \bar{\mathbf{x}}_{N-r}), \qquad r = 1, 2, \ldots, N-1, \tag{3.6.9}$$

Define $\hat{r}$ to be the value such that

$$T_{\hat{r}}^2 = \max_{1 \le r \le N-1} T_r^2. \tag{3.6.10}$$

Then $\hat{r}$ is the maximum likelihood estimate of the point of change r.

From a computational point of view, it may be convenient to rewrite T_r^2. Let

$$\mathbf{y}_r = N^{-1/2} r^{1/2} (N-r)^{1/2} (\bar{\mathbf{x}}_r - \bar{\mathbf{x}}_{n-r}), \tag{3.6.11}$$

$$V = \sum_{i=1}^{N} (\mathbf{x}_i - \bar{\mathbf{x}})(\mathbf{x}_i - \bar{\mathbf{x}})', \qquad \bar{\mathbf{x}} = N^{-1} \sum_{i=1}^{N} \mathbf{x}_i, \tag{3.6.12}$$

and

$$S_r = \mathbf{y}_r' V^{-1} \mathbf{y}_r. \tag{3.6.13}$$

Then it can be shown that

$$T_r = S_r (1 - S_r)^{-1}. \tag{3.6.14}$$

It may be somewhat easier to compute S_r first and then obtain T_r. A likelihood ratio test for taking the null hypothesis of no shift in mean is based on $T_{\hat{r}}$. Such tests have been shown to have high power (Sen and Srivastava 1973, 1975a–c) for $p=1$. However, for $p>1$, it is inferior to many competitive tests (Sen and Srivastava, 1982). In the next section, we describe a test with reasonable power.

3.6.2 *A Test for a Shift in the Mean*

Let $\mathbf{x}_1, \mathbf{x}_2, \ldots, \mathbf{x}_N$ be independently distributed random p-vectors, and

$$u^{(j)} = \sum_{i=1}^{N-1} \frac{i\left(x_{i+1}^{(j)} - \bar{x}^{(j)}\right)}{(\mathbf{a}'\mathbf{a})^{1/2}}, \qquad j=1,2,\ldots,p, \tag{3.6.15}$$

where

$$\mathbf{a}' = \tfrac{1}{2}(-(N-1), -(N-3), -(N-5), \ldots, (N-5), (N-3), (N-1)), \tag{3.6.16}$$

and $\mathbf{x}$ and V are defined in (3.6.12). Let $\mathbf{u} = (u^{(1)}, \ldots, u^{(p)})$. Then we propose a test statistic

$$Q = \mathbf{u}'V^{-1}\mathbf{u} \tag{3.6.17}$$

for testing the hypothesis H against the alternative A defined in (3.6.5) and (3.6.6), respectively. Thus the hypothesis H is rejected if

$$Q \geq c_\alpha, \tag{3.6.18}$$

where

$$c_\alpha = 1 - \left[1 + p(N-p-1)^{-1} F_{p,N-p-1,\alpha}\right]^{-1}, \tag{3.6.19}$$

and $F_{p,N-p-1,\alpha}$ is the upper $\alpha 100\%$ point of the F-distribution with p and $N-p-1$ df.

It may be noted that $u^{(j)}$ is the test statistic used by Chernoff and Zacks (1964) for testing one-sided change when the variance is known. It may also be noted that a nonparametric multivariate generalization of the Bhattacharya and Johnson (1968) test can also be carried out on the same lines.

Example 3.6.1 Consider the Illinois traffic data of Table 3.7. The increase per year in deaths, injuries, accidents and deaths per 10^8 vehicle miles is given in the following table. The values of T_r, obtained by the SHIFT 1 program in Appendix II, were 2.64, 3.92, 6.21, 6.44, 1.71, 0.66, 1.55, and 2.20. A change in mean was estimated to occur after the fourth interval. That is, there was a shift in the annual increase of traffic casualties after 1966. The value of the test statistic Q is 0.8750. With $c_{0.05} = 1 - [1+4(4)^{-1}6.39]^{-1} = 0.865$, we reject H and claim that there was a significant

Table 3.7

	1963	1964	1965	1966	1967	1968	1969	1970	1971
Deaths	1.38	1.79	0.49	2.66	−0.29	0.06	0.34	−1.87	0.54
Injuries	7.58	14.27	11.38	3.6	0.37	−0.78	8.72	2.43	−11.05
Accidents	4.66	24.48	42.91	5.35	8.14	13.51	54.44	3.66	−15.6
Deaths per 10^8 vehicle miles	0.2	0.1	−0.1	0.2	−0.2	−0.2	−0.2	−0.5	0

shift in the rise of traffic casualties in Illinois with the shift estimated to have occurred in 1966.

3.7 Normal Probability Plots and Transformations

The methods of Sections 3.1–3.6 have all required the assumption of normality. In this section we discuss methods of assessing normality and methods of transforming the response variables to obtain data that are

Table 3.8[a]

x_1	x_2	x_1	x_2	
59	40.9	102	77.0	
80	42.0	67	23.0	
17	42.7	74	302.0	
11	306.0	30	48.8	
134	146.5	3	30.1	
76	101.0	25	16.2	
50	21.0	35	8.7	
85	22.7	70	152.5	
139	72.0	67	36.4	
175	73.0	55	151.5	
55	7.7	31	25.0	
30	28.1	57	303.0	
115	305.0	54	11.9	
152	48.8	68.4	90.5	Arithmetic means
		52.4	52.1	Geometric means

[a]Data from Huntingford (1976).

Table 3.9

	Mean	Variance	Skewness	Kurtosis	Shapiro–Wilk W	p
x_1	68.4	1928.1	0.816	0.200	0.936	0.112
x_2	90.5	9944	1.475	0.862	0.730	<0.01

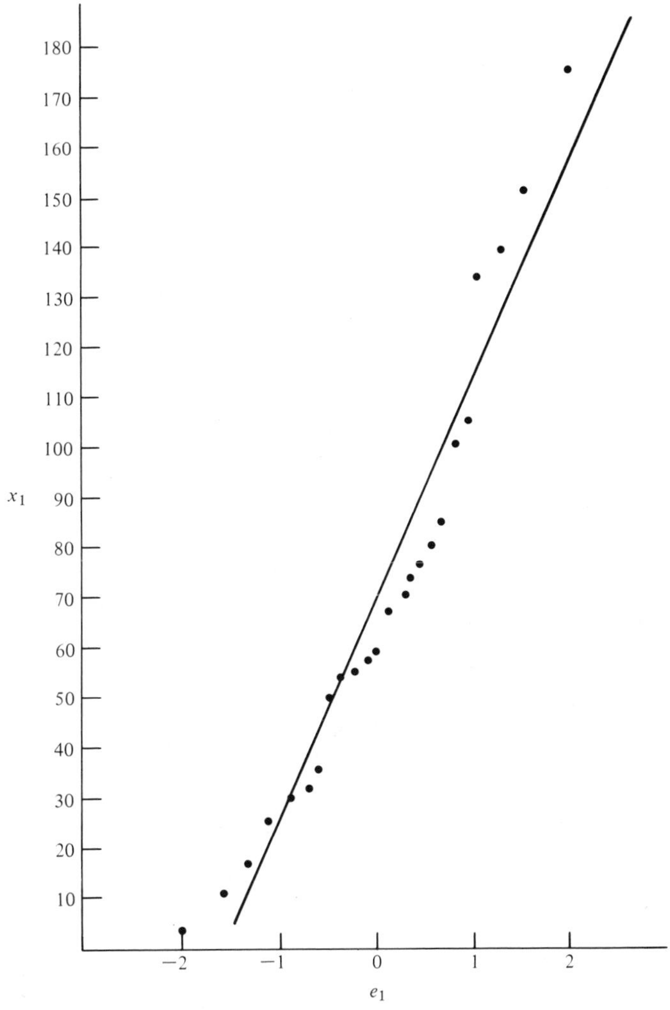

Figure 3.2 Normal probability plot for x_1.

approximately normally distributed. We consider the univariate situation first.

3.7.1 Univariate Tests of Normality

To test for normality the Kolmogorov–Smirnov test and the chi-square goodness of fit test are standard. These are used to test the null hypothesis that the distribution of the data follows a given distribution. Two other tests, used exclusively for testing normality, are the D'Agostino (1971) and

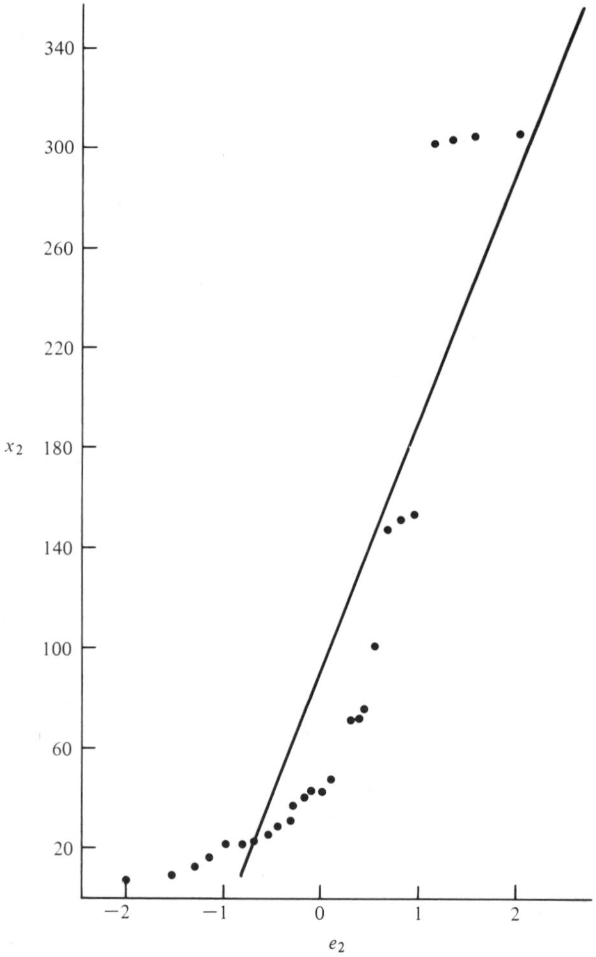

Figure 3.3 Normal probability plot for x_2.

the Shapiro–Wilk (1965) tests. We shall give the Shapiro–Wilk procedure, which exhibits reasonable power. Consider univariate data $x_1,\ldots,x_N$. Define descriptive statistics $\bar{x}$, s^2, α, and κ by

$$\bar{x}=N^{-1}\sum_{i=1}^{N}x_i, \qquad s^2=(N-1)^{-1}\sum_{i=1}^{N}(x_i-\bar{x})^2,$$
$$\alpha=\sum_{i=1}^{N}\frac{(x_i-\bar{x})^3}{(N-1)s^3}, \qquad \kappa=-3+\sum_{i=1}^{N}\frac{(x_i-\bar{x})^4}{(N-1)s^4}, \tag{3.7.1}$$

called the *mean*, *variance*, *skewness*, and *kurtosis*, respectively. The skewness and kurtosis are zero for the normal distribution, so these two measures are

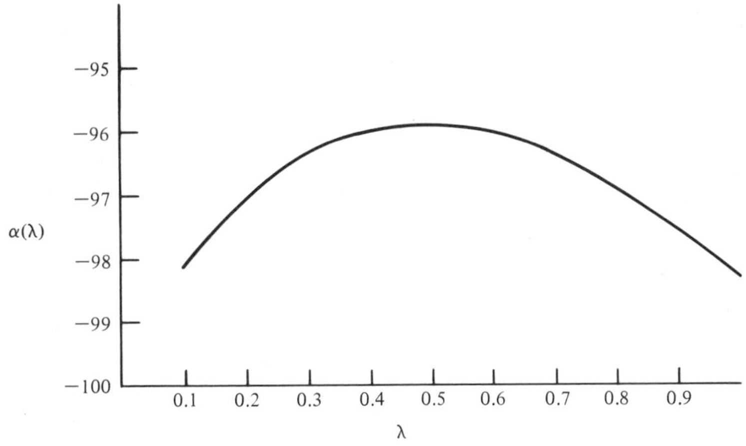

Figure 3.4 Likelihood plot for variable 1.

used as measures of normality of data. The Shapiro–Wilk test statistic is given by

$$W=\sum_{i=1}^{N} a_i x_{(i)}/s \tag{3.7.2}$$

where $a_1,\ldots,a_N$ depend on the expected values of the order statistics from a standard normal distribution and are given by Shapiro and Wilk (1965) and where $x_{(1)}<\cdots<x_{(N)}$ are the ordered values of the observations $x_1,\ldots,x_N$.

Figure 3.5 Likelihood plot for variable 2.

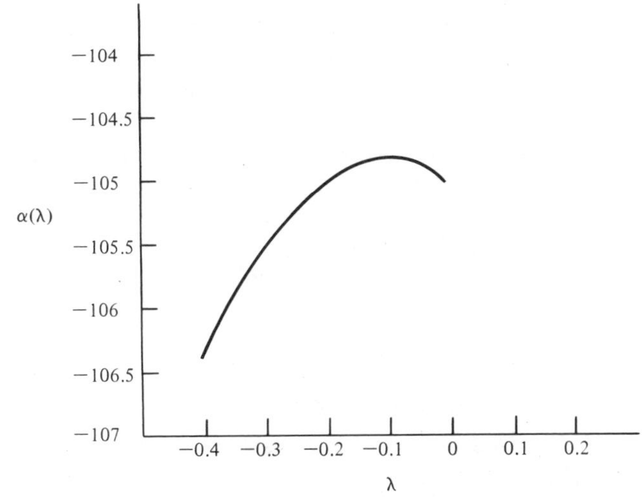

Table 3.10

y_1	e_1	y_2	e_2
96.727	0.0000	239.845	−0.1850
115.014	0.5834	241.260	−0.0921
45.215	−1.3027	242.139	0.0000
33.539	−1.5579	337.144	1.9966
153.112	1.1122	303.782	0.6963
111.735	0.4775	285.980	0.5834
87.894	−0.4775	203.051	−0.9553
118.999	0.6963	207.475	−0.8189
156.210	1.3027	269.194	0.2796
177.043	1.9966	269.889	0.3767
92.891	−0.2323	142.837	−1.9966
64.819	−0.8871	219.429	−0.4775
140.777	0.9553	337.001	1.5579
164.014	1.5579	249.189	0.1386
131.739	0.8189	272.569	0.4775
104.026	0.1386	208.218	−0.6963
110.063	0.3767	336.569	1.1122
64.819	−0.8871	249.189	0.1386
10.598	−1.9966	223.226	−0.3767
57.910	−1.1122	188.051	−1.1122
71.173	−0.5834	150.492	−1.5579
106.650	0.2796	305.664	0.9553
104.026	0.1386	233.587	−0.2796
92.891	−0.2323	305.356	0.8189
66.130	−0.6963	212.913	−0.5834
94.826	−0.0921	336.714	1.3027
91.910	−0.3767	169.707	−1.3027

The null hypothesis of normality is rejected if $W \leq W_\alpha$, where W are the tabulated percentage points, also given by Shapiro and Wilk (1965). For small departures from zero in the skewness and kurtosis of a distribution, we can still use the Student's t-test, which is robust to slight departures from normality (Scheffé, 1959).

We can make a graphical assessment of normality by normal probability plots of $x_{(1)},\ldots,x_{(N)}$ against $\Phi^{-1}((i-\frac{3}{8})/(N+\frac{1}{4}))$, where $\Phi^{-1}(p)$ is defined by $P(z \leq \Phi^{-1}(p)) = p$, $z \sim N(0,1)$. If the data are normal, the plot should result in a reasonably straight line. The analysis presented here is available using procedures in the SAS computer package.

3.7.2 *Transformations*

The application of transformations to data to achieve normality has been suggested and discussed by several authors, including Bartlett (1947b), Box and Cox (1964), Draper and Hunter (1969a, b), and John and Draper (1980). We give now two families of transformations, a family of power

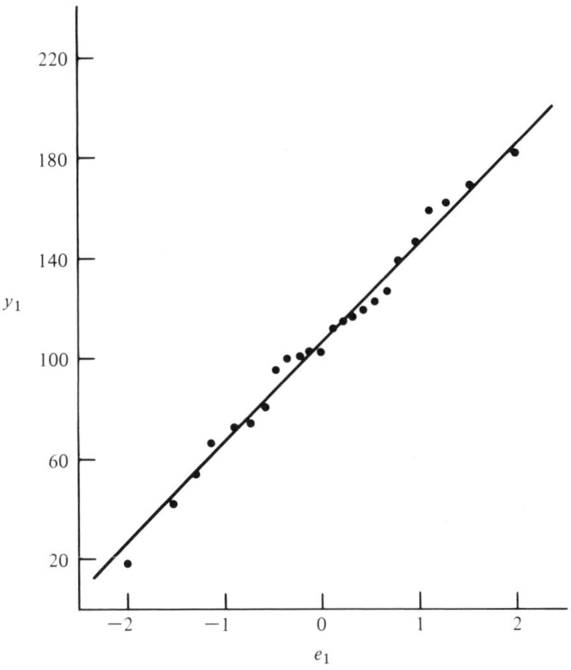

Figure 3.6 Normal probability plot for y_1.

transformations suggested by Box and Cox (1964) and a family of modulus transformations suggested by John and Draper (1980). For data points $x_1,\ldots,x_N$ let $g_1=\prod_{i=1}^{N}(x_i)^{1/N}$ be the geometric mean. Then we define the power transformation by

$$\begin{aligned} y_i^{(\lambda)} &= \left(x_i^{\lambda}-1\right)/\left(\lambda g_1^{\lambda-1}\right), \qquad \lambda \neq 0 \\ &= g_1 \log x_i, \qquad \lambda = 0, \end{aligned} \tag{3.7.3}$$

$i=1,\ldots,N$, if the data are positive. If some of the observations are negative, we consider the family of transformations

$$\begin{aligned} y_i^{(\lambda)} &= \left[(x_i+a)^{\lambda}-1\right]/\left(\lambda g_2^{\lambda-1}\right), \qquad \lambda \neq 0 \\ &= g_2 \log x_i, \qquad \lambda = 0, \end{aligned} \tag{3.7.4}$$

$i=1,\ldots,N$, where $g_2=\prod_{i=1}^{N}(x_i+a)^{1/N}$ and a is chosen such that $x_i+a>0$, $i=1,\ldots,N$. The transformation in (3.7.3) and (3.7.4) tends to eliminate the skewness of data. For data tending to be symmetrical but with a nonzero kurtosis John and Draper (1980) proposed the modulus transformation

$$\begin{aligned} z_j^{(\lambda)} &= \operatorname{sign}\left[\left(|x_j-b|+1\right)^{\lambda}-1\right]/\left(\lambda g_3^{\lambda-1}\right), \qquad \lambda \neq 0, \\ &= \operatorname{sign} g_3 \log\left(|x_j-b|+1\right), \qquad \lambda = 0, \end{aligned} \tag{3.7.5}$$

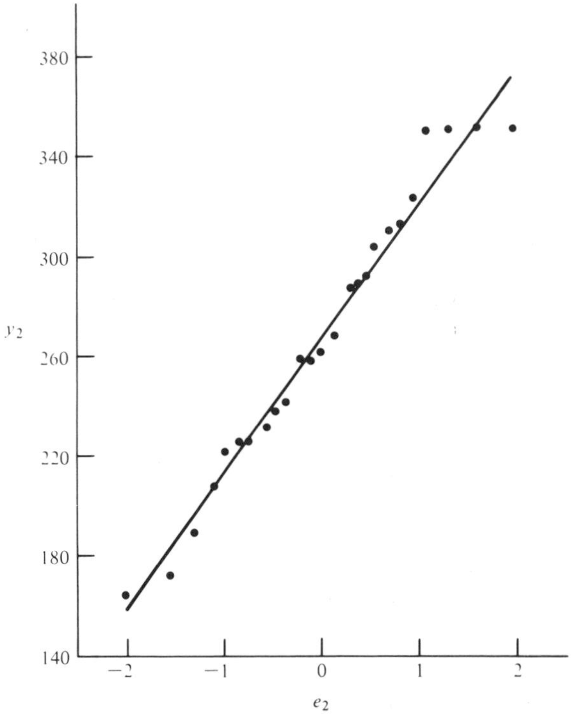

Figure 3.7 Normal probability plot for y_2.

$j=1,\ldots,N$, where $g_3=\prod_{j=1}^{N}(|(x_j-b)|+1)^{1/N}$, sign = sign$(x_j-b)$, and b is a preselected value such as the arithmetic or geometric mean. In Example 3.7.1 we shall use the geometric mean for b. Having chosen the family of transformations we then find the maximum likelihood estimate of λ. If one of the members of the family does produce normal data, then the log of the maximum likelihood function attains a maximum value of

$$\max_{\lambda} -(N/2)\log\left[N^{-1}(N-1)s_\lambda^2\right],$$

where s_λ^2 is the pooled variance of the transformed data. The maximum likelihood estimate $\hat{\lambda}$ is the value of λ that maximizes the log likelihood function. For convenience we use an equivalent, the modified likelihood ratio function, which takes the form

$$l(\lambda)=-(f/2)\log s_\lambda^2 \tag{3.7.6}$$

where f is the number of degrees of freedom associated with s_λ^2, the sample variance of the transformed data. For the one-sample situation, $f=N-1$. This method is used again in the analysis of variance in Section 4.6. The maximum likelihood estimate of λ is obtained by plotting $l(\lambda)$ or by

Table 3.11

	Mean	Variance	Skewness	Kurtosis	Shapiro–Wilk W	p
y_1	98.69	1597.18	-0.050	-0.079	0.983	0.918
y_2	249.65	3183.24	-0.012	-0.712	0.957	0.397

methods of steepest ascent. Note that for $\lambda=1$ the data are not transformed.

3.7.3 Multivariate Normality

The multivariate situation is complex. Generalizations of univariate tests for normality exist and are given by Andrews et al. (1973), Malkovich and Afifi (1973), Mardia (1975), and Cox and Small (1978), but their power is still to be shown. We suggest looking at univariate tests and probability plots: If we believe that the data are not normal, then we may apply univariate transformations to each variable separately. Although this method may only achieve marginal normality, it has the advantage of simplicity. Note that if we are testing p characteristics for normality, each test should be performed at a significance level of α/p to maintain an overall level of significance of α.

One multivariate method that requires extensive programming is to transform each characteristic of the data set by one of the family of transformations (3.7.3)–(3.7.5) and then choose the p values of λ, $\boldsymbol{\lambda}=(\lambda_1,\ldots,\lambda_p)'$, such that

$$l(\boldsymbol{\lambda})=(-f/2)\log|S_{(\lambda)}| \tag{3.7.7}$$

is maximized, where $S_{(\lambda)}$ is the sample covariance matrix of the transformed data with f degrees of freedom.

As an alternative to this extensive programming, we suggest separate transformations for each variable to achieve marginal normality. A graphical method is given in this section to test for joint normality. Another elaborate graphical method based on principal components, due to Srivastava (1982), is given in Section 10.5. Some other graphical methods have been given by Andrews et al. (1973) and Gnanadesikan (1977). Mardia's (1970, 1974, 1975) tests based on multivariate skewness and kurtosis lack tables for $p>2$.

Suppose we have observations $x_1,\ldots,x_N$ each independently distributed as $N_p(\boldsymbol{\mu},\Sigma)$. Then

$$c_i=N(N-1)^{-2}(\mathbf{x}_i-\bar{\mathbf{x}})'S^{-1}(\mathbf{x}_i-\bar{\mathbf{x}}) \tag{3.7.8}$$

has a beta distribution with parameters $a=p/2$ and $b=(f-p)/2$, where $\bar{\mathbf{x}}$ is the sample mean and S is the sample covariance matrix with $f=N-1$

Table 3.12

c		
Transformed data	Original data	d expected
0.0012	0.0014	0.0031
0.0052	0.0108	0.0063
0.0096	0.0117	0.0096
0.0175	0.0136	0.0130
0.0264	0.0155	0.0166
0.0273	0.0183	0.0203
0.0287	0.0207	0.0241
0.0309	0.0239	0.0282
0.0326	0.0254	0.0324
0.0343	0.0269	0.0369
0.0396	0.0270	0.0416
0.0443	0.0292	0.0466
0.0465	0.0346	0.0518
0.0700	0.0415	0.0575
0.0739	0.0416	0.0635
0.0819	0.0447	0.0699
0.0829	0.0550	0.0769
0.0888	0.0594	0.0845
0.0956	0.0949	0.0930
0.1069	0.0961	0.1023
0.1126	0.1088	0.1129
0.1158	0.1614	0.1251
0.1252	0.1803	0.1395
0.1493	0.1921	0.1571
0.1550	0.2109	0.1801
0.1949	0.2439	0.2136
0.2794	0.2859	0.2794

degrees of freedom. c_i, $i=1,\ldots,N$, ordered as $c_{(1)}<\cdots<c_{(N)}$, is then plotted against d_i, which is calculated from the inverse beta distribution. Thus suppose v has a beta distribution with parameters a and b. Then d_i satisfies

$$P(v\leq d_i)=(i-\alpha)/(N-\alpha-\beta+1), \qquad (3.7.9)$$

where $\alpha=\frac{1}{2}(a-1)/a$ and $\beta=\frac{1}{2}(b-1)/b$, chosen such that the expected value of $c_{(i)}$ comes close to d_i in the tails of the distribution (Small, 1978). If the plot results in approximately a straight line through the origin with slope 1, then we may assume that the data are approximately normal. Note that we can never be sure that the data are normal (although we may in some cases be sure that they are not).

For data that deviate slightly from normality in skewness and kurtosis, Hopkins and Clay (1963), Ito (1969), Mardia (1971), and Mardia (1975) have shown that tests of location such as Hotelling's T^2 are reasonably

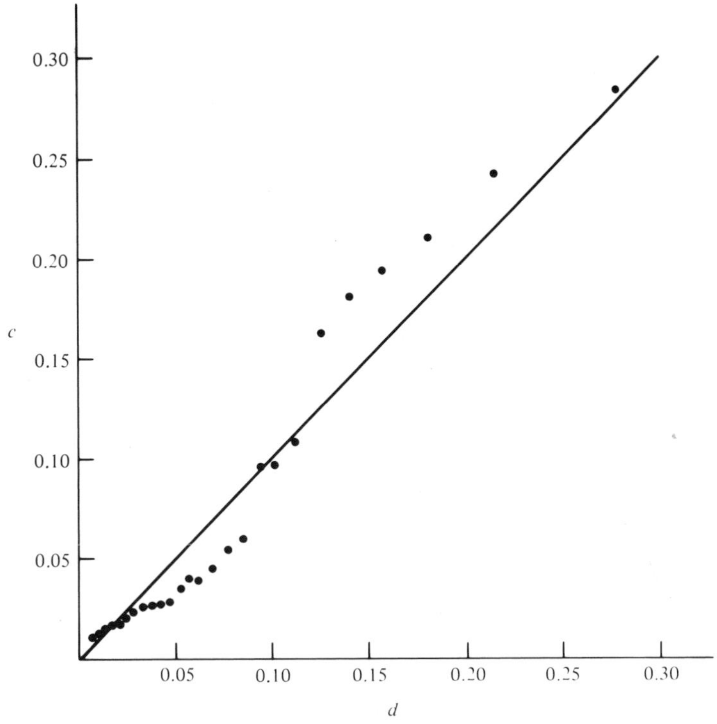

Figure 3.8 c_i vs d_i for the original data.

robust, so that the significance level under the assumption of normality is reasonable. Hence we may use the T^2-procedure with data only approximately normally distributed. Other forms of nonnormality may have a different effect on the significance level of the T^2-test. Srivastava and Awan (1982) have shown that the Hotelling's T^2-procedure is not robust when the data come from a mixture of two normal populations. Hence the data should be plotted to check at least the approximate normality.

Example 3.7.1 Consider two measurements on sticklebacks (Table 3.8): x_1, the number of bites, and x_2, the mean bout length of facing from less than 10 and more than 20 cm. The covariance matrix is

$$S = \begin{pmatrix} 1928.1 & 522.7 \\ 522.7 & 9944 \end{pmatrix}.$$

Descriptive statistics were obtained by the use of the proc univariate and proc rank procedures in SAS (Table 3.9). From Table 3.9 and the probability plots in Figures 3.2 and 3.3 we can conclude that x_1 may be normal but x_2 is definitely not normally distributed. Both x_1 and x_2 take on only positive values, so we use the power transformation $y^{(\lambda)}$ on both of these

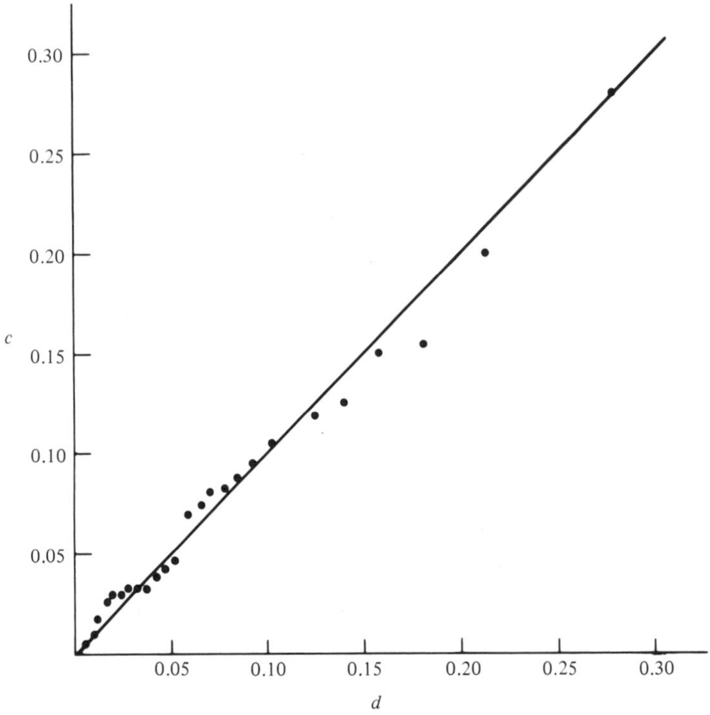

Figure 3.9 c_i vs d_i for the transformed data.

variables. The plots of their likelihood function in Figures 3.4 and 3.5 produce maxima at $\lambda = 0.50$ for x_1 and $\lambda = -0.1$ for x_2. Note that λ is farther from 1 for x_2 than for x_1.

The two new variables

$$y_1 = (x_1^{0.5} - 1)/(0.5 \times 52.39^{0.5} - 1)$$

and

$$y_2 = (x_2^{-0.1} - 1)/(-0.1 \times 52.09^{-0.1} - 1)$$

take on the values in Table 3.10, which shows approximate expected values of the order statistics from a standard normal distribution (e_1 and e_2) for the orders of the two variables y_1 and y_2. Plots of y_1 vs e_1 and y_2 vs e_2 are given in Figures 3.6 and 3.7. Both yield fairly straight lines, and hence normality may exist. Descriptive statistics for y_1 and y_2 are given in Table 3.11. The skewness in the second variable has been reduced from 1.475 to -0.012, and the Shapiro–Wilk test for normality is insignificant for both variables. At this point we can assume marginal normality.

We now look at multivariate plots of c_i vs d_i, $i = 1, \ldots, n$, defined in (3.7.7)–(3.7.9). For this example $N = 27$, $p = 2$, $a = 1$, $b = 13$, $\alpha = 0$, and

$\beta = \frac{6}{13}$. Hence, from (3.7.9), the values of d_i, $i=1,\ldots,27$, are given implicitly by

$$P(V \le d_i) = i/27.54,$$

where V is a beta random variable with parameters 1 and 13. Values of c_i for the transformed and original data and the values of d_i, $i=1,\ldots,27$, are given in Table 3.12. Plots of c_i vs d_i are given in Figure 3.9 for the transformed data and in Figure 3.8 for the original data.

The transformed data tend to produce a straight-line plot, whereas the original data deviate considerably from a 45° line. Hence we may assume multivariate normality for the transformed data.

3.8 Missing Observations

The problem of missing data arises frequently in practice. It occurs most frequently in large-scale surveys. For example, consider a survey of families conducted in 1975 and a follow-up survey conducted in 1977. It is very likely that many families studied in 1975 could not be located in 1977, resulting in missing observations. Here the missing observations are *unintentional*, but the case where they are *intentional* is not uncommon. For example, in successive sampling stages new observations are added and some old ones dropped. Every branch of the social sciences and economics, medical trials, and other fields as well have problems involving missing observations. The available results are far from complete since the solution presents many difficulties. For simplicity of presentation, consider the bivariate situation first. Let $(x_1, x_2)'$ be a bivariate vector normally distributed with mean vector $\boldsymbol{\mu} = (\mu_1, \mu_2)'$ and covariance matrix

$$\Sigma = \begin{pmatrix} \sigma_1^2 & \sigma_{12} \\ \sigma_{12} & \sigma_2^2 \end{pmatrix}. \tag{3.8.1}$$

Without loss of generality, the missing data in the bivariate situation may be arranged as follows:

$$\begin{aligned} &x_{11},\ldots,x_{1n_1}, x_{1,n_1+1},\ldots,x_{1,n_1+n_2}, \\ &x_{21},\ldots,x_{2n_1}, x_{2,n_1+1},\ldots,x_{2,n_1+n_3}. \end{aligned} \tag{3.8.2}$$

That is, we have n_1 pairs of observations $(x_{1i}, x_{2i})'$, $i=1,2,\ldots,n_1$, n_2 observations on x_1 (the corresponding observations on x_2 are missing), and n_3 observations on x_2 (the corresponding observations on x_1 are missing). This is the most general situation in the bivariate case. We assume that the observations $(x_{1i}, x_{2i})'$, $i=1,2,\ldots,n_1$, x_{1,n_1+j}, $j=1,2,\ldots,n_2$, and x_{2,n_1+k}, $k=1,2,\ldots,n_3$, are independently distributed. Now we shall obtain the maximum likelihood estimates of $\boldsymbol{\mu}$ and Σ. An implicit estimate of $\boldsymbol{\mu}$ and Σ was given by Rao (1952). In order to give explicit estimates, we introduce some notation keeping in mind the generalization to the p-response case. Thus in many places a matrix is used to denote a vector and a vector to denote a scalar. Let

$$B_1 = I_2, \qquad B_2 = (1,0), \qquad B_3 = (0,1),$$

$$\mathbf{z}_{1\alpha} = \begin{pmatrix} x_{1\alpha} \\ x_{2\alpha} \end{pmatrix}, \qquad \alpha = 1,2,\ldots,n_1, \tag{3.8.3}$$

$$\mathbf{z}_{2\beta} = x_{1\beta}, \qquad \beta = n_1+1,\ldots,n_1+n_2,$$

$$\mathbf{z}_{2\gamma} = x_{2\gamma}, \qquad \gamma = n_1+1,\ldots,n_1+n_3,$$

$$\bar{\mathbf{z}}_1 = n_1^{-1} \sum_{\alpha=1}^{n_1} \mathbf{z}_{1\alpha},$$

$$\bar{\mathbf{z}}_2 = n_2^{-1} \sum_{\beta=n_1+1}^{n_1+n_2} \mathbf{z}_{2\beta}, \tag{3.8.4}$$

and

$$\bar{\mathbf{z}}_3 = n_3^{-1} \sum_{\gamma=n_1+1}^{n_1+n_3} \mathbf{z}_{2\gamma}.$$

Then

$$\begin{aligned} E(\bar{\mathbf{z}}_1) &= E(\mathbf{z}_{1\alpha}) = B_1\boldsymbol{\mu} = \boldsymbol{\mu}, \qquad & \alpha &= 1,\ldots,n_1, \\ E(\bar{\mathbf{z}}_2) &= E(\mathbf{z}_{2\beta}) = B_2\boldsymbol{\mu} = \mu_1, \qquad & \beta &= n_1+1,\ldots,n_1+n_2, \\ E(\bar{\mathbf{z}}_3) &= E(\mathbf{z}_{3\gamma}) = B_3\boldsymbol{\mu} = \mu_2, \qquad & \gamma &= n_1+1,\ldots,n_1+n_3, \end{aligned} \tag{3.8.5}$$

and $\mathbf{z}_{1\alpha}$, $\mathbf{z}_{2\beta}$, and $\mathbf{z}_{3\gamma}$ are independently distributed. Let

$$\Sigma_j = B_j \Sigma B_j', \qquad j = 1,2,3. \tag{3.8.6}$$

Then the maximum likelihood estimates of $\boldsymbol{\mu}$ and Σ are the solutions of the following two equations:

$$\mu = \left[\sum_{j=1}^{3} n_j \left(B_j'\Sigma_j^{-1}B_j \right) \right]^{-1} \left(\sum_{j=1}^{3} n_j B_j'\Sigma_j^{-1}\bar{\mathbf{z}}_j \right) \tag{3.8.7}$$

and

$$\sum_{j=1}^{3} n_j B_j'\Sigma_j^{-1}B_j = \sum_{j=1}^{3} B_j'\Sigma_j^{-1}U_jU_j'\Sigma_j^{-1}B_j, \tag{3.8.8}$$

where

$$U_j = Z_j - B_j\boldsymbol{\mu}\mathbf{e}_{n_j}', \qquad \mathbf{e}_{n_j}' = (1,\ldots,1) : 1 \times n_j \tag{3.8.9}$$

and

$$Z_j = \left(\mathbf{z}_{j1},\ldots,\mathbf{z}_{j,n_j} \right), \qquad j = 1,2,3. \tag{3.8.10}$$

Let us test the hypothesis

$$H\colon \boldsymbol{\mu} = \boldsymbol{\mu}_0 \qquad \text{vs} \qquad A\colon \boldsymbol{\mu} \neq \boldsymbol{\mu}_0.$$

The likelihood function under the alternative hypothesis A is given by

$$L(\hat{\boldsymbol{\mu}}, \hat{\Sigma}) = (2\pi)^{-(2n_1+n_2+n_3)/2} \prod_{j=1}^{3} |\hat{\Sigma}_j|^{-n_j/2}$$

$$\times \exp -\frac{1}{2} \sum_{j=1}^{3} \sum_{i=1}^{n_j} (\mathbf{z}_{ji} - B_j\hat{\boldsymbol{\mu}})'\hat{\Sigma}_j^{-1}(\mathbf{z}_{ji} - B_j\hat{\boldsymbol{\mu}}).$$

The likelihood function under H is given by

$$L(\boldsymbol{\mu}_0, \tilde{\Sigma}) = (2\pi)^{-(2n_1+n_2+n_3)/2} \prod_{j=1}^{3} |\tilde{\Sigma}_j|^{-n_j/2}$$

$$\times \exp -\frac{1}{2} \sum_{j=1}^{3} \sum_{i=1}^{n_j} (\mathbf{z}_{ji} - B_j\boldsymbol{\mu}_0)'\tilde{\Sigma}_j^{-1}(\mathbf{z}_{ji} - B_j\boldsymbol{\mu}_0),$$

where $\tilde{\Sigma}$ is the solution of (3.8.8) and (3.8.9) with $\boldsymbol{\mu} = \boldsymbol{\mu}_0$. The likelihood test statistic is now given by

$$\lambda = L(\boldsymbol{\mu}_0, \tilde{\Sigma})/L(\hat{\boldsymbol{\mu}}, \hat{\Sigma}).$$

The null hypothesis is then rejected at the α-level if

$$-2\ln\lambda \geq \chi^2_{2,\alpha}.$$

Example 3.8.1 Consider the following data on the improvement of reading and arithmetic scores after a certain teaching program for slow learners:

Reading	5	−1	2	3	0	6	5	4	2	1	2	3	5	—	—	—
Arithmetic	1	2	−2	3	5	4	−1	7	5	4	—	—	—	2	−1	4

Incomplete data can arise from the failure of a student to be present at the time of the particular test. The use of the program INCOMPLETE in Appendix II yielded as estimates of the mean and covariance matrix

$$\hat{\boldsymbol{\mu}} = \begin{pmatrix} 2.86 \\ 2.52 \end{pmatrix}, \qquad \hat{\Sigma} = \begin{pmatrix} 4.125 & -0.564 \\ -0.564 & 6.70 \end{pmatrix},$$

and under the hypothesis H: $\boldsymbol{\mu} = 0$ yielded the following estimate of Σ:

$$\tilde{\Sigma} = \begin{pmatrix} 11.84 & 6.15 \\ 6.15 & 13.20 \end{pmatrix}.$$

The value of $-2\ln\lambda$ is then calculated to be $22.0 \geq \chi^2_{2,0.05} = 5.99$. Hence H is rejected, and we claim that there is an improvement in scores with this teaching program. 95% confidence intervals for the means of the two characteristics are obtained by the use of univariate confidence intervals and Bonferroni bounds. For change in reading scores the confidence interval is given by

$$2.85 \pm t_{12,0.0125}\sqrt{4.48/13} = (1.1, 4.6),$$

where 2.85 is the sample mean and 4.48 is the sample variance of the improvements in reading scores. For the improvement in arithmetic scores the interval is given by

$$2.54 \pm t_{12,0.0125}\sqrt{7.26/13} = (0.3, 4.7).$$

Hence we would conclude that both scores improved with the teaching program.

The generalization of the above results to the p-response case is quite straightforward. It can be described as follows. Note that in the bivariate situation ($p=2$) we have three subsets of responses $\mathbf{z}_{1\alpha}$, $\mathbf{z}_{2\beta}$, and $\mathbf{z}_{3\gamma}$; the first represents the observations on both the variables, the second observations on the first variable only, and the third observations on the second variable only. In the case of p responses we have $K = 2^p - 1$ subsets R_k, $k = 1, 2, \ldots, K$. For example, R_1 represents the response on all of the variables. Thus by properly defining $B_1, \ldots, B_K$, we still have the estimates given in (3.8.7) and (3.8.8), except that 3 is now replaced by K and $n_1, \ldots, n_K$ denote the number of observations in the subsets $R_1, \ldots, R_K$, where $n_i \geq 0$, $i = 1, 2, \ldots, K$. If there are p_i characteristics, $j_1, \ldots, j_{pi}$, say in the ith set of observations, then B_i would be a $p_i \times p$ matrix with a 1 in the (s, j_s) position, $s = 1, \ldots, p_i$, and zeros elsewhere. For example, suppose we have $p = 5$ and there are four sets of observations as follows:

Set	Variables included	n_i
R_1	1,2,3,4,5	63
R_2	1,3,5	5
R_3	2,3,4	4
R_4	1	10

Then

$$B_1 = I_5, \qquad B_2 = \begin{pmatrix} 1 & 0 & 0 & 0 & 0 \\ 0 & 0 & 1 & 0 & 0 \\ 0 & 0 & 0 & 0 & 1 \end{pmatrix},$$

$$B_3 = \begin{pmatrix} 0 & 1 & 0 & 0 & 0 \\ 0 & 0 & 1 & 0 & 0 \\ 0 & 0 & 0 & 1 & 0 \end{pmatrix}, \qquad B_4 = (1 \quad 0 \quad 0 \quad 0 \quad 0).$$

Estimates of $\boldsymbol{\mu}$ and Σ are given by (3.8.7)–(3.8.10). If we wish to test H:

$\boldsymbol{\mu}=\boldsymbol{\mu}_0$ vs A: $\boldsymbol{\mu}\neq\boldsymbol{\mu}_0$, we calculate the likelihood functions

$$L(\hat{\boldsymbol{\mu}},\hat{\Sigma})=(2\pi)^{-1/2\sum_{j=1}^{K}n_jp_j}\prod_{j=1}^{K}|\hat{\Sigma}_j|^{-n_j/2}$$
$$\times\exp-\frac{1}{2}\sum_{j=1}^{K}\sum_{i=1}^{n_j}(\mathbf{z}_{ji}-B_j\hat{\boldsymbol{\mu}})'\hat{\Sigma}_j^{-1}(\mathbf{z}_{ji}-B_j\hat{\boldsymbol{\mu}}),$$

$$L(\boldsymbol{\mu}_0,\tilde{\Sigma})=(2\pi)^{-1/2\sum_{j=1}^{K}n_jp_j}\prod_{j=1}^{K}|\tilde{\Sigma}_j|^{-n_j/2}$$
$$\times\exp-\frac{1}{2}\sum_{j=1}^{K}\sum_{i=1}^{n_j}(\mathbf{z}_{ji}-B_j\boldsymbol{\mu}_0)'\tilde{\Sigma}_j^{-1}(\mathbf{z}_{ji}-B_j\boldsymbol{\mu}_0),$$

where $\hat{\boldsymbol{\mu}}$ and $\hat{\Sigma}$ are the maximum likelihood estimates obtained from (3.8.7)–(3.8.10) and $\tilde{\Sigma}$ is obtained from (3.8.8)–(3.8.9) with $\boldsymbol{\mu}=\boldsymbol{\mu}_0$. The null hypothesis is rejected if

$$-2\ln\lambda>\chi^2_{p,\alpha},\qquad \lambda=L(\boldsymbol{\mu}_0,\tilde{\Sigma})/L(\hat{\boldsymbol{\mu}},\hat{\Sigma}).$$

For details, the reader is referred to Srivastava (1980a), where results for other models, including regression and growth curve models, are also given. For classification problems, see Srivastava and Zaatar (1972) and Leung (1977). Macdonald (1971) gives a heuristic method for estimating the missing observations by minimizing the sample covariance matrix. Kleinbaum (1973) gives best asymptotic normal estimators for the mean and covariance matrix but does not derive the maximum likelihood estimators and provides no guarantee that the estimated covariance matrix is positive definite. Woolson (1980) gives estimators for the mean and covariance matrix when the missing observations follow structured patterns such as the generalizations of the monotone sample discussed in Section 3.8.1.

The distribution of the estimates proposed is known only as $n\to\infty$. Asymptotic expansions for the maximum likelihood estimates are given by Carter (1982). An alternative computational procedure to obtain maximum likelihood estimates, known as the EM algorithm, has been given by Dempster et al. (1977); these estimates will obviously be the same as those given here. However, it is not clear how to use their method to obtain estimates in complicated cases, such as growth curve models, nor is it known how to use these estimates in testing hypotheses. Also, the properties of these estimates cannot be obtained without the use of the preceding expression from Srivastava (1980a).

In the next sections we shall consider what is called the monotone sample case, where some exact tests are available; see Bhargava (1962) for more details.

3.8.1 *Monotone Sample — Bivariate Case*

For simplicity of presentation we first give the bivariate case, also considered by Anderson (1957). The data are of the form

$$\begin{aligned} &x_{11},\dots,x_{1N_2},x_{1,N_2+1},\dots,x_{1,N_1},\\ &x_{21},\dots,x_{2N_2} \end{aligned} \tag{3.8.11}$$

Comparing this with the general case, we have $n_1 \to N_2$, $n_2 \to N_1 - N_2$, and $n_3 = 0$. Data of this form are called a *monotone sample*.

In this case it is simple to write down the likelihood function, and tests and estimates can be obtained. The distribution problems, however, present some difficulties. The asymptotic distributions of the chi-square type (or their linear combinations) under the null hypotheses have been given by Bhargava (1962).

To give an idea how to write down the likelihood function, we consider the bivariate case. Let $f(x_1, x_2)$ denote the joint pdf of x_1 and x_2; $g(x_2 | x_1)$ the conditional pdf of x_2, given x_1; and $h(x_1)$ the marginal pdf of x_1. Then we have

$$f(x_1, x_2) = g(x_2 | x_1)h(x_1).$$

Under the assumption of bivariate normality of (x_1, x_2), $x_1 \sim N(\mu_1, \sigma_{11})$ and $x_2 | x_1 \sim N[\mu_2 + \beta(x_1 - \mu_1), \sigma_{2\cdot1}]$, where $\beta = \sigma_{12}/\sigma_{11}$ and $\sigma_{2\cdot1} = \sigma_{22} - (\sigma_{12}^2/\sigma_{11})$. Hence the likelihood function based on all the available observations can be written down explicitly.

3.8.2 *Tests and Estimates Based on a Bivariate Monotone Sample*

In this case explicit expressions for the maximum likelihood estimates of μ_1, μ_2, σ_{11}, σ_{22}, and σ_{12} can be given. Let

$$\bar{x}_{1N_1} = N_1^{-1}\sum_{\alpha=1}^{N_1} x_{1\alpha}, \qquad \bar{x}_{1N_2} = N_2^{-1}\sum_{\alpha=1}^{N_2} x_{1\alpha}, \qquad \bar{x}_{2N_2} = N_2^{-1}\sum_{\alpha=1}^{N_2} x_{2\alpha},$$

$$\hat{\sigma}_{11} = N_1^{-1}\sum_{\alpha=1}^{N_1}(x_{1\alpha} - \bar{x}_{1N_1})^2,$$

$$\hat{\sigma}_{2\cdot1} = N_2^{-1}\left[\sum_{\alpha=1}^{N_2}(x_{2\alpha} - \bar{x}_{2N_2})^2 - \hat{\beta}^2\sum_{\alpha=1}^{N_2}(x_{1\alpha} - \bar{x}_{1N_2})^2\right], \tag{3.8.12}$$

and

$$\hat{\beta} = \sum_{\alpha=1}^{N_2}(x_{1\alpha} - \bar{x}_{1N_2})(x_{2\alpha} - \bar{x}_{2N_2}) \Big/ \sum_{\alpha=1}^{N_2}(x_{1\alpha} - \bar{x}_{1N_2})^2.$$

Then the maximum likelihood estimates of μ_1, μ_2, σ_{11}, σ_{22}, and σ_{12} are given, respectively, by

$$\hat{\mu}_1=\bar{x}_{1N_1}, \qquad \hat{\mu}_2=\bar{x}_{2N_2}-\hat{\beta}\,(\bar{x}_{1N_2}-\bar{x}_{1N_1}),$$

$$\hat{\sigma}_{11}=N_1^{-1}\sum_{\alpha=1}^{N_1}(x_{1\alpha}-\bar{x}_{1N})^2, \qquad \hat{\sigma}_{22}=\hat{\sigma}_{2\cdot 1}+\hat{\beta}^2\hat{\sigma}_{11}, \qquad \hat{\sigma}_{12}=\hat{\beta}\hat{\sigma}_{11}. \tag{3.8.13}$$

Having obtained the maximum likelihood estimates, we now give the likelihood ratio test for the hypothesis

$$H\colon \boldsymbol{\mu}=\mathbf{0},\quad \Sigma>0, \qquad \text{vs} \qquad A\colon \boldsymbol{\mu}\neq\mathbf{0},\quad \Sigma>0.$$

It is based on the statistic

$$\lambda=\left(N_1\hat{\sigma}_{11}\Big/\sum_{\alpha=1}^{N_1}x_{1\alpha}^2\right)^{N_1/2}\left[N_2\hat{\sigma}_{2\cdot 1}\Big/\left(\sum_{\alpha=1}^{N_2}x_{2\alpha}^2-\hat{\beta}_H^2\sum_{\alpha=1}^{N_2}x_{1\alpha}^2\right)\right]^{N_2/2}, \tag{3.8.14}$$

where

$$\hat{\beta}_H=\left(\sum_{\alpha=1}^{N_2}x_{1\alpha}x_{2\alpha}\Big/\sum_{\alpha=1}^{N_2}x_{1\alpha}^2\right). \tag{3.8.15}$$

The asymptotic cumulative distribution function (cdf) of $M=-2\ln\lambda$ under the hypothesis H (Bhargava, 1962) is given by

$$P\{M\le M_0\}=P\{\rho M\le\rho M_0\}=(1-w_2)P\{\chi_2^2\le\rho M_0\}+w_2P\{\chi_6^2\le\rho M_0\}+O(N^{-3}), \tag{3.8.16}$$

where

$$N=N_1+N_2, \qquad \rho=1-\tfrac{1}{4}(3/N_1+5/N_2), \qquad N_1\ge N_2>2,$$

$$w_2=\frac{1}{2\rho^2}\sum_{k=1}^{2}\frac{h_k(2h_k-1)}{N_k^2}, \qquad \text{and} \qquad h_k=\tfrac{1}{2}[N_k(1-\rho)-k]. \tag{3.8.17}$$

Hence, as a first approximation, $-2\rho\ln\lambda$ is asymptotically distributed as a chi-square with 2 df. Thus for large N the hypothesis is rejected if

$$-2\rho\ln\lambda\ge\chi_{2,\alpha}^2, \tag{3.8.18}$$

where χ_2^2 is the upper $\alpha100\%$ point of the chi-square distribution. To obtain p-values one uses (3.8.16) with M_0, the observed value for M.

Example 3.8.2 For comparison purposes, consider the following subset of Table 3.1.2:

systolic	−8	7	−2	0	−2	0	−2	1
diastolic	−1	6	4	2	5	3	—	—

The program MONOTONE 1 gives estimates of the mean and covariance matrices:

$$\hat{\Sigma}=\begin{pmatrix} 15.188 & 6.218 \\ 6.218 & 4.421 \end{pmatrix}, \qquad \hat{\mu}=\begin{pmatrix} -0.75 \\ 3.20 \end{pmatrix}.$$

The test of H: $\boldsymbol{\mu}=0$ vs A: $\boldsymbol{\mu}\neq 0$ has an observed χ^2-value of 8.548 on 2 df. Using (3.8.16), we find

$$P(\rho M\geq 8.548)=(1-w_2)P(\chi_2^2\geq 8.548)+w_2P(\chi_6^2\geq 8.548).$$

Using (3.8.17), w_2 is calculated to be 0.0067, so

$$P(\rho M\geq 8.548)=(1-0.0067)(0.018)+(0.0067)(0.223)=0.019.$$

Hence we reject H at $\alpha=0.05$ and claim that $\boldsymbol{\mu}\neq 0$. The observed significance level was actually 0.019.

3.8.3 *Test of the Equality of Means Based on the Bivariate Monotone Sample*

Consider the bivariate monotone sample given in Section 3.8.1. Suppose we wish to test the hypothesis

$$H: \mu_1=\mu_2, \quad \Sigma>0, \qquad \text{vs} \qquad A: \mu_1\neq\mu_2, \quad \Sigma>0. \tag{3.8.19}$$

Let

$$z_\alpha=x_{2\alpha}-\left(N_2N_1^{-1}\right)^{1/2}x_{1\alpha}+\left(N_1N_2\right)^{-1/2}\sum_{\beta=1}^{N_2}x_{1\beta}$$
$$-N_1^{-1}\sum_{\beta=1}^{N_1}x_{1\beta}, \qquad \alpha=1,2,\ldots,N_2. \tag{3.8.20}$$

Then

$$E(z_\alpha)=\mu_2-\mu_1\equiv\delta,$$
$$\operatorname{var}(z_\alpha)=\sigma_{22}+\left(N_2N_1^{-1}\right)\sigma_{11}-2\left[\left(N_2/N_1\right)^{1/2}-\left(N_1N_2\right)^{-1/2}+N_1^{-1}\right]\sigma_{12}$$
$$=\tau^2, \qquad \alpha=1,2,\ldots,N_2 \tag{3.8.21}$$

and

$$\operatorname{cov}(z_\alpha,z_\beta)=N_1^{-1}\left[\left(N_1/N_2\right)^{1/2}-1\right]\sigma_{12}, \qquad \alpha\neq\beta.$$

Thus, asymptotically, $z_1,\ldots,z_{N_2}$ are independently normally distributed with mean δ and variance τ^2. Hence the hypothesis H is rejected if

$$N_2^{1/2}|\bar{z}/s_z| \geq t_{N_2-1,\alpha/2}, \tag{3.8.22}$$

where $t_{N_2-1,\alpha/2}$ is the upper $(\alpha/2)100\%$ point of the Student's t-distribution with N_2-1 df, and

$$s_z^2=(N_2-1)^{-1}\sum_{\alpha=1}^{N_2}(z_\alpha-\bar{z})^2, \qquad \bar{z}=N_2^{-1}\sum_{\alpha=1}^{N_2} z_\alpha. \tag{3.8.23}$$

This procedure is based on one method of solution of a Behrens–Fisher problem. Other solutions are given by Mehta and Gurland (1969, 1973), Lin (1971, 1973), and Morrison (1973).

Example 3.8.3 Consider the scores (out of 20) on a comprehension test at the beginning and end of a remedial reading program:

subject	1	2	3	4	5	6	7	8	9	10
beginning	15	13	11	8	10	17	20	19	17	15
end	17	14	10	10	10	18	20	20	—	—

Two subjects left the program, and hence the final test was not completed. Using the MONOTONE 2 program to test H: $\mu_1=\mu_2$ vs A: $\mu_1 \neq \mu_2$ (i.e., no change in scores) yielded a t-value of $t_7=2.672>t_{7,0.025}=2.36$. Hence we reject H and claim that there is a change in comprehension test scores over the course of the program. Note that in this instance a one-sided t-test could be performed if we are only interested in seeing if there was an improvement in scores. In this case the critical value is $t_{7,0.05}=1.895$.

3.8.4 Monotone Sample—General Case (Estimation)

Let $\mathbf{x} \sim N_p(\boldsymbol{\mu},\Sigma)$, $\Sigma>0$, where $\mathbf{x}=(x_1,\ldots,x_p)'$. Suppose we have N_1 observations on x_1, N_2 observations on x_2, N_3 observations on x_3, and so on such that $N_1 \geq N_2 \geq N_3 \geq \cdots \geq N_p > p$, where N_p observations are on x_p. Such a sample is called a *monotone sample* and can be presented in tabular form as follows:

$$\left\{\begin{matrix} x_{11},\ldots,x_{1N_p},x_{1N_p+1},\ldots,x_{1Np-1},\ldots,x_{1N_2},x_{1N_2+1},\ldots,x_{1N_1} \\ x_{21},\ldots,x_{2N_p},x_{2N_p+1},\ldots,x_{2Np-1},\ldots,x_{2N_2} \\ \vdots \quad \vdots \quad \vdots \quad \vdots \\ x_{p-11},\ldots,x_{p-1N_p},x_{p-1,N_p+1},\ldots,x_{p-1N_{p-1}} \\ x_{p1},\ldots,x_{pN_p} \end{matrix}\right\}. \tag{3.8.24}$$

Let

$$\boldsymbol{\mu} = (\mu_1, \ldots, \mu_p), \qquad \text{and} \qquad \Sigma = (\sigma_{ij}). \tag{3.8.25}$$

From multivariate regression theory (see Chapter 5), there is a one-to-one and onto correspondence between the $p + \frac{1}{2}p(p+1)$ parameters of $\boldsymbol{\mu}$ and Σ and $\boldsymbol{\nu}$, β_{kj}, $k = 1, 2, \ldots, p$, $j = 1, 2, \ldots, k-1$, and $\sigma_k^2(e)$, $k = 1, 2, \ldots, p$, where $\beta_{k0} = 0$ and

$$\begin{aligned}
\mu_k &= \nu_k + \sum_{j=1}^{k-1} \beta_{kj}\mu_j, \qquad k = 1, 2, \ldots, p, \\
\sigma_{ik} &= \sum_{j=1}^{k-1} \beta_{kj}\sigma_{ij}, \qquad i = 1, 2, \ldots, k-1, \quad k = 1, 2, \ldots, p, \\
\sigma_{kk} &= \sigma_k^2(e) + \sum_{i=1}^{k-1}\sum_{j=1}^{k-1} \beta_{ki}\beta_{kj}\sigma_{ij}, \qquad k = 1, 2, \ldots, p.
\end{aligned} \tag{3.8.26}$$

Thus, if we have the estimates of $\boldsymbol{\nu}$, β_{kj}, and $\sigma_k^2(e)$, we can obtain the estimates of $\boldsymbol{\mu}$ and Σ. Let

$$\begin{aligned}
\mathbf{x}'_{j,N_k} &= (x_{j1}, \ldots, x_{jN_k}), \qquad j = 1, 2, \ldots, k, \\
X_{k-1,N_k} &= (\mathbf{x}_{1,N_k}, \ldots, \mathbf{x}_{k-1,N_k}) : N_k \times (k-1), \\
e'_{N_k} &= (1, 1, \ldots, 1) : 1 \times N_k, \\
Z_{k,N_k} &= (e_{N_k}, X_{k-1,N_k}) : N_k \times k,
\end{aligned} \tag{3.8.27}$$

and

$$\boldsymbol{\beta}'_{(k)} = (\beta_{k1}, \ldots, \beta_{k,k-1}) : 1 \times (k-1). \tag{3.8.28}$$

Then the maximum likelihood estimates of ν_k, $\boldsymbol{\beta}_{(k)}$, and $\sigma_k^2(e)$ are given by

$$\begin{pmatrix} \hat{\nu}_k \\ \hat{\boldsymbol{\beta}}_{(k)} \end{pmatrix} = \left[Z'_{k,N_k} Z_{k,N_k}\right]^{-1} Z'_{k,N_k}\mathbf{x}_{k,N_k},$$

and

$$N_k\hat{\sigma}_k^2(e) = \mathbf{x}'_{k,N_k}\left[I - Z_{k,N_k}\left(Z'_{k,N_k}Z_{k,N_k}\right)^{-1} Z'_{k,N_k}\right]\mathbf{x}_{k,N_k}, \qquad k = 1, 2, \ldots, p. \tag{3.8.29}$$

3.8.5 Monotone Sample—General Case (Testing for Location)

Let $\mathbf{x} \sim N_p(\boldsymbol{\mu}, \Sigma)$, $\Sigma > 0$, and suppose we have a monotone sample on the vector $\mathbf{x}$ as described in Section 3.8.4. Consider the problem of testing the hypothesis

$$H: \boldsymbol{\mu} = 0, \quad \Sigma > 0, \qquad \text{vs} \qquad A: \boldsymbol{\mu} \neq \mathbf{0}, \quad \Sigma > 0. \tag{3.8.30}$$

The likelihood ratio test for H vs A is based on the statistic

$$\lambda = \prod_{k=1}^{p} \left[\frac{\hat{\sigma}_k^2(e)}{\tilde{\sigma}_k^2(e)} \right]^{N_k/2}, \tag{3.8.31}$$

where $\hat{\sigma}_k(e)$ is as defined in (3.8.29) and $\tilde{\sigma}_k^2(e)$ is given by

$$N_k\tilde{\sigma}_k^2(e) = \mathbf{x}'_{k,N_k}\Big[I - X_{k-1,N_k}\big(X'_{k-1,N_k}X_{k-1,N_k}\big)^{-1}X_{k-1,N_k}\Big]\mathbf{x}'_{k,N_k}. \tag{3.8.32}$$

The asymptotic cdf of $M = -2\ln\lambda$ is given by

$$P\{M \le M_0\} = P\{\rho M \le \rho M_0\} = (1-w_2)P\{\chi_p^2 \le \rho M_0\} + w_2 P\{\chi_{p+4}^2 \le \rho M_0\} + O(N^{-3}),$$

where

$$N = \sum_{k=1}^{p} N_k, \qquad \rho = 1 - \frac{1}{2p}\sum_{k=1}^{p}\frac{2k+1}{N_k},$$

$$w_2 = \frac{1}{2\rho^2}\sum_{k=1}^{p}\frac{h_k(2h_k-1)}{N_k^2}, \qquad h_k = \tfrac{1}{2}\big[N_k(1-\rho)-k\big].$$

Thus, as a first approximation, $-2\rho\ln\lambda$ has a chi-square distribution with p df. This result is due to Bhargava (1962).

3.8.6 *Monotone Sample—Two-Sample Case*

Let the observation matrix with missing observations of monotone type from the first population, distributed as $N_p(\boldsymbol{\mu}, \Sigma)$, be given by

$$X = \begin{pmatrix} x_{11} & \cdots & x_{1M_1} \\ x_{21} & \cdots & x_{2M_2} \\ \vdots & & \vdots \\ x_{p1} & \cdots & x_{pM_p} \end{pmatrix}, \qquad M_1 \ge M_2 \ge \cdots \ge M_p,$$

and that from the second population, distributed as $N_p(\boldsymbol{\nu}, \Sigma)$, be given by

$$Y = \begin{pmatrix} y_{11} & \cdots & y_{1N_1} \\ y_{21} & \cdots & y_{2N_2} \\ \vdots & & \vdots \\ y_{p1} & \cdots & y_{pN_p} \end{pmatrix}, \qquad N_1 \ge N_2 \ge \cdots \ge N_p.$$

We wish to test the hypothesis

$$H: \boldsymbol{\mu} = \boldsymbol{\nu}, \quad \Sigma > 0, \qquad \text{vs} \qquad A: \boldsymbol{\mu} \ne \boldsymbol{\nu}, \quad \Sigma > 0.$$

SOLUTION: Let k be a positive integer, $k \leq p$. Define

$$\mathbf{x}'_{j,M_k}=(x_{j1},\dots,x_{jM_k}):1\times M_k, \qquad j=1,2,\dots,k,$$

$$\mathbf{y}'_{j,N_k}=(y_{j1},\dots,y_{jN_k}):1\times N_k, \qquad j=1,2,\dots,k,$$

$$\mathbf{z}'_{j,T_k}=(\mathbf{x}'_{j,M_k},\mathbf{y}'_{j,N_k}):1\times T_k, \qquad T_k=M_k+N_k, \quad j=1,\dots,p,$$

$$Z_{k-1,T_k}=(\mathbf{z}_{1,T_k},\dots,\mathbf{z}_{k-1,T_k}):T_k\times(k-1),$$

$$e'_{T_k}=(1,1,\dots,1):1\times T_k,$$

$$V_{k,T_k}=(e_{T_k},Z_{k-1,T_k}):T_k\times k,$$

$$A_2(k)=\begin{pmatrix}\mathbf{e}_{M_k} & \mathbf{0}_{M_k}\\ \mathbf{0}_{N_k} & \mathbf{e}_{N_k}\end{pmatrix}:T_k\times 2,$$

$$W_{k+1,T_k}=(A_2(k),Z_{k-1,T_k}):T_k\times(k+1),$$

$$=\begin{pmatrix}\mathbf{e}_{M_k} & \mathbf{0}_{M_k} & \mathbf{x}_{1,M_k} & \mathbf{x}_{2,M_k} & \cdots & \mathbf{x}_{k-1,M_k}\\ \mathbf{0}_{N_k} & \mathbf{e}_{N_k} & \mathbf{y}_{1,N_k} & \mathbf{y}_{2,N_k} & \cdots & \mathbf{y}_{k-1,N_k}\end{pmatrix},$$

$$T_k\hat{\sigma}_k^2(e)=\mathbf{z}'_{k,T_k}\Big[I_{T_k}-W_{k+1,T_k}\big(W'_{k+1,T_k}W_{k+1,T_k}\big)^{-1}W'_{k+1,T_k}\Big]\mathbf{z}_{k,T_k},$$

$$T_k\tilde{\sigma}_k^2(e)=\mathbf{z}'_{k,T_k}\Big[I_{T_k}-V_{k,T_k}\big(V'_{k,T_k}V_{k,T_k}\big)^{-1}V'_{k,T_k}\Big]\mathbf{z}_{k,T_k}.$$

Then the likelihood ratio test is given by

$$\lambda=\prod_{k=1}^{p}\left[\frac{\hat{\sigma}_k^2(e)}{\tilde{\sigma}_k^2(e)}\right]^{T_k/2}.$$

Let

$$M=-2\ln\lambda=-\sum_{k=1}^{p}T_k\ln\frac{\hat{\sigma}_k^2(e)}{\tilde{\sigma}_k^2(e)}.$$

Then Bhargava (1962) has shown that

$$P\{M\leq M_0\}=P\{\rho M\leq\rho M_0\}=(1-w_2)P\{\chi_p^2\leq\rho M_0\}$$
$$+w_2P\{\chi_{p+4}^2\leq\rho M_0\}+O(T^{-3}),$$

where

$$T=\sum_{k=1}^{p}T_k,$$

$$0<\rho=1-\frac{1}{2p}\sum_{k=1}^{p}\frac{2k+3}{T_k},$$

$$h_k=\tfrac{1}{2}(1-\rho)T_k,$$

Table 3.13[a]

Haltica oleracea ($M_1=16$, $M_2=12$, $T_1=26$, $T_2=20$)		*Haltica carduorum* ($N_1=10$, $N_2=8$)	
X_1	X_2	X_1	X_2
137	245	184	305
132	260	133	237
141	276	166	300
142	299	162	273
128	239	163	297
147	262	160	308
136	278	166	301
128	255	141	208
128	244	146	—
146	276	171	—
128	242		
147	263		
121	—		
138	—		
138	—		
150	—		

[a]Data edited from Lubischew (1962).

and

$$w_2 = \frac{1}{6\rho^2}\sum_{k=1}^{p}\frac{12h_k(h_k-1)}{T_k^2}.$$

Example 3.8.4 Consider (Table 3.13) data on the length of the second antennal joint (in microns), X_1, and the elytra length (in 0.01 mm), X_2, for two species of flea beetle, genus *Halticus*. To test for the equality of means for the two species of beetle, we use the program MONOTONE 3 in Appendix II. The following values were obtained:

$$\hat{\sigma}_1^2 = 121.5, \qquad \hat{\sigma}_2^2 = 257.8,$$

$$\tilde{\sigma}_1^2 = 241.56, \qquad \tilde{\sigma}_2^2 = 262.5,$$

$$\rho = 1 - 0.25\left(\tfrac{5}{26} + \tfrac{7}{20}\right) = 0.889,$$

$$h_1 = 1.443, \qquad h_2 = 1.11, \qquad \text{and} \qquad w_2 = 0.01,$$

$$-2\rho\ln(121.5/241.46)^{26}(257.8/262.5)^{20} = 32.27 = \rho M_0.$$

The p-value for testing the equality of means is therefore given by

$$(1-0.01)P\left(\chi_2^2 > 32.27\right) + 0.01P\left(\chi_6^2 > 32.27\right) = 0.000.$$

Hence there is a significant difference in the two species with respect to these two measurements at the 1% significance level.

3.9 Computational Procedures

Some programs are given in Appendix II for problems of a shift in mean, bivariate monotone samples, and the two-sample T^2-test. Programs for one- and two-sample T^2-tests are available in such packages as BMDP 3D.

The following cards were used to obtain the T^2-value for Example 3.3.1.

```
/PROBLEM     TITLE IS 'DOG DATA'.
/INPUT       VARIABLES ARE 3.
             FORMAT IS '(F3.0,2F2.0)'.
/VARIABLES   NAMES ARE SUGAR, SYST, DIAS.
/TEST        VARIABLES ARE SUGAR, SYST, DIAS.
             HOTELLING.
/PRINT       DATA.
             COVARIANCE.
             CORRELATION.
/END
```

The problem paragraph specifies the title. The input paragraph specifies the number of variables and the format of the input. For a one-sample test that the population mean is zero, the variables paragraph lists the names of the input variables, and the test paragraph gives the variables to be tested using Hotelling's T^2-procedure. Optional output, such as the data, the covariance matrix, and the correlation matrix, is obtained by the use of the print paragraph. The following output was obtained:

```
CORRELATION MATRIX FOR GROUP 0
              SUGAR 1   SYST 2   DIAS 3
SUGAR    1    1.0000
SYST     2    0.6055    1.0000
DIAS     3    0.2393    0.7081   1.0000

MAHALANOBIS D SQUARE      9.8830
HOTELLING T SQUARE       79.0637
F VALUE                  18.8247   P VALUE  0.004
  DEGREES OF FREEDOM 3,   5

DIFFERENCES ON SINGLE VARIABLES
* * * * * * * *
*  SUGAR  *    VARIABLE NUMBER 1
* * * * * * * *

                                      MEAN            31.2500
T STATISTIC   P VALUE   D. F.         STD DEV         32.7054
                                      S.E.M.          11.5631
   2.70        0.031      7           SAMPLE SIZE           8
                                      MAXIMUM         90.0000            H
                                      MINIMUM        -10.0000    HH H    HH H       H
                                                                 - - - - - - - - - - -
                                                              MIN                     MAX
                                                                 AN  H=    1  CASES

* * * * * * * *
*  SYST  *     VARIABLE NUMBER 2
* * * * * * * *
```

```
                                        MEAN          -0.7500
T STATISTIC  P VALUE  D. F.   STD DEV        4.1662
                                        S.E.M.         1.4730
  -0.51         0.626      7    SAMPLE SIZE         8                H
                                        MAXIMUM      7.0000                H H
                                        MINIMUM     -8.0000      H        H HH        H
                                                                 - - - - - - - - - - -
                                                              MIN                      MAX
                                                                    AN  H=    1 CASES

* * * * * * * *
*  DIAS  *   VARIABLE NUMBER 3
* * * * * * * *

                                        MEAN           3.1250
T STATISTIC  P VALUE  D. F.   STD DEV        2.1671
                                        S.E.M.         0.7662
   4.08         0.005      7    SAMPLE SIZE         8
                                        MAXIMUM      6.0000                 H    H
                                        MINIMUM     -1.0000      H         H H  H  H H
                                                                 - - - - - - - - - - -
                                                              MIN                      MAX
                                                                    AN  H=    1 CASES
```

For a two-sample problem, such as Example 3.3.2, the following program cards were used.

```
/PROBLEM      TITLE IS 'DOG DATA'.
/INPUT        VARIABLES ARE 3.
              FORMAT IS '(F5.1,F4.1,F1.0)'.
/VARIABLES    NAMES ARE STRIDE, STRAIN, TREAT.
              GROUPING IS TREAT.
/GROUP        CODES(3) ARE 1,2.
              NAMES(3) ARE CONTROL, TREAT.
/TEST         VARIABLES ARE STRIDE, STRAIN.
              GROUPS ARE 1,2.
              HOTELLING.
/PRINT        DATA.
              COVARIANCE.
              CORRELATION.
/END
```

In addition to the two characteristics STRIDE and STRAIN, the variable TREAT is used. This is a dummy variable to distinguish the two groups of dogs. It is designated a grouping variable in the variable paragraph. A group paragraph is included, giving the codes or values for the third variable, the grouping variable, and their names. The groups to be tested are then given in the test paragraph. The following output resulted:

```
CORRELATION MATRIX FOR GROUP 1 CONTROL
                         STRIDE1      STRAIN2
STRIDE          1        1.0000
STRAIN          2        0.6741       1.0000
CORRELATION MATRIX FOR GROUP 2 TREAT
                         STRIDE1      STRAIN2
STRIDE          1        1.0000
STRAIN          2        0.5057       1.0000
```

```
DIFFERENCES AMONG GROUP MEANS USING ALL VARIABLES
FOR THE FOLLOWING GROUPS
* * * * * * * * * * * *
*   CONTROL        *
*   TREAT          *
* * * * * * * * * * * *

MAHALANOBIS D SQUARE      58.3541
HOTELLING T SQUARE       116.7081
F VALUE                   48.6284   P VALUE   0.001
  DEGREES OF FREEDOM 2,     5

DIFFERENCES ON SINGLE VARIABLES
* * * * * * * * * *
*  STRIDE   * VARIABLE NUMBER 1              GROUP          1 CONTROL   2 TREAT
* * * * * * * * * *                          MEAN            141.8750   65.1250
           STATISTICS          P VALUE  D. F. STD DEV           7.8356   23.6304
                                              S.E.M.            3.9178   11.8152
T (SEPARATE)         6.17      0.005     3.7  SAMPLE SIZE            4         4
T (POOLED)           6.17      0.001     6    MAXIMUM          150.0000   90.0000
                                              MINIMUM          131.5000   40.5000
F(FOR VARIANCES)     9.09      0.103     3,3
* * * * * * * * * *
*  STRAIN   * VARIABLE NUMBER 2              GROUP          1 CONTROL   2 TREAT
* * * * * * * * * *                          MEAN             21.7500   63.3750
           STATISTICS          P VALUE  D. F. STD DEV          13.2759    8.5866
                                              S.E.M.            6.6380    4.2933
T (SEPARATE)        -5.27      0.003     5.1  SAMPLE SIZE            4         4
T (POOLED)          -5.27      0.002     6    MAXIMUM           36.0000   74.5000
                                              MINIMUM            9.0000   54.0000
F(FOR VARIANCES)     2.39      0.493     3,3
```

The SAS package may also be used to produce F values for one- and two-sample problems. The GLM procedure will handle the k-sample problem discussed in the next chapter.

Examples 3.3.1 and 3.3.2 were set up as follows:

```
DATA BLOOD;
INPUT BLOOD SYST DIAS;
A = 1.0;
CARDS;
30 -8 2
:   : :
PROC GLM;
CLASS A;
MODEL BLOOD SYST DIAS = A/NOINT;
MANOVA H = A;
```

The DATA BLOOD; statement defines the name of the data set. INPUT BLOOD SYST DIAS; names the three variables being inputed. The card A = 1.0 defines a dummy variable needed later on along with CLASS A. The model being defined is

```
BLOOD SYST DIAS = A
```

That is, the three response variables are BLOOD, SYST, and DIAS, and the class variable A indicates from which group the responses were taken. In this case there is only one group. A always equals 1.0.

NOINT indicates that we wish to test the means equal to zero. The MANOVA statement indicates that a multivariate test should be performed.

```
DATA DOGS;
INPUT STRIDE STRAIN TREAT;
CARDS;
131.5   9.01  1
145.0  12.01  1
141.0  30.01  1
150.0  36.01  1
 40.5  54.02  2
 80.0  74.52  2
 50.0  64.52  2
 90.0  60.52  2
PROC GLM;
CLASS TREAT;
MODEL STRIDE STRAIN = TREAT;
MANOVA H = TREAT;
```

For this second example the variable TREAT indicates to which group the subjects belong, control (1) or plate (2). The statement MANOVA H= TREAT; indicates that we wish a multivariate test for equality of treatment means. This procedure GLM is discussed more fully in the next two chapters.

The values of λ in Example 3.7.1 were calculated by running the GLM procedure and computing the error sum of squares. For given values of λ the following file was used to produce statistics and values to test for normality.

```
DATA ONE;
INPUT X1 X2;
A = 52.4;
B = 52.1
Y1 = (X1**.5 - 1)/(A** - .5*.5);
Y2 = (X2** - .1 - 1)/(B** - 1.1* - .1);
CARDS;
59  40.9
:    :
:    :
PROC RANK NORMAL = BLOM TIES = MEAN;
VAR X1 X2;
RANKS R1 R2;

PROC PRINT;
VAR X1 Y1 R1 X2 Y2 R2;

PROC UNIVARIATE NORMAL;
VAR X1 X2 Y1 Y2;
```

Problems

3.1 Let $\mathbf{x} \sim N_p(\boldsymbol{\mu}, \Sigma)$, $\Sigma > 0$. Suppose an independent sample $\mathbf{x}_1, \mathbf{x}_2, \ldots, \mathbf{x}_N$ of size N on $\mathbf{x}$ is given. Let C be a $k \times p$ matrix of rank $k \times p$. Obtain a test for testing the hypothesis H: $C\boldsymbol{\mu} = C\boldsymbol{\mu}_0$ vs A: $C\boldsymbol{\mu} \neq C\boldsymbol{\mu}_0$, where $\boldsymbol{\mu}_0$ is specified. *Hint*: Note that $C\mathbf{x} \sim N_k(C\boldsymbol{\mu}, C\Sigma C')$.

Table 3.14[a]

F_1 Period			F_2 Period		
1	2	3	1	2	3
194	192	141	239	127	90
208	188	165	189	105	85
233	217	171	224	123	79
241	222	201	243	123	110
265	252	207	243	117	100
269	283	191	226	125	75

[a]Data modified from Box (1950).

3.2 Show that the tests in (3.2.1) and (3.2.2) are equivalent.

3.3 Tires were measured for their wear during the first 1000 revolutions, the second 1000 revolutions, and the third 1000 revolutions (Table 3.14). Two types of filler in the tires were used, F_1 and F_2. Is there a significant difference in the two fillers? If so, which periods differ?

3.4 Glycine in the spinal cord of cats with local tetanus rigidity (Table 3.15): The left sides of the cats were considered a control, and the right sides have local tetanus rigidity. The amount of glycine present in the gray and white matter was recorded. Is there a difference between control and tetanus for the two characteristics of gray and white matter?

3.5 The sample means data of Table 3.16 give the means and covariances of observations from two groups of patients. The 28 subjects in the first group

Table 3.15[a]

	Gray matter		White matter	
	Control (L)	Tetanus (R)	Control (L)	Tetanus (R)
1	5.7	4.6	3.0	3.6
2	6.1	5.9	3.2	2.7
3	5.6	5.3	2.7	2.9
4	6.1	5.8	3.3	3.4
5	6.7	6.6	2.9	3.8
6	5.4	5.3	3.0	2.8
7	5.9	5.5	3.3	3.1
8	5.9	5.4	3.9	3.5
9	5.7	5.2	2.8	3.6
10	4.8	4.4	2.6	2.6
11	5.8	4.9	2.7	3.2

[a]From Sema and Kano (1969) with permission of the editors of *Science*. © 1969 by the American Association for the Advancement of Science.

Table 3.16[a]

Group	Days to union	Age	Lymphocytes	Polymorphonucleors	Esinophils
single	48.00	36.61	37.04	58.36	1.54
multiple	46.36	29.86	32.09	62.91	1.50

[a]From Woodard (1931).

had sustained single fractures of the jaw, and the 23 subjects of the second group had sustained multiple fractures of the jaw. The covariance matrix is

$$\begin{pmatrix} 489.19 & -24.21 & 62.38 & -57.03 & 0.15 \\ & 104.07 & 0.76 & -2.97 & 2.50 \\ & & 45.68 & -39.38 & -0.39 \\ & & & 39.92 & -1.99 \\ & & & & 1.93 \end{pmatrix}.$$

(a) Is there a significant difference between the two groups with respect to the five characteristics measured?

(b) Jensen (1972) suggests analyzing the first two characteristics and the last three characteristics separately, each at $\alpha = 0.05$, with an overall error of 10%.

(c) Assuming that the population means for the two characteristics "days to union" and "age" are the same in each group, test for the equality of the population means for the remaining three characteristics of the two groups.

3.6 The first- and second-year grade point averages (GPA) were studied for students in their first and second years of university. 406 students were assessed, 252 males and 154 females. The means and covariances of Table 3.17 resulted.

(a) Test for a significant difference in GPAs between male and female students, assuming equal covariances.

(b) If there is a significant difference, does it occur in first year? In the second year?

Table 3.17[a]

	Males			Females		
	Means	Covariance matrix		Means	Covariance matrix	
first year	2.61	0.260	0.181	2.54	0.303	0.206
second year	2.63	0.181	0.203	2.55	0.206	0.194

[a]From Hosseini (1978).

3.7 The following data (courtesy of A. D. Genova) give the change in Durrell test scores over one year of a remedial reading program:

oral reading	0.8	1.2	0.9	1.4	1.0	1.5	0.8	2.0	0.4	0.5	1.3
	1.0	1.0	0.8	1.2							
silent reading	1.3	1.3	0.7	0.6	1.2	1.7	0.8	0.7	1.3	0	1.3
	—	—	—	—							

(a) Estimate the mean and covariance matrix of the data ignoring the four incomplete responses.
(b) Test the null hypothesis that there is no improvement in scores after one year of the program.
(c) Repeat **(a)** and **(b)**, including the four incomplete responses.

3.8 Table 3.18 gives the binding of Warfarin to albumin A/A and albumin A/Me variants.
(a) Perform a two-sample T^2-test on the above data, both with and without using the incomplete data. If there is a significant difference, for which characteristic does it occur?
(b) Because x_2 is actually a proportion, a variance stabilizing transform such as $y = \arcsin(x_2/100)^{1/2}$ generally eliminates the dependence between the mean and the variance of x_2. Use a T^2-test based on x_1 and y to test for differences between the two albumin variants, both with and without using the incomplete data. What characteristics are different, if any, for the two variants?

Table 3.18

	A/A		A/Me	
	x_1	x_2	x_1	x_2
	2.04	98.7	1.69	98.3
	1.70	98.2	1.00	97.8
	1.72	98.5	1.16	97.5
	1.01	98.5	1.18	97.7
	1.21	98.2	0.92	96.7
	1.22	98.1	1.06	96.7
	1.69	97.2	1.24	96.3
	1.40	96.7	1.05	96.9
	1.88	97.2	0.68	97.6
	1.92	97.6	1.69	97.4
	2.07	97.8	1.68	97.7
	1.41	97.9	1.38	96.8
	1.42	97.2	0.87	96.9
	2.05	—	1.76	97.7
			1.61	—
mean	1.62	97.8	1.26	97.2

x_1 is the binding constant ($10^5 M^{-1}$) for isolated charcoal treated albumin, x_2 the percentage of Warfarin bound on whole plasma. From Wilding et al. (1977).

3.9 Three measurements were obtained on Pacific ocean perch (Bernard, 1981). The variables x_1, the asymptotic length (in mm), x_2, the coefficient of growth, and x_3, the time (in years) at which length is zero were measured by fitting von Bertalanffy growth curves to measurements over time for each fish. The means and covariance matrices for three groups of 76 fish were as follows: For males off the Columbia River,

$$\mathbf{x}_1 = \begin{pmatrix} 441.16 \\ 0.13 \\ -3.36 \end{pmatrix}, \quad S_1 = \begin{pmatrix} 294.74 & -0.60 & -32.57 \\ -0.60 & 0.0013 & 0.073 \\ -32.57 & 0.073 & 4.23 \end{pmatrix};$$

for females off the Columbia River,

$$\mathbf{x}_2 = \begin{pmatrix} 505.97 \\ 0.09 \\ -4.57 \end{pmatrix}, \quad S_3 = \begin{pmatrix} 1596.18 & -1.19 & -91.05 \\ -1.19 & 0.0009 & 0.071 \\ -91.05 & 0.071 & 5.76 \end{pmatrix};$$

for males off Vancouver Island,

$$\mathbf{x}_3 = \begin{pmatrix} 432.51 \\ 0.14 \\ -3.31 \end{pmatrix}, \quad S_3 = \begin{pmatrix} 182.67 & -0.42 & -22.00 \\ -0.42 & 0.001 & 0.056 \\ -22.00 & 0.056 & 3.14 \end{pmatrix}.$$

(a) Assuming equal covariance matrices for the two groups of male perch, test for the equality of means between male perch off the Columbia river and off Vancouver Island at $\alpha = 0.01$. If there is a significant difference, which variables differ in means?

(b) Assuming unequal covariance matrices test for a difference in means between male and female perch off the Columbia river at $\alpha = 0.01$. If there is a significant difference, which variables differ in means?

Chapter 4
Multivariate Analysis of Variance

4.1 Introduction

In Chapter 3 inferences on the means of one and two populations were discussed. These problems will now be generalized to three or more populations. As in the univariate model, analysis of variance, regression analysis, and covariance analysis can all be included in a study of the general linear model, as we show in Chapter 5. First, however, we analyze a few basic designs.

4.2 Completely Randomized Design

Suppose $N_1,\dots,N_t$ subjects have been randomly assigned to t treatments. For each treatment p variables are measured on each subject. Assuming that the observations are normally distributed with common covariance matrix Σ, we would like to test the hypothesis that the treatment mean responses $\boldsymbol{\mu}_1,\dots,\boldsymbol{\mu}_t$ are the same. If they are significantly different, we must decide where the difference lies.

Let $\mathbf{y}_{ij}$ be independently distributed as $N_p(\boldsymbol{\mu}_j,\Sigma)$, $i=1,\dots,N_j$, $j=1,\dots,t$. We define $\bar{\mathbf{y}}_{\cdot j}$ to be the mean of the jth treatment group and $\bar{\mathbf{y}}_{\cdot\cdot}$ to be the overall mean; that is,

$$\bar{\mathbf{y}}_{\cdot j}=\frac{1}{N_j}\sum_{i=1}^{N_j}\mathbf{y}_{ij} \qquad \text{and} \qquad \bar{\mathbf{y}}_{\cdot\cdot}=\frac{1}{N}\sum_{j=1}^{t}\sum_{i=1}^{N_j}\mathbf{y}_{ij}, \tag{4.2.1}$$

where $N=\Sigma_{j=1}^{t}N_j$. The following sums of squares are as in the univariate case:

$$\begin{aligned}\mathrm{SS(TR)}&=\sum_{j=1}^{t}N_j\bar{\mathbf{y}}_{\cdot j}\bar{\mathbf{y}}'_{\cdot j}-N\bar{\mathbf{y}}_{\cdot\cdot}\bar{\mathbf{y}}'_{\cdot\cdot},\\ \mathrm{SST}&=\sum_{j=1}^{t}\sum_{i=1}^{N_j}\mathbf{y}_{ij}\mathbf{y}'_{ij}-N\bar{\mathbf{y}}_{\cdot\cdot}\bar{\mathbf{y}}'_{\cdot\cdot},\\ \mathrm{SSE}&=\mathrm{SST}-\mathrm{SS(TR)}.\end{aligned} \tag{4.2.2}$$

Table 4.1

Source	df	SS	$U_{p,m,n}$	p	m	n
treatments	$t-1$	SS(TR)	$\dfrac{\lvert \text{SSE} \rvert}{\lvert \text{SS}(E)+\text{SS(TR)} \rvert}$	p	$t-1$	$N-t$
error	$N-t$	SSE				
total	$N-1$	SST				

In order to test H: $\boldsymbol{\mu}_1 = \cdots = \boldsymbol{\mu}_t$ vs A: $\boldsymbol{\mu}_i \neq \boldsymbol{\mu}_j$ for at least one i and j, $1 \le i < j \le t$, we set up the MANOVA (Table 4.1). The likelihood ratio test rejects H if

$$|\text{SSE}|/|\text{SSE}+\text{SS(TR)}| = U_{p,t-1,N-t} < c, \tag{4.2.3}$$

where c is chosen such that the error of the first kind is at a specified level α. We define the α percentage point of $U_{p,m,n}$ to be $U_{p,m,n,\alpha}$. Table values are given by Schatzoff (1966), Pillai and Gupta (1969), and Lee (1972).

The following asymptotic expansion gives a good approximation even for moderate sample sizes. Here $O(n^{-\gamma})$ represents terms that go to zero at the rate of $n^{-\gamma}$; that is, $n^{\gamma}O(n^{-\gamma})$ tends to a constant.

Theorem 4.2.1

$$P\left\{-\left[n-\tfrac{1}{2}(p-m+1)\right]\ln U_{p,m,n} > z\right\} = P\left\{\chi_f^2 \ge z\right\} + n^{-2}\gamma_2\left(P\left\{\chi_{f+4}^2 \ge z\right\} - P\left\{\chi_f^2 \ge z\right\}\right) + O(n^{-4}),$$

where $f = pm$, $\gamma_2 = f(p^2+m-5)/48$, *and* χ_f^2 *denotes a chi-square random variable with* f *df*.

Corollary 4.2.1 *As* $n \to \infty$,

$$P\left\{-\left[n-\tfrac{1}{2}(p-m+1)\right]\ln U_{p,m,n} \ge z\right\} = P\left\{\chi_f^2 \ge z\right\}.$$

Thus, as a first approximation, we shall use the values from a chi-square table. However, to obtain the p-values, we shall use Theorem 4.2.1 to get a better approximation.

As the next theorem points out, the tabulated values in the references cited above are given for $m > p$.

Theorem 4.2.2 *The distribution of* $U_{p,m,n}$ *is the same as that of* $U_{m,p,n+m-p}$.

For a proof, refer to Srivastava and Khatri (1979). Another important result is contained in the following.

Theorem 4.2.3 *When $p=2$,*

$$(n-1)\left(1-U_{2,m,n}^{1/2}\right)/mU_{2,m,n}^{1/2}=F_{2m,2(n-1)}.$$

That is, we obtain the percentage points for $U_{2,m,n}$ from the tables of the F-distribution.

Corollary 4.2.2 *When $m=2$, the distribution of $U_{p,2,n}$ is the same as that of $U_{2,p,n+2-p}$. Hence*

$$\left[(n-p+1)\left(1-U_{p,2,n}^{1/2}\right)/pU_{p,2,n}^{1/2}\right]=F_{2p,2(n-p+1)}.$$

Theorem 4.2.4 *When $p=1$,*

$$\left[n\left(1-U_{1,m,n}\right)/mU_{1,m,n}\right]=F_{m,n}.$$

Corollary 4.2.3 *When $m=1$, the distribution of $U_{p,1,n}$ is the same as that of $U_{1,p,n+1-p}$. Hence*

$$\left[(n+1-p)\left(1-U_{p,1,n}\right)/pU_{p,1,n}\right]=F_{p,n+1-p}.$$

Perhaps more important than testing is the problem of placing simultaneous confidence bounds on linear combinations of variables for various contrasts of the populations. If we are interested in the linear combination of variables $\Sigma_{j=1}^{t}\mathbf{a}'\boldsymbol{\mu}_{j}b_{j}$, where $\mathbf{a}'=(a_1,\dots,a_p)$ and where $\mathbf{b}'=(b_1,\dots,b_t)$ is such that $\Sigma_{j=1}^{t}b_j=0$, then a simultaneous $(1-\alpha)100\%$ confidence interval for $\Sigma_{j=1}^{t}\mathbf{a}'\boldsymbol{\mu}_{j}b_{j}$ for any $\mathbf{a}$ and $\mathbf{b}$ is given by

$$\sum_{j=1}^{t}\mathbf{a}'\bar{\mathbf{y}}_{\cdot j}b_j\pm\left[\left(\sum_{j=1}^{t}\frac{b_j^2}{N_j}\right)\mathbf{a}'V\mathbf{a}\left(\frac{x_\alpha}{1-x_\alpha}\right)\right]^{1/2}, \tag{4.2.4}$$

where $V=\text{SSE}$ and x_α can be obtained from Table A.5. The values of the parameters specified in the tables are

$$\begin{aligned} p_0&=p, &&\text{the number of characteristics,}\\ m_0&=t-1, &&\text{df treatments,}\\ n&=N-t, &&\text{df error.}\end{aligned} \tag{4.2.5}$$

For the case of $p_0=1$, $x_\alpha/(1-x_\alpha)$ reduces to $m_0n^{-1}F_{m_0,n,\alpha}$, and the procedure reduces to that given by Scheffé (1953). If only k contrasts are to be performed, then simultaneous confidence intervals of the form

$$\sum_{j=1}^{t}\mathbf{a}'\boldsymbol{\mu}_{j}b_{j}$$

are given from Bonferroni's inequalities:

$$\sum_{j=1}^{t} \mathbf{a}'\bar{\mathbf{y}}_{.j}b_j \pm t_{N-t,\alpha/2k}\left[\left(\sum_{j=1}^{t} \frac{b_j^2}{N_j}\right)(\mathbf{a}'S\mathbf{a})\right]^{1/2}, \tag{4.2.6}$$

where $S=(N-t)^{-1}V$, the sample covariance matrix. Bonferroni confidence intervals are shorter than those in (4.2.4) if $t_{N-t,\alpha/2k}^2(N-t)^{-1} \le x_\alpha/(1-x_\alpha)$.

More complicated designs are written in terms of a linear model. For the test for equality of means given in this section, this model can be written

$$\mathbf{y}_{ij}=\boldsymbol{\mu}_j+\boldsymbol{\varepsilon}_{ij}, \qquad i=1,\ldots,N_j, \quad j=1,\ldots,t, \tag{4.2.7}$$

where the $\boldsymbol{\varepsilon}_{ij}$s are independent errors assumed to be identically and independently distributed as $N_p(\mathbf{0},\Sigma)$. Section 4.7 treats the situation in which these assumptions need not be met. An alternative model, which will be generalized in Sections 4.3–4.6, can be written

$$\mathbf{y}_{ij}=\boldsymbol{\mu}+\boldsymbol{\tau}_j+\boldsymbol{\varepsilon}_{ij}, \qquad i=1,\ldots,N_j, \quad j=1,\ldots,t, \tag{4.2.8}$$

where $\boldsymbol{\mu}$ represents an overall mean effect and $\boldsymbol{\tau}_j=\boldsymbol{\mu}_j-\boldsymbol{\mu}$ represents the deviation of the treatment effect from the overall mean. Since the average treatment effect is zero, we impose the condition $\Sigma_{j=1}^{t}N_j\boldsymbol{\tau}_j=\mathbf{0}$. This restriction is obtained by looking at

$$E(\mathbf{y}_{..})=N^{-1}E\left(\sum_{j=1}^{t}\sum_{i=1}^{N_j}\mathbf{y}_{ij}\right)=\boldsymbol{\mu}+N^{-1}\sum_{j=1}^{t}\sum_{i=1}^{N_j}\boldsymbol{\tau}_j$$
$$=\boldsymbol{\mu}+N^{-1}\sum_{j=1}^{t}N_j\boldsymbol{\tau}_j.$$

If we wish to define $\boldsymbol{\mu}$ to be the overall mean effect, then $E(\mathbf{y}_{..})=\boldsymbol{\mu}$. Hence we must have $\Sigma_{j=1}^{t}N_j\boldsymbol{\tau}_j=\mathbf{0}$. The hypothesis of equal means can now be written H: $\boldsymbol{\tau}_1=\cdots=\boldsymbol{\tau}_t=0$ vs A: $\boldsymbol{\tau}_i\neq\mathbf{0}$ for some $i=1,\ldots,t$.

Example 4.2.1 Suppose four varieties of rice are sown in 20 plots, where each variety of rice is assigned at random to five plots. Two variables were measured six weeks after transplanting: y_1, the height of the plant, and y_2, the number of tillers per plant. The observations are shown in Table 4.2. The means for the four groups are as follows: A—58.4, 5.8; B—50.6, 5.6; C—55.2, 6.0; D—53.0, 4.6. Overall means are 54.3 and 5.50. The analysis of variance table is shown in Table 4.3. Using the approximation for $U_{p,m,n}$ with $n=16$, we obtain from Theorem 4.2.1, with $p=2$, $m=3$, and $n=16$,

$$P\{-[n-\tfrac{1}{2}(p-m+1)]\ln U > -[n-\tfrac{1}{2}(p-m+1)]\ln 0.384\}$$
$$=P\{X_6^2\ge 15.31\}+0.25\times(16^{-2})\left(P\{X_{10}^2>15.31\}-P\{X_6^2>15.31\}\right)$$
$$=0.02,$$

Table 4.2

Variety							
A		B		C		D	
y_1	y_2	y_1	y_2	y_1	y_2	y_1	y_2
58	4	49	7	54	6	58	4
62	6	55	6	52	6	55	4
60	7	48	4	51	7	51	5
54	6	52	4	58	5	49	6
58	6	49	7	61	6	52	4

since $f=2\times3=6$, and $\gamma_2=6\times(4+3-5)/48$. Hence we reject the null hypothesis and claim a difference in varieties at $\alpha=0.05$ with an observed p-value of 0.02. Using Theorem 4.2.3 with $p=2$ we can obtain exact p-values. That is,

$$15\times\left[1-(0.384)^{1/2}\right]/\left[3(0.384)^{1/2}\right]=3.07=F_{6,30}.$$

Hence, $P\{F_{6,30}\geq 3.07\}=0.0183$, and we obtain a p-value of 0.0183: The approximation does yield reasonable results.

Simultaneous 95% confidence intervals for differences between varieties in height and number of tillers can be obtained from (4.2.5) with $x_\alpha=0.49$ ($p_0=2$, $m_0=5$, and $n=16$). Shorter intervals can be obtained from (4.2.6) for these 12 specific contrasts. The value of $t_{16,0.025/12}$ is found by interpolation from Table A.2 to be 3.31. Hence, since $(3.31)^2/16=0.69\leq 0.49/(1-0.49)=0.96$, we use Bonferroni's bounds and (4.2.6) to obtain confidence intervals (Table 4.4).

In this case, however, one interval does not contain zero—that for the difference in mean height between varieties A and B. We can conclude at the 5% significance level that there is a difference in height between the two varieties.

Table 4.3

Source	df	SS	$U_{p,m,n}$	p	m	n
treatments	3	$\begin{pmatrix}165 & 12.4\\ 12.4 & 5.8\end{pmatrix}$	0.384	2	3	16
error	16	$\begin{pmatrix}189.2 & -15.4\\ -15.4 & 19.2\end{pmatrix}$				
total	19	$\begin{pmatrix}354.2 & -3\\ -3 & 25\end{pmatrix}$				

Table 4.4

a′	b′	Variable	Varieties contrasted	Confidence interval
(1,0)	(1, −1,0,0)	height	A, B	(0.99, 14.61)
(1,0)	(1,0, −1,0)	height	A, C	(−3.61, 10.01)
(1,0)	(1,0,0, −1)	height	A, D	(−1.41, 12.21)
(1,0)	(0,1, −1,0)	height	B, C	(−11.41, 2.21)
(1,0)	(0,1,0, −1)	height	B, D	(−9.21, 4.41)
(1,0)	(0,0,1, −1)	height	C, D	(−4.61, 9.01)
(0,1)	(1, −1,0,0)	tillers	A, B	(−2.1, 2.5)
(0,1)	(1,0, −1,0)	tillers	A, C	(−2.5, 2.1)
(0,1)	(1,0,0, −1)	tillers	A, D	(−1.1, 3.5)
(0,1)	(0,1, −1,0)	tillers	B, C	(−2.7, 1.9)
(0,1)	(0,1,0, −1)	tillers	B, D	(−1.3, 3.3)
(0,1)	(0,0,1, −1)	tillers	C, D	(−0.9, 3.7)

4.2.1 Derivation of the Likelihood Ratio Test

The maximum likelihood estimates of Σ under A and H are given by

$$N\hat{\Sigma}_A = \sum_{j=1}^{t}\sum_{i=1}^{N_j}(\mathbf{y}_{ij}-\bar{\mathbf{y}}_j)(\mathbf{y}_{ij}-\bar{\mathbf{y}}_{\cdot j})' = \text{SSE}$$

and

$$N\hat{\Sigma}_H = \sum_{j=1}^{t}\sum_{i=1}^{N_j}(\mathbf{y}_{ij}-\bar{\mathbf{y}}_{\cdot\cdot})(\mathbf{y}_{ij}-\bar{\mathbf{y}}_{\cdot\cdot})' = \text{SST}$$

$$= N\hat{\Sigma}_A + \sum_{j=1}^{t} N_j(\bar{\mathbf{y}}_{\cdot j}-\bar{\mathbf{y}}_{\cdot\cdot})(\bar{\mathbf{y}}_{\cdot j}-\bar{\mathbf{y}}_{\cdot\cdot})'$$

$$= \text{SSE} + \text{SS(TR)},$$

respectively. Hence the likelihood statistic is simply the ratio of the determinants of the two estimates of Σ, namely,

$$\frac{|\hat{\Sigma}_A|}{|\hat{\Sigma}_H|} = \frac{|N\hat{\Sigma}_A|}{|N\hat{\Sigma}_H|} = \frac{|\text{SSE}|}{|\text{SSE}+\text{SS(TR)}|}.$$

This test does not change if we relabel the observations as $\mathbf{y}_{ij} \to A\mathbf{y}_{ij} + \mathbf{b}$, where A is a $p \times p$ nonsingular matrix and $\mathbf{b}$ is a vector of constants. To check this property, we first consider nonsingular linear transformations of the kind $\mathbf{y}_{ij} \to A\mathbf{y}_{ij}$, where A is a $p \times p$ nonsingular matrix. Then $\text{SSE} \to A\text{SSE}A'$ and $\text{SS(TR)} \to A(\text{SS(TR)})A'$. Since $|A(\text{SS(TR)})A'| = |A||\text{SSE}||A'|$,

$$|A(\text{SSE})A' + A(\text{SS(TR)})A'| = |A(\text{SSE}+\text{SS(TR)})A'|$$

$$= |A||\text{SSE}+\text{SS(TR)}||A'|,$$

the likelihood ratio test remains unchanged by relabeling of observations of the above kind. Similarly, it can be shown that if $\mathbf{y}_{ij} \to \mathbf{y}_{ij} + \mathbf{b}$, the value remains unchanged. For $p = 1$, A and $\mathbf{b}$ are scalars, and it is well known that for one-way classifications, the analysis of variance test remains unchanged under the relabeling $y_{ij} \to ay_{ij} + b$. This property is known as invariance.

4.2.2 Some Other Possible Tests

In the univariate case there is usually only one reasonable test. In multivariate situations, however, several are possible. To limit their number tests we may wish to restrict our attention to only those tests which are invariant—possessing the property of symmetry. The likelihood ratio test is invariant, but there are other several possible tests which are also invariant. It can be shown that the characteristic roots $l_1, l_2, \ldots, l_s$ of $(\mathrm{SSE})^{-1}\mathrm{SS(TR)}$, where $s = \min(p, t-1)$, are also invariant. Without loss of generality, let us assume that these roots are ordered such that $l_1 > l_2 > \cdots > l_s$. Then the following tests are in use:

i. Wilks's likelihood ratio test, where the acceptance region is $\prod_{i=1}^{s}(1 + l_i) \le c_1$, $c_1 \ge 1$;
ii. The Lawley–Hotelling trace test, where the acceptance region is $\Sigma_{i=1}^{s} l_1 \le c_2$, $c_2 \ge 0$;
iii. Roy's maximum roots test, where the acceptance region is $l_1 \le c_3$, $c_3 \ge 0$; and
iv. Pillai's test, where the acceptance region is $\Sigma l_i(1 + l_i)^{-1} \le c_4$, $0 \le c_4 \le \min(p, t-1)$.

Let $\lambda_1 \ge \lambda_2 \ge \cdots \ge \lambda_r$ be the ordered ch roots of

$$\Sigma^{-1}\left[\sum_{i=1}^{t} N_i(\boldsymbol{\mu}_i - \bar{\boldsymbol{\mu}})(\boldsymbol{\mu}_i - \bar{\boldsymbol{\mu}}')\right], \qquad \text{where} \quad \bar{\boldsymbol{\mu}} = N^{-1}\sum_{i=1}^{t} N_i \boldsymbol{\mu}_i,$$
$$r \le \min(p, t-1).$$

Then the power functions of these tests depend on $\lambda_1, \ldots, \lambda_r$. Mardia (1971) has shown that Pillai's test is robust to nonnormality so long as the distribution is symmetric, and John (1971) has shown that under some conditions Pillai's test is also locally most powerful. However, in the power comparisons of Pillai and Jayachandran (1967) no single test proves superior, although the Roy's test should clearly be used if $\lambda_2 = \cdots = \lambda_r$ and λ_1 is far away from the rest. All four tests are unbiased (provided $0 \le c_4 \le 1$) and have monotone power functions (Das Gupta et al., 1964; J. N. Srivastava, 1964; Perlman, 1974).

4.3 Randomized Complete Block Design

Suppose that in Example 4.2.1 the four varieties of rice are assigned randomly within five blocks of land. Then if we are to test for differences among the varieties of rice, we must account for any differences between the blocks. In general, we let $\mathbf{y}_{ij}$ be the p measurements on the jth variety of rice in the ith block. Then

$$\mathbf{y}_{ij} = \boldsymbol{\mu} + \boldsymbol{\tau}_j + \boldsymbol{\beta}_i + \boldsymbol{\varepsilon}_{ij}, \quad \sum \boldsymbol{\tau}_j = \sum \boldsymbol{\beta}_i = \mathbf{0}, \qquad i = 1, \ldots, b, \qquad j = 1, \ldots, t, \tag{4.3.1}$$

where the $\boldsymbol{\tau}_j$ are the treatment effects, $\boldsymbol{\beta}_i$ are the block effects, and $\boldsymbol{\varepsilon}_{ij}$ are independent $N_p(\mathbf{0}, \Sigma)$. As in the univariate case, we are assuming that the errors $\boldsymbol{\epsilon}_{ij}$ have a common covariance matrix Σ. We define the following:

$$\bar{\mathbf{y}}_{i\cdot} = \frac{1}{t} \sum_{j=1}^{t} \mathbf{y}_{ij}, \qquad \bar{\mathbf{y}}_{\cdot j} = \frac{1}{b} \sum_{i=1}^{b} \mathbf{y}_{ij}, \qquad \bar{\mathbf{y}}_{\cdot\cdot} = \frac{1}{bt} \sum_{i,j} \mathbf{y}_{ij},$$

$$\mathrm{SST} = \sum_{i,j} \mathbf{y}_{ij}\mathbf{y}'_{ij} - bt\bar{\mathbf{y}}_{\cdot\cdot}\bar{\mathbf{y}}'_{\cdot\cdot},$$

$$\mathrm{SS(TR)} = b \sum_{j=1}^{t} \bar{\mathbf{y}}_{\cdot j}\bar{\mathbf{y}}'_{\cdot j} - bt\bar{\mathbf{y}}_{\cdot\cdot}\bar{\mathbf{y}}'_{\cdot\cdot}, \tag{4.3.2}$$

$$\mathrm{SS(BL)} = t \sum_{i=1}^{b} \bar{\mathbf{y}}_{i\cdot}\bar{\mathbf{y}}'_{i\cdot} - bt\bar{\mathbf{y}}_{\cdot\cdot}\bar{\mathbf{y}}'_{\cdot\cdot},$$

$$\mathrm{SSE} = \mathrm{SST} - \mathrm{SS(TR)} - \mathrm{SS(BL)}.$$

An analysis of the variance table is given in Table 4.5, where $U_{p,m,n}$ is the likelihood ratio statistic for testing treatment or block effects as defined in Section 4.2.

Example 4.3.1 Return to the data of Example 4.2.1, but assuming a randomized complete block design. We have the analysis of Table 4.6. With $p = 2$ one obtains an observed F-value from Theorem 4.2.3:

$$F_{6,22} = 2.46.$$

Table 4.5

Source	df	SS	$U_{p,m,n}$	p	m	n
treatments	$t-1$	SS(TR)	$\lvert\mathrm{SSE}\rvert / \lvert\mathrm{SSE}+\mathrm{SS(TR)}\rvert$	p	$t-1$	$(b-1)(t-1)$
blocks	$b-1$	SS(BL)		p	$b-1$	$(b-1)(t-1)$
error	$(b-1)(t-1)$	SSE				
total	$bt-1$	SST				

Table 4.6

Source	df	SS	$U_{p,m,n}$	p	m	n
variety	3	$\begin{pmatrix} 165 & 12.4 \\ 12.4 & 5.8 \end{pmatrix}$	0.34	2	3	12
blocks	4	$\begin{pmatrix} 31.7 & -0.5 \\ -0.5 & 1.0 \end{pmatrix}$				
error	12	$\begin{pmatrix} 157.4 & -14.9 \\ -14.9 & 18.2 \end{pmatrix}$				
total	19	$\begin{pmatrix} 354.2 & -3 \\ -3 & 25 \end{pmatrix}$				

This gives an exact p-value from Table A.4 of $P(F_{6,22} \geq 2.46) = 0.046$. Here we reject the null hypothesis at $\alpha = 0.05$ and claim a difference in varieties of rice. Simultaneous confidence bands for contrasts in treatment effects $\boldsymbol{\tau}_j$ can be obtained as in the completely randomized designs. That is, $(1-\alpha)100\%$ simultaneous confidence intervals for $\mathbf{a}'\Sigma_{j=1}^{t}\boldsymbol{\tau}_j c_j$ are given by

$$\mathbf{a}'\sum \bar{\mathbf{y}}_{\cdot j}c_j \pm \left[\left(\sum c_j^2/b\right)\mathbf{a}'\,\mathrm{SSE}\,\mathbf{a}\,x_\alpha/(1-x_\alpha)\right]^{1/2}, \tag{4.3.3}$$

where b is the number of blocks, $\mathbf{a}$ can be any p-vector, and $c_1, \ldots, c_t$ satisfy the restraint $\Sigma_{j=1}^{t} c_j = 0$. The values of x_α can be obtained from Table A.5 with

$$p_0 = p, \quad \text{the number of characteristics}$$

$$m_0 = t-1, \quad \text{df treatments,}$$

and

$$n = (b-1)(t-1), \quad \text{df error.}$$

Table 4.7

$\mathbf{a}'$	$\mathbf{c}'$	Variable	Varieties contrasted	Confidence interval
(1,0)	(1, −1,0,0)	height	A, B	(−0.23, 15.83)
(1,0	(1,0, −1,0)	height	A, C	(−4.83, 11.23)
(1,0)	(1,0,0, −1)	height	A, D	(−2.63, 13.43)
(1,0)	(0, 1, −1,0)	height	B, C	(−12.63, 3.43)
(1,0)	(0, 1,0, −1)	height	B, D	(−10.43, 5.63)
(1,0)	(0,0, 1, −1)	height	C, D	(−5.83, 10.23)
(0, 1)	(1, −1,0,0)	tillers	A, B	(−2.53, 2.93)
(0, 1)	(1,0, −1,0)	tillers	A, C	(−2.93, 2.53)
(0, 1)	(1,0,0, −1)	tillers	A, D	(−1.53, 3.93)
(0, 1)	(0, 1, −1,0)	tillers	B, C	(−3.13, 2.33)
(0, 1)	(0, 1,0, −1)	tillers	B, D	(−1.73, 3.73)
(0, 1)	(0,0, 1, −1)	tillers	C, D	(−1.33, 4.13)

If k Bonferroni-type intervals are used, then $x_\alpha/(1-x_\alpha)$ is replaced by $n^{-1}t^2_{n,\alpha/2k}$ if the latter is smaller.

In Example 4.3.1 we have four treatments or varieties and two characteristics. If we wish confidence intervals for the six differences between the four varieties for both characteristics of height and tillers, then we shall need 12 confidence intervals. For this example $p_0=2$, $m_0=3$, $n=12$, $p=2$, $t=4$, and $b=5$. The value of $x_{0.05}$, from Table A.5, is 0.59. The value of $t_{0.0021,12}$, by interpolation from Table A.2, is 3.51. Because $(3.51)^2/12=1.027<1.44=x_{0.05}/(1-x_{0.05})$, we shall use the Bonferroni inequalities to obtain confidence intervals. The intervals in Table 4.7 were calculated. Note that all 12 intervals contain zero. Therefore we cannot claim a significant difference for these comparisons at an overall error rate of $\alpha=0.05$.

4.3.1 Randomized Complete Block Design with Interaction

Equation (4.3.1) presented the randomized block design without interaction. However, the assumption that the treatment and block effects are additive may not be correct. Since there are no replicates, we cannot consider all types of nonadditivity as interaction. Hence, we consider the special kind of nonadditivity considered by Tukey (1949) in the univariate case.

Let

$$y_{ijk}=\mu_k+\beta_{ki}+\tau_{kj}+\alpha_k\beta_{ki}\tau_{kj}+\varepsilon_{ijk},\qquad k=1,2,\ldots,p,\quad j=1,2,\ldots,t,\quad i=1,2,\ldots,b, \tag{4.3.4}$$

where $\Sigma_{i=1}^b\beta_{ki}=\Sigma_{j=1}^t\tau_{kj}=0$, $k=1,2,\ldots,p$.

Let $D_c=\operatorname{diag}(c_1,\ldots,c_p)$. Then

$$D_{\beta_i}\boldsymbol{\tau}_j=D_{\tau_j}\boldsymbol{\beta}_i=\begin{pmatrix}\beta_{1i}\tau_{1j}\\ \vdots\\ \beta_{pi}\tau_{pj}\end{pmatrix},$$

a $(p\times1)$-vector. Hence, in matrix notation, the model (4.3.4) can be written

$$\mathbf{y}_{ij}=\boldsymbol{\mu}+\boldsymbol{\beta}_i+\boldsymbol{\tau}_j+D_\alpha D_{\beta_i}\boldsymbol{\tau}_j+\boldsymbol{\varepsilon}_{ij}, \tag{4.3.5}$$

where $i=1,2,\ldots,b$, $j=1,2,\ldots,t$, $\Sigma_{i=1}^b\boldsymbol{\beta}_i=\mathbf{0}$, $\Sigma_{j=1}^t\boldsymbol{\tau}_j=\mathbf{0}$, and the $\boldsymbol{\varepsilon}_{ij}$ are independently distributed as $N_p(\mathbf{0},\Sigma)$. Thus to test for the absence of interaction we need to test the hypothesis

$$H: \alpha_1=\alpha_2=\cdots=\alpha_p=0 \qquad \text{vs} \qquad A: \alpha_i\neq0 \tag{4.3.6}$$

for at least one i, $i=1,2,\ldots,p$.

Let

$$\hat{\boldsymbol{\beta}}_{i\cdot} = \mathbf{y}_{i\cdot} - \mathbf{y}_{\cdot\cdot}, \qquad \hat{\boldsymbol{\tau}}_j = \mathbf{y}_{\cdot j} - \mathbf{y}_{\cdot\cdot},$$
$$\hat{F} = \left(D_{\hat{\beta}_1}\hat{\boldsymbol{\tau}}_1, D_{\hat{\beta}_1}\hat{\boldsymbol{\tau}}_2, \ldots, D_{\hat{\beta}_1}\hat{\boldsymbol{\tau}}_t, \ldots, D_{\hat{\beta}_b}\hat{\boldsymbol{\tau}}_t\right) : p \times bt, \tag{4.3.7}$$
$$Z = (\mathbf{z}_{11}, \mathbf{z}_{12}, \ldots, \mathbf{z}_{1t}, \ldots, \mathbf{z}_{bt}) : p \times bt,$$

where

$$\mathbf{z}_{ij} = \mathbf{y}_{ij} - \mathbf{y}_{i\cdot} - \mathbf{y}_{\cdot j} + \mathbf{y}_{\cdot\cdot}.$$

Then a test for the hypothesis in (4.3.6) is based on

$$U_{p,p,m} = |S_{\mathrm{e}}| / |S_{\mathrm{e}} + S_{\mathrm{h}}|,$$

where

$$m = bt - t - b - p + 1,$$
$$S_{\mathrm{h}} = Z\hat{F}'(\hat{F}\hat{F}')^{-1}\hat{F}Z',$$

and

$$S_{\mathrm{e}} = ZZ' - S_{\mathrm{h}}.$$

Note that the above analysis assumes that $\hat{F}\hat{F}'$ is of full rank p. In case $\hat{F}\hat{F}'$ is not of full rank, we use a generalized inverse $(\hat{F}\hat{F}')^{-}$ in place of $(\hat{F}\hat{F}')^{-1}$, and $U_{p,p,m}$ will reduce to $U_{p,r,m}$, where $r = \rho(\hat{F}\hat{F}')$.

Example 4.3.2 Return to the data of Examples 4.2.1 and 4.3.1 with the rows considered as blocks. To test for the presence of interaction between blocks and varieties (Tukey type), we use (4.3.4)–(4.3.7). Estimates of the treatment effects are

$$\hat{\boldsymbol{\tau}}_1 = \begin{pmatrix} 0.45 \\ -0.25 \end{pmatrix}, \qquad \hat{\boldsymbol{\tau}}_2 = \begin{pmatrix} 1.7 \\ 0.0 \end{pmatrix}, \qquad \hat{\boldsymbol{\tau}}_3 = \begin{pmatrix} -1.80 \\ 0.25 \end{pmatrix},$$
$$\hat{\boldsymbol{\tau}}_4 = \begin{pmatrix} -1.05 \\ -0.25 \end{pmatrix}, \qquad \text{and} \qquad \hat{\boldsymbol{\tau}}_5 = \begin{pmatrix} 0.70 \\ 0.25 \end{pmatrix}.$$

Estimates of the block effects are

$$\hat{\boldsymbol{\beta}}_1 = \begin{pmatrix} 4.1 \\ 0.3 \end{pmatrix}, \qquad \hat{\boldsymbol{\beta}}_2 = \begin{pmatrix} -3.7 \\ 0.1 \end{pmatrix}, \qquad \hat{\boldsymbol{\beta}}_3 = \begin{pmatrix} 0.9 \\ 0.5 \end{pmatrix}, \qquad \text{and} \qquad \hat{\boldsymbol{\beta}}_4 = \begin{pmatrix} -1.3 \\ -0.9 \end{pmatrix}.$$

For example, $D_{\hat{\beta}_1}$ takes the form

$$D_{\hat{\beta}_1} = \begin{pmatrix} 4.1 & 0.0 \\ 0.0 & 0.3 \end{pmatrix}.$$

We may now calculate $\hat{F}$ and Z, which are displayed in Table 4.8 with

$$S_{\mathrm{h}} = \begin{pmatrix} 4.00 & 3.31 \\ 3.31 & 4.18 \end{pmatrix}, \qquad S_{\mathrm{e}} = \begin{pmatrix} 157.4 & -14.93 \\ -14.93 & 18.175 \end{pmatrix},$$

and

$$m = 20 - 5 - 4 - 2 + 1 = 10.$$

Table 4.8

$\hat{F}'$		Z	
1.845	−0.075	−0.85	−1.55
6.970	0.000	1.9	0.2
−7.380	0.075	3.4	0.95
−4.305	−0.075	−3.35	0.45
2.870	0.075	−1.1	−0.05
−1.665	−0.025	−2.05	1.65
−6.290	0.000	2.7	0.4
6.660	0.025	−0.8	−1.85
3.885	−0.025	2.45	−1.35
−2.590	0.025	−2.3	1.15
0.405	−0.125	−1.65	0.25
1.530	0.000	−4.9	0.00
−1.620	0.125	−2.4	0.75
−0.945	−0.125	3.85	−0.75
0.630	0.125	5.1	−0.25
−0.585	−0.225	4.55	−0.35
−2.210	0.000	0.3	−0.6
2.340	0.225	−0.2	0.15
1.365	−0.225	−2.95	1.65
−0.910	0.225	−1.7	−0.85

Hence the test statistic for testing H: (no interaction) vs A: (an interaction of Tukey type) is given by

$$U_{2,2,10} = |S_e| / |S_e + S_h| = 0.7595.$$

From Theorem 4.2.3, when $p = 2$,

$$9(1 - U_{2,2,10}^{1/2}) / 2U_{2,2,10}^{1/2} = 0.664 = F_{4,18} < F_{4,18,0.05} = 2.93.$$

We conclude that there is no interaction between blocks and varieties.

4.4 Latin Square Design

A certain region of land was to be used for testing t varieties of corn $C_1, \ldots, C_t$. It was felt that due to the slope of the land that there would be differences from north to south (B) and from east to west (A). The region was divided into t^2 plots, and a Latin square was chosen from all possible Latin squares of order t. For example, for $t = 4$ one such design could be

N-S \ $E-W$	A_1	A_2	A_3	A_4
B_1	C_2	C_3	C_4	C_1
B_2	C_3	C_2	C_1	C_4
B_3	C_1	C_4	C_2	C_3
B_4	C_4	C_1	C_3	C_2

(4.4.1)

It becomes difficult randomly to select Latin squares of large order ($t \geq 6$). One simply chooses as randomly as possible. [See Fisher and Yates (1948) and Kempthorne (1952).] For each of these plots of land, p characteristics were measured, for example, the yield, the height of a typical plant, the percentage of plants infected, and a measure of the sweetness of the corn. The model for this design is given by

$$\mathbf{y}_{ijk} = \boldsymbol{\mu} + \boldsymbol{\alpha}_i + \boldsymbol{\beta}_j + \boldsymbol{\tau}_k + \boldsymbol{\varepsilon}_{ijk}, \qquad i=1,\dots,t, \quad j=1,\dots,t, \quad k=k(i,j), \tag{4.4.2}$$

where $\mathbf{y}_{ijk}$ is the vector of observations for the p characteristics for the ith E–W plot, the jth N–S plot, and the kth variety assigned to the ijth plot. The assumptions are that $\Sigma\boldsymbol{\alpha}_i = \Sigma\boldsymbol{\beta}_j = \Sigma\boldsymbol{\tau}_k = \mathbf{0}$ and that $\varepsilon_{ijk} \sim N_p(\mathbf{0}, \Sigma)$. The hypothesis of interest is simply H: $\boldsymbol{\tau}_1 = \cdots = \boldsymbol{\tau}_t = 0$ vs A: $\boldsymbol{\tau}_k \neq \mathbf{0}$ for at least one k. To perform the analysis, we need the following expressions:

$$\bar{\mathbf{y}}_{i\cdot\cdot} = t^{-1}\sum_{j,k}\mathbf{y}_{ijk}, \quad i=1,\dots,t, \qquad \bar{\mathbf{y}}_{\cdot j\cdot} = t^{-1}\sum_{i,k}\mathbf{y}_{ijk}, \quad j=1,\dots,t,$$

$$\bar{\mathbf{y}}_{\cdot\cdot k} = t^{-1}\sum_{i,j}\mathbf{y}_{ijk}, \quad k=1,\dots,t, \qquad \bar{\mathbf{y}}_{\cdot\cdot\cdot} = \sum_{ijk}\mathbf{y}_{ijk}/t^2.$$

$$\mathrm{SST} = \sum_{i,j,k}\mathbf{y}_{ijk}\mathbf{y}'_{ijk} - t^2\bar{\mathbf{y}}_{\cdot\cdot\cdot}\bar{\mathbf{y}}'_{\cdot\cdot\cdot}, \qquad \mathrm{SSA} = t\sum_{i=1}^{t}\bar{\mathbf{y}}_{i\cdot\cdot}\bar{\mathbf{y}}'_{i\cdot\cdot} - t^2\bar{\mathbf{y}}_{\cdot\cdot\cdot}\bar{\mathbf{y}}'_{\cdot\cdot\cdot},$$

$$\mathrm{SSB} = t\sum_{j=1}^{t}\bar{\mathbf{y}}_{\cdot j\cdot}\bar{\mathbf{y}}'_{\cdot j\cdot} - t^2\bar{\mathbf{y}}_{\cdot\cdot\cdot}\bar{\mathbf{y}}'_{\cdot\cdot\cdot}, \qquad \mathrm{SSC} = t\sum_{k=1}^{t}\bar{\mathbf{y}}_{\cdot\cdot k}\bar{\mathbf{y}}'_{\cdot\cdot k} - t^2\bar{\mathbf{y}}_{\cdot\cdot\cdot}\bar{\mathbf{y}}'_{\cdot\cdot\cdot},$$

$$\mathrm{SSE} = \mathrm{SST} - \mathrm{SSA} - \mathrm{SSB} - \mathrm{SSC}. \tag{4.4.3}$$

The resulting MANOVA table is given in Table 4.9. Simultaneous $(1-\alpha)100\%$ confidence intervals for the difference in varieties $C_1,\dots,C_t$ can be obtained using Table A.5. That is, simultaneous confidence intervals for contrasts of the form $\Sigma_{k=1}^{t} d_k \mathbf{a}'\boldsymbol{\tau}_k$, $\Sigma_{k=1}^{t} d_k = 0$, where $\mathbf{a}$ chooses the linear

Table 4.9

Source	df	SS	$U_{p,m,n}$	p	m	n
A	$t-1$	SSA	—	—	—	—
B	$t-1$	SSB	—	—	—	—
C (varieties)	$t-1$	SSC	$\lvert\mathrm{SSE}\rvert / \lvert\mathrm{SSE}+\mathrm{SSC}\rvert$	p	$t-1$	$(t-1)(t-2)$
error	$(t-1)(t-2)$	SSE				
total	t^2-1	SST				

combination of characteristics of interest, are given by

$$\sum_{k=1}^{t} d_k \mathbf{a}'\bar{\mathbf{y}}_{\cdot\cdot k} \pm \left[x_\alpha (1-x_\alpha)^{-1} \sum_{k=1}^{t} d_k t^{-1} \mathbf{a}'\mathrm{SSE}\mathbf{a} \right]^{1/2}. \tag{4.4.4}$$

The value of x_α is obtained from Table A.5 with $p_0 = p$, the number of characteristics, $m_0 = t-1$, and $n = (t-1)(t-2)$. To obtain shorter intervals, $x_\alpha/(1-x_\alpha)$ can be replaced by $t^2_{n,\alpha/2k} n^{-1}$ for k comparisons if the latter yields the smaller number.

Example 4.4.1 Four varieties of corn were planted according to the Latin square design (4.4.2), and the height of a typical plant Y_1 and the yield Y_2 were measured for each plot. The following data resulted:

B \ A		A_1	A_2	A_3	A_4
B_1	height	65	68	67	68
	yield	24	21	26	27
B_2	height	66	63	64	67
	yield	20	23	25	24
B_3	height	65	67	64	63
	yield	24	25	19	20
B_4	height	65	64	64	68
	yield	26	25	25	22

Calculations yield

$$\bar{\mathbf{y}}_{1\cdot\cdot} = \begin{pmatrix} 65.25 \\ 23.5 \end{pmatrix}, \quad \bar{\mathbf{y}}_{2\cdot\cdot} = \begin{pmatrix} 65.5 \\ 23.5 \end{pmatrix}, \quad \bar{\mathbf{y}}_{3\cdot\cdot} = \begin{pmatrix} 64.75 \\ 23.75 \end{pmatrix}, \quad \bar{\mathbf{y}}_{4\cdot\cdot} = \begin{pmatrix} 66.5 \\ 23.25 \end{pmatrix},$$

$$\bar{\mathbf{y}}_{\cdot 1\cdot} = \begin{pmatrix} 67 \\ 24.5 \end{pmatrix}, \quad \bar{\mathbf{y}}_{\cdot 2\cdot} = \begin{pmatrix} 65.0 \\ 23.0 \end{pmatrix}, \quad \bar{\mathbf{y}}_{\cdot 3\cdot} = \begin{pmatrix} 64.75 \\ 22.0 \end{pmatrix}, \quad \bar{\mathbf{y}}_{\cdot 4\cdot} = \begin{pmatrix} 65.25 \\ 24.5 \end{pmatrix},$$

$$\bar{\mathbf{y}}_{\cdot\cdot 1} = \begin{pmatrix} 65.25 \\ 25.25 \end{pmatrix}, \quad \bar{\mathbf{y}}_{\cdot\cdot 2} = \begin{pmatrix} 65 \\ 22 \end{pmatrix}, \quad \bar{\mathbf{y}}_{\cdot\cdot 3} = \begin{pmatrix} 65.25 \\ 21.5 \end{pmatrix}, \quad \bar{\mathbf{y}}_{\cdot\cdot 4} = \begin{pmatrix} 66.50 \\ 25.25 \end{pmatrix},$$

$$\bar{\mathbf{y}}_{\cdots} = \begin{pmatrix} 65.5 \\ 23.5 \end{pmatrix}, \qquad t^2 \bar{\mathbf{y}}_{\cdots} \bar{\mathbf{y}}'_{\cdots} = \begin{pmatrix} 68644 & 24625 \\ 24625 & 8826 \end{pmatrix}.$$

The MANOVA table (Table 4.8) now becomes Table 4.10. For $p = 2$, we use the formula (from Theorem 4.2.3)

$$F_{2m,2(n-1)} = \frac{1-\sqrt{U}}{\sqrt{U}} \frac{n-1}{m} \tag{4.4.5}$$

Table 4.10

Source	df	SS	$U_{p,m,n}$	p	m	n
A	3	$\begin{pmatrix} 6.5 & -1.75 \\ -1.75 & 0.5 \end{pmatrix}$	—	—	—	—
B	3	$\begin{pmatrix} 12.5 & 10.5 \\ 10.5 & 18.0 \end{pmatrix}$	—	—	—	—
C	3	$\begin{pmatrix} 5.5 & 10.25 \\ 10.25 & 49.5 \end{pmatrix}$	0.2100	2	3	6
error	6	$\begin{pmatrix} 23.5 & -7 \\ -7 & 20 \end{pmatrix}$				
total	15	$\begin{pmatrix} 48 & 12 \\ 12 & 88 \end{pmatrix}$				

to find the percentage points of $U_{p,m,n}$. With $U = 0.2100$

$$F_{6,10} = \frac{1-(0.2100)^{1/2}}{(0.2100)^{1/2}} \times \frac{5}{3} = 1.97.$$

Because $F_{6,10,0.05} = 3.22$, we do not reject H_0 and claim that there is no significant difference in varieties of corn.

4.5 Factorial Experiments

In Sections 4.2–4.4 we assumed t treatments with no clear relationships among them. A factorial experiment has t treatment combinations of levels of factors. That is, if two drugs A and B are given in dosages of 0, 5, and 10 ml, there will be nine treatment combinations. Each dosage of one drug is given in combination with each dosage of the other. If each of these nine treatment combinations is assigned randomly to four experimental units, the experiment is designated a 3×3 factorial experiment on a completely randomized design with four replications.

The analysis of a multivariate factorial design is analogous to that of a univariate factorial design. That is, the treatment sum of squares matrix is broken down into main effects and interaction matrices. We give only the analysis of a three-factor multivariate fixed effects model on a completely randomized design. Two-factor fixed effects, random effects, and mixed effects designs are analyzed in much the same way. Note, however, that any matrix used as a divisor in a test statistic must have at least as many degrees of freedom as there are characteristics.

Example 4.5.1 Box (1950) gives the wear on tires during three periods: the first 1000 revolutions, the second 1000 revolutions, and the third 1000

Table 4.11[a] Wear of Coated Fabrics (mg)

Surface treatment	Filler	Q_1 (25%) Period 1	2	3	Q_2 (50%) Period 1	2	3	Q_3 (75%) Period 1	2	3
	F_1	194	192	141	233	217	171	265	252	207
		208	188	165	241	222	201	269	283	191
T_1	F_2	239	127	90	224	123	79	243	117	100
		187	105	85	243	123	110	226	125	75
	F_1	155	169	151	198	187	176	235	225	166
		173	152	141	177	196	167	229	270	183
T_2	F_2	137	82	77	129	94	78	155	76	91
		160	82	83	98	89	48	132	105	67

[a]Data from Box (1950).

revolutions. The data are reproduced in Table 4.11. Three factors were of interest to the experimenter: the type of road surface ($i=1,2$), the type of filler ($j=1,2$), and the proportion of filler ($k=1,2,3$). Two replications ($l=1,2$) were made for each combination of surface, filler, and proportion of filler, giving a total of $N=24$ observations for each period of rotation. The model is given by

$$\mathbf{y}_{ijkl}=\boldsymbol{\mu}+\boldsymbol{\alpha}_i+\boldsymbol{\beta}_j+\boldsymbol{\gamma}_k+\boldsymbol{\delta}_{ij}+\boldsymbol{\lambda}_{ik}+\boldsymbol{\zeta}_{jk}+\boldsymbol{\xi}_{ijk}+\boldsymbol{\varepsilon}_{ijkl},$$

where $i=1,\ldots,a$, $j=1,\ldots,b$, $k=1,\ldots,c$, and $l=1,\ldots,r$, with $a=b=r=2$, $c=3$, and with the restrictions $\Sigma_i\boldsymbol{\alpha}_i=\Sigma_j\boldsymbol{\beta}_j=\Sigma_k\boldsymbol{\gamma}_k=\mathbf{0}$,

$$\sum_i\boldsymbol{\delta}_{ij}=\mathbf{0}\quad\text{for}\quad j=1,2,\qquad \sum_j\boldsymbol{\delta}_{ij}=\mathbf{0},\quad i=1,2,$$

$$\sum_i\boldsymbol{\lambda}_{ik}=\mathbf{0}\quad\text{for}\quad k=1,2,3,\qquad \sum_k\boldsymbol{\lambda}_{ik}=\mathbf{0},\quad i=1,2,$$

$$\sum_j\boldsymbol{\zeta}_{jk}=\mathbf{0}\quad\text{for}\quad k=1,2,3,\qquad \sum_k\boldsymbol{\zeta}_{jk}=\mathbf{0},\quad j=1,2,$$

$$\sum_i\boldsymbol{\xi}_{ijk}=\mathbf{0},\qquad k=1,2,3,\quad j=1,2,$$

$$\sum_j\boldsymbol{\xi}_{ijk}=\mathbf{0},\qquad k=1,2,3,\quad i=1,2,$$

$$\sum_k\boldsymbol{\xi}_{ijk}=\mathbf{0},\qquad i,\,j=1,2;$$

the $\boldsymbol{\varepsilon}_{ijkl}$ are independently distributed as $N_p(\mathbf{0},\Sigma)$. The sum-of-squares matrices are now defined as follows (the average over a particular index is indicated by replacing the index with a period and overscoring the y), where the factors road surface, filler type, and proportion of filler are denoted A,

Table 4.12[a]

Source	df	SS	$U_{p,m,n}$ (Wilk's criterion)
A	$a-1$	SSA	$\lvert SSE\rvert / \lvert SSA+SSE\rvert$
B	$b-1$	SSB	$\lvert SSE\rvert / \lvert SSB+SSE\rvert$
C	$c-1$	SSC	$\lvert SSE\rvert / \lvert SSC+SSE\rvert$
AB	$(a-1)(b-1)$	SSAB	$\lvert SSE\rvert / \lvert SSAB+SSE\rvert$
AC	$(a-1)(c-1)$	SSAC	$\lvert SSE\rvert / \lvert SSAC+SSE\rvert$
BC	$(b-1)(c-1)$	SSBC	$\lvert SSE\rvert / \lvert SSBC+SSE\rvert$
ABC	$(a-1)(b-1)(c-1)$	SSABC	$\lvert SSE\rvert / \lvert SSABC+SSE\rvert$
error	$(r-1)abc$	SSE	
total	$rabc$	SST	

[a]p is the number characteristics, $m=$ df, and $n=(r-1)abc$.

B, and C, respectively:

$$\text{SST}=\sum \mathbf{y}_{ijkl}\mathbf{y}'_{ijkl}-N\bar{\mathbf{y}}_{....}\mathbf{y}'_{....}, \qquad N=24,$$

$$\text{SSE}=\text{SST}-r\sum \bar{\mathbf{y}}_{ijk\cdot}\bar{\mathbf{y}}'_{ijk\cdot}+N\bar{\mathbf{y}}_{....}\bar{\mathbf{y}}'_{....}, \qquad r=2,$$

$$\text{SSA}=rbc\sum \bar{\mathbf{y}}_{i...}\bar{\mathbf{y}}'_{i...}-N\bar{\mathbf{y}}_{....}\bar{\mathbf{y}}'_{....}, \qquad b=2, \quad c=3,$$

$$\text{SSB}=rac\sum \bar{\mathbf{y}}_{\cdot j\cdot\cdot}\bar{\mathbf{y}}'_{\cdot j\cdot\cdot}-N\bar{\mathbf{y}}_{....}\bar{\mathbf{y}}'_{....}, \qquad a=2,$$

$$\text{SSC}=rbc\sum \bar{\mathbf{y}}_{\cdot\cdot k\cdot}\bar{\mathbf{y}}'_{\cdot\cdot k\cdot}-N\bar{\mathbf{y}}_{....}\bar{\mathbf{y}}'_{....},$$

$$\text{SSAB}=rc\sum \bar{\mathbf{y}}_{ij\cdot\cdot}\bar{\mathbf{y}}'_{ij\cdot\cdot}-\text{SSA}-\text{SSB}+N\bar{\mathbf{y}}_{....}\bar{\mathbf{y}}'_{....},$$

$$\text{SSBC}=ra\sum \bar{\mathbf{y}}_{\cdot jk\cdot}\bar{\mathbf{y}}'_{\cdot jk\cdot}-\text{SSB}-\text{SSC}+N\bar{\mathbf{y}}_{....}\bar{\mathbf{y}}'_{....},$$

$$\text{SSAC}=rb\sum \bar{\mathbf{y}}_{i\cdot k\cdot}\bar{\mathbf{y}}'_{i\cdot k\cdot}-\text{SSA}-\text{SSC}+N\bar{\mathbf{y}}_{....}\bar{\mathbf{y}}'_{....},$$

and

$$\text{SSABC}=\text{SST}-\text{SSE}-\text{SSA}-\text{SSB}-\text{SSC}-\text{SSAB}-\text{SSAC}-\text{SSBC}.$$

Table 4.13 First-Period ANOVA

Source	df	SS	F	$P(>F)$
A	1	26268	97.74	0.0001
B	1	6800	25.30	0.0003
C	2	5968	11.10	0.0019
AB	1	3953	14.71	0.0024
AC	2	1186	2.21	0.1527
BC	2	3529	6.57	0.0119
ABC	2	479	0.89	0.4360
error	12	3225		
total	23	51408		

Table 4.14 Second-Period ANOVA

Source	df	SS	F	$P(>F)$
A	1	5017	25.03	0.0003
B	1	70959	353.99	0.0001
C	2	7969	19.88	0.0002
AB	1	57	0.28	0.6035
AC	2	44	0.11	0.8962
BC	2	6031	15.04	0.0005
ABC	2	14	0.04	0.9650
error	12	2406		
total	23	90092		

The MANOVA table for a general three-factor experiment is given in Table 4.12. It is assumed that either $n \geq p$ or else $U_{p,m,n} = 0$ or is undefined. The calculations for Example 4.5.1 may be obtained by simple programming or by the use of Wilk's criterion with statistical packages such as BMD12V or SAS. Further details on statistical packages are given in Section 4.7.

From Tables 4.13–4.16 there appears to be no three-way interaction and no interaction between road surface and filler quantity. All main effects are significant, and in addition the filler type interacts with road surface and quantity of filler.

Since no three-way interaction exists, we should look at two-way interaction diagrams. This requires plotting the average value of treatment combinations taken two factors at a time for each characteristic. The mean values of road surface and filler proportion combinations averaged over replications and filler type for time period 1 are

	Q_1	Q_2	Q_3
T_1	207.00	235.25	250.75
T_2	156.25	150.50	187.75

Table 4.15 Third-Period ANOVA

Source	df	SS	F	$P(>F)$
A	1	1457	6.57	0.0249
B	1	48330	217.83	0.0001
C	2	1397	3.15	0.0796
AB	1	0.38	0.00	0.9679
AC	2	251	0.56	0.5829
BC	2	1740	3.92	0.0489
ABC	2	272	0.61	0.5576
error	12	2663		
total	23	56110.38		

Table 4.16 MANOVA 0

Source	df	p, m, n	$U_{p,m,n}$	F	(df)	$P(>F)$
A	1	3, 1, 12	0.080	38.31	(3, 10)	0.0001
B	1	3, 1, 12	0.019	170.27	(3, 10)	0.0001
C	2	3, 2, 12	0.137	5.66	(6, 20)	0.0014
AB	1	3, 1, 12	0.356	6.02	(3, 10)	0.0131
AC	2	3, 2, 12	0.711	0.62	(6, 20)	0.7116
BC	2	3, 2, 12	0.178	4.56	(6, 20)	0.0045
ABC	2	3, 2, 12	0.755	0.50	(6, 20)	0.7979
error	12					
total	23					

Figure 4.1 Interaction diagrams.

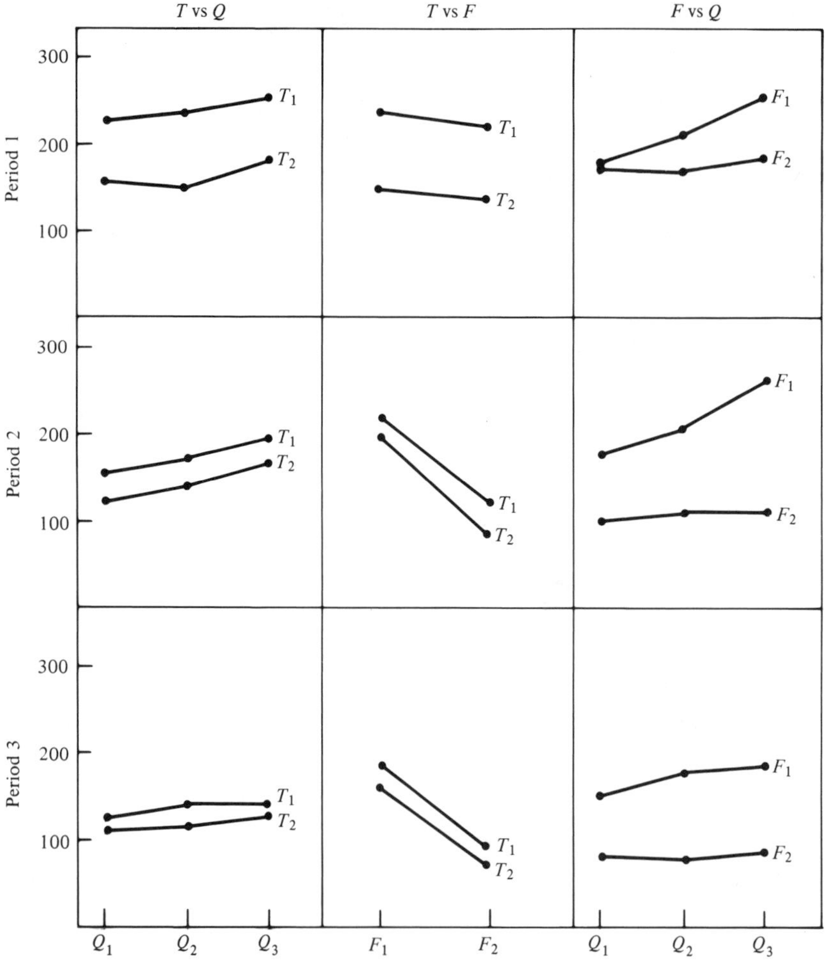

The corresponding interaction diagram is given in Figure 4.1 along with eight other diagrams.

From Figure 4.1 and Table 4.16 we can conclude that filler 2 produces less wear for both road surfaces and for all three filler proportions and should be used, preferably in proportion Q_1 or Q_2. We also find that road surface T_2 produces greater wear than T_1; however, this information does not help in the design of the tire.

4.6 Analysis of Covariance

Consider the following example, from Box (1950).

Example 4.6.1 In a biological experiment 27 rats were divided at random into three groups containing 10, 7, and 10 rats, respectively. The first group was kept as a control, the second group had thyroxin, and the third group thiouracil added to its drinking water. The initial weight and weight gain after 1, 2, 3, and 4 weeks are shown in Table 4.17. We may be interested to find out whether the gain in weight (x_1, x_2, x_3, x_4) for the three treatments (control, thyroxin, and thiouracil) is the same, taking into account the effect of initial weight on weight gain.

Since the rats were randomly assigned to the three groups, the expected initial weight for each group is the same. If the initial weight is correlated with weight gain, we may use this relationship to reduce the error variance. The initial weight is labeled a *covariable*, or *concomitant variable*. The testing procedure is the same whether the covariable is a controlled (or fixed variable) or a random variable. That is, the initial weights could be preselected values or randomly selected. The general model is analyzed in Section 5.5. Consider the random variable situation. We let $\mathbf{x}_{ij}$ be independently distributed as $N_p(\boldsymbol{\mu}_j, \Sigma)$, $j=1,\ldots,t$, $i=1,\ldots,N_j$. In Example 4.6.1,

Table 4.17 Initial Weight and Weekly Gains in Weight for 27 Rats[a]

Group 1 (control)						Group 2 (thyroxin)						Group 3 (thiouracil)					
Rat	x_0	x_1	x_2	x_3	x_4	Rat	x_0	x_1	x_2	x_3	x_4	Rat	x_0	x_1	x_2	x_3	x_4
1	57	29	28	25	33	11	59	26	36	35	35	18	61	25	23	11	9
2	60	33	30	23	31	12	54	17	19	20	28	19	59	21	21	10	11
3	52	25	34	33	41	13	56	19	33	43	38	20	53	26	21	6	27
4	49	18	33	29	35	14	59	26	31	32	29	21	59	29	12	11	11
5	56	25	23	17	30	15	57	15	25	23	24	22	51	24	26	22	17
6	46	24	32	29	22	16	52	21	24	19	24	23	51	24	17	8	19
7	51	20	23	16	31	17	52	18	35	33	33	24	56	22	17	8	5
8	63	28	21	18	24							25	58	11	24	21	24
9	49	18	23	22	28							26	46	15	17	12	17
10	57	25	28	29	30							27	53	19	17	15	18

[a] x_0 is the initial weight of the rat, x_1 the gain in the first week, x_2 the gain in the second week, x_3 the gain in the third week, and x_4 the gain in the fourth week.

$t=3$, $N_1=10$, $N_2=7$, and $N_3=10$. Suppose that $\mathbf{x}_{ij}$ is partitioned

$$\mathbf{x}_{ij}=\begin{pmatrix}\mathbf{y}_{ij}\\ \mathbf{z}_{ij}\end{pmatrix},$$

where $\mathbf{y}_{ij}$ is a q-vector and $\mathbf{z}_{iJ}$ is a $(p-q)$-vector. Partition $\boldsymbol{\mu}_j$ and Σ similarly:

$$\boldsymbol{\mu}_j=\begin{pmatrix}\boldsymbol{\gamma}_j\\ \boldsymbol{\nu}_j\end{pmatrix} \quad \text{and} \quad \Sigma=\begin{pmatrix}\Sigma_{11} & \Sigma_{12}\\ \Sigma_{12}' & \Sigma_{22}\end{pmatrix},$$

where $\boldsymbol{\gamma}_j$ is a q-vector, Σ_{11} is a $q\times q$ symmetric matrix, and Σ_{12} is a $q\times(p-q)$ matrix. Then the hypothesis to be tested is

$$H\colon \mathbf{v}_1=\cdots=\mathbf{v}_t \quad \text{and} \quad \boldsymbol{\gamma}_1=\cdots=\boldsymbol{\gamma}_t$$

vs (4.6.1)

$$A\colon \mathbf{v}_i\neq\mathbf{v}_j \quad \text{and} \quad \boldsymbol{\gamma}_1=\cdots=\boldsymbol{\gamma}_t$$

for some $1\le i<j\le t$.

We first define

$$N=\sum_{j=1}^{t}N_j,\qquad \bar{\mathbf{x}}_{\cdot j}=N_j^{-1}\sum_{i=1}^{N_j}\mathbf{x}_{ij},\qquad \bar{\mathbf{x}}_{\cdot\cdot}=N^{-1}\sum_{j=1}^{t}\sum_{i=1}^{N_j}\mathbf{x}_{ij}.$$

The sum of squares due to the error matrix is calculated from

$$V=\text{SSE}=\sum_{j=1}^{t}\sum_{i=1}^{N_j}(\mathbf{x}_{ij}-\bar{\mathbf{x}}_{\cdot j})(\mathbf{x}_{ij}-\bar{\mathbf{x}}_{\cdot j})', \tag{4.6.2}$$

which may be partitioned

$$V=\begin{pmatrix}V_{11} & V_{12}\\ V_{12}' & V_{22}\end{pmatrix}$$

as for Σ. The sum of squares due to the treatments is

$$W=\text{SS(TR)}=\sum_{j=1}^{t}N_j(\bar{\mathbf{x}}_{\cdot j}-\bar{\mathbf{x}}_{\cdot\cdot})(\mathbf{x}_{\cdot j}-\bar{\mathbf{x}}_{\cdot\cdot})', \tag{4.6.3}$$

which may also be partitioned

$$W=\begin{pmatrix}W_{11} & W_{12}\\ W_{12}' & W_{22}\end{pmatrix}.$$

Then the U-statistic for testing (4.6.1) is

$$U_{p-q,t-1,N-t-q}=|V_{11}|^{-1}|V_{11}+W_{11}||V+W|^{-1}|V|. \tag{4.6.4}$$

H is rejected if $U_{p-q,t-1,N-p-q}\ge U_{p-q,t-1,N-t-q,\alpha}$. It should be noted that the U-statistic in (4.6.4) is the ratio of the two U-statistics for testing the two hypotheses

$$H: \boldsymbol{\mu}_1 = \cdots = \boldsymbol{\mu}_t \tag{4.6.5}$$

and

$$H: \gamma_1 = \cdots = \gamma_t. \tag{4.6.6}$$

For Example 4.6.1 the values of the U-statistics were obtained from the BMDP7M program explained in Section 4.8. For the hypothesis of (4.6.5) the value of the U-statistic was 0.2431, and for the hypothesis of (4.6.6) the value of the U-statistic was 0.9807. Hence the value of the U-statistic in (4.6.4) is

$$U_{4,2,23} = (0.2431)/(0.9807) = 0.2479.$$

Using the fact that $U_{r,m,n} = U_{m,r,n+m-r}$, we have from Theorem 4.2.3 that for $m=2$, $r=4$, and $n=23$

$$\frac{1-\sqrt{U}}{\sqrt{U}}\,\frac{n+m-r-1}{r} = F_{2r,2(n+m-r-1)}.$$

Hence

$$\frac{1-(0.2479)^{1/2}}{(0.2479)^{1/2}}\,\frac{20}{4} = 5.04 \geq F_{8,40,0.05} = 2.18.$$

We therefore claim that there is a significant difference in groups. Without adjustment for the covariable, the U-statistic is $U_{4,2,24} = 0.2655$. Hence

$$\frac{1-(0.2655)^{1/2}}{(0.2655)^{1/2}} \times \frac{21}{4} = 4.939 \geq F_{8,42,0.05} = 2.17.$$

Although the F-value is slightly lower it still is significant.

The general case of analysis of covariance is discussed in Section 5.5 for both random and fixed covariables. Confidence intervals for comparisons of $\boldsymbol{\mu}_j$, $j=1,\ldots,t$, are also given in that section. Statistical computation of the test statistics in this section can be performed by taking the ratio of the two U-statistics for (4.6.5) and (4.6.6). The U-statistics can be obtained from the GLM procedure of SAS or by BMDP7M or BMD11V (see Section 4.8). An alternative procedure is discussed in Section 5.7.

4.7 Transformations

To assess the normality of a set of observations we must eliminate the effect of the model, which for a randomized block design (Section 4.3) is given by

$$\mathbf{y}_{ij} = \boldsymbol{\mu} + \boldsymbol{\beta}_i + \boldsymbol{\tau}_j + \boldsymbol{\varepsilon}_{ij}, \qquad i=1,\ldots,b, \quad j=1,\ldots,t.$$

Table 4.18

Control	25–50*r*	75–100*r*	125–250*r*
223	53	206	202
72	45	208	126
171	47	224	54
138	107	119	158
22	193	144	175
172	91	170	147
	115	93	105
	32	237	213
	38	208	258
	66	187	257
	210	95	257
	167	46	
	23	95	
	234	59	
		186	

Figure 4.2 *X*-normal plot.

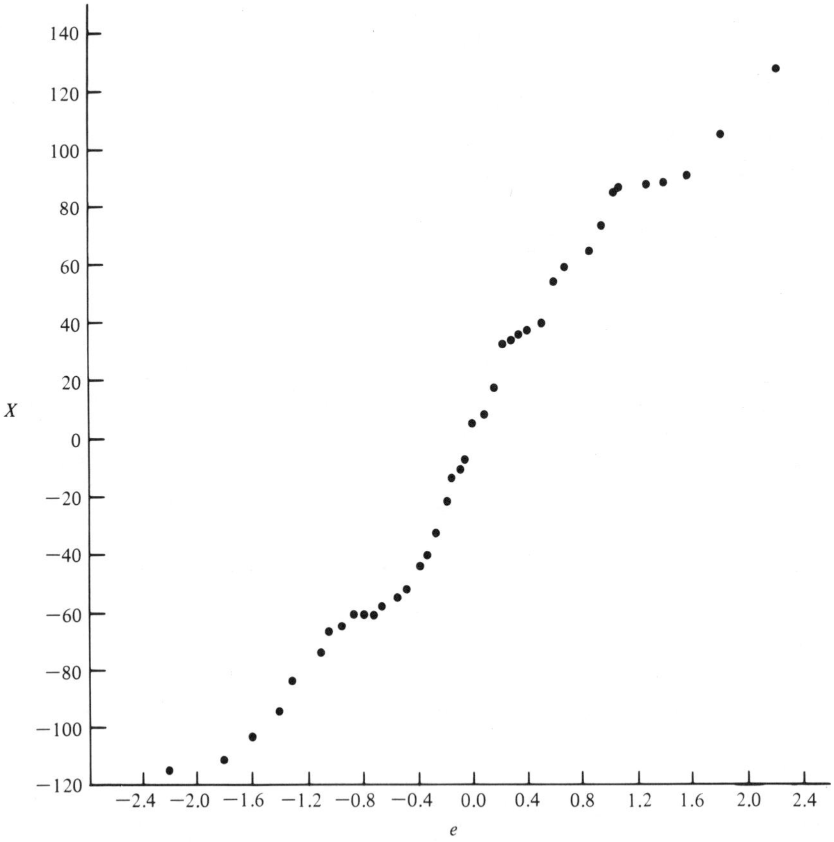

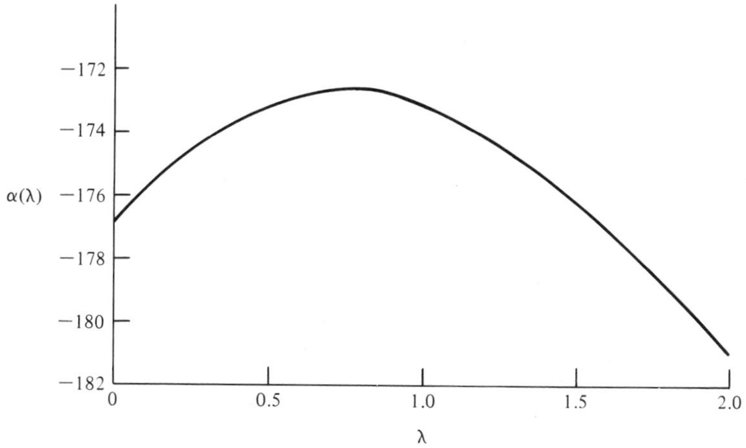

Figure 4.3 Power transformation.

where $\boldsymbol{\varepsilon}_{ij} \sim N_p(0, \Sigma)$. Estimates of the parameters are

$$\hat{\boldsymbol{\mu}} = \bar{\mathbf{y}}_{..}, \qquad \hat{\boldsymbol{\beta}}_i = \bar{\mathbf{y}}_{i.} - \bar{\mathbf{y}}_{..}, \qquad \hat{\boldsymbol{\tau}}_j = \bar{\mathbf{y}}_{.j} - \bar{\mathbf{y}}_{..}.$$

Hence the variables

$$\mathbf{z}_{ij} = \mathbf{y}_{ij} - \hat{\boldsymbol{\beta}}_i - \hat{\boldsymbol{\tau}}_j = \mathbf{y}_{ij} - \bar{\mathbf{y}}_{i.} - \bar{\mathbf{y}}_{.j} + \bar{\mathbf{y}}_{..} \tag{4.7.1}$$

are normally distributed with the same covariance matrix and mean zero.

Figure 4.4 Modulus transformation.

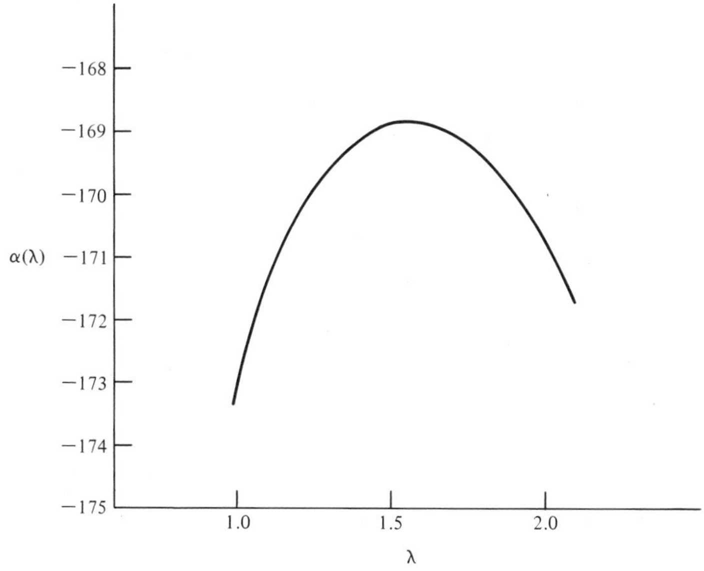

Table 4.19

Transformation	Skewness	Kurtosis	Wilk–Shapiro p
original	−0.023	−1.149	0.063
power	−0.199	−1.057	0.124
modulus	0.072	−0.727	0.503

Although they are not independent, the dependency can be ignored for most practical situations (Anscombe and Tukey, 1963).

The residuals can then be handled with regard to normal probability plots and transformations as described in Section 3.7. That is, for each characteristic we choose a transformation from one of the families given in (3.7.3)–

Figure 4.5 Normal probability plot for power-transformed data.

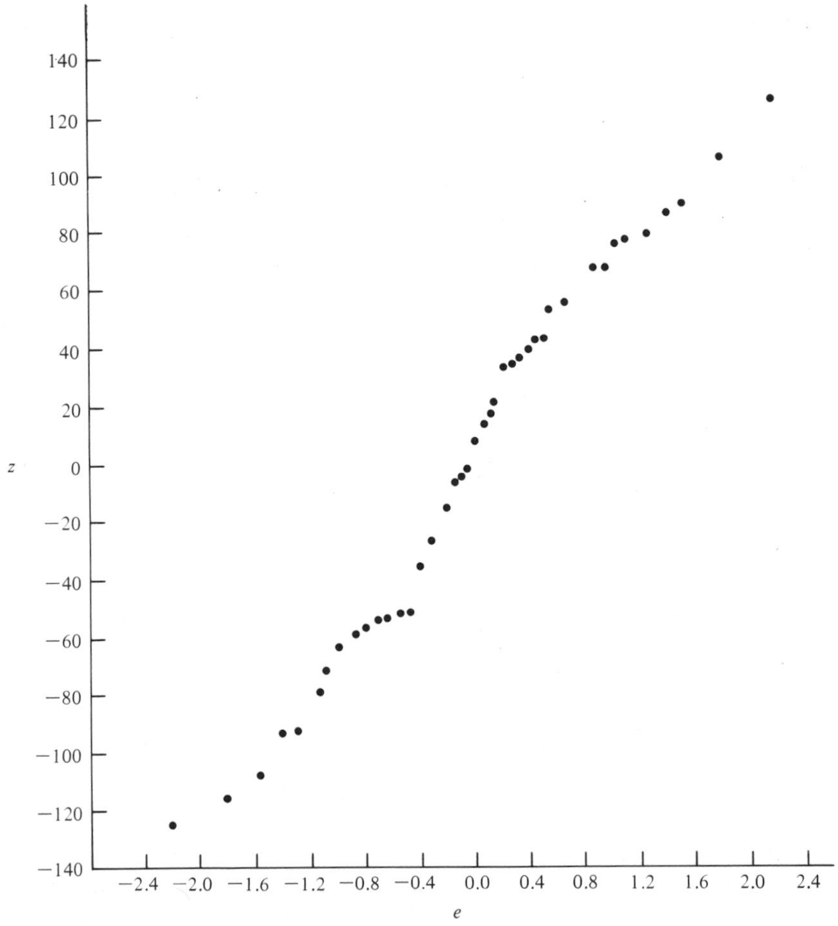

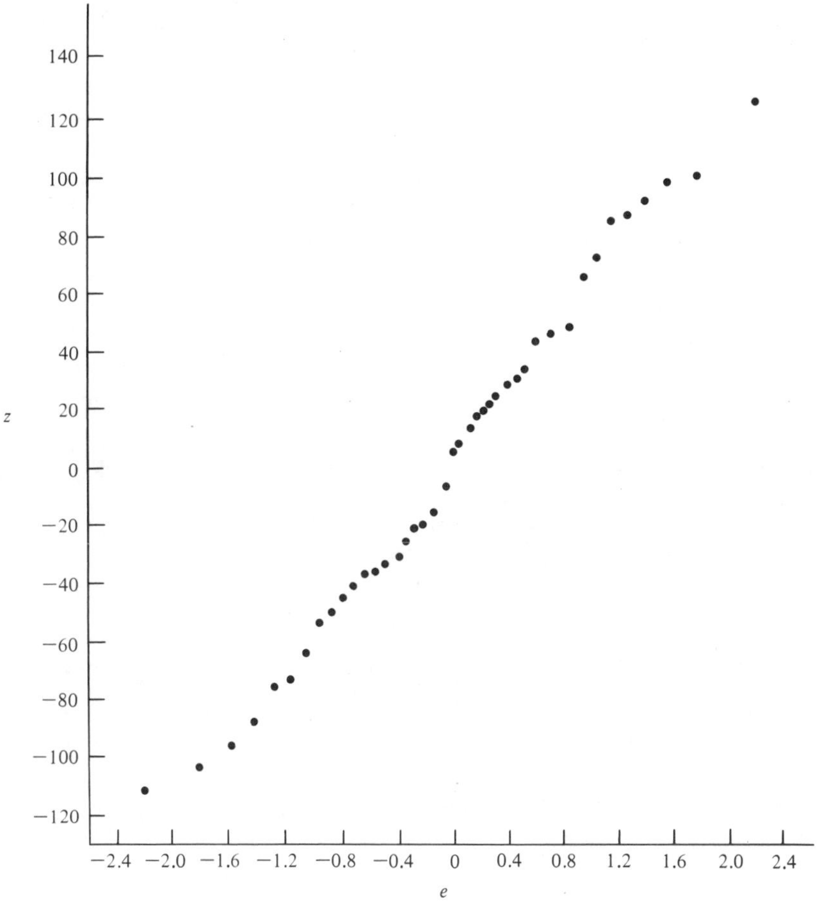

Figure 4.6 Normal probability plot for modulus-transformed data.

(3.7.5) that maximizes $l(\lambda) = -f/2\log s_\lambda^2$, where f is the number of degrees of freedom for error and s_λ^2, the mean sum of squares due to error, is obtained by dividing the sum of squares due to error for the transformed characteristic by f. The plots of $c_{(i)}$ vs d_i given in Section 3.7.3 can also be made for the residuals of a multivariate analysis of variance. We calculate

$$c_{ij}^2 = f\mathbf{z}_{ij}'(\mathrm{SSE})^{-1}\mathbf{z}_{ij}. \tag{4.7.2}$$

These values are ordered as $c_{(1)} < \cdots < c_{(N)}$, where N is the sample size, and $c_{(i)}$ is then plotted vs d_i, defined in (3.7.9). Multivariate normality can alternatively be assessed using the principal component method, which we shall discuss in Section 10.5.

Example 4.7.1 Consider the univariate data of Table 4.18 (given in detail in Problem 7.4). The data consist of psychomotor scores for four treatment groups one day postradiation. The normal probability plot, given in Figure 4.2, does not yield a straight line, indicating that the data are not normal. Plots of the likelihood function as a function of λ are given in Figures 4.3 and 4.4 for the power and modulus families of transformations, respectively. The maximum value of the likelihood function occurred at $\lambda = 0.7$ for the power transformation and at $\lambda = 1.6$ for the modulus transformation. From the summary statistics, given in Table 4.19, we can see that the data are symmetrical; hence the modulus transformation should be used to eliminate kurtosis. We may notice that the power transformation has little effect on the kurtosis, whereas the modulus transformation has a large effect. The *p*-values for the Wilk–Shapiro test for normality are also much higher with the modulus transformation. Normal probability plots for the power and modulus transformed variables are given in Figures 4.5 and 4.6, respectively. The plot for the modulus transformed data tends to give the straighter line, although none of the three probability plots in Figures 4.2, 4.5, and 4.6 produces a perfectly straight line.

4.8 Computational Procedures

Example 4.5.1 was run using the BMD12V and SAS GLM procedures. The BMD package has generally been replaced by BMDP and is not as accessible as SAS.

The program language to run the problem in SAS is given by

```
DATA GROWTH;
INPUT A B C Y1 Y2 Y3;
CARDS;
1 1 1 194 192 141
1 1 1 208 188 165
1 1 2 233 217 171
1 1 2 241 222 201
1 1 3 265 252 207
1 1 3 269 283 191
1 2 1 239 127  90
1 2 1 187 105  85
1 2 2 224 123  79
1 2 2 243 123 110
1 2 3 243 117 100
1 2 3 226 125  75
2 1 1 155 169 151
2 1 1 173 152 141
2 1 2 198 187 176
2 1 2 177 196 167
2 1 3 235 225 166
2 1 3 229 270 183
2 2 1 137  82  77
2 2 1 160  82  83
2 2 2 129  94  78
```

```
2 2 2  98  89  48
2 2 3 155  76  91
2 2 3 132 105  67
PROC GLM;
CLASSES A B C;
MODEL Y1 Y2 Y3 = A B C A*B A*C B*C A*B*C;
MANOVA H = A B C A*B A*C B*C A*B*C/PRINTH PRINTE;
```

The input data consisted of the levels of the factors and the values of the characteristics observed. The GLM procedure needs the model specified. (In this case, the full model was specified, although third-order interaction could be omitted if desired.) The MANOVA line specifies the hypothesis to be tested using multivariate techniques. Univariate summary tables are given by default.

The ANOVA tables for the individual characteristics were given in Tables 4.13–4.15. These tables were the output of the SAS GLM program. The summary for multivariate tests based on Wilks's U-statistic was given in Table 4.16. A typical printout for testing the main effect of A, for example, is summarized in Table 4.20.

Note that SAS gives all four test statistics: the criteria of Hotelling and Lawley, Pillai, Wilks, and Roy. One procedure should be chosen prior to the analysis and adhered to throughout. In the absence of any other information, the likelihood ratio statistic (Wilks's) is preferred.

For the case of unbalanced designs, the various treatment effects $A, B, C, A \times B, \ldots, A \times B \times C$ are not independent. SAS gives four types of sum of squares matrices. The type-1 SS adjusts the sum of squares for all other effects already in the model. Type-4 SS adjusts the sum of squares for all other effects including terms further down the ANOVA table.

We now give the program control language for the remaining examples of this chapter. For the case of Example 4.2.1 the following file was used:

```
DATA;
INPUT BLOCK VAR HEIGHT TILLERS;
CARDS;
1 1 58 4
2 1 62 6
3 1 60 7
4 1 54 6
5 1 58 6
1 2 49 7
2 2 55 6
3 2 48 4
4 2 52 4
5 2 49 7
1 3 54 6
2 3 52 6
3 3 51 7
4 3 58 5
5 3 61 6
1 4 58 4
2 4 55 4
```

Table 4.20 MANOVA Test Criteria for the Hypothesis of No Overall A Effect

Let

H = type IV SS and CP matrix for A,

E = error SS and CP matrix,

P = dependent variables = 3,

Q = hypothesis df = 1,

NE = df of E = 12,

S = min(P, Q) = 1,

$M = 0.5(ABS(P-Q)-1) = 0.5$,

$N = 0.5(\text{NE} - P - 1) = 4.0$.

Hotelling–Lawley trace[a]

TR(E** − 1*H) = 11.49374000

with an F-approximation

2(S*N + L)*Tr(E** − 1*H)/(S*S*(2M + S + 1))

for $S(2M+S+1)$ and $S(2N+S+1)$ df, $F(3,10) = 38.31$, probability > F = 0.0001.

Pillai's trace[b]

V = Tr(H*INV(H + E)) = 0.91995992

with F-approximation

(2N + S + 1)/(2M + S + 1)*V*(S − V)

for $S(2M+S+1)$ and $S(2N+S+1)$ df, $F(3,10) = 38.31$, probability > F = 0.0001.

Wilks's criterion[c]

L = DET(E)/DET(H + E) = 0.08004008

and F (exactly)

(1 − L)/L*(NE + Q − P)/P

for P and NE + $Q - P$ df, $F(3,10) = 38.31$, probability > 0.0001.

Roy's maximum root criterion[d]

11.49374000

The first canonical variable yields an upper bound for F of $F(1,12) = 137.92$.

---

[a] Pillai (1960, Table 3).
[b] Pillai (1960, Table 2).
[c] Rao (1973, p. 555).
[d] Heck (1960, p. 625).

```
3 4 51 5
4 4 49 6
5 4 52 4
PROC GLM;
CLASSES BLOCK VAR;
MODEL HEIGHT TILLERS = VAR;
MANOVA H = VAR/PRINTE PRINTH;
MEANS A;
```

If the data were analyzed as a randomized complete block design then the model would be

```
MODEL HEIGHT TILLERS = VAR BLOCK;
```

For Example 4.4.1 the following file was used:

```
DATA;
INPUT A B C Y1 Y2;
CARDS;
1 1 2 65 24
1 2 3 66 20
1 3 1 65 24
1 4 4 65 26
2 1 3 68 21
2 2 2 63 23
2 3 4 67 25
2 4 1 64 25
3 1 4 67 26
3 2 1 64 25
3 3 2 64 19
3 4 3 64 25
4 1 1 68 27
4 2 4 67 24
4 3 3 63 20
4 4 2 68 22
PROC GLM;
CLASSES A B C;
MODEL Y1 Y2 = A B C;
MANOVA H = A B C/PRINTE PRINTH;
MEANS A;
```

The following data file was used in the calculations of Example 4.7.1:

```
DATA ONE;
INPUT Y A;
Z = 49.38;
X = Y - 138.91;
X1 = SIGN(X);
X2 = ABS(X) + 1;
Y1 = X1*(X2**1.25 - 1)/(1.25*Z**.25);
Y2 = X1*(X2**2 - 1)/(2*Z);
Y3 = X1*(X2**1.9 - 1)/(1.9*Z*.9);
Y4 = X1*(X2**1.8 - 1)/(1.8*Z**.8);
Y5 = X1*(X2**1.7 - 1)/(1.7*Z**.7);
Y6 = X1*(X2**1.6 - 1)/(1.6*Z**.6);
Y7 = X1*(X2**1.5 - 1)/(1.5*Z**.5);
Y8 = X1*(X2**1.4 - 1)/(1.4*Z**.4);
Y9 = X1*(X2**1.3 - 1)/(1.3*Z**.3);
Y10 = X1*(X2**1.2 - 1)/(1.2*Z**.2);
```

```
CARDS;
223 1
72  1
172 1
171 1
 :  :
 :  :
258 4
257 4
PROC GLM;
CLASS A;
MODEL Y Y1-Y10=A;
OUTPUT OUT=RES RESIDUAL=RE RE1-RE10;
PROC UNIVARIATE PLOT NORMAL;
VAR RE RE1-RE10;
PROC RANK TIES=MEAN NORMAL=BLOM;
RANKS RA RA1-RA10;
VAR RE RE1-RE10;
PROC PRINT;
VAR RE RE1-RE10 RA RA1-RA10;
```

The GLM procedure is used to calculate the sum of squares due to error, and the residuals are then analyzed by the PROC UNIVARIATE and PROC RANK procedures.

Problems

4.1 Clifford (1972) studied the effects of competition in the classroom. 66 classes were randomly divided into three treatments groups of 22 each. The first group was a control, the second group was rewarded with lifesavers for their efforts, and the third group was rewarded with advantages in ensuing games. The students were scored on three characteristics: performance, interest, and retention on vocabulary tasks. The resulting data (SSE divided by the degrees of freedom) for the means of the groups and the pooled covariance matrix S are shown in Table 4.21.

$$S=\begin{pmatrix} 0.84 & -0.03 & 0.18 \\ -0.03 & 0.16 & -0.009 \\ 0.18 & -0.009 & 0.05 \end{pmatrix}.$$

(a) Is there a significant difference among the three groups?

(b) Of interest to the experimenters were two planned comparisons:

$$\mu_3+\mu_2-2\mu_1, \qquad \mu_3-\mu_2.$$

Test each hypothesis at the $\alpha=0.05$ level, and find out which characteristics caused a significant difference among the groups.

Table 4.21[a]

	Means of treatment groups		
Characteristic	M_1 (Control)	M_2 (Reward)	M_3 (Game)
performance	5.72	5.92	5.90
interest	2.41	2.63	2.57
retention	30.85	31.55	31.19

Table 4.22[a]

Reasonable				Fairly reasonable				Unreasonable			
R_1	R_2	R_3	R_4	R_1	R_2	R_3	R_4	R_1	R_2	R_3	R_4
				Unassertive							
11	15	21	21	16	15	16	20	23	16	23	23
12	11	21	17	11	12	10	11	18	21	11	12
16	12	21	12	15	23	20	20	22	24	12	13
11	11	17	11	14	13	16	15	19	19	15	17
6	14	15	13	12	11	13	11	16	23	14	18
12	11	18	14	8	16	15	14	20	16	16	15
8	10	18	11	11	9	10	13	14	15	17	14
19	10	21	8	18	19	11	17	9	19	8	17
11	9	15	18	9	14	21	15	21	24	17	24
10	9	14	12	12	15	12	18	21	21	14	24
				Fairly assertive							
14	8	21	9	17	21	21	14	24	24	8	24
12	10	21	7	18	20	21	18	24	24	18	22
16	14	21	23	15	14	15	17	23	24	15	24
6	11	15	12	11	19	7	12	21	13	19	10
13	18	15	16	15	21	11	15	16	24	20	18
4	3	10	9	13	15	8	5	13	15	10	19
15	12	18	12	16	18	!3	15	18	17	15	19
19	9	22	15	18	22	21	17	21	21	18	24
10	10	14	20	16	12	13	7	21	20	19	18
16	20	19	16	21	18	19	13	18	19	14	22
				Highly assertive							
12	23	21	16	13	21	11	12	24	23	11	23
6	11	6	11	10	13	5	7	13	5	18	8
12	15	15	18	18	22	18	15	22	21	23	15
21	9	10	18	21	21	20	13	24	24	19	21
9	12	12	7	10	17	8	13	22	21	24	24
8	24	24	15	18	17	10	13	24	24	11	12
7	8	9	10	13	20	10	16	24	23	18	19
14	19	21	24	20	21	20	17	24	24	14	21
10	6	11	8	20	16	9	20	22	22	23	24
21	21	22	23	24	23	24	20	24	24	24	20

[a]Data from Josefawitz (1979).

4.2 Table 4.22 shows partial data for the response to a reasonable, fairly reasonable, and unreasonable request under four situations R_1, R_2, R_3, R_4. Scores ranged from 3 to 24, with 3 indicating a response of definitely comply and 24 indicating a response of definitely refuse. The subjects were classified into 3 classes prior to the testing: unassertive, fairly assertive, and highly assertive.

Is there a difference in the mean responses for the three classes of subjects: unassertive, fairly assertive, and highly assertive? Note that each response vector consists of the 12 responses by each subject. Hence the analysis is that of a CRD with three treatment groups.

4.3 The data in Tables 4.23 and 4.24 analyzed by Horton et al. (1968), give the clinical analysis of soil (nine characteristics) for three contours and four depths

Table 4.23 Original Data[a]

		Variable[b]								
Block	Group	1	2	3	4	5	6	7	8	9
1	1	5.40	0.188	0.92	215	16.35	7.65	0.72	1.14	1.09
2	1	5.65	0.165	1.04	208	12.25	5.15	0.71	0.94	1.35
3	1	5.14	0.260	0.95	300	13.02	5.68	0.68	0.60	1.41
4	1	5.14	0.169	1.10	248	11.92	7.88	1.09	1.01	1.64
1	2	5.14	0.164	1.12	174	14.17	8.12	0.70	2.17	1.85
2	2	5.10	0.094	1.22	129	8.55	6.92	0.81	2.67	3.18
3	2	4.70	0.100	1.52	117	8.74	8.16	0.39	3.32	4.16
4	2	4.46	0.112	1.47	170	9.49	9.16	0.70	3.76	5.14
1	3	4.37	0.112	1.07	121	8.85	10.35	0.74	5.74	5.73
2	3	4.39	0.058	1.54	115	4.73	6.91	0.77	5.85	6.45
3	3	4.17	0.078	1.26	112	6.29	7.95	0.26	5.30	8.37
4	3	3.89	0.070	1.42	117	6.61	9.76	0.41	8.30	9.21
1	4	3.88	0.077	1.25	127	6.41	10.96	0.56	9.67	10.64
2	4	4.07	0.046	1.54	91	3.82	6.61	0.50	7.67	10.07
3	4	3.88	0.055	1.53	91	4.98	8.00	0.23	8.78	11.26
4	4	3.74	0.053	1.40	79	5.86	10.14	0.41	11.04	12.15
1	5	5.11	0.247	0.94	261	13.25	7.55	0.61	1.86	2.61
2	5	5.46	0.298	0.96	300	12.30	7.50	0.68	2.00	1.98
3	5	5.61	0.145	1.10	242	9.66	6.76	0.63	1.01	0.76
4	5	5.85	0.186	1.20	229	13.78	7.12	0.62	3.09	2.85
1	6	4.57	0.102	1.37	156	8.58	9.92	0.63	3.67	3.24
2	6	5.11	0.097	1.30	139	8.58	8.69	0.42	4.70	4.63
3	6	4.78	0.122	1.30	214	8.22	7.75	0.32	3.07	3.67
4	6	6.67	0.083	1.42	132	12.68	9.56	0.55	8.30	8.10
1	7	3.96	0.059	1.53	98	4.80	10.00	0.36	6.52	7.72
2	7	4.00	0.050	1.50	115	5.06	8.91	0.28	7.91	9.78
3	7	4.12	0.086	1.55	148	6.16	7.58	0.16	6.39	9.07
4	7	4.99	0.048	1.46	97	7.49	9.38	0.40	9.70	9.13
1	8	3.80	0.049	1.48	108	3.82	8.80	0.24	9.57	11.57
2	8	3.96	0.036	1.28	103	4.78	7.29	0.24	9.67	11.42
3	8	3.93	0.048	1.42	109	4.93	7.47	0.14	9.65	13.32
4	8	4.02	0.039	1.51	100	5.66	8.84	0.37	10.54	11.57

Table 4.23 (*continued*)

		Variable[b]								
Block	Group	1	2	3	4	5	6	7	8	9
1	9	5.24	0.194	1.00	445	12.27	6.27	0.72	1.02	0.75
2	9	5.20	0.256	0.78	380	11.39	7.55	0.78	1.63	2.20
3	9	5.30	0.136	1.00	259	9.96	8.08	0.45	1.97	2.27
4	9	5.67	0.127	1.13	248	9.12	7.04	0.55	1.43	0.67
1	10	4.46	0.087	1.24	276	7.24	9.40	0.43	4.17	5.08
2	10	4.91	0.092	1.47	158	7.37	10.57	0.59	5.07	6.37
3	10	4.79	0.047	1.46	121	6.99	9.91	0.30	5.15	6.82
4	10	5.36	0.095	1.26	195	8.59	8.66	0.48	4.17	3.65
1	11	3.94	0.054	1.60	148	4.85	9.62	0.18	7.20	10.14
2	11	4.52	0.051	1.53	115	6.34	9.78	0.34	8.52	9.74
3	11	4.35	0.032	1.55	82	5.99	9.73	0.22	7.02	8.60
4	11	4.64	0.065	1.46	152	4.43	10.54	0.22	7.61	9.09
1	12	3.82	0.038	1.40	105	4.65	9.85	0.18	10.15	12.26
2	12	4.24	0.035	1.47	100	4.56	8.95	0.33	10.51	11.29
3	12	4.22	0.030	1.56	97	5.29	8.37	0.14	8.27	9.51
4	12	4.41	0.058	1.58	130	4.58	9.46	0.14	9.28	12.69

[a]From Russel et al. (1967).
[b]Variables and units are as follows:

1. pH
2. Total nitrogen (%)
3. Bulk density (gm/cm^3)
4. Total phosphorus (ppm)
5. Exchangeable + soluble calcium (me/100 gm)
6. Exchangeable + soluble magnesium (me/100 gm)
7. Exchangeable + soluble potassium (me/100 gm)
8. Exchangeable + soluble sodium (me/100 gm)
9. Conductivity (mmhos/cm at 25°C)

of soil. The area in question was divided into four blocks, and samples were taken randomly at depths of 0–10, 10–30, 30–60, and 60–90 cm for the three contours of top, slope, and depression.

(a) Test for a significant difference among the 12 groups at $\alpha = 0.10$.
(b) The design is a two-factor design; test for interaction and main effects.
(c) For the variables pH and total nitrogen calculate the residuals and perform univariate probability plots. If the data do not appear normal, do power or modulus transformations to produce approximately normal data.

Table 4.24 Group Numbers in Relation to Depth and Microtopographic Position

Microtopographic	Soil layer			
position	0–10 cm	10–30 cm	30–60 cm	60–90 cm
top: above 60-cm contour	1	2	3	4
slope: between 30- and 60-cm contours	5	6	7	8
depression: below 30-cm contour	9	10	11	12

Table 4.25 Activity Remaining[a]

	Blocks I and II		Blocks III and IV	
Treatment[b]	α	β	α	β
$P_1B_1A_1C_1$	881	1437	834	1236
$P_1B_1A_1C_2$	650	1024	494	1009
$P_1B_1A_2C_1$	191	574	257	837
$P_1B_1A_2C_2$	183	716	193	720
$P_1B_2A_1C_1$	289	579	178	563
$P_1B_2A_1C_2$	188	498	163	537
$P_1B_2A_2C_1$	225	527	370	762
$P_1B_2A_2C_2$	135	507	156	577
$P_2B_1A_1C_1$	1180	1364	1193	1528
$P_2B_1A_1C_2$	1039	1433	1146	1306
$P_2B_1A_2C_1$	466	906	890	1144
$P_2B_1A_2C_2$	781	1130	775	1078
$P_2B_2A_1C_1$	298	639	273	512
$P_2B_2A_1C_2$	238	475	254	499
$P_2B_2A_2C_1$	420	664	429	700
$P_2B_2A_2C_2$	350	686	389	660

[a] In tens of count per minute per milliliter of filtrate. Data from Barnett and Mead (1956).
[b] A subscript 1 denotes the lower level of the factor, and a subscript 2 indicates the higher level of the factor.

4.4 The data in Table 4.25 show a 2^4 experiment performed in four blocks of 8. Each treatment combination is represented twice. Both alpha and beta counts of radioactive waste were recorded as response variables. The four treatment factors were four types of chemicals: A, B, C, and P, added to the unfiltered waste. Factor A was an aluminum reagent, B a barium chloride solution, C a carbon slurry, and P either sodium hydroxide or hydrochloric acid. A and B were given in amounts of 0.4 and 2.5 g; C was given in amounts of 0.08 and 0.4 g. Finally, sodium chloride or hydrochloric acid was added to produce a pH of 6 or 10 (P).

(a) Ignoring the presence of blocks, analyze the data as a 2^4 factorial experiment with two replications.

(b) The following blocked design was actually used; eight tests could be run on a given day and PBAC interaction was confounded with blocks:

	Block I		
$P_1B_1A_2C_2$	$P_2B_2A_2C_2$	$P_1B_2A_1C_2$	$P_1B_1A_1C_1$
$P_1B_2A_2C_1$	$P_2B_2A_1C_1$	$P_2B_1A_1C_2$	$P_2B_1A_2C_1$
	Block II		
$P_1B_1A_1C_2$	$P_2B_1A_1C_1$	$P_2B_2A_1C_2$	$P_1B_1A_2C_1$
$P_2B_2A_2C_1$	$P_1B_2A_1C_1$	$P_1B_2A_2C_2$	$P_2B_1A_2C_2$

	Block III		
$P_2B_2A_1C_1$	$P_2B_1A_2C_1$	$P_1B_2A_2C_1$	$P_1B_1A_1C_1$
$P_1B_1A_2C_2$	$P_2B_2A_2C_2$	$P_1B_2A_1C_2$	$P_2B_1A_1C_2$
	Block IV		
$P_2B_1A_1C_1$	$P_1B_2A_2C_2$	$P_1B_1A_2C_1$	$P_1B_2A_1C_1$
$P_2B_2A_1C_2$	$P_2B_1A_2C_2$	$P_1B_1A_1C_2$	$P_2B_2A_2C_1$

Analyze the data, adjusting for the presence of blocks.

(c) Calculate the residuals for these data and perform a normal probability plot. Do the data appear normal? If not, use a power or modulus transform to obtain data that are approximately normally distributed. Analyze the transformed data.

Table 4.26

Subject	Left ear	Right ear	Subject	Left ear	Right ear
		Ordinary murderers			
1	59	59	11	66	63
2	60	65	12	56	56
3	58	62	13	62	64
4	59	59	14	66	68
5	50	48	15	65	66
6	59	65	16	61	60
7	62	62	17	60	64
8	63	62	18	60	57
9	68	72	19	58	60
10	63	66	20	58	59
		Lunatic criminals			
1	70	69	11	60	57
2	69	68	12	53	55
3	65	65	13	66	65
4	62	60	14	60	53
5	59	56	15	59	58
6	55	58	16	58	54
7	60	58	17	60	56
8	58	64	18	54	59
9	65	67	19	62	66
10	67	62	20	59	61
		Other criminals			
1	63	63	11	65	70
2	56	57	12	64	64
3	62	62	13	65	65
4	59	58	14	67	67
5	62	58	15	55	55
6	50	57	16	56	56
7	63	63	17	65	67
8	61	62	18	62	65
9	55	59	19	55	61
10	63	63	20	58	58

4.5 The data of Table 4.26 represent ear length measurements on "ordinary murderers," "lunatic criminals," and "other criminals" at Parkurst prison. Test at $\alpha = 0.05$ whether there is a difference in ear measurements for the three groups of criminals.

4.6 In Table 4.27 the response variables consist of average psychomotor scores for three days postradiation for different radiation strengths. (The data are given in greater detail in Problem 7.4.)

(a) Test for difference in the four treatment groups, ignoring the preradiation psychomotor ability.

(b) Test for differences in the four treatment groups adjusting for preradiation psychomotor ability. Justify any discrepencies.

(c) Compute the values of the observations within a group minus the mean of that group for the four variables measured. Perform four univariate normal probability plots. Make a power or modulus transformation to the data. Note that in this case, because the variables all measure psychomotor

Table 4.27

	Preradiation	Day 1	Day 2	Day 3
		Control group		
1	191	223	242	248
2	64	72	81	66
3	206	172	214	239
4	155	171	191	203
5	85	138	204	213
6	15	22	24	24
		25–50 r		
1	53	53	102	104
2	33	45	50	54
3	16	47	45	34
4	121	167	188	209
5	179	193	206	210
6	114	91	154	152
		75–100 r		
1	181	206	199	237
2	178	208	222	237
3	190	224	224	261
4	127	119	149	196
5	94	144	169	164
6	148	170	202	181
		125–250 r		
1	201	202	229	232
2	113	126	159	157
3	86	54	75	75
4	115	158	168	175
5	183	175	217	235
6	131	147	183	181

Table 4.28[a]

Diet	Data base	x_1	x_2	x_3	Diet	Data base	x_1	x_2	x_3
1	1	0.25	1.50	9.82	6	1	1.14	2.12	22.06
	2	0.61	1.82	9.97		2	2.04	3.03	22.88
	3	0.68	1.19	14.64		3	2.45	3.44	30.38
	4	0.69	2.19	14.64		4	2.29	3.14	24.41
	5	0.52	1.03	11.02		5	2.08	3.30	28.73
	6	0.63	1.23	9.70		6	1.59	2.56	18.20
2	1	1.86	2.86	49.66	7	1	3.48	0.66	16.99
	2	3.66	4.02	50.62		2	5.17	2.16	17.63
	3	1.84	3.16	31.00		3	2.41	3.43	21.49
	4	1.90	3.20	29.36		4	5.73	2.60	21.33
	5	2.02	3.33	30.47		5	1.87	2.69	20.12
	6	3.21	3.82	45.00		6	1.92	2.91	18.50
3	1	0.78	0.98	15.68	8	1	0.00	1.11	8.19
	2	1.24	1.88	16.25		2	0.66	1.81	8.95
	3	1.78	2.31	17.40		3	0.87	1.90	7.27
	4	1.58	1.90	16.74		4	1.16	1.92	6.14
	5	0.98	1.54	14.22		5	0.66	1.65	5.82
	6	1.21	2.18	16.80		6	0.64	1.81	8.80
4	1	0.59	1.22	19.56	9	1	0.38	0.74	10.81
	2	1.45	1.79	19.97		2	0.86	1.28	10.80
	3	1.66	2.07	23.26		3	1.02	1.59	12.88
	4	1.54	1.79	21.23		4	1.03	1.25	11.95
	5	1.69	1.99	22.64		5	0.80	1.20	11.11
	6	1.19	1.54	16.40		6	0.86	1.46	1.22
5	1	0.63	2.10	8.61	10	1	0.56	0.20	5.24
	2	1.60	3.28	9.56		2	0.60	0.43	5.24
	3	1.07	2.91	5.89		3	1.20	0.52	12.44
	4	1.32	2.96	6.64		4	0.44	0.32	3.55
	5	0.91	2.36	4.69		5	1.56	0.66	7.29
	6	0.85	2.66	6.60		6	0.63	0.43	4.30

[a] Data courtesy of Dr. B. Kazlowski, Nisonger Center, Columbus, Ohio.

scores, we should use a common value of λ in the transformation. Test for a difference in treatment means, adjusting for the covariable with the transformed data.

4.7 Table 4.28 gives the nutritional content of ten diets as analyzed by six data bases. The variables measured were x_1 (thiamin), x_2 (riboflavin), and x_3 (niacin).

(a) Test at $\alpha = 0.05$ for a difference in the analyses for the six data bases with diets considered as blocks. If there is a significant difference, which data bases differ and for which variables?

(b) Test for a Tukey-type interaction between diets and data bases.

Chapter 5
Multivariate Regression

5.1 Introduction

Simple linear regression investigates the linear relationship between a controlled variable x, called the independent variable, and a response variable y, called the dependent variable. The controlled variable x is often used to predict a response y. For example, x could be the log concentration of a toxic insecticide, and y could be the weight of trout. If k insecticides are used in different concentrations, one tries to find a linear relationship between y and their observed log concentrations $x_1,\ldots,x_k$. Such a procedure is called *multivariable* linear regression. *Multivariate* regression takes the problem one step further. Here many dependent variables $y_1,\ldots,y_p$ are measured and then regressed against the independent variables $x_1,\ldots,x_k$.

For example, p of the independent variables $x_1,\ldots,x_k$ could denote the concentration of p chemicals in a tank, and $y_1,\ldots,y_p$ could denote the amount of these chemicals in a fish in the tank after 2 weeks. In the next section the case of simple linear regression is studied. The results are then generalized to multivariable and multivariate regression.

5.2 Linear Regression

5.2.1 Parametric Regression

The simplest linear model takes the form

$$y=\beta_0+x\beta_1+\varepsilon, \tag{5.2.1}$$

where $\varepsilon \sim N(0,\sigma^2)$. In (5.2.1) x is designated the independent variable and the response y the dependent variable. This model can be generalized to the case of many independent variables and written in vector notation

$$y=\mathbf{x}'\boldsymbol{\beta}+\varepsilon,$$

where $\mathbf{x}'=(x_1,\ldots,x_k)$ represents the independent variables (possibly dummy variables), $\boldsymbol{\beta}$ is a k-vector of regression parameters, and $\varepsilon \sim N(0,\sigma^2)$. If a sample of N independent observations is taken with observations $y_1,\ldots,y_N$ and $\mathbf{x}_1,\ldots,\mathbf{x}_N$, then the model can be written

$$\mathbf{y}=X'\boldsymbol{\beta}+\boldsymbol{\varepsilon}, \tag{5.2.2}$$

where $\mathbf{y}'=(y_1,\ldots,y_N)$, $X=(\mathbf{x}_1,\ldots,\mathbf{x}_N)$, and $\boldsymbol{\varepsilon}'=(\varepsilon_1,\ldots,\varepsilon_N)$. The least squares solution for $\boldsymbol{\beta}$ in (5.2.2) is given by $\hat{\boldsymbol{\beta}}=(XX')^{-1}X\mathbf{y}$. The statistic has the properties

$$E(\hat{\boldsymbol{\beta}})=\boldsymbol{\beta} \qquad \text{and} \qquad \operatorname{cov}(\hat{\boldsymbol{\beta}})=\sigma^2(XX')^{-1}. \tag{5.2.3}$$

Since $\hat{\boldsymbol{\beta}}$ is a linear combination of $\mathbf{y}$, it is normally distributed with mean $\boldsymbol{\beta}$ and covariance matrix $\sigma^2(XX')^{-1}$. (See the appendix to this chapter.) To test hypotheses concerning $\boldsymbol{\beta}$, such as $\boldsymbol{\beta}=\mathbf{0}$ or $\beta_2=\beta_3=0$, we consider the more general hypothesis $C\boldsymbol{\beta}=C\boldsymbol{\beta}_0$, where C is $r\times k$ and $\boldsymbol{\beta}$ is $k\times 1$. To test $\boldsymbol{\beta}=\mathbf{0}$, we set $C=I_k$, and to test $\beta_i=0$ for some specified i, we set $C=(0,\ldots,0,1,0,\ldots,0)$ with the 1 in the ith position. For testing $\beta_1=\cdots=\beta_q=0$, $q\le k$, we get $C_{q\times k}=(I_q,0)$. To test H: $C\boldsymbol{\beta}=C\boldsymbol{\beta}_0$ vs A: $C\boldsymbol{\beta}\neq C\boldsymbol{\beta}_0$, we reject H if

$$\frac{(C\hat{\boldsymbol{\beta}}-C\boldsymbol{\beta}_0)'\left[C(XX')^{-1}C'\right]^{-1}(C\hat{\boldsymbol{\beta}}-C\boldsymbol{\beta}_0)}{\mathbf{y}'\left[I-X'(XX')^{-1}X\right]\mathbf{y}}\ge\frac{r}{N-k}F_{r,N-k,\alpha}, \tag{5.2.4}$$

where $F_{r,N-k,\alpha}$ is the upper $\alpha 100\%$ point of the F-distribution on r and $N-k$ df. Note that in (5.2.4) the numerator is the sum of squares due to the null hypothesis and the denominator is the sum of squares due to error. A $(1-\alpha)100\%$ confidence interval for a (single) linear combination of the regression parameters, such as $\mathbf{a}'\boldsymbol{\beta}$, is given by

$$\mathbf{a}'\hat{\boldsymbol{\beta}}\pm\left[(N-k)^{-1}\mathbf{a}'(XX')^{-1}\mathbf{a}\,\mathrm{SSE}\,F_{1,N-k,\alpha}\right]^{1/2}, \tag{5.2.5}$$

where $\mathrm{SSE}=\mathbf{y}'[\mathrm{I}-X(XX')^{-1}X]\mathbf{y}$. *Simultaneous* $(1-\alpha)100\%$ confidence intervals for $\mathbf{a}'\boldsymbol{\beta}$ for all possible values of $\mathbf{a}$ are given by

$$\mathbf{a}'\hat{\boldsymbol{\beta}}\pm\left[\mathbf{a}'(XX')^{-1}\mathbf{a}\,\mathrm{SSE}\,r/(N-k)F_{k,N-k,\alpha}\right]^{1/2}. \tag{5.2.6}$$

Consider a model

$$y=\beta_0+\beta_1x_1+\beta_2x_2+\varepsilon. \tag{5.2.7}$$

If we are interested in confidence intervals for survival time of fish for all weights and amounts of toxic substance, then we use (5.2.6) with $k=3$. In this case $\mathbf{a}$ may be represented $\mathbf{a}'=(1,x_1,x_2)$, and we are interested in confidence intervals for $\mathbf{a}'\boldsymbol{\beta}=\beta_0+\beta_1x_1+\beta_2x_2$. However, if we are interested in survival time for all amounts of toxic material but only for a typical fish of weight 5 oz, then we should be able to obtain shorter confidence intervals. In this case $\mathbf{a}$ can be represented $\mathbf{a}'=(1,5,x_2)$ for all values of x_2, or equivalently,

$$\mathbf{a}=\begin{pmatrix}1 & 0\\ 5 & 0\\ 0 & 1\end{pmatrix}\begin{pmatrix}1\\ x_2\end{pmatrix}. \tag{5.2.8}$$

In linear algebra terms, although $\mathbf{a}$ is a vector of length 3 the space spanned by all vectors of the form $\mathbf{a}'=(1,5,x_2)$ is only of dimension 2 because the second component is always 5 times the first. Statistically, if the vector $\mathbf{a}$ is restricted to the form

$$\mathbf{a}'=\mathbf{b}'C \tag{5.2.9}$$

for some fixed matrix $C: r\times k$ and an arbitrary vector $\mathbf{b}: r\times 1$, then $(1-\alpha)100\%$ simultaneous confidence intervals for $\mathbf{a}'\boldsymbol{\beta}$ are given by

$$\mathbf{a}'\hat{\boldsymbol{\beta}}\pm\left[\mathbf{a}'(XX')^{-1}\mathbf{a}\mathrm{SSE}(r/N-k)F_{r,N-k,\alpha}\right]^{1/2}. \tag{5.2.10}$$

For our example

$$C'=\begin{pmatrix}1 & 0\\ 5 & 0\\ 0 & 1\end{pmatrix} \quad\text{and}\quad \mathbf{b}=\begin{pmatrix}1\\ x_2\end{pmatrix}$$

are 3×2 and 2×1 respectively.

Hence $(1-\alpha)100\%$ confidence intervals for the survival time of fish weighing 5 oz for all values of amount of toxic substance are given by (5.2.10) with $r=2$, $k=3$, and $\mathbf{a}'=(1,5,x_2)$.

Example 5.2.1 Suppose we have the model of (5.2.1) with observations $(x_1,y_1),\ldots,(x_N,y_N)$. The model can be written

$$\mathbf{y}=X'\boldsymbol{\beta}+\boldsymbol{\epsilon},$$

where $\mathbf{y}'=y_1,\ldots,y_N$, $\boldsymbol{\beta}'=(\beta_0,\beta_1)$,

$$X'=\begin{pmatrix}1 & x_1\\ \vdots & \vdots\\ 1 & x_N\end{pmatrix}$$

and $\boldsymbol{\varepsilon}'=(\varepsilon_1,\ldots,\varepsilon_N)$. We then have

$$\begin{pmatrix}\hat{\beta}_0\\ \hat{\beta}_1\end{pmatrix}=(XX')^{-1}X\mathbf{y}=\begin{pmatrix}N & \Sigma x_i\\ \Sigma x_i & \Sigma x_i^2\end{pmatrix}^{-1}\begin{pmatrix}\Sigma y_i\\ \Sigma x_i y_i\end{pmatrix},$$

which can be simplified to

$$\hat{\beta}_1=\frac{\Sigma x_i y_i-(\Sigma x_i)(\Sigma y_i)/N}{\Sigma x_i^2-(\Sigma x_i)(\Sigma x_i)/N} \quad\text{and}\quad \hat{\beta}_0=\bar{y}-\hat{\beta}_1\bar{x},$$

where $\bar{y}=N^{-1}\Sigma_{i=1}^{N}y_i$ and $\bar{x}=N^{-1}\Sigma_{i=1}^{N}x_i$. These are the well-known least squares estimates of β_0 and β_1. The distributions of $\hat{\beta}_0$ and $\hat{\beta}_1$ are given by

$$\begin{pmatrix}\hat{\beta}_0\\ \hat{\beta}_1\end{pmatrix}\sim N_2\left(\begin{pmatrix}\beta_1\\ \beta_2\end{pmatrix},\ \frac{\sigma^2}{N\Sigma x_i^2-(\Sigma x_i)^2}\begin{pmatrix}\Sigma x_i^2 & -\Sigma x_i\\ -\Sigma x_i & N\end{pmatrix}\right).$$

To obtain a $(1-\alpha)100\%$ confidence interval for one linear combination of β_0 and β_1, such as $\beta_0+\beta_1 x_0$ for a given x_0, the expression for the confidence interval given in (5.2.5) can be simplified to

$$\hat{\beta}_0+\hat{\beta}_1 x_0 \pm \left\{\left[\frac{1}{N}+\frac{(x_0-\bar{x})^2}{\Sigma(x_i-\bar{x})^2}\right]\mathrm{SSE}\,F_{1,N-2,\alpha}\right\}^{1/2}.$$

However, if we want a $(1-\alpha)100\%$ confidence interval for the line $\beta_0+\beta_1 x$, that is, for all values of x, then the expression for the confidence interval given in (5.2.6) becomes

$$\hat{\beta}_0+\hat{\beta}_1 x \pm \left\{\left[\frac{1}{N}+\frac{(x-\bar{x})^2}{\Sigma(x_i-\bar{x})^2}\right]\mathrm{SSE}\frac{2}{N-2}F_{2,N-2,\alpha}\right\}^{1/2}.$$

Further generalizations will be made in Chapter 6.

Example 5.2.2 The following data represent the mean development per day y of the pupae of apple maggots (*Rhagoletis pumonella*) at different breeding temperatures x in degrees Celsius:

x	10	12	14	16	18	20	22	24	26	28	30
y	0.15	0.24	0.52	0.61	0.82	1.1	1.4	1.6	1.8	2.4	2.7

Fitting a quadratic model of the form

$$y=\beta_0+\beta_1 x+\beta_2 x^2+\varepsilon, \qquad \varepsilon\sim N(0,\sigma^2)$$

yields $\hat{\beta}_0=-0.0518$, $\hat{\beta}_1=-0.014$, $\hat{\beta}_2=0.0035$, and $\mathrm{SSE}=0.048$. The estimate of σ^2 is given by

$$\hat{\sigma}^2=\mathrm{SSE}/(11-3)=0.006,$$

and the F-test for H: $\beta_1=\beta_2=0$ vs A: (at least one $\beta_i\neq 0$, $i=1,2$) has a value of

$$F_{2,8}=601.9\geq F_{2,8,0.05}=4.07.$$

Hence there is certain to be a relationship between the development of maggots and their breeding temperature.

The design matrix X may be written

$$X=\begin{pmatrix}1&1&1&1&1&1&1&1&1&1&1\\ 10&12&14&16&18&20&22&24&26&28&30\\ 100&144&196&256&324&400&484&576&676&784&900\end{pmatrix}.$$

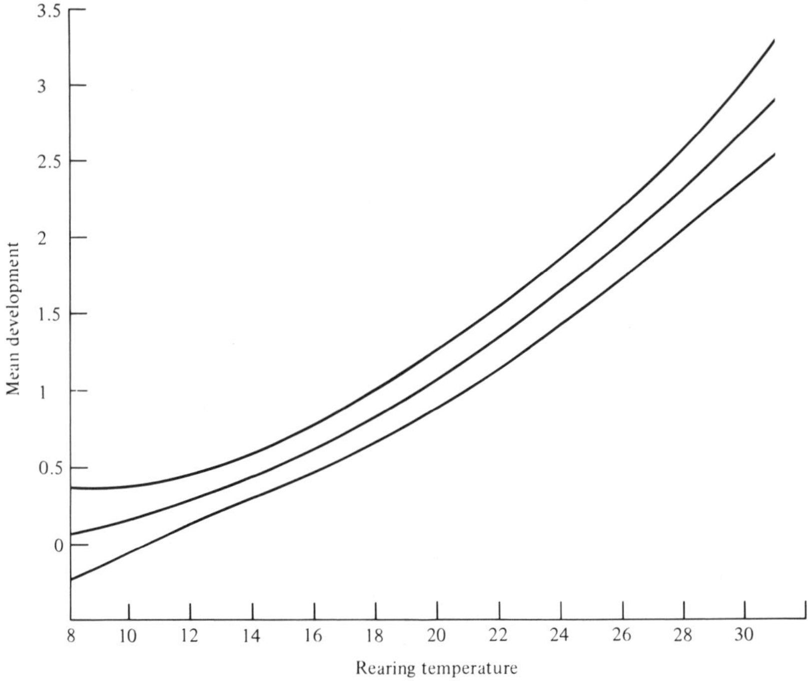

Figure 5.1 Confidence bands.

If we wish to plot the estimated regression line $\hat{y}=\hat{\beta}_0+\hat{\beta}_1x+\hat{\beta}_2x^2$ vs x on a graph and impose a 95% confidence band on the curve, then we are actually interested in finding simultaneous confidence intervals for $\mathbf{a}'\boldsymbol{\beta}$, where $\mathbf{a}'=(1, x, x^2)$ as x varies. From (5.2.6) such bands are given by

$$\hat{\beta}_0+\hat{\beta}_1x+\hat{\beta}_2x^2 \pm \left[(1, x, x^2)(XX')^{-1}\begin{pmatrix}1\\x\\x^2\end{pmatrix}\text{SSE}\frac{3}{8}F_{3,8,0.05}\right]^{1/2},$$

or

$$-0.0518-0.014x+0.0035x^2$$

$$\pm\left[(1, x, x^2)\begin{pmatrix}10.44 & -1.094 & 0.0262\\ -1.094 & 0.119 & -0.0029\\ 0.0262 & -0.0029 & 0.0000728\end{pmatrix}\begin{pmatrix}1\\x\\x^2\end{pmatrix}(0.07326)\right]^{1/2}.$$

Plots of the estimated regression line and the simultaneous 95% confidence bands are given in Fig. 5.1. Note that the bands are narrower in the neighborhood of the mean value of the breeding temperature (20°C).

5.3 Multivariate Regression Model

In sample surveys many variables are assigned values according to responses to a questionnaire. Variables such as age, socioeconomic status, religion, and marital status are designated independent, whereas variables measuring quantities such as authority patterns, desire for children, and homemaking skills are designated dependent. The relationship between one dependent variable and a single independent variable can be measured by a linear regression analysis. If there is more than one dependent variable, however, a multivariate regression must be performed; this model can be expressed in the general form

$$\mathbf{y} = \beta\mathbf{x} + \boldsymbol{\varepsilon},$$

where $\mathbf{y}$ is a vector of p dependent variables, $\mathbf{x}$ is a vector of q independent variables, $\boldsymbol{\varepsilon} \sim N_p(0, \Sigma)$, and β is a $p \times q$ matrix of regression coefficients. Given the observations $Y_{p\times N} = (\mathbf{y}_1, \ldots, \mathbf{y}_N)$ and $X_{q\times N} = (\mathbf{x}_1, \ldots, \mathbf{x}_N)$, the least squares estimate of β is given by

$$\hat{\beta} = YX'(XX')^{-1}. \tag{5.3.1}$$

The estimates $\hat{\beta}_{ij}$ of β_{ij} have the property of being jointly normally distributed with

$$E(\hat{\beta}_{ij}) = \beta_{ij}, \qquad \operatorname{var}(\hat{\beta}_{ij}) = \sigma_{ii}a_{jj}, \qquad \operatorname{cov}(\beta_{ij}\beta_{kl}) = \sigma_{ik}a_{jl}, \tag{5.3.2}$$

where $(a_{kl}) = (XX')^{-1}$ and $(\sigma_{ij}) = \Sigma$. An unbiased estimate of Σ is given by $(N-q)^{-1}Y[I - X'(XX')^{-1}X]Y'$; the maximum likelihood estimate has the divisor N in place of $N-q$. If we want to test for relationships between the dependent and independent variables, then we test the hypothesis H: $C_1\beta C_2 = 0$ vs A: $C_1\beta C_2 \neq 0$ for some $r \times p$ matrix C_1 and some $q \times t$ matrix C_2. It will be assumed that C_1 and C_2 are of rank $r \leq p$ and $t \leq q$. The estimate of $C_1\beta C_2$ is given by

$$C_1\hat{\beta}C_2 = C_1YX'(XX')^{-1}C_2. \tag{5.3.3}$$

Under H: $C_1\beta C_2 = 0$, the statistic

$$W = (C_1\hat{\beta}C_2)\left[C_2'(XX')^{-1}C_2\right]^{-1}C_2'\hat{\beta}'C_1' \tag{5.3.4}$$

is an unbiased estimate of $tC_1\Sigma C_1'$. The statistic

$$V = C_1Y\left(I - X'(XX')^{-1}X\right)Y'C_1' \tag{5.3.5}$$

is also an unbiased estimate of $(N-r)C_1\Sigma C_1'$. On comparing V and W, we reject H if $V^{-1}W$ is too large. More exactly, the likelihood ratio test rejects H if

$$|V|/|V+W| = U_{r,t,n} \leq U_{r,t,n,\alpha}, \tag{5.3.6}$$

where $n = N-q$ and where $U_{r,t,n,\alpha}$ is the $\alpha100\%$ point of the U-statistic as given in Section 4.2. The asymptotic expansion for $U_{r,t,n}$ is given by

$$P\{-[n-\tfrac{1}{2}(r-t+1)]\ln U_{r,t,n}\geq z\}$$
$$=P\{\chi_f^2\geq z\}+n^{-2}\gamma_2\left(P\{\chi_{f+4}^2\geq z\}-P\{\chi_f^2\geq z\}\right)+O(n^{-3}),\tag{5.3.7}$$

where $f=rt$ and $\gamma_2=f(r^2+t-5)/48$. This approximation can be used to obtain p-values. If the observed value of the likelihood ratio or Wilks's statistic is $U=U_0$, the p-value would be obtained from (5.3.7) with $z=-[n-\frac{1}{2}(r-t+1)]\ln U_0$.

Let us consider some special cases of the hypothesis H: $C_1BC_2=0$. Let $C_1=I_p$ and $C_2=I_q$; then H becomes H: $\beta=0$. Further, let $\beta=(\beta_1\vdots\beta_2)$ where β_1 is $(p\times q_1)$, β_2 is $(p\times q_2)$, and $q_1+q_2=q$. Choosing

$$C_1=I_p \qquad \text{and} \qquad C_2=\begin{pmatrix} I_{q_1} \\ 0_{q_2,q_1} \end{pmatrix}$$

where $0_{q_2,q_1}$ is a zero matrix of order $q_2\times q_1$, we obtain H: $\beta_1=0$. Thus by a proper choice of C_1 and C_2 any hypothesis of interest can be generated.

Simultaneous $(1-\alpha)100\%$ confidence limits for $\mathbf{a}'\hat{\beta}\mathbf{b}$, where $\mathbf{a}$ and $\mathbf{b}$ can be written $\mathbf{a}'=\mathbf{a}^{*\prime}C_1$ and $\mathbf{b}=C_2\mathbf{b}^*$ for some $\mathbf{a}^*$ and $\mathbf{b}^*$, are given by

$$\mathbf{a}'\hat{\beta}\mathbf{b}\pm\left[x_\alpha(1-x_\alpha)^{-1}\mathbf{a}'V\mathbf{a}\mathbf{b}'(XX')^{-1}\mathbf{b}\right]^{1/2},\tag{5.3.8}$$

where the values of x_α are obtained from Table A.5 with the parameters

$$p_0=r, \qquad m_0=t, \qquad \text{and} \qquad n=N-q,\tag{5.3.9}$$

for C_1 an $r\times p$ matrix and C_2 a $q\times t$ matrix.

Note that by restricting our choices of $\mathbf{a}$ and $\mathbf{b}$ we reduce the width of the confidence intervals through the value of x_α, the only quantity in (5.3.8) involving the restrictions.

The testing and estimating problems of the previous sections can also be analyzed using the general linear model. For example, the model for the completely randomized design of Section 4.2 can be written

$$\mathbf{y}_{ij}=\boldsymbol{\mu}+\boldsymbol{\tau}_j+\boldsymbol{\varepsilon}_{ij}, \qquad j=1,\ldots,t, \quad i=1,\ldots,N_j,$$

where $N_t\boldsymbol{\tau}_t=-\Sigma_{j=1}^{t-1}N_j\boldsymbol{\tau}_j$. Equivalently we can write

$$Y_{p\times N}=\beta_{p\times t}X_{p\times N}+\varepsilon_{p\times N},$$

where

$$Y=\left(\mathbf{y}_{11},\ldots,\mathbf{y}_{N_1 1},\ldots,\mathbf{y}_{1t},\ldots,\mathbf{y}_{N_t t}\right), \qquad \beta=(\boldsymbol{\mu},\boldsymbol{\tau}_1,\ldots,\boldsymbol{\tau}_{t-1}),\tag{5.3.10}$$

and

$$X=\left(\begin{array}{c|c|c|c|c} \mathbf{e}'_{N_1} & \mathbf{e}'_{N_2} & \cdots & \mathbf{e}'_{N_{t-1}} & \mathbf{e}'_{N_t} \\ \hline \mathbf{e}'_{N_1} & 0 & \cdots & 0 & -\mathbf{e}'_{N_t}\left(\dfrac{N_1}{N_t}\right) \\ \hline 0 & \mathbf{e}'_{N_2} & \cdots & 0 & -\mathbf{e}'_{N_t}\left(\dfrac{N_2}{N_t}\right) \\ \hline 0 & 0 & \cdots & 0 & \vdots \\ \hline 0 & 0 & \cdots & \mathbf{e}'_{N_{t-1}} & -\mathbf{e}'_{N_t}\left(\dfrac{N_{t-1}}{N_t}\right) \end{array}\right).$$

The problem of testing for equal treatment effects can be summarized as

$$H\colon \boldsymbol{\tau}_1=\cdots=\boldsymbol{\tau}_{t-1}=0 \qquad \text{vs} \qquad A\colon \boldsymbol{\tau}_j\neq 0, \qquad 1\leq j\leq t-1,$$

or, in terms of the parameter matrix β,

$$H\colon \beta C=0 \qquad \text{vs} \qquad A\colon \beta C\neq 0,$$

where

$$C_{t\times(t-1)}=\begin{pmatrix} \mathbf{0}' \\ I_{t-1} \end{pmatrix}.$$

The test statistic for this problem in the general linear model can be shown to reduce to the statistic given in Section 4.2. for this special case of a completely randomized design. Similar models can be obtained for the randomized complete block and Latin square designs.

Example 5.3.1 Ranek et al. (1975) investigated the presence of diploid, tetraploid, and octaploid nuclei in liver cells. Table 5.1 gives the estimated frequencies of each of these three types of liver nuclei for ten patients along with the patients' age and sex. Ranek et al. showed that the percentage of diploid nuclei is affected by age. A regression of the percentage of diploid nuclei y vs age x produced the following regression curve:

$$y=\hat{\beta}_0+\hat{\beta}_1 x,$$

where $\hat{\beta}_0=99.2$ and $\hat{\beta}_1=-0.20$. To test $H\colon \beta_1=0$ vs $A\colon \beta_1\neq 0$, one rejects H if

$$F_{1,N-2}=\hat{\beta}_1^2\left[\Sigma(x_i-\bar{x})^2\right]\left[\text{SSE}/(N-2)\right]^{-1}\geq F_{1,N-2,\alpha}=F_{1,8,0.05}=5.32.$$

The observed value of F is 5.78. Hence the hypothesis that the percentage of diploid nuclei is not affected by age is rejected. A 95% confidence interval for the expected percentage of diploid nuclei for a patient of age 50 is given

Table 5.1

Patient	Age (years)	Sex	Estimated frequencies		
			Diploid	Tetraploid	Octaploid
1	22	F	93.6	5.7	0.7
2	30	F	91.8	7.4	0.8
3	36	F	99.4	0.3	0.3
4	50	F	85.2	14.2	0.6
5	57	F	90.4	9.2	0.4
6	58	F	86.4	12.2	1.4
7	19	M	95.6	4.4	0.0
8	50	M	89.0	10.8	0.2
9	53	M	84.3	14.7	1.0
10	62	M	90.9	8.4	0.7

by

$$\hat{y} = \hat{\beta}_0 + \hat{\beta}_1 x_0 \pm \left[F_{1,8,0.05} \left(\frac{1}{N} + \frac{(x_0 - \bar{x})^2}{\Sigma (x_i - \bar{x})^2} \right) s_e^2 \right]^{1/2}$$

$$= 99.2 - 0.20 \times 50 \pm \left\{ 5.32 \left[\tfrac{1}{10} - (50 - 43.7)^2 / 2210 \right] 14.66 \right\}^{1/2}$$

$$= 89.2 \pm 3.03,$$

where $s_e^2 = \text{SSE}/(N-2)$.

At this point we caution the reader about the indiscriminate use of multivariate techniques. Suppose we perform a multivariate regression of the percent diploid and percent tetraploid cells vs the age of the patient. The repsonse matrix Y may be written

$$Y = \begin{pmatrix} 93.6 & 91.8 & 99.4 & 85.2 & 90.4 & 86.4 & 95.6 & 89.0 & 84.3 & 90.9 \\ 5.7 & 7.4 & 0.3 & 14.2 & 9.2 & 12.2 & 4.4 & 10.8 & 14.7 & 8.4 \end{pmatrix}.$$

The design matrix is written

$$X = \begin{pmatrix} 1 & 1 & 1 & 1 & 1 & 1 & 1 & 1 & 1 & 1 \\ 22 & 30 & 36 & 50 & 57 & 58 & 19 & 50 & 53 & 62 \end{pmatrix},$$

and the regression parameters are

$$\beta = \begin{pmatrix} \beta_{11} & \beta_{12} \\ \beta_{21} & \beta_{22} \end{pmatrix}.$$

We wish to test the hypothesis of a linear relationship between age and tetraploid and diploid nuclei. That is, we wish to test H: $\beta_{12} = \beta_{22} = 0$ vs A: (at least one of β_{12} or β_{22} is zero). Equivalently we test H: $C_1 \beta C_2 = 0$ vs $C_1 \beta C_2 \neq 0$, where

$$C_1 = \begin{pmatrix} 1 & 0 \\ 0 & 1 \end{pmatrix} \quad \text{and} \quad C_2 = \begin{pmatrix} 0 \\ 1 \end{pmatrix}.$$

The value of the U-statistic from (5.3.6) is calculated to be $U_{2,1,8}=0.579$. From Corollary 4.2.3 we have that $7(1-0.579)/(2\times0.579)=2.55=F_{2,7}$. Hence the observed significance level is $P(F_{2,7}>2.55)=0.1475$. Hence we do not reject H. The discrepancy between the univariate and multivariate results occurs because we are performing multivariate tests on two variables that are essentially measuring the same quantity. That is, the percent tetraploid nuclei is 100 minus the percent of diploid nuclei. The correlation between these two variables is -0.995. Hence it is not an advantage to perform multivariate tests on as many variables as possible. The experimenter should choose only those variables of interest or use data reducing techniques such as principal component analysis (to be discussed in Chapter 10).

Example 5.3.2 The data in Table 5.2 were obtained from an experiment in the Department of Zoology at the University of Guelph. Twenty-five tanks of 20 trout were given various dosages of copper in milligrams per liter. The average weight of the fish was also recorded. The proportions (p_i, $i=1,\ldots,5$) of fish dead after 8, 14, 24, 36, and 48 hours were recorded. The response variables p_i were then transformed by an arcsine transformation,

$$Y_i=\arcsin\sqrt{p_i}\,,$$

to stabilize the variance. The following model was estimated:

$$Y_{5\times25}=\beta_{5\times3}X_{3\times25}+\varepsilon_{5\times25}.$$

The estimated matrix of regression coefficients was obtained from program 5 in Appendix II. The estimated values of β were

$$\hat{\beta}=YX'(XX')^{-1}=\begin{pmatrix} -94.601 & 19.275 & -24.683\\ -229.176 & 43.667 & -20.027\\ -258.803 & 48.990 & 3.322\\ -246.048 & 47.327 & 5.224\\ -232.958 & 45.098 & 8.239 \end{pmatrix}.$$

The within-covariance matrix is

$$\begin{pmatrix} 52.543 & 19.752 & -12.174 & -35.589 & -31.567\\ 19.752 & 59.479 & 36.992 & 35.111 & 33.614\\ -12.174 & 36.992 & 183.144 & 182.365 & 162.527\\ -35.589 & 35.111 & 182.365 & 216.453 & 196.113\\ -31.567 & 33.614 & 162.527 & 196.113 & 182.694 \end{pmatrix},$$

and the correlation matrix is

$$\begin{pmatrix} 1 & 0.353 & -0.124 & -0.334 & -0.322\\ 0.353 & 1 & 0.354 & 0.309 & 0.322\\ -0.124 & 0.354 & 1 & 0.916 & 0.889\\ -0.334 & 0.309 & 0.916 & 1 & 0.986\\ -0.322 & 0.322 & 0.889 & 0.986 & 1 \end{pmatrix}.$$

Table 5.2

		Y'		
Y_1	Y_2	Y_3	Y_4	Y_5
0	0	30	30	30
0	18.435	33.211	33.211	33.211
0	45	60	71.565	71.565
22.786	53.729	90	90	90
42.130	90	90	90	90
0	12.921	26.565	26.565	26.565
12.921	18.435	33.211	33.211	33.211
12.921	42.130	77.079	90	90
18.435	56.789	90	90	90
26.565	67.214	90	90	90
0	0	0	0	12.921
0	12.921	22.786	30	33.211
0	22.786	77.079	77.079	77.079
0	47.869	77.079	90	90
18.435	67.213	90	90	90
0	0	0	12.921	18.435
0	12.921	22.786	26.565	30
18.435	42.130	77.079	77.079	77.079
18.435	56.789	90	90	90
36.271	77.079	90	90	90
0	12.921	26.565	26.565	26.565
0	0	22.786	30	30
0	39.231	71.565	90	90
12.921	53.729	90	90	90
33.211	67.213	90	90	90

As can be seen from the estimates of β and the correlation matrix, observations after 36 and 48 hours are highly correlated, and the regression coefficients are roughly the same. This is owing to the fishes' acclimatization to the toxic substance. It would be of interest to test the effect on the death rate of the average weight of the fish in the tank. Writing $\beta = (\boldsymbol{\beta}_1, \boldsymbol{\beta}_2, \boldsymbol{\beta}_3)$, we are interested in testing the hypothesis H: $\boldsymbol{\beta}_3 = 0$ vs A: $\boldsymbol{\beta}_3 \neq 0$. This is equivalent to testing the hypothesis

$$H: \mathbf{C}_1 \beta \mathbf{C}_2 = 0 \qquad \text{vs} \qquad A: \mathbf{C}_1 \beta \mathbf{C}_2 \neq 0,$$

where

$$\mathbf{C}_1 = \mathbf{I}_5 \qquad \text{and} \qquad \mathbf{C}_2 = \begin{pmatrix} 0 \\ 0 \\ 1 \end{pmatrix},$$

Table 5.2 (*continued*)

	X′	
	log dose X_1	Average weight (g) X_2
1	5.5984	0.6695
1	6.0161	0.6405
1	6.4134	0.729
1	6.8459	0.77
1	7.2793	0.5655
1	5.5984	0.782
1	6.0161	0.812
1	6.4135	0.8215
1	6.8459	0.869
1	7.2793	0.8395
1	5.5984	0.8615
1	6.0161	0.9045
1	6.4135	1.028
1	6.8459	1.0445
1	7.2793	1.0455
1	5.5984	0.6195
1	6.0161	0.5305
1	6.4135	0.597
1	6.8459	0.6385
1	7.2793	0.6645
1	5.5984	0.5685
1	6.0162	0.604
1	6.4135	0.6325
1	6.8459	0.6845
1	7.2793	0.723

The U-statistic has a value of $U_{r,t,n}=U_{5,1,22}=0.34$. One could use an exact F-value for U with $t=1$. That is,

$$\frac{n+1-r}{r}\times\frac{1-U_{r,1,n}}{U_{r,1,n}}=F_{r,n+1-r}.$$

In this case

$$F_{5,18}=\frac{18}{5}\times\frac{0.34}{0.66}=1.90\leq F_{5,18,0.05}=2.77.$$

Hence we can conclude that the average weight of the fish did not have an effect on death rate. The problem of also fitting a reduced polynomial in *time* to this data is discussed in Section 6.3.

Table 5.3[a]

Group 1		Control diet (0.1% P)						Treatment diet (0.25% P)					
Rat	Initial weight	1	2	3 (days)	4	5	6	1	2	3 (days)	4	5	6
1	254	12.7	10.4	5.1	8.6	7.1	9.7	3.0	4.9	3.8	5.5	6.3	5.3
2	262	7.2	8.4	8.5	6.8	6.3	4.5	7.9	7.0	7.7	8.7	7.1	11.4
3	301	14.8	13.9	8.4	7.3	8.0	10.4	6.0	7.4	7.2	10.5	12.7	8.9
4	311	5.6	10.2	7.8	6.1	6.4	16.5	16.7	12.8	11.6	15.9	10.6	4.5
5	290	13.9	12.1	8.8	8.8	8.1	7.8	4.8	6.5	4.1	7.1	7.0	7.6
6	300	10.4	11.2	12.5	7.0	6.9	6.9	6.0	7.3	1.2	9.0	10.8	8.1
7	306	16.6	17.8	14.0	6.8	5.9	5.3	3.7	2.7	1.0	7.7	7.4	11.5
8	286	13.9	14.3	5.9	7.7	9.2	5.7	7.3	6.1	9.2	9.7	7.3	9.0

Group 2		Control diet (0.1% P)						Treatment diet (0.65% P)					
Rat	Initial weight	1	2	3 (days)	4	5	6	1	2	3 (days)	4	5	6
1	275	11.9	7.0	5.9	6.1	0.8	5.1	4.2	9.7	10.1	7.7	15.3	10.9
2	282	10.7	11.3	4.4	3.9	4.7	5.3	6.6	8.1	9.9	8.8	11.4	10.1
3	256	10.1	6.9	7.8	6.4	9.5	7.9	5.7	9.8	4.7	5.9	4.5	7.6
4	276	10.8	5.2	1.3	1.3	2.1	6.3	6.6	8.9	12.9	15.0	13.0	8.6
5	337	14.7	14.4	11.6	7.4	7.8	14.8	9.7	9.9	8.9	15.5	11.0	5.0
6	296	9.7	12.1	5.2	9.1	9.7	5.2	8.8	6.5	7.3	5.7	6.0	12.2
7	309	5.5	7.1	7.8	3.1	1.5	8.4	12.6	13.9	4.2	11.1	16.0	8.9
8	296	13.1	6.5	1.3	0.9	0.8	0.5	0.9	10.7	13.4	7.9	9.8	12.9

Group 3		Control diet						Treatment diet (1.3% P)					
Rat	Initial weight	1	2	3 (days)	4	5	6	1	2	3 (days)	4	5	6
1	275	8.8	17.7	11.5	6.6	5.4	12.0	2.3	0.6	1.0	7.0	9.8	4.8
2	292	8.3	3.2	5.2	8.9	4.3	4.4	8.1	11.8	7.3	7.0	10.3	9.5
3	338	16.2	11.9	10.2	15.6	15.3	13.9	5.6	2.0	0.0	0.8	3.6	8.9
4	248	7.7	4.9	11.7	12.7	13.2	10.7	9.0	7.5	0.7	1.4	0.4	0.3
5	315	14.5	14.0	16.9	8.4	13.1	9.8	6.6	6.0	9.7	9.8	5.1	6.0
6	295	11.6	2.5	5.5	4.5	5.8	8.6	6.0	16.4	9.9	8.8	9.4	8.4
7	312	5.3	6.1	1.5	4.1	6.2	2.1	11.7	14.7	13.4	7.0	4.2	13.3
8	286	11.2	11.0	5.7	8.1	10.0	8.1	11.2	12.1	9.2	6.6	8.9	10.4

Group 4		Control diet						Treatment diet (1.71% P)					
Rat	Initial weight	1	2	3 (days)	4	5	6	1	2	3 (days)	4	5	6
1	275	13.5	9.7	12.3	13.4	14.0	6.1	3.3	1.2	1.9	1.3	2.8	5.8
2	270	11.6	2.4	9.7	14.0	10.8	10.3	5.7	15.5	3.8	1.8	6.3	1.5
3	290	10.0	14.8	9.1	9.6	8.2	9.3	7.2	4.1	7.0	3.4	6.5	1.3
4	260	12.3	16.2	6.6	9.2	8.3	12.6	8.1	4.8	7.9	8.2	9.9	5.2
5	302	13.6	14.9	9.3	10.2	11.5	15.8	2.7	5.6	2.7	4.3	3.0	1.3
6	284	12.8	13.2	11.6	11.5	11.1	10.5	6.2	6.4	2.5	4.1	4.9	4.0
7	280	10.9	14.3	10.8	9.6	13.2	10.0	6.0	2.2	2.0	7.0	4.1	4.2
8	329	8.3	10.5	7.5	10.6	8.5	6.2	13.1	4.1	5.1	6.2	8.3	6.0

[a] Data courtesy of M. Hedley, Department of Nutrition, University of Guelph.

Example 5.3.3 Consider the amount of food eaten from two diets, a control and a treatment diet, containing phosphorus, over six days by 32 rats randomly assigned to four groups (Table 5.3). It was of interest to relate the amount of each diet eaten by a rat to the amount of phosphorus in the treatment diet. In this example we are measuring two characteristics (the amount of food consumed for two diets) over six time intervals; the MANOVA model can handle this type of situation. The transpose of the observation matrix consists of 32 vectors of the 12 responses for each rat; that is,

$$
Y' = \begin{bmatrix}
12.7 & 10.4 & 5.1 & 8.6 & 7.1 & 9.7 & 3.0 & 4.9 & 3.8 & 5.5 & 6.3 & 5.3 \\
7.2 & 8.4 & 8.5 & 6.8 & 8.3 & 4.5 & 7.9 & 7.0 & 7.7 & 8.7 & 7.1 & 11.4 \\
14.8 & 13.9 & 8.4 & 7.3 & 8.0 & 10.4 & 6.0 & 7.4 & 7.2 & 10.5 & 12.7 & 8.9 \\
5.6 & 10.2 & 7.8 & 6.1 & 6.4 & 16.5 & 16.7 & 12.8 & 11.6 & 15.9 & 10.6 & 4.5 \\
13.9 & 12.1 & 8.8 & 8.8 & 8.1 & 7.8 & 4.8 & 6.5 & 4.1 & 7.1 & 7.0 & 7.6 \\
10.4 & 11.2 & 12.5 & 7.0 & 6.9 & 6.9 & 6.0 & 7.3 & 1.2 & 9.0 & 10.8 & 8.1 \\
16.6 & 17.8 & 14.0 & 6.8 & 5.9 & 5.3 & 3.7 & 2.7 & 1.0 & 7.7 & 7.4 & 11.5 \\
13.9 & 14.3 & 5.9 & 7.7 & 9.2 & 5.7 & 7.3 & 6.1 & 9.2 & 9.7 & 7.3 & 9.0 \\
11.9 & 7.0 & 5.9 & 6.1 & 0.8 & 5.1 & 4.2 & 9.7 & 10.1 & 7.7 & 15.3 & 10.9 \\
10.7 & 11.3 & 4.4 & 3.9 & 4.7 & 5.3 & 6.6 & 8.1 & 9.9 & 8.8 & 11.4 & 10.1 \\
10.1 & 6.9 & 7.8 & 6.4 & 9.5 & 7.9 & 5.7 & 9.8 & 4.7 & 5.9 & 4.5 & 7.6 \\
10.8 & 5.2 & 1.3 & 1.3 & 2.1 & 6.3 & 6.6 & 8.9 & 12.9 & 15.0 & 13.0 & 8.6 \\
14.7 & 14.4 & 11.6 & 7.4 & 7.8 & 14.8 & 9.7 & 9.9 & 8.9 & 5.5 & 11.0 & 5.0 \\
9.7 & 12.1 & 5.2 & 9.1 & 9.7 & 5.2 & 8.8 & 6.5 & 7.3 & 5.7 & 6.0 & 12.2 \\
5.5 & 7.1 & 7.8 & 3.1 & 1.5 & 8.4 & 12.6 & 13.9 & 4.2 & 11.1 & 16.0 & 8.9 \\
13.1 & 6.5 & 1.3 & 0.9 & 0.8 & 0.5 & 0.9 & 10.7 & 13.4 & 7.9 & 9.8 & 12.9 \\
8.8 & 17.7 & 11.5 & 6.6 & 5.4 & 12.0 & 2.3 & 0.6 & 1.0 & 7.0 & 9.8 & 4.8 \\
8.3 & 3.2 & 5.2 & 8.9 & 4.3 & 4.4 & 8.1 & 11.8 & 7.3 & 7.0 & 10.3 & 9.5 \\
16.2 & 11.9 & 10.2 & 15.6 & 15.3 & 13.9 & 5.6 & 2.0 & 0.0 & 0.8 & 3.6 & 8.9 \\
7.7 & 4.9 & 11.7 & 12.7 & 13.2 & 10.7 & 9.0 & 7.5 & 0.7 & 1.4 & 0.4 & 0.3 \\
14.5 & 14.0 & 16.9 & 8.4 & 13.1 & 9.8 & 6.6 & 6.0 & 9.7 & 9.8 & 5.1 & 6.0 \\
11.6 & 2.5 & 5.5 & 4.5 & 5.8 & 8.6 & 6.0 & 9.7 & 9.8 & 8.8 & 9.4 & 8.4 \\
5.3 & 6.1 & 1.5 & 4.1 & 6.2 & 2.1 & 11.7 & 14.7 & 13.4 & 7.0 & 4.2 & 13.3 \\
11.2 & 11.0 & 5.7 & 8.1 & 10.0 & 8.1 & 11.3 & 12.1 & 9.2 & 6.6 & 8.9 & 10.4 \\
13.5 & 9.7 & 11.3 & 13.4 & 14.0 & 6.1 & 3.3 & 1.2 & 1.9 & 1.3 & 2.8 & 5.8 \\
11.6 & 2.4 & 9.7 & 14.0 & 10.8 & 10.3 & 5.7 & 15.5 & 3.8 & 1.8 & 6.3 & 1.5 \\
10.0 & 14.8 & 9.1 & 9.6 & 8.2 & 9.3 & 7.2 & 4.1 & 7.0 & 3.4 & 6.5 & 1.3 \\
12.3 & 16.2 & 6.6 & 9.2 & 8.3 & 12.6 & 8.1 & 4.8 & 7.9 & 8.2 & 9.9 & 5.2 \\
13.6 & 14.9 & 9.3 & 10.2 & 11.5 & 15.8 & 2.7 & 5.6 & 2.7 & 4.3 & 3.0 & 1.3 \\
12.8 & 13.2 & 11.6 & 11.5 & 11.1 & 10.5 & 6.2 & 6.4 & 2.5 & 4.1 & 4.9 & 4.0 \\
10.9 & 14.3 & 10.8 & 9.6 & 13.2 & 10.0 & 6.0 & 2.2 & 2.0 & 7.0 & 4.1 & 4.2 \\
8.3 & 10.5 & 7.5 & 10.6 & 8.5 & 6.2 & 13.1 & 4.1 & 5.1 & 6.2 & 8.3 & 6.0
\end{bmatrix}
$$

The model we fit is given by

$$E(y_i) = \beta_{i0} + \beta_{i1}x + \beta_{i2}x^2 + \beta_{i3}x^3 + \beta_{i4}w, \qquad i = 1, \ldots, 12,$$

where x represents the amount of phosphorus given and w is the covariable initial weight. The transpose of the design matrix for this model is given by

$$X' = \begin{bmatrix} 1 & 0.25 & 0.0625 & 0.015625 & 254 \\ 1 & 0.25 & 0.0625 & 0.015625 & 262 \\ 1 & 0.25 & 0.0625 & 0.015625 & 301 \\ 1 & 0.25 & 0.0625 & 0.015625 & 311 \\ 1 & 0.25 & 0.0625 & 0.015625 & 290 \\ 1 & 0.25 & 0.0625 & 0.015625 & 300 \\ 1 & 0.25 & 0.0625 & 0.015625 & 306 \\ 1 & 0.25 & 0.0625 & 0.015625 & 286 \\ 1 & 0.65 & 0.4225 & 0.274625 & 275 \\ 1 & 0.65 & 0.4225 & 0.274625 & 282 \\ 1 & 0.65 & 0.4225 & 0.274625 & 256 \\ 1 & 0.65 & 0.4225 & 0.274625 & 276 \\ 1 & 0.65 & 0.4225 & 0.274625 & 337 \\ 1 & 0.65 & 0.4225 & 0.274625 & 296 \\ 1 & 0.65 & 0.4225 & 0.274625 & 309 \\ 1 & 0.65 & 0.4225 & 0.274625 & 296 \\ 1 & 1.3 & 1.69 & 2.197 & 275 \\ 1 & 1.3 & 1.69 & 2.197 & 192 \\ 1 & 1.3 & 1.69 & 2.197 & 338 \\ 1 & 1.3 & 1.69 & 2.197 & 248 \\ 1 & 1.3 & 1.69 & 2.197 & 315 \\ 1 & 1.3 & 1.69 & 2.197 & 295 \\ 1 & 1.3 & 1.69 & 2.197 & 312 \\ 1 & 1.3 & 1.69 & 2.197 & 286 \\ 1 & 1.71 & 2.9241 & 5.000211 & 275 \\ 1 & 1.71 & 2.9241 & 5.000211 & 270 \\ 1 & 1.71 & 2.9241 & 5.000211 & 290 \\ 1 & 1.71 & 2.9241 & 5.000211 & 260 \\ 1 & 1.71 & 2.9241 & 5.000211 & 302 \\ 1 & 1.71 & 2.9241 & 5.000211 & 284 \\ 1 & 1.71 & 2.9241 & 5.000211 & 280 \\ 1 & 1.71 & 2.9241 & 5.000211 & 329 \end{bmatrix}$$

It was felt that the response would be quadratic in x, the amount of phosphorus as in the treatment diet. Hence with the model

$$E(Y) = \beta X$$

we test the hypothesis of no cubic term:

$$H: \beta\mathbf{C} = 0 \qquad \text{vs} \qquad A: \beta\mathbf{C} \neq 0$$

where

$$\mathbf{C} = \begin{pmatrix} 0 \\ 0 \\ 0 \\ 1 \\ 0 \end{pmatrix}.$$

The U-statistic, using the formulas in (5.3.4)–(5.3.6), has a value of

$$U_{12,1,27} = 0.648.$$

From Corollary 4.2.3,

$$U_{12,1,27} = U_{1,12,16},$$

and $\frac{16}{12}(1 - U_{1,12,16})/U_{1,12,16} = F_{12,16}$. Hence we have an observed F-value of

$$F_{12,16} = 0.724 < 1 \leq F_{12,16,0.05} = 2.42,$$

and we do not reject the hypothesis H that a cubic term is zero. We therefore fit only a quadratic model in x. The design matrix now becomes

$$A' = \begin{bmatrix} 1 & 0.25 & 0.0625 & 254 \\ 1 & 0.25 & 0.0625 & 262 \\ 1 & 0.25 & 0.0625 & 301 \\ 1 & 0.25 & 0.0625 & 311 \\ 1 & 0.25 & 0.0625 & 290 \\ 1 & 0.25 & 0.0625 & 300 \\ 1 & 0.25 & 0.0625 & 306 \\ 1 & 0.25 & 0.0625 & 286 \\ 1 & 0.65 & 0.4225 & 275 \\ 1 & 0.65 & 0.4225 & 282 \\ 1 & 0.65 & 0.4225 & 256 \\ 1 & 0.65 & 0.4225 & 276 \\ 1 & 0.65 & 0.4225 & 337 \\ 1 & 0.65 & 0.4225 & 296 \\ 1 & 0.65 & 0.4225 & 309 \\ 1 & 0.65 & 0.4225 & 296 \\ 1 & 1.3 & 1.69 & 275 \\ 1 & 1.3 & 1.69 & 192 \\ 1 & 1.3 & 1.69 & 338 \\ 1 & 1.3 & 1.69 & 248 \\ 1 & 1.3 & 1.69 & 315 \\ 1 & 1.3 & 1.69 & 295 \\ 1 & 1.3 & 1.69 & 312 \\ 1 & 1.3 & 1.69 & 286 \\ 1 & 1.71 & 2.9241 & 275 \\ 1 & 1.71 & 2.9241 & 270 \\ 1 & 1.71 & 2.9241 & 290 \\ 1 & 1.71 & 2.9241 & 260 \\ 1 & 1.71 & 2.9241 & 302 \\ 1 & 1.71 & 2.9241 & 284 \\ 1 & 1.71 & 2.9241 & 280 \\ 1 & 1.71 & 2.9241 & 329 \end{bmatrix}.$$

Suppose we check that, for this design matrix, the linear and quadratic terms are significant. For the model

$$E(Y)=\beta A, \qquad \beta=(\beta_{ij}), \quad i=1,\ldots,12, \quad j=1,2,3,4,$$

we test H: $\beta C=0$ vs A: $\beta C\neq 0$, where

$$C=\begin{pmatrix} 0 & 0 \\ 1 & 0 \\ 0 & 1 \\ 0 & 0 \end{pmatrix}.$$

The U-statistic in this case is

$$U_{12,2,28}=0.1566.$$

Asymptotically, we have that

$$-\left[28-\tfrac{1}{2}(12-2+1)\right]\ln 0.1566=41.716\geq\chi^2_{24}=36.4.$$

For p-values, we have

$$\begin{aligned} P\left[-(22.5\ln U_{12,2,28}\right]>41.716&=P\left(\chi^2_{24}>41.716\right) \\ &\quad+(28)^{-2}\gamma_2 P\left(\chi^2_{28}>41.716\right)-P\left(\chi^2_{24}>41.716\right) \\ &=0.0177+0.090(0.0478-0.0177) \\ &=0.02. \end{aligned}$$

Hence $\beta C\neq 0$ at the 2% level. Alternatively, using Corollary 4.2.2, we have

$$U_{12,2,28}=U_{2,12,18},$$

and from Theorem 4.2.3,

$$17\left[1-(0.1566)^{1/2}\right]/12(0.1566)^{1/2}=2.1632\geq F_{24,34,0.05}=1.90.$$

Hence we have a relationship between amount of diet eaten and amount of phosphorus in the treatment diet, a relationship that is quadratic. The significance of the quadratic terms can be obtained by testing

$$H: \beta\mathbf{C}=0 \qquad \text{vs} \qquad A: \beta\mathbf{C}\neq 0,$$

where

$$\mathbf{C}=\begin{pmatrix} 0 \\ 0 \\ 1 \\ 0 \end{pmatrix}.$$

This produces test value $U_{12,1,28}=0.346$. The observed F-value is then $17(1-0.346)/[12(0.346)]=2.677>\mathrm{F}_{12,17,0.05}=2.38$: A quadratic model is indeed necessary.

Example 5.3.4 Let us illustrate the use of dummy variables in a multivariate regression model. Consider the data of Example 5.3.2. This time, however, let us suppose that the five distinct dosage levels were given over five weeks to each of five tanks of fish. Thus, a different dosage was administered to each tank in week 1, a procedure that was replicated over the next four weeks.

We would like to model the the possible week-to-week variations. One way is to write the model

$$\mathbf{y}_{jk}=\boldsymbol{\beta}_1+\boldsymbol{\beta}_2 x_j+\boldsymbol{\gamma}_k+\boldsymbol{\varepsilon}_{jk}, \qquad j=1,\ldots,5, \quad k=1,\ldots,5,$$

where $\boldsymbol{\beta}_1$ and $\boldsymbol{\beta}_2$ are the regression coefficients, the γ_k are parameters reflecting differences among the five weeks, and $x_1,\ldots,x_5$ are the five different dosages. The observation $\mathbf{y}_{jk}$ represents the vector of five response values for the jth dosage and the kth week or block. (The variable mean weight is omitted in this example. However, it may be added to the model using the techniques of this chapter.)

To avoid singularities in the regression model we must assume that the average block effect is zero, or

$$\sum_{k=1}^{5} \gamma_k = 0.$$

Hence, we write $\gamma_5 = -\sum_{k=1}^{4}\gamma_k$.

In matrix notation we may write the model

$$Y = \beta\mathbf{X} + \varepsilon$$

where

$$Y = (\mathbf{y}_{11},\ldots,\mathbf{y}_{51},\ldots,\mathbf{y}_{15},\ldots,\mathbf{y}_{55}),$$
$$\beta = (\boldsymbol{\beta}_1, \boldsymbol{\beta}_2, \gamma_1, \gamma_2, \gamma_3, \gamma_4),$$
$$\varepsilon = (\boldsymbol{\varepsilon}_{11},\ldots,\boldsymbol{\varepsilon}_{51},\ldots,\boldsymbol{\varepsilon}_{15},\ldots,\boldsymbol{\varepsilon}_{55}),$$

and

$$X' = \begin{pmatrix}
1 & 5.598421959 & 1 & 0 & 0 & 0 \\
1 & 6.01615716 & 1 & 0 & 0 & 0 \\
1 & 6.413458957 & 1 & 0 & 0 & 0 \\
1 & 6.845879875 & 1 & 0 & 0 & 0 \\
1 & 7.279318835 & 1 & 0 & 0 & 0 \\
1 & 5.598421959 & 0 & 1 & 0 & 0 \\
1 & 6.01615716 & 0 & 1 & 0 & 0 \\
1 & 6.413458957 & 0 & 1 & 0 & 0 \\
1 & 6.845879875 & 0 & 1 & 0 & 0 \\
1 & 7.279318835 & 0 & 1 & 0 & 0 \\
1 & 5.598421959 & 0 & 0 & 1 & 0 \\
1 & 6.01615716 & 0 & 0 & 1 & 0 \\
1 & 6.413458957 & 0 & 0 & 1 & 0 \\
1 & 6.845879875 & 0 & 0 & 1 & 0 \\
1 & 7.279318835 & 0 & 0 & 1 & 0 \\
1 & 5.598421959 & 0 & 0 & 0 & 1 \\
1 & 6.01615716 & 0 & 0 & 0 & 1 \\
1 & 6.413458957 & 0 & 0 & 0 & 1 \\
1 & 6.845879875 & 0 & 0 & 0 & 1 \\
1 & 7.279318835 & 0 & 0 & 0 & 1 \\
1 & 5.598421959 & -1 & -1 & -1 & -1 \\
1 & 6.01615716 & -1 & -1 & -1 & -1 \\
1 & 6.413458957 & -1 & -1 & -1 & -1 \\
1 & 6.845879875 & -1 & -1 & -1 & -1 \\
1 & 7.279318835 & -1 & -1 & -1 & -1
\end{pmatrix}$$

The covariance matrix is given by

$$S=\begin{pmatrix} 55.677 & 17.165 & -28.202 & -53.302 & -46.288 \\ 17.165 & 59.315 & 27.578 & 28.344 & 29.979 \\ -28.202 & 27.578 & 195.674 & 196.664 & 179.781 \\ -53.302 & 28.344 & 196.664 & 236.758 & 219.052 \\ -46.288 & 29.979 & 179.781 & 219.052 & 207.353 \end{pmatrix},$$

and the estimate of the regression coefficient matrix is

$$\hat{\beta}=\begin{pmatrix} -104.075 & 17.885 & 3.690 & -1.712 & 2.045 & 3.230 \\ -236.863 & 42.540 & 1.086 & -2.079 & 4.734 & 2.799 \\ -257.528 & 49.177 & -2.739 & 1.472 & 1.930 & 4.659 \\ -244.043 & 47.621 & -2.877 & 3.123 & 0.765 & 3.765 \\ -229.796 & 45.561 & -2.091 & 2.119 & -0.238 & 2.762 \end{pmatrix}.$$

If we wish to test that the dosage has no effect on mortality we test

$$H:\ \boldsymbol{\beta}_2=0 \quad \text{vs} \quad A:\ \boldsymbol{\beta}_2\neq 0,$$

or

$$H:\ \mathbf{C}_1\beta\mathbf{C}_2=0 \quad \text{vs} \quad A:\ \mathbf{C}_1\beta\mathbf{C}_2\neq 0$$

where $\mathbf{C}_2=(0,1,0,0,0,0)$ and $\mathbf{C}_1=\mathbf{I}_5$. We obtain the hypothesis sum of squares matrix

$$W=\begin{pmatrix} 2810.757 & 6685.396 & 7728.371 & 7483.848 & 7160.183 \\ 6685.396 & 15901.236 & 18381.955 & 17800.356 & 17030.519 \\ 7728.371 & 18381.955 & 21249.687 & 20577.354 & 19687.416 \\ 7483.848 & 17800.356 & 20577.354 & 19926.294 & 19064.513 \\ 7160.183 & 17030.519 & 19687.416 & 19064.513 & 18240.002 \end{pmatrix}$$

and the error sum of squares matrix

$$V=\begin{pmatrix} 1057.868 & 326.140 & -535.833 & -1012.744 & -879.464 \\ 326.140 & 1126.980 & 523.991 & 538.529 & 569.605 \\ -535.833 & 523.991 & 3717.814 & 3736.625 & 3415.836 \\ -1012.744 & 538.529 & 3736.625 & 4498.399 & 4161.990 \\ -879.464 & 569.605 & 3415.836 & 4161.990 & 3939.711 \end{pmatrix}.$$

Hence the value of the test statistic is $U_{p,m,n}=|V|/|V+W|=0.0519$ with $p=5$, $m=1$, and $n=25-6=19$. From Corollary 4.2.3 we have

$$F_{5,15}=\tfrac{15}{5}(1-0.0519)/0.0519=54.8.$$

As $54.8\geq F_{5,15,0.05}=2.9$, we reject H and claim that the copper had an effect on the mortality rate of the fish.

One can also modify the model tested here. For example, a term quadratic in x could be added and the result tested to see whether the term linear in x was adequate. Other variables, such as water hardness or temperature, may also be included in the model.

5.4 A Test for Nonadditivity

Let

$$E(Y)=\beta X+D_{\alpha}F,$$

where Y: $p\times N$ is the observation matrix; β: $p\times q$ is the matrix of regression parameters; X: $q\times N$ is the design matrix, which is assumed to be known; D_{α}: $p\times p$ is a diagonal matrix diag$(\alpha_1,\ldots,\alpha_p)$; and F: $p\times N$ is a matrix whose elements are functions of βX. $D_{\alpha}F$ represents the interaction part and is not known except for the assumption that f_{ij} is a real-valued function of βX, where $F=(f_{ij})$. It is assumed that

$$XF'=0.$$

Also, without loss of generality, we assume that the rank of (X',F) is $\rho(X',F)=p+q$. We wish to test the hypothesis

$$H\colon \alpha_1=\cdots=\alpha_p=0 \qquad \text{vs} \qquad A\colon \alpha_i\neq 0$$

for at least one i, $i=1,2,\ldots,p$. The test statistic for this hypothesis is

$$|S_{\text{e}}|/|S_{\text{e}}+S_{\text{h}}|=U_{p,p,N-q-p},$$

where

$$S_{\text{h}}=Z\hat{F}'(\hat{F}\hat{F}')^{-1}\hat{F}Z', \qquad Z=Y\left[I-X'(XX')^{-1}X\right],$$

and

$$S_{\text{e}}=ZZ'-S_{\text{h}}.$$

Note that for simplicity in the above development we have assumed that

$$XF'=0, \qquad \rho(X)=q, \qquad \text{and} \qquad \rho(X',F)=p+q.$$

However, we may just assume that

$$XF'=0, \qquad \rho(X)=s\leq q, \qquad \rho(X',F)=r, \qquad q<r\leq p+q.$$

Then

$$S_{\text{h}}=Z\hat{F}'\hat{F}'^{-}Z', \qquad Z=Y[I-X'X'^{-}], \qquad S_{\text{e}}=ZZ'-S_{\text{h}},$$

and

$$|S_{\text{e}}|/|S_{\text{e}}+S_{\text{h}}|=U_{p,r-q,N-r},$$

where $\hat{F}'^{-}$ and X'^{-} are the generalized inverses of $\hat{F}'$ and X', respectively.

This result for $p=1$ is given by Milliken and Graybill (1970). An extension to the multivariate case is given by McDonald (1972), who suggests using Roy's maximum root test.

5.5 Analysis of Covariance

Section 4.6 described the analysis of a CRD with covariables. The procedure can be extended to an RCBD or a Latin square design. In this section we analyze the general model.

5.5.1 The General Model

The general model for the analysis of covariance can be written

$$Y_{p\times N}=\beta_{1_{p\times q}}X_{1_{q\times N}}+\beta_{2_{p\times u}}X_{2_{u\times N}}+\varepsilon_{p\times N}, \tag{5.5.1}$$

where Y is the matrix of observations; ε is the corresponding error matrix, whose columns are independently distributed as $N_p(\mathbf{0},\Sigma)$; β_1 is the regression coefficients of the design matrix X_1; and X_2 is the matrix of observed covariables with coefficients β_2. Note that

$$Y=\left(\beta_1\vdots\beta_2\right)\begin{pmatrix}X_1\\ \hline X_2\end{pmatrix}+\varepsilon \tag{5.5.2}$$

is the form of the general linear model of Section 5.3. We are only interested in testing hypotheses concerning the treatment effects or regression coefficients β_1, so we limit our tests to

$$H: D_1\beta_1D_2=0 \qquad \text{vs} \qquad A: D_1\beta_1D_2\neq 0, \tag{5.5.3}$$

where D_1 and D_2 are $r\times p$ and $q\times s$ and of ranks r and s, respectively. Equivalently, we may write the hypothesis

$$H: C_1\beta C_2=0 \qquad \text{vs} \qquad A: C_1\beta C_2\neq 0,$$

where $\beta=(\beta_1,\beta_2)$, $C_1=D_1$, and

$$C_2=\begin{pmatrix}D_2\\ 0_{u\times s}\end{pmatrix}.$$

Note that D_1 specifies the characteristics of interest and D_2 specifies the contrasts of the treatments or specifies the regression parameters. The null hypothesis is rejected if

$$U_{r,s,n}\le U_{r,s,n,\alpha},$$

where $n=N-q-u$ and

$$U_{r,s,n}=|V|/|V+W|$$

with

$$V=C_1Y\left(I-X'(XX')^{-1}X\right)Y'C_1', \tag{5.5.4}$$

$$W=C_1\hat{\beta}C_2\left(C_2'(XX')^{-1}C_2\right)^{-1}C_2'\hat{\beta}C_1', \tag{5.5.5}$$

$$\hat{\beta}=YX'(XX')^{-1}. \tag{5.5.6}$$

This test procedure can be written in the usual form for analysis of covariance, for the case of u covariables and a CRD with t treatments. Let $\mathbf{x}_{ij}$ be the vector of observations for the covariables from the ith replication of the jth treatment groups. Let $\mathbf{y}_{ij}$ be the corresponding p-vector of response variables. Then the model may be written

$$\mathbf{y}_{ij} = \boldsymbol{\mu}_j + \gamma \mathbf{x}_{ij} + \boldsymbol{\varepsilon}_{ij}, \tag{5.5.7}$$

where ε_{ij} is independently distributed as $N_p(\mathbf{0}, \Sigma)$, $j=1,\ldots,t$, $i=1,\ldots,N_j$, and γ is the $p\times u$ matrix of coefficients of the covariables. It is assumed that γ is the same for all groups.

Then let

$$Y_{p\times N} = \left(\mathbf{y}_{11},\ldots,\mathbf{y}_{N_1 1},\ldots,\mathbf{y}_{1t},\ldots,\mathbf{y}_{N_t t}\right),$$

$$X_{2_{u\times N}} = \left(\mathbf{x}_{11},\ldots,\mathbf{x}_{N_1 1},\ldots,\mathbf{x}_{1t},\ldots,\mathbf{x}_{N_t t}\right),$$

$\beta_1 = (\boldsymbol{\mu}_1,\ldots,\boldsymbol{\mu}_t)$, and $\beta_2 = \gamma$. The design matrix X_1 may be written

$$X_1 = \begin{pmatrix} \mathbf{e}'_{N_1} & & & 0 \\ & \mathbf{e}'_{N_2} & & \\ & & \ddots & \\ 0 & & & \mathbf{e}'_{N_t} \end{pmatrix}. \tag{5.5.8}$$

With this model the hypothesis

$$H: \boldsymbol{\mu}_1 = \cdots = \boldsymbol{\mu}_t \qquad \text{vs} \qquad A: \boldsymbol{\mu}_i \neq \boldsymbol{\mu}_j$$

for some i, j, $1 \le i < j \le t$, can be written

$$H: D_1\beta_1 D_2 = 0 \qquad \text{vs} \qquad A: D_1\beta_1 D_2 \neq 0,$$

where $D_1 = I_p$ and

$$D_2 = \begin{pmatrix} 1 & & & & 0 \\ -1 & 1 & & & \\ & -1 & 1 & & \\ & & -1 & \ddots & \\ & & & \ddots & 1 \\ 0 & & & & -1 \end{pmatrix}. \tag{5.5.9}$$

One may then test the hypothesis H: $\boldsymbol{\mu}_1 = \cdots = \boldsymbol{\mu}_t$ using the statistic $U_{r,s,n} = |V|/|V+W|$. In problems involving complex designs it may be easier to use the following analysis of variance method.

$$\begin{aligned} N &= \sum_{j=1}^{t} N_j, \qquad \bar{\mathbf{y}}_{\cdot j} = N_j^{-1}\sum_{i=1}^{N_j}\mathbf{y}_{ij}, \qquad \bar{\mathbf{y}}_{\cdot\cdot} = N^{-1}\sum_{j=1}^{t}\sum_{i=1}^{N_j}\mathbf{y}_{ij}, \\ \bar{\mathbf{x}}_{\cdot j} &= N_j^{-1}\sum_{i=1}^{N_j}\mathbf{x}_{ij}, \qquad \bar{\mathbf{x}}_{\cdot\cdot} = N^{-1}\sum_{j=1}^{t}\sum_{i=1}^{N_j}\mathbf{x}_{ij}. \end{aligned} \tag{5.5.10}$$

Then define

$$
\begin{aligned}
T_{xx} &= \sum_{j=1}^{t}\sum_{i=1}^{N_j} \mathbf{x}_{ij}\mathbf{x}'_{ij} - N\bar{\mathbf{x}}_{..}\mathbf{x}'_{..},\\
S_{xx} &= \sum_{j=1}^{t} N_j\bar{\mathbf{x}}_{.j}\bar{\mathbf{x}}'_{.j} - N\bar{\mathbf{x}}_{..}\bar{\mathbf{x}}'_{..},\\
E_{xx} &= T_{xx} - S_{xx};\\
T_{yy} &= \sum_{j=1}^{t}\sum_{i=1}^{N_j} \mathbf{y}_{ij}\mathbf{y}'_{ij} - N\bar{\mathbf{y}}_{..}\bar{\mathbf{y}}'_{..},\\
S_{yy} &= \sum_{j=1}^{t} N_j\bar{\mathbf{y}}_{.j}\bar{\mathbf{y}}'_{.j} - N\bar{\mathbf{y}}_{..}\bar{\mathbf{y}}'_{..},\\
E_{yy} &= T_{yy} - S_{yy};\\
T_{xy} &= \sum_{j=1}^{t}\sum_{i=1}^{N_j} \mathbf{x}_{ij}\mathbf{y}'_{ij} - N\bar{\mathbf{x}}_{..}\bar{\mathbf{y}}'_{..},\\
S_{xy} &= \sum_{j=1}^{t} N_j\bar{\mathbf{x}}_{.j}\bar{\mathbf{y}}'_{.j} - N\bar{\mathbf{x}}_{..}\bar{\mathbf{y}}'_{..},\\
E_{xy} &= T_{xy} - S_{xy}.
\end{aligned}
\tag{5.5.11}
$$

The appropriate sum of squares matrices can be calculated from

$$
\mathrm{SS}(\mathbf{x}) = T'_{xy}T_{xx}^{-1}T_{xy}, \qquad \mathrm{SS}(E\,|\,\mathbf{x}) = E_{yy} - E'_{xy}E_{xx}^{-1}E_{xy},
$$

$$
\mathrm{SST} = T_{yy}, \qquad \mathrm{SS}(\mathrm{TR}\,|\,\mathbf{x}) = \mathrm{SST} - \mathrm{SS}(E\,|\,\mathbf{x}) - \mathrm{SS}(\mathbf{x}).
$$

We can now write the MANOVA table (Table 5.4). Both it and (5.5.3)–(5.5.6) yield the same value of the U-statistic. In Table 5.4 the sum of squares and products matrices SS(TR|**x**) and SS(E|x) are "adjusted" for **x** (the covariables). Hence **x** is included in the notation. The matrix SS(TR|**x**) does not equal the raw SS(TR) matrix of (4.2.2) unless the means of the covariables are the same for each group.

Table 5.4

Source	df	SS	$U_{r,m,n}$	r, m, n
(**x**)	u	SS(**x**)		
(TR|**x**)	$t-1$	SS(TR|**x**)	$\dfrac{\lvert \mathrm{SS}(E\,\lvert\,\mathbf{x}) \rvert}{\lvert \mathrm{SS}(E\,\lvert\,\mathbf{x}) + \mathrm{SS}(\mathrm{TR}\,\lvert\,\mathbf{x}) \rvert}$	$p, t-1, N-u-t$
(E|**x**)	$N-u-t$	SS(E|**x**)		
total	$N-1$	SST		

Note that even for the case of one dependent variable and one covariable SS(TR|**x**) can be larger than SS(TR). This is because in adjusting SS(TR) for a covariable treatment differences can produce a larger sum of squares. However, the error sum of squares will always be smaller.

Example 5.5.1 Consider again the data of Example 4.6.1 (Box, 1950). The test of equality of the treatment means can be obtained by the methods of this section. Using (5.5.4)–(5.5.6) and the model in (5.5.7) with $C_1 = I_4$ and

$$C_2' = \begin{pmatrix} 1 & -1 & 0 & 0 \\ 0 & 1 & -1 & 0 \end{pmatrix},$$

we obtain $U_{4,2,23} = 0.248$. Alternatively, if we use the ANOVA technique we define the following matrices:

$$T_{xx} = 528, \qquad \mathbf{T}_{xy} = (247.00, -27.67, -24.00, -120.33),$$

$$T_{yy} = \begin{pmatrix} 664.00 & 79.67 & -44.00 & 38.33 \\ 79.67 & 1085.85 & 1409.22 & 1131.93 \\ -44.00 & 1409.22 & 2362.67 & 1719.11 \\ 38.33 & 1131.93 & 1719.11 & 2186.96 \end{pmatrix};$$

$$S_{xx} = 10.19, \qquad \mathbf{S}_{xy} = (-27.66, 6.83, 28.94, -7.16),$$

$$S_{yy} = \begin{pmatrix} 81.67 & 37.17 & 11.47 & 112.92 \\ 37.17 & 476.85 & 782.72 & 787.43 \\ 11.47 & 782.72 & 1315.94 & 1260.10 \\ 112.92 & 7807.43 & 1260.10 & 1334.01 \end{pmatrix};$$

$$E_{xx} = T_{xx} - S_{xx} = 517.81,$$

$$\mathbf{E}_{xy} = \mathbf{T}_{xy} - \mathbf{S}_{xy} = (274.66, -34.50, -52.94, -113.17),$$

$$E_{yy} = T_{yy} - S_{yy} = \begin{pmatrix} 582.33 & 42.50 & -55.47 & -74.59 \\ 42.50 & 609.00 & 626.50 & 344.50 \\ -55.47 & 626.50 & 1046.73 & 459.01 \\ -74.59 & 344.50 & 459.01 & 852.96 \end{pmatrix};$$

$$\mathrm{SS}(\mathbf{x}) = \mathbf{T}_{xy}' T_{xx}^{-1} \mathbf{T}_{xy}$$

$$= \begin{pmatrix} 115.54 & -12.94 & -11.23 & -56.29 \\ -12.94 & 1.45 & 1.26 & 6.31 \\ -11.23 & 1.26 & 1.10 & 5.47 \\ -56.29 & 6.31 & 5.47 & 27.42 \end{pmatrix},$$

$$\mathrm{SS}(E\,|\,\mathbf{x}) = E_{yy} - \mathbf{E}_{xy}' E_{xx}^{-1} \mathbf{E}_{xy}$$

$$= \begin{pmatrix} 436.65 & 60.80 & -27.39 & -14.56 \\ 60.80 & 606.70 & 622.97 & 336.96 \\ -27.39 & 622.97 & 1041.31 & 447.44 \\ -14.56 & 336.96 & 447.44 & 828.22 \end{pmatrix},$$

$$\mathrm{SS}(\mathrm{TR}\,|\,\mathbf{x}) = T_{yy} - \mathrm{SS}(E\,|\,\mathbf{x}) - \mathrm{SS}(\mathbf{x})$$

$$= \begin{pmatrix} 111.81 & 31.81 & -5.38 & 109.18 \\ 31.81 & 477.70 & 784.99 & 788.66 \\ -5.38 & 784.99 & 1320.26 & 1266.20 \\ 109.18 & 788.66 & 1266.20 & 1331.32 \end{pmatrix}.$$

Calculations for the U-statistics (of Wilks) yield

$$U_{4,2,23} = \frac{|\mathrm{SS}(E\,|\,\mathbf{x})|}{|\mathrm{SS}(E\,|\,\mathbf{x}) + \mathrm{SS}(\mathrm{TR}\,|\,\mathbf{x})|} = 0.248.$$

Hence we obtain

$$\tfrac{20}{4}(1 - U^{1/2})/U^{1/2} = 5.04 = F_{8,40}.$$

Because $5.04 \geq F_{8,40,0.05} = 2.18$, we claim a significant difference in treatments. The problem of finding which treatments differ and for which weeks is discussed in Section 5.5.3.

5.5.2 Random and Fixed Covariables

In Section 4.6 we discussed the case of random covariables, whereas the models of Section 5.5.1 involve fixed covariables. Nevertheless, the testing and estimation procedures yield identical results, and from a practical point of view any difference between the two types of variables can be ignored. Theoretically, the power of the test procedures in the random covariable situation depends on the population covariance of the covariables and not on the values of the covariables. For fixed covariables the power of the tests depends on the actual chosen or fixed values.

The following example illustrates an extremely important application of covariables. Suppose we wish to examine the effect of three teaching methods on the final grades of the students in a statistics course. If possible, the students should be randomly assigned to the methods, but suppose scheduling problems prevent this. In that case one method would probably have the better students, and a significant difference in final grades between methods could not be attributed to the methods themselves. However, using past performance in a statistics course as a covariable, we can adjust the students' scores so that the three teaching methods can be compared.

In effect, the use of covariables enables us to compare treatments that could not otherwise be compared. Moreover, the addition of covariables reduces the error sum of squares and products matrices.

5.5.3 Confidence Intervals in the Presence of Covariables

In analysis of covariance a significant difference in treatment groups does not tell which groups differ and for which characteristics. We would like to have a method simultaneously to test linear contrasts in treatment groups.

We shall first give a general formula for the model of (5.5.1) using the results of Section 5.3. We then present more simplistic formulas for a one-factor completely randomized design.

Simultaneous $(1-\alpha)100\%$ confidence intervals for $\mathbf{a}'\beta_1\mathbf{b}$, for all $a: p\times 1$ and $b: q\times 1$, are

$$\mathbf{a}'\hat{\beta}_1\mathbf{b}\pm\left[x_\alpha(1-x_\alpha)^{-1}\mathbf{a}'V\mathbf{a}\mathbf{c}'(XX')^{-1}\mathbf{c}\right]^{1/2} \tag{5.5.12}$$

where

$$\hat{\beta}=YX'(XX')^{-1}, \qquad X=\left(X_1 \vdots X_2\right),$$

$$\beta=\left(\beta_1 \vdots \beta_2\right) \qquad \text{for} \quad \beta_1: p\times q,$$

$$V=Y\left[I-X'(XX')^{-1}X\right]Y',$$

$$\mathbf{c}'=(\mathbf{b}',\mathbf{0}'_u)=(b_1,\ldots,b_q,0,\ldots,0),$$

and x_α is obtained from Table A.5 with parameters

$$p_0=p, \qquad m_0=q, \qquad n=N-q-u. \tag{5.5.13}$$

If, for example, the vectors **a** and **b** are restricted to the form $\mathbf{a}'=\mathbf{a}'_1C_1$ and $\mathbf{b}'=\mathbf{b}'_1C_2$, where C_1 is of rank $r\leq p$, C_2 is of rank $s\leq q$, $\mathbf{a}_1$ is an r-vector, and $\mathbf{b}_1$ is an s-vector, then the parameters become

$$p_0=r, \qquad m_0=s, \qquad n=N-q-u. \tag{5.5.14}$$

However, (5.5.12) remains unchanged except for the value of x_α. One example of this type of restriction is the case in which all characteristics are of importance (that is, **a** is unrestricted) but the columns of β_1 are treatment means. In this case a comparison of treatment means restricts **b** to a *contrast vector* (one orthogonal to the unit vector). The parameters for Table A.5 are then

$$p_0=p, \qquad m_0=q-1, \qquad n=N-q-u. \tag{5.5.15}$$

If only a few comparisons are to be made, then Bonferroni bounds can be used, so that for k confidence intervals we obtain k simultaneous $(1-\alpha)100\%$ confidence intervals:

$$\mathbf{a}'\hat{\beta}\mathbf{b}\pm\left[\left(t^2_{\alpha/2k,N-u-q}\right)\mathbf{a}'V\mathbf{a}\mathbf{c}'(XX')^{-1}\mathbf{c}/(N-u-q)\right]^{1/2}. \tag{5.5.16}$$

Once again, for $t^2_{\alpha/2k}/(N-u-q)\leq x_\alpha(1-x_\alpha)^{-1}$, (5.5.16) produces shorter intervals. In general, (5.5.12) is used when the experimenter has no hypothesis of interest before the experiment and requires all comparisons. It is advisable, therefore, to perform a trial run of the experiment, or pretest, to obtain a few well-defined hypotheses.

For a CRD with one factor, simultaneous confidence intervals can be written in simplified form. The general model, given by (5.5.7)–(5.5.9), is $\mathbf{y}_{ij}=\boldsymbol{\mu}_j+\gamma\mathbf{x}_{ij}+\boldsymbol{\varepsilon}_{ij}$, $j=1,\ldots,t$, $i=1,\ldots,N_j$. The least squares estimates of the treatment means $\boldsymbol{\mu}_j, j=1,\ldots,t$, are given by

$$\hat{\boldsymbol{\mu}}_j = \bar{\mathbf{y}}_{\cdot j} + \hat{\gamma}(\bar{\mathbf{x}}_{\cdot j} - \bar{\mathbf{x}}_{\cdot\cdot}), \tag{5.5.17}$$

where

$$\hat{\gamma} = E'_{xy}E_{xx}^{-1} \tag{5.5.18}$$

with E_{xy} and E_{xy} defined in (5.5.11) and $\bar{\mathbf{y}}_{\cdot j}$ and $\bar{\mathbf{x}}_{\cdot j}$ the mean vectors of response and covariable, respectively, for the jth treatment group. Simultaneous confidence intervals for the means of all treatment groups for each characteristic are then given by

$$\hat{\mu}_{lj} \pm \left\{x_\alpha(1-x_\alpha)^{-1}v_{ll}\left[N_j^{-1} + (\bar{\mathbf{x}}_{\cdot j} - \bar{\mathbf{x}}_{\cdot\cdot})'E_{xx}^{-1}(\bar{\mathbf{x}}_{\cdot j} - \bar{\mathbf{x}}_{\cdot\cdot})\right]\right\}^{1/2} \tag{5.5.19}$$

where x_α is given in Table A.5 with

$$p_0 = p, \qquad m_0 = t, \qquad n = N - t - u. \tag{5.5.20}$$

$\hat{\mu}_{lj}$ is obtained from $\hat{\boldsymbol{\mu}}_j = (\hat{\mu}_{1j},\ldots,\hat{\mu}_{pj})'$.

The values v_{ll}, $l=1,\ldots,p$, are obtained as the diagonal entries of the error sum of squares matrix, $V=(v_{ij})=\mathrm{SS}(E|\mathbf{x})$. N_j is the sample size at the jth treatment group with $\Sigma_{j=1}^{t}N_j = N$, $\bar{\mathbf{x}}_{\cdot\cdot}$ is the vector of the means of the covariables from the jth treatment group, and $\bar{\mathbf{x}}_{\cdot\cdot}$ is the corresponding overall mean vector of the covariables. If Bonferroni bounds are to be used, (5.5.19) is replaced by

$$\hat{\mu}_{lj} \pm \left[t^2_{\alpha/2pt,n}(v_{ll}/n)\left[N_j^{-1} + (\bar{\mathbf{x}}_{\cdot j} - \bar{\mathbf{x}}_{\cdot\cdot})'E_{xx}^{-1}(\bar{\mathbf{x}}_{\cdot j} - \bar{\mathbf{x}}_{\cdot\cdot})\right]\right\}^{1/2}. \tag{5.5.21}$$

The shorter bounds are obtained by using the smaller of $x_\alpha(1-x_\alpha)^{-1}$ and $t^2_{\alpha/2pt,n}/n$. Note that the number of comparisons is pt in this particular case.

Suppose that instead of looking at the treatment means separately we compose the treatment means two at a time. In other words, we wish to obtain simultaneous confidence intervals for pairwise differences in treatments for each characteristic or response variable. We then require confidence intervals for parameters of the form $\mu_{li} - \mu_{lj}$, $l=1,\ldots,p$, $1 \le i < j \le t$. These intervals are given by

$$\hat{\mu}_{li} - \hat{\mu}_{lj} \pm \left\{x_\alpha(1-x_\alpha)^{-1}v_{ll}\left[N_i^{-1} + N_j^{-1} + (\bar{\mathbf{x}}_{\cdot i} - \bar{\mathbf{x}}_{\cdot j})'E_{xx}^{-1}[\bar{\mathbf{x}}_{\cdot i} - \bar{\mathbf{x}}_{\cdot j})\right]\right\}^{1/2} \tag{5.5.22}$$

where x_α can be obtained from Table A.5 with

$$p_0 = p, \qquad m_0 = t-1, \qquad n = N - p - u.$$

With Bonferroni bounds (5.5.22) becomes

$$\hat{\mu}_{li}-\hat{\mu}_{lj}\pm\left\{t^2_{\alpha/2k,n}(v_{ll}/n)\left[N_i^{-1}+N_j^{-1}+(\bar{\mathbf{x}}_{.i}-\bar{\mathbf{x}}_{.j})'E_{xx}^{-1}(\bar{\mathbf{x}}_{.i}-\bar{\mathbf{x}}_{.j})\right]\right\}^{1/2}, \tag{5.5.23}$$

where k, the number of comparison, is now $\frac{1}{2}pt(t-1)$.

Note that $\hat{\mu}_{li}$ and $\hat{\mu}_{lj}$ usually are not independent statistics. Hence the variance of $\hat{\mu}_{li}-\hat{\mu}_{lj}$ is not the sum of the individual variances of $\hat{\mu}_{li}$ and $\hat{\mu}_{lJ}$. Calculations to obtain confidence interval and the values of test statistics are described at the end of Section 5.6.

Example 5.5.2 Consider the data of Example 4.6.1 from Box (1950). The experiment consisted of three treatments applied randomly to 27 rats. The first treatment, a control, was given to ten rats; the second treatment, the addition of thyroxin to the drinking water, was administered to seven rats; and the third treatment, the addition of thiouracil to the drinking water, was given to ten rats. The initial weight of a rat was used as a covariable, and the weight gains of a rat for four weeks were considered as the four response or dependent variables. The test for the equality of the treatment effects was discussed in Section 4.6. The methods of this section can determine which treatments differ and for which weeks. We have $p=4$, $u=1$, $t=3$, and $N=27$. (Example 5.5.1 contains detailed calculations.)

Calculations yield $\hat{\gamma}=(-0.5304, 0.0666, 0.1022, 0.2186)'$, $E_{xx}=517.81$. Table 5.5 shows the means for each group and least squares estimates of the treatment means (adjusted for the covariables). The overall mean for the covariable initial weight was $\bar{x}=54.67$. From these estimates we find the simultaneous confidence intervals for the treatment means μ_{lj}, $l=1,\ldots,4$,

Table 5.5

	Group		
Week	Control ($N_j=10$)	Thyroxin ($N_j=7$)	Thiouracil ($N_j=10$)
		(means)	
0	54.00	55.57	54.70
1	24.50	20.29	21.60
2	27.50	29.00	19.50
3	24.10	29.29	12.40
4	30.50	30.14	15.80
		(least squares estimates)	
1	24.85	19.81	21.58
2	27.46	29.06	19.50
3	24.03	29.38	12.40
4	30.35	30.34	15.81

$j=1,\ldots,3$, at $\alpha=0.05$. We use $t_{0.05/24,23}=t_{0.002,23}=3.2$ with $p_0=4$, $m_0=3$, and $n=27-3-1=23$. The value of $x_{0.05}$ is given as 0.552 in Table A.5, from which values are obtained by linear interpolation. Thus $x_{0.05}$ for $n=26$ is 0.472, $x_{0.05}=0.536$ for $n=21$, and $x_{0.05}$ on 23 df is calculated to be $0.536+[(21-23)/(26-21)](0.536-0.472)=0.511$. Now,

$$t^2_{0.002,23}/23=0.45\leq x_{0.05}/(1-x_{0.05})=0.511/0.489=1.05,$$

so we use Bonferroni's bounds given in (5.5.21) for simultaneous confidence intervals for $\boldsymbol{\mu}_j$, $j=1,\ldots,t$.

The value of the error sum of squares matrix was calculated in Example 5.5.1 to be

$$V=\mathrm{SS}(E\,|\,\mathbf{x})=\begin{pmatrix} 436.65 & 60.80 & -27.39 & -14.56 \\ 60.80 & 606.70 & 622.97 & 336.96 \\ -27.39 & 622.97 & 1041.31 & 447.44 \\ -14.56 & 336.96 & 447.44 & 828.22 \end{pmatrix},$$

and from (5.5.21) we obtain the simultaneous 95% intervals for the treatment means, which appear in Table 5.6. A sample calculation is given now for μ_{11}. The confidence interval, from (5.5.21), is

$$24.85\pm\left\{(3.2)^2(436.65/23)\left[0.1+(54-54.67)^2/518\right]\right\}^{1/2}.$$

Equivalently, it can be represented in the form of a least squares estimate plus-or-minus the t-value times the standard error, where the least squares estimate of μ_{11} is 24.85, the t-value is 3.2, and the standard error of the estimate is $\{(436.65/23)[0.1+154-54-67)^2/518]\}^{1/2}$. This formula is useful when using statistical packages such as SAS in which the least squares estimates and their standard errors are given (see Section 5.6).

Table 5.6

Group	Parameter	Estimate	Interval
control	μ_{11}	24.85	(20.42, 29.28)
	μ_{21}	27.46	(22.24, 32.68)
	μ_{31}	24.03	(17.19, 30.87)
	μ_{41}	30.35	(24.25, 36.45)
thyroxin	μ_{12}	19.81	(14.50, 25.12)
	μ_{22}	29.06	(22.82, 35.30)
	μ_{32}	29.38	(21.20, 37.56)
	μ_{42}	30.34	(23.04, 37.64)
thiouacil	μ_{13}	21.58	(17.17, 25.99)
	μ_{23}	19.50	(14.30, 24.64)
	μ_{33}	12.40	(5.59, 19.21)
	μ_{43}	15.81	(9.74, 21.88)

Table 5.7

Parameter	Estimate	Interval
$\mu_{11}-\mu_{12}$	5.04	(−1.90, 11.98)
$\mu_{21}-\mu_{22}$	−1.60	(−9.78, 6.58)
$\mu_{31}-\mu_{32}$	−5.35	(16.06, 5.36)
$\mu_{41}-\mu_{42}$	0.01	(−9.55, 9.57)
$\mu_{11}-\mu_{13}$	3.27	(−2.98, 9.52)
$\mu_{21}-\mu_{23}$	7.96	(0.59, 15.33)
$\mu_{31}-\mu_{33}$	11.63	(1.98, 21.28)
$\mu_{41}-\mu_{43}$	14.54	(5.93, 23.15)
$\mu_{12}-\mu_{13}$	−1.77	(−8.66, 5.12)
$\mu_{22}-\mu_{23}$	9.56	(1.44, 17.68)
$\mu_{32}-\mu_{33}$	16.98	(6.34, 27.62)
$\mu_{42}-\mu_{43}$	14.53	(5.04, 24.02)

For contrasts among the three treatments we have $p_0=4$, $m_0=2$, and $n=23$. In this case $x_\alpha=0.455$. Bonferroni bounds, from (5.5.23), use $t_{0.05/24,23}=t_{0.002,23}=3.2$. Again in this instance

$$t^2_{0.002,23}/23=0.45\leq x_{0.05}/(1-x_{0.05})=0.455/0.545=0.835.$$

Hence Bonferroni bounds yield shorter intervals. Note that if we are interested only in comparing the treatment groups with the control, then we use $t_{\alpha/2p(t-1),n}=t_{0.05/16,23}=t_{0.003,23}$. This t-value produces even shorter intervals because we are restricting the number of comparisons.

Table 5.7 gives the difference in means for each treatment group and each of the four weeks. From these confidence intervals we can draw the following conclusions. The thyroxin group and the control group have the same mean weight gain for all four weeks. The thiouracil group, however, has a significantly lower weight gain for weeks 2–4 compared with both the control and thyroxin group. We can conclude that thiouracil decreases weight gain after an initial period of one week.

5.6 Computational Procedures

Multivariate regression can be handled by many programs, such as BMDP6R and the GLM procedure of SAS. The examples in this chapter were run on the SAS system. One advantage of the SAS system is that it can easily handle the analysis of class variables, such as levels of a factor or treatment, as well as continuous independent variables such as covariables. The BMDP6R procedure requires that all the entries of the design matrix be entered as data. Hence we shall give only the SAS procedures in this section.

For Example 5.2.2 the following program statements were used:

```
DATA;
INPUT X Y;
CARDS;
10 0.15
14 0.24
16 0.61
18 0.82
20 1.1
22 1.4
24 1.6
26 1.8
28 2.4
30 2.7
PROC GLM;
 MODEL Y = X X*X/XPX I;
```

The DATA; step creates a data file consisting of two variables X and Y. The PROC GLM; step analyzes the general linear model of the form $Y=\beta_0+\beta_1 X+\beta_2 X^2+\varepsilon$. The terms XPX and I are included so that $X'X$ and $(X'X)^{-1}$ are printed out. Terms after a slash generally ask for additional information and are not essential to the program. For Example 5.3.1 the following program statements were used.

```
DATA;
INPUT AGE DI TETRA;
CARDS;
22 93.6  5.7
30 91.8  7.4
36 99.4  0.3
50 85.2 14.2
57 90.4  9.2
58 86.4 12.2
19 95.6  4.4
50 89   10.8
53 84.3 14.7
62 90.9  8.4
PROC GLM;
MODEL DI TETRA = AGE;
 MANOVA H = AGE/PRINTE PRINTH;
```

In the program a data set is created consisting of the three variables AGE, DI, and TETRA. The two dependent or response variables DI and TETRA are then regressed against age. The MANOVA statement asks for the multivariate statistics for testing no relationship between age and DI and Tetraploid nuclei. PRINTE and PRINTH indicate that the error and hypothesis sum of squares matrices are to be printed. The univariate F tests are printed by default. For Example 5.3.2 the following program statements were used:

```
DATA;
INPUT Y1 Y2 Y3 Y4 Y5 X1 X2;
CARDS;
```

```
0.000 0.000 30.000 30.000 30.000 5.5984 0.6695
.
.
.
33.211 67.213 90.000 90.000 90.000 7.2793 0.7230
PROC GLM;
MODEL Y1 Y2 Y3 Y4 Y5 = X1 X2;
MANOVA H = X1 X2/PRINTE PRINTH;
```

The DATA; procedure creates a data set of the seven variables $Y_1, \ldots, Y_5$, X_1 and X_2. The PROC GLM; statement analyzes the data as a general linear model with the MODEL specified as five dependent or response variables $Y_1, \ldots, Y_5$ with two independent variables X_1 and X_2. Univariate results and tests are printed by default for all response variables. The MANOVA statements asks for multivariate tests for the effects of X_1 and the effect of X_2. PRINTE PRINTH signifies that the error and hypothesis matrices are to be printed out. (The error matrix E is called V and the hypothesis matrix H is labeled W in the formulas of this chapter.) For Example 5.5.2 the following statements were used in an SAS file:

```
DATA;
INPUT    Y0    Y1    Y2    Y3    Y4    A;
CARD;
57  29  28  25  33  1
60  33  30  23  31  1
52  25  34  33  41  1
49  18  33  29  35  1
56  25  23  17  30  1
46  24  32  29  22  1
51  20  23  16  31  1
63  28  21  18  24  1
49  18  23  22  28  1
57  25  28  29  30  1
59  26  36  35  35  2
54  17  19  20  28  2
56  19  33  43  38  2
59  26  31  32  29  2
57  15  25  23  24  2
52  21  24  19  24  2
52  18  35  33  33  2
61  25  23  11   9  3
59  21  21  10  11  3
53  26  21   6  27  3
59  29  12  11  11  3
51  24  26  22  17  3
51  24  17   8  19  3
56  22  17   8   5  3
58  11  24  21  24  3
46  15  17  12  17  3
53  19  17  15  18  3
PROC GLM;
CLASSES A;
MODEL Y1 Y2 Y3 Y4 = Y0 A;
MANOVA H = A/PRINTE PRINTH;
MEANS A;
LSMEANS A/STDERR PDIFF;
```

In this file a data set is created consisting of six variables. The PROC GLM; statement calls the subprogram on general linear models to analyze the data. The variable A is a dummy variable consisting of three levels of treatments. To differentiate A from an explanatory continuous variable the CLASSES A; statement is included. The MODEL consists of four response variables Y_1, Y_2, Y_3 and Y_4, the covariable Y_0, and the class variable A. The MANOVA statement asks for the multivariate test statistics for testing the equality of treatment means. The PRINTE PRINTH statement asks for the error and hypothesis sum of squares matrices. The MEANS A statement asks for the means of all the treatments. However, for the confidence intervals discussed in Section 5.3.3 the least squares means are required. Hence the LSMEANS A statement asks for the treatment means adjusted for the covariable [see (5.5.17)]. The STDERR statement asks for the standard error of the adjusted treatment means. These values can be used for simultaneous confidence intervals (see Example 5.5.2). The PDIFF statements requests the significance level of the p-value for testing H: $\mu_{li}=\mu_{lj}$ vs A: $\mu_{li}\neq\mu_{lj}$, $l=1,\ldots,4$, $1\leq i<j\leq 3$. Note that in the test the p-values should be compared with α/k, where k is the number of desired comparisons, in order to obtain an overall significance level for α.

Appendix

Let $X=(x_{ij})$ be a $p\times N$ matrix of random variables with mean $\mu=(\mu_{ij})$. Writing all the elements of X as a vector,

$$\mathbf{y}'=(x_{11},\ldots,x_{1N};\ x_{21},\ldots,x_{2N};\ \ldots;\ x_{p1},\ldots,x_{pN}),$$

we find that the mean of $\mathbf{y}'$ is

$$E(\mathbf{y}')\equiv\boldsymbol{\delta}'=(\mu_{11},\ldots,\mu_{1N};\ \mu_{21},\ldots,\mu_{2N};\ \ldots;\ \mu_{p1},\ldots,\mu_{pN}),$$

and the covariance matrix of $\mathbf{y}$ is a $pN\times pN$ symmetric, positive, semidefinite matrix. We shall, however, consider the case when the covariance matrix of $\mathbf{y}$ can be written $\Sigma\otimes B=(\sigma_{ij}B)$, where Σ is a $p\times p$ matrix and B is an $N\times N$ matrix. If $\mathbf{y}$ is normally distributed, then we can write $\mathbf{y}\sim N_{pN}(\delta,\Sigma\otimes B)$, and its pdf can be written

$$(2\pi)^{-pN/2}|\Sigma|^{-N/2}|B|^{-p/2}\operatorname{etr}-\tfrac{1}{2}\Sigma^{-1}(X-\mu)B^{-1}(X-\mu)'.$$

This implies that the elements of X are normally distributed. We shall write this $X\sim N_{p,N}(\mu,\Sigma,B)$. Thus if $B=I$, the columns of X are independently normally distributed with covariance matrix Σ.

Definition A.5.1 $X\sim N_{p,N}(\mu,\Sigma,B)$ if $\mathbf{y}\sim N_{pN}(\boldsymbol{\delta},\Sigma\otimes B)$.

Theorem A.5.1 *If $X\sim N_{p,n}(\mu,\Sigma,B)$, then $LXM\sim N_{l,m}(L\mu M, L\Sigma L', M'BM)$, where $L: l\times p$ and $M: N\times m$ are real nonnull matrices.*

Corollary A.5.1 *Let $X\sim N_{p,N}(\mu,\Sigma,B)$. Then $\mathbf{a}'XM\sim N_{1,m}(\mathbf{a}'\mu M,\mathbf{a}'\Sigma\mathbf{a}, M'BM)$ where $\mathbf{a}': 1\times p$ and $m: N\times m$. That is, $M'X\mathbf{a}\sim N_m(M\mu\mathbf{a},(\mathbf{a}'\Sigma\mathbf{a})(M'BM))$.*

Problems

5.1 Lohnes (1972) analyzes the data of the second grade phase of the Cooperative Reading Studies Project. The mean m, variance s, showness g, and kurtosis h of 219 classes were calculated for four tests: the Pintner–Cunningham primary general abilities test (P–C) and the Durell–Murphy Total Letters test (D–ML), given the first week of the first grade, and the Standford Paragraph Meaning tests (SPM 1 and 2), given at the end of the second grade. It was of interest to relate the scores in the second grade tests to those for the first grade tests. The data are shown in Table 5.8.

Note: The matrix of regression coefficient of $\mathbf{x}_1$ and $\mathbf{x}_2$, where $\mathbf{x}_1$ and $\mathbf{x}_2$ are p_1- and p_2-vectors, is given by $S_{12}'S_{11}^{-1}$, where the sample covariance matrix

$$S=\begin{pmatrix} S_{11} & S_{12} \\ S_{12}' & S_{22} \end{pmatrix}$$

and S_{11} and S_{22} are $p_1 \times p_1$ and $p_2 \times p_2$ pd matrices. Using the terms of the correlation matrix, we can write

$$S_{12}'S_{11}^{-1}=D_2R_{12}R_{11}^{-1}D_1,$$

where the correlation matrix R is partitioned

$$R=\begin{pmatrix} R_{11} & R_{12} \\ R_{12}' & R_{22} \end{pmatrix},$$

D_1 is a diagonal matrix of the standard deviations of the independent variables $\mathbf{x}_1$, and D_2 is a diagonal matrix of the standard deviation of the dependent variable $\mathbf{x}_2$. The intercept is given by $\bar{\mathbf{x}}_2 - S_{12}'S_{11}^{-1}\bar{\mathbf{x}}_1$, where $\bar{\mathbf{x}}_1$ and $\bar{\mathbf{x}}_2$ are the means of the two groups of variables.

5.2 Let

$$\mathbf{y}_{ij}=\boldsymbol{\mu}_j+\boldsymbol{\varepsilon}_{ij}, \qquad j=1,2,\ldots,t, \quad i=1,2,\ldots,N_j.$$

Let $\beta=(\boldsymbol{\mu}_1,\ldots,\boldsymbol{\mu}_t)$ and $Y=(\mathbf{y}_{11},\mathbf{y}_{12},\ldots,\mathbf{y}_{t,N_t})$, where $N=\Sigma_{j=1}^{t}N_j$. Write this model $Y=\beta X+\varepsilon$. Show that $\boldsymbol{\mu}_1=\boldsymbol{\mu}_2=\cdots=\boldsymbol{\mu}_t=\mathbf{0}$ is equivalent to testing $C\beta'=0$, where $C=(I_{t-1},-\mathbf{e}):(t-1)\times t$ and $\mathbf{e}=(1,\ldots,1)'$, a $(t-1)$-vector of 1.

5.3 The data in Table 5.9 give the respiratory quotient of each of five male, queen, and worker pupae of *Formica polyctera* at various stages of development.

(a) Given a general linear regression model assuming that the three groups have different means but the temperature effect is the same for all three groups.

(b) Estimate the regression parameters.

(c) Give the estimated covariance matrix and the correlation matrix.

Table 5.8[a]

Test and descriptor	2	3	4	5	6	7	8
1 P–C *m*	−0.33	−0.24	0.13	0.75	−0.08	−0.58	−0.05
2 P–C *s*		−0.03	0.02	−0.29	0.41	0.28	−0.19
3 P–C *g*			−0.30	−0.19	0.00	0.18	0.05
4 P–C *h*				0.11	0.03	−0.08	−0.06
5 D–ML *m*					−0.29	−0.82	0.16
6 D–ML *s*						0.31	−0.54
7 D–ML *g*							−0.34
8 D–ML *h*							
9 SPM 1 *m*							
10 SPM 1 *s*							
11 SPM 1 *g*							
12 SPM 1 *h*							
13 SPM 2 *m*							
14 SPM 2 *s*							
15 SPM 2 *g*							
16 SPM 2 *h*							

Table 5.9[a]

T	g_1	g_2	g_3	g_4	g_5	g_6	g_7
				Male			
16	0.79	0.78	0.83	0.83	0.75	0.80	0.75
20	0.78	0.86	0.84	0.77	0.77	0.64	0.80
24	0.77	0.83	0.79	0.85	0.82	0.79	0.82
28	0.87	0.79	0.81	0.95	0.79	0.82	0.82
32	0.82	0.89	0.82	0.78	0.78	0.84	0.85
				Worker			
16	0.96	1.13	0.76	0.73	0.76	0.83	0.91
20	0.80	0.92	0.85	0.87	0.82	0.82	0.87
24	0.97	0.94	0.87	0.82	0.84	0.86	0.87
28	0.96	0.96	0.98	0.94	0.90	0.97	0.89
32	0.97	1.06	0.99	0.93	0.89	0.92	0.90
				Queen			
16	0.72	0.75	0.81	0.78	0.80	0.72	0.74
20	0.80	0.83	0.84	0.72	0.75	0.80	0.76
24	0.78	0.85	0.79	0.77	0.77	0.74	0.76
28	0.82	0.83	0.75	0.86	0.79	0.74	0.82
32	0.83	0.89	0.87	0.84	0.82	0.81	0.83

[a]Data altered slightly from Brian (1978). Key: T, temperature; g_1, prepupae; g_2, pupation; g_3, beginning of eye pigmentation; g_4, well-pigmented eyes; g_5, beginning of body pigmentation; g_6, well-pigmented body; g_7, at emergence.

Table 5.8 (*continued*)

Correlations									
9	10	11	12	13	14	15	16	m	s.d.
0.64	0.37	−0.25	−0.09	0.68	0.09	−0.38	0.02	37.8	6.3
−0.30	0.16	0.26	−0.03	−0.25	0.37	0.18	−0.14	7.1	1.9
−0.14	−0.03	0.06	0.09	−0.19	0.04	0.09	0.02	−3.1	1.62
0.05	−0.02	−0.06	−0.05	0.01	−0.01	0.03	−0.04	0.10	1.4
0.61	0.24	−0.28	0.00	0.66	−0.04	−0.28	0.05	33.1	8.4
−0.17	0.21	0.21	0.06	−0.15	0.44	0.08	−0.06	11.5	2.8
−0.51	−0.20	0.27	0.03	−0.60	0.10	0.34	−0.08	−0.41	0.82
0.12	−0.17	−0.26	−0.03	0.14	−0.28	−0.09	0.06	−0.08	1.8
	0.10	−0.66	−0.02	0.85	−0.21	−0.53	0.13	20.7	6.5
		0.11	−0.30	0.22	0.39	−0.15	−0.24	7.7	2.2
			0.00	−0.52	0.39	0.41	−0.16	0.01	0.76
				−0.08	−0.03	0.08	0.29	−0.26	1.4
					−0.19	−0.56	0.12	31.6	8.2
						0.01	−0.21	9.9	2.3
							−0.19	−0.15	0.58
								−0.34	0.96

Problem 5.3 (*continued*)

(d) Test the null hypothesis that temperature has no effect on respiratory quotient.

(e) For the males, plot on one graph the responses vs temperature at each stage of development.

5.4 **(a)** For the data of Example 5.3.2, test the hypothesis of no linear relationship between the proportion of dead fish and the log dose.

(b) Estimate the regression parameters, and give the estimated model.

(c) For each time period, estimate the amount of copper necessary to kill 50% of the fish in a tank for fish of weight 1 gm. That is, for $Y = 45°$ (arcsin $\sqrt{0.5}$, $X_2 = 1$, find X_1. This value is called the LD50.

(d) Plot LD50 vs time for the five time periods.

Chapter 6
Analysis of Growth Curves

6.1 Introduction

In order to test for the effect of various dosages of a drug on a limited number of experimental subjects, one can randomly assign the dosages to the subjects and perform the analysis of a randomized complete block design (RCB).

Example 6.1.1 Six rats were each given dosages of 1, 10, or 100 units of a drug, and a response y was recorded, as shown in Table 6.1, which also displays the ANOVA table for this RCB design.

Observing that $245.1 \geq F_{2,10,0.05} = 4.10$, one concludes that there is a change in response with a change in dosage. The model for this design is given by

$$y_{ij} = \mu + \tau_j + \beta_i + \varepsilon_{ij}, \qquad \sum \tau_j = \sum \beta_i = 0,$$

where μ is the overall mean, τ_j the treatment effect, and β_i the block effect. The errors ε_{ij} are assumed to be normally distributed. It is not necessary, however, that the ε_{ij} be independent. A sufficient condition for the analysis to be correct is that the correlations between ε_{1j}, ε_{2j}, and ε_{3j}, $j = 1, \ldots, 6$, be the same. This condition is usually satisfied by the process of randomly assigning treatments to the subject.

Example 6.1.2 A data set similar to that of Example 6.1.1 but for which the process of randomization cannot be applied is as follows:

Subject time (min)	1	2	3	4	5	6
0	6	5	6	7	6	3
1	12	9	11	13	10	9
2	20	15	21	25	20	16

Table 6.1

$\log_{10}$(subject dose)	1	2	3	4	5	6	*total*
0	5	4	5	5	6	3	28
1	10	9	9	8	11	9	56
2	16	13	14	14	15	12	84
total	31	26	28	27	32	24	

Source	df	SS	MS	F	F^*
treatments	2	261.33	130.57	245.1	4.10
subjects	5	15.33	3.07	5.76	—
error	10	5.33	0.533		
total	17	282			

In this case, a treatment was given to the subjects, and the responses at various time intervals were recorded. As time cannot be randomly assigned to the subjects, the assumption of equal correlation between responses at time 0, 1, and 2 minutes may not be valid. In many cases, where growth is measured over time, equal correlation between responses at different times cannot be assumed. By considering the responses over time as characteristics of the subjects, one has six independent observations $\mathbf{y}_1,\ldots,\mathbf{y}_6$, where $\mathbf{y}_i$ is the vector of observations in time on the ith subject, $\mathbf{y}_i \sim N_p(\boldsymbol{\mu}, \Sigma)$, $\Sigma > 0$, and $\boldsymbol{\mu} = (\mu_1, \mu_2, \mu_3)$. This situation is shown in Table 6.2. The null hypothesis for testing for a change in response over time is given by

$$H\colon \mu_1 = \mu_2 = \mu_3 \qquad \text{vs} \qquad A \neq H.$$

We now describe the solution to this problem.

6.1.1 Equality of Means ($\Sigma > 0$)

Let $\mathbf{y} \sim N_p(\boldsymbol{\mu}, \Sigma)$, $\Sigma > 0$, and let $\mathbf{y}_1,\ldots,\mathbf{y}_N$ be an independent sample of size N on $\mathbf{y}$. The sufficient statistics are

$$\bar{\mathbf{y}} = N^{-1} \sum_{\alpha=1}^{N} \mathbf{y}_\alpha \qquad \text{and} \qquad V = \sum_{\alpha=1}^{N} (\mathbf{y}_\alpha - \bar{\mathbf{y}})(\mathbf{y}_\alpha - \bar{\mathbf{y}})'.$$

Table 6.2

	Observations						
Time	y_1	y_2	y_3	y_4	y_5	y_6	Population mean
0	6	5	6	7	6	3	μ_1
1	12	9	11	13	10	9	μ_2
2	20	13	21	25	20	16	μ_3

We are interested in testing the equality of the means μ_i. Thus we wish to test the hypothesis

$$H: \boldsymbol{\mu} = \mathbf{e}\gamma, \qquad \mathbf{e} = (1,\ldots,1)', \quad \gamma \text{ unknown},$$

vs

$$A: \mu_i \neq \mu_j \qquad \text{for at least one pair } (i, j), \quad i \neq j, \quad i, j = 1,2,\ldots,p. \tag{6.1.1}$$

The likelihood ratio test (Srivastava and Khatri, 1979) for this problem is to reject H if

$$\frac{n-p+1}{n(p-1)} N\bar{\mathbf{y}}'\left(S^{-1} - \frac{S^{-1}\mathbf{ee}'S^{-1}}{\mathbf{e}'S^{-1}\mathbf{e}} \right)\bar{\mathbf{y}} \geq F_{p-1,n-p+2,\alpha}, \tag{6.1.2}$$

where $n = N-1$, $S = n^{-1}V$, and $F_{p-1,n-p+1,\alpha}$ is the upper $\alpha 100\%$ point of the F-distribution with $p-1$ and $n-p+2$ df.

It is not convenient to use (6.1.2) with statistical packages. Instead we reconsider (6.1.1) and give an equivalent form. Let C be a $(p-1) \times p$ matrix of rank $p-1$ such that

$$C\mathbf{e} = \mathbf{0}. \tag{6.1.3}$$

The reader familiar with the analysis of variance (see Chapter 4) will immediately recognize that C is a matrix of $p-1$ independent linear contrasts of the means. Thus (6.1.1) is equivalent to the problem (Rao, 1948)

$$H: C\boldsymbol{\mu} = \mathbf{0} \qquad \text{vs} \qquad A: C\boldsymbol{\mu} \neq \mathbf{0}. \tag{6.1.4}$$

Hence the likelihood ratio test (6.1.2) may be restated: Reject H if

$$\frac{n-p+2}{n(p-1)} N\bar{\mathbf{y}}'C'(CSC')^{-1}C\bar{\mathbf{y}} \geq F_{p-1,n-p+2,\alpha}, \tag{6.1.5}$$

where $F_{p-1,n-p+2}$ is the upper $\alpha 100\%$ point of the F-distribution with $p-1$ and $n-p+2$ degrees of freedom. Note that the statistic in (6.1.5) does not depend on the choice of C so long as $C\mathbf{e} = \mathbf{0}$. Also, if we use a pooled estimate S_p on f df, then in (6.1.5) $n-p+2$ changes to $f-p+2$, and S changes to S_p.

Simultaneous $(1-\alpha)100\%$ confidence intervals for $\mathbf{a}'\boldsymbol{\mu}$, where the vector $\mathbf{a}$ is such that

$$\mathbf{a}'\mathbf{e} \equiv \sum_{i=1}^{p} a_i = 0,$$

are given by

$$\mathbf{a}'\bar{\mathbf{y}} - T_\alpha N^{-1/2}(\mathbf{a}'S\mathbf{a})^{1/2} \leq \mathbf{a}'\boldsymbol{\mu} \leq \mathbf{a}'\bar{\mathbf{y}} + T_\alpha N^{-1/2}(\mathbf{a}'S\mathbf{a})^{1/2}, \tag{6.1.6}$$

where

$$T_\alpha^2 = \frac{n(p-1)}{n-p+2} F_{p-1,n-p+2,\alpha}. \tag{6.1.7}$$

These confidence bounds for $\mathbf{a}'\boldsymbol{\mu}$ are narrower than those given in Section 3.3.1, because the values of $\mathbf{a}$ are restricted to the form $\mathbf{a}'\mathbf{e}=0$. Note that if we are interested in a finite number of contrasts, say k, then Bonferroni's inequality may give shorter intervals. Thus, if $t_{n,\alpha/2k}^2 \le T_\alpha^2$, then Bonferroni's bounds should be used with T_α replaced by $t_{n,\alpha/2k}$ in (6.1.6).

It may, however, be the case that we are not interested in the confidence intervals for $\mathbf{a}'\boldsymbol{\mu}$. If the hypothesis H: $\boldsymbol{\mu}=\mathbf{e}\gamma$ is accepted, we are instead interested in the estimates and confidence intervals of γ—testing about γ rather than $\boldsymbol{\mu}$. The maximum likelihood estimate of γ is given by

$$\hat{\gamma}=\mathbf{e}'S^{-1}\bar{\mathbf{y}}/\mathbf{e}'S^{-1}\mathbf{e}, \qquad \mathbf{e}=(1,1,\ldots,1)'. \tag{6.1.8}$$

Now $E(\bar{\mathbf{y}})=\mathbf{e}\gamma$, and we are minimizing the sum of squares

$$(\bar{\mathbf{y}}-\gamma\mathbf{e})'S^{-1}(\bar{\mathbf{y}}-\gamma\mathbf{e}) \tag{6.1.9}$$

with respect to γ, since $\boldsymbol{\mu}=\gamma\mathbf{e}$ is a regression model for the correlated variables. A $(1-\alpha)100\%$ confidence interval for γ is given by

$$\begin{aligned}\hat{\gamma}-N^{-1/2}t_{n-p+1,\alpha/2}\left(1+n^{-1}T_{p-1}^2\right)^{1/2}(\mathbf{e}'S^{-1}\mathbf{e})^{-1/2}\\ \le\gamma\le\hat{\gamma}+N^{-1/2}t_{n-p+1,\alpha/2}\left(1+n^{-1}T_{p-1}^2\right)^{1/2}(\mathbf{e}'S^{-1}\mathbf{e})^{-1/2},\end{aligned} \tag{6.1.10}$$

where

$$T_{p-1}^2=N\bar{\mathbf{y}}'C'(CSC')^{-1}C\bar{\mathbf{y}}, \tag{6.1.11}$$

and $t_{n-p+1,\alpha/2}$ is the upper $(\alpha/2)100\%$ point of the t-distribution with $n-p+1$ df. Note that $t_{n-p+1,\alpha/2}^2=F_{1,n-p+1,\alpha}$. Thus, assuming that the model $E(\bar{\mathbf{y}})=\mathbf{e}\gamma$ is correct, (6.1.10) can be used to test any hypothesis about γ. If the hypothesis value lies in the interval (6.1.10), the hypothesis is accepted; otherwise, it is rejected. Note that (6.1.10) *differs* from several available in the literature; it gives a shorter confidence interval and uses the results of Section 3.5.

For Example 6.1.2, the null hypothesis becomes

$$H\colon C\boldsymbol{\mu}=\mathbf{0},$$

where C is a 2×3 matrix orthogonal to $(1,1,1)$. For example, C could be

$$\begin{pmatrix}1 & -1 & 0\\ 0 & 1 & -1\end{pmatrix}.$$

From (6.1.5) we have

$$F=44.25\ge 6.94=F_{2,4,0.05}.$$

$$\bar{\mathbf{y}}=\begin{pmatrix}5.5\\ 10.67\\ 19.17\end{pmatrix}$$

Table 6.3

	Cow							
Time (minutes)	1	2	3	4	5	6	7	8
0	20	28	22	20	28	36	30	28
5	36	20	24	36	32	44	32	38
10	28	52	16	18	30	52	28	28
15	20	44	26	12	30	48	30	20

is the sample mean and the sample covariance matrix is

$$S=\begin{pmatrix} 1.9 & 1.8 & 4.3 \\ 1.8 & 2.67 & 6.07 \\ 4.3 & 6.07 & 17.37 \end{pmatrix}.$$

Hence one concludes that there is a change in response with time.

Example 6.1.3 The data of Table 6.3 represent respiratory rates of eight cows over 15 min after administration of the drug Ketamin. For the data

$$\bar{\mathbf{x}}=\begin{pmatrix} 26.5 \\ 32.8 \\ 31.5 \\ 28.8 \end{pmatrix}, \qquad S=\begin{pmatrix} 30.6 & 12.7 & 59.4 & 53.9 \\ 12.7 & 59.4 & 5.0 & -12.1 \\ 55.1 & 5.0 & 186.0 & 147.3 \\ 53.9 & -12.1 & 147.3 & 149.5 \end{pmatrix}.$$

The hypothesis of interest is

$$H\colon \mu_1=\mu_2=\mu_3=\mu_4 \qquad \text{vs} \qquad A\colon \mu_i\neq\mu_j$$

for some $i\neq j$, i, j, $=1,2,3,4$. This hypothesis can be written in the form of (6.1.4) with

$$C=\begin{pmatrix} 1 & -1 & 0 & 0 \\ 1 & 0 & -1 & 0 \\ 1 & 0 & 0 & -1 \end{pmatrix}.$$

The T^2-value is 15.3, with $[(N-p+1)/(p-1)n]T^2=3.64\leq F_{3,5,0.05}=5.41$. Hence the respiratory rate does not change significantly with time after the injection of Ketamin.

Example 6.1.4 The data of Table 6.4 consist of weights (in centigrams) of cork borings reported by Rao (1948) north (N), east (E), south (S), and west (W) of the trunk for 28 trees in a block of plantations. The problem is to test whether the bark deposit varies in thickness, and hence in weight, in the four directions.

It was suggested by Mahalanobis that the bark deposit is likely to be uniform in all directions but less so E and W, so that N−E−W+S can be taken as the best contrast.

Table 6.4 Weights of Cork Borings for the Four Directions for 28 Trees

N	E	S	W	N	E	S	W
72	66	76	77	91	79	100	75
60	53	66	63	56	68	47	50
56	57	64	58	79	65	70	61
41	29	36	38	81	80	68	58
32	32	35	36	78	55	67	60
30	35	34	26	46	38	37	38
39	39	31	27	39	35	34	37
42	43	31	25	32	30	30	32
37	40	31	25	60	50	67	54
33	29	27	36	35	37	48	39
32	30	34	28	39	36	39	31
63	45	74	63	50	34	37	40
54	46	60	52	43	37	39	50
47	51	52	43	48	54	57	43

The mean vector is given by

$$\bar{\mathbf{y}}' = (50.535, 46.179, 49.679, 45.179),$$

and the sample covariance matrix is given by

$$S = \begin{pmatrix} 290.41 & 223.75 & 288.49 & 226.27 \\ 223.75 & 219.93 & 229.06 & 171.37 \\ 288.49 & 229.06 & 350.00 & 259.54 \\ 226.27 & 171.37 & 259.54 & 226.00 \end{pmatrix}.$$

It is of interest to test the hypothesis

$$H: \mu_1 = \mu_2 = \mu_3 = \mu_4 \qquad \text{vs} \qquad A: \mu_i \neq \mu_j$$

for some i, j, $i \neq j$, or equivalently

$$H: C\boldsymbol{\mu} = 0 \qquad \text{vs} \qquad A: C\boldsymbol{\mu} \neq 0,$$

where

$$C = \begin{pmatrix} 1 & -1 & 0 & 0 \\ 0 & 1 & -1 & 0 \\ 0 & 0 & 1 & -1 \end{pmatrix}.$$

The T_α^2-value is

$$T_\alpha^2 = \frac{n(p-1)}{N-p+1} F_{p-1, N-p+1, \alpha} = \frac{3 \times 27}{25} \times 2.99 = 9.7.$$

The calculated value of T^2 is

$$T^2 = N(C\bar{\mathbf{y}})'(CSC')^{-1}(C\bar{\mathbf{y}}) = 20.7420.$$

Hence we conclude that there is a significant difference in the weights from the four directions. A 95% confidence interval for the best contrast $\mu_1-\mu_2+\mu_3-\mu_4$ is given, with $\mathbf{a}=(1,-1,1,-1)$, $N=28$, by

$$\mathbf{a}'\bar{\mathbf{y}}\pm t_{N-1,\alpha/2}\sqrt{\mathbf{a}'S\mathbf{a}/N}=8.857\pm 2.052\sqrt{128.72/28}=8.857\pm 4.39.$$

This confidence interval is obtained by using Bonferroni's inequality with the number of contrasts $k=1$. For this case the confidence interval is exactly 95% (for one contrast), whereas if (6.1.6) is used, the confidence interval is

$$8.837\pm 6.6734,$$

a much wider interval. However, zero is included in the interval in both cases.

6.2 Polynomial Regression

Suppose that N subjects receive a treatment and that their responses over p time intervals are recorded. The model can be written

$$Y_{p\times N}=\boldsymbol{\mu}_{p\times 1}\mathbf{e}_{1\times N}+\boldsymbol{\varepsilon}_{p\times N}, \tag{6.2.1}$$

where y_{ij} is the response of the jth subject at time i, $\boldsymbol{\mu}$ is the population mean vector of the responses at each time period, $\mathbf{e}'=(1,\dots,1)$, and ε is a $p\times N$ matrix whose columns are independently distributed as $N(\mathbf{0},\Sigma)$. Let

$$Y=(\mathbf{y}_1,\dots,\mathbf{y}_N)\qquad\text{and}\qquad \varepsilon=(\boldsymbol{\varepsilon}_1,\dots,\boldsymbol{\varepsilon}_N); \tag{6.2.2}$$

the model can now be rewritten

$$\mathbf{y}_j=\boldsymbol{\mu}+\boldsymbol{\varepsilon}_j,\qquad j=1,2,\dots,N. \tag{6.2.3}$$

To test that there is a response over time, one tests the hypothesis H: $C\boldsymbol{\mu}=0$ vs A: $C\boldsymbol{\mu}\neq\mathbf{0}$, where C is chosen such that $C\mathbf{e}=\mathbf{0}$. If there is a significant change in time, what order of polynomial should be fitted to the data? If a polynomial of order $r-1$, $r\leq p$, is to be fitted, the model (6.2.3) becomes

$$E(\mathbf{y}_j)=B\boldsymbol{\xi},\qquad j=1,2,\dots,N, \tag{6.2.4}$$

where B is the $p\times r$ matrix

$$B=\begin{pmatrix}1 & t_1 & \cdots & t_1^{r-1}\\ 1 & t_2 & \cdots & t_2^{r-1}\\ \vdots & \vdots & & \vdots\\ 1 & t_p & \cdots & t_p^{r-1}\end{pmatrix}\qquad\text{and}\qquad \boldsymbol{\xi}=\begin{pmatrix}\xi_0\\ \xi_1\\ \vdots\\ \xi_{r-1}\end{pmatrix}. \tag{6.2.5}$$

6.2.1 Test of the Adequacy of the Model

To check the adequacy of the model (6.2.4), we need to test the hypothesis

$$H: \boldsymbol{\mu} = B\boldsymbol{\xi} \qquad \text{vs} \qquad A: \boldsymbol{\mu} \neq B\boldsymbol{\xi}, \tag{6.2.6}$$

where B is a known $p \times r$ matrix of rank r. Let C be a $(p-r)\times p$ matrix of full rank such that

$$CB = 0. \tag{6.2.7}$$

For example, if we let

$$D = \begin{pmatrix} t_1^r & \cdots & t_1^{p-1} \\ \vdots & & \vdots \\ t_p^r & \cdots & t_p^{p-1} \end{pmatrix},$$

then one choice of C could be

$$C = D'\left[I - B(B'B)^{-1}B'\right].$$

It may be noted that D can be any $p\times(p-r)$ matrix as long as $(B: D)$ is nonsingular. (See Remark 6.3.1.)

If equal-time points are chosen, then C can be obtained from tables of orthogonal polynomials (Pearson and Hartley, 1948). The likelihood ratio test for H vs A [see Srivastava and Khatri (1979) or Rao (1959)] rejects H if

$$\frac{n-p+r+1}{(N-1)(p-r)} N\bar{\mathbf{y}}'C'(CSC')^{-1}C\bar{\mathbf{y}} \geq F_{p-r,\, n-p+r-1}, \tag{6.2.8}$$

where

$$nS = \sum_{\alpha=1}^{N} (\mathbf{y}_\alpha - \bar{\mathbf{y}})(\mathbf{y}_\alpha - \bar{\mathbf{y}})', \quad \bar{\mathbf{y}} = N^{-1}\sum_{\alpha=1}^{N} \mathbf{y}_\alpha, \qquad n = N-1, \tag{6.2.9}$$

and $F_{p-r,\, n-p+r+1,\, \alpha}$ is the upper $\alpha100\%$ of the F-distribution with $p-r$, $n-p+r+1$ df. If we use a pooled estimate S_p on f degrees of freedom in place of S, then in the above formula (as well as in the formula for confidence intervals given below) $n-p+r+1$ changes to $f-p+r+1$.

Note again that the formula in (6.2.8) does not depend upon the choice of C so long as C satisfies (6.2.7). A $(1-\alpha)100\%$ confidence bound for $\mathbf{a}'\boldsymbol{\mu}$, where $\mathbf{a} \neq \mathbf{0}$ such that $\mathbf{a}'B = \mathbf{0}'$, is

$$\mathbf{a}'\bar{\mathbf{y}} - T_\alpha N^{-1/2}(\mathbf{a}'S\mathbf{a})^{1/2} \leq \mathbf{a}'\boldsymbol{\mu} \leq \mathbf{a}'\bar{\mathbf{y}} + T_\alpha N^{-1/2}(\mathbf{a}'S\mathbf{a})^{1/2}, \tag{6.2.10}$$

where

$$T_\alpha = \frac{(N-1)(p-r)}{n-p+r+1} F_{p-r,\, n-p+r+1,\, \alpha}. \tag{6.2.11}$$

6.2.2 Estimates and Confidence Intervals

Suppose the hypothesis H of (6.2.6) is accepted. Then we are interested in the estimate and confidence intervals of $\boldsymbol{\xi}$. The maximum likelihood estimate of $\boldsymbol{\xi}$ is given by

$$\hat{\boldsymbol{\xi}}=(B'S^{-1}B)^{-1}B'S^{-1}\bar{\mathbf{y}}. \tag{6.2.12}$$

Simultaneous $(1-\alpha)100\%$ confidence intervals for $\mathbf{a}'\boldsymbol{\xi}$ are

$$\mathbf{a}'\hat{\boldsymbol{\xi}}-N^{-1/2}T_{\alpha}\left(1+n^{-1}T_{p-r}^{2}\right)^{1/2}\left[\mathbf{a}'(B'S^{-1}B)^{-1}\mathbf{a}\right]^{1/2}$$
$$\leq \mathbf{a}'\boldsymbol{\xi}\leq \mathbf{a}'\hat{\boldsymbol{\xi}}+N^{-1/2}T_{\alpha}\left(1+n^{-1}T_{p-r}^{2}\right)^{1/2}\left[\mathbf{a}'(B'S^{-1}B)^{-1}\mathbf{a}\right]^{1/2}, \tag{6.2.13}$$

where

$$T_{\alpha}^{2}=\frac{nr}{n-p+1}F_{r,n-p+1,\alpha}$$

and

$$T_{p-r}^{2}=N\bar{\mathbf{y}}'C'(CSC')^{-1}C\bar{\mathbf{y}}, \qquad n=N-1. \tag{6.2.14}$$

6.2.3 Test of the General Hypotheses

It will be interesting to test a hypothesis of the kind

$$H\colon U\boldsymbol{\xi}=\boldsymbol{\eta}_0 \qquad \text{vs} \qquad A\colon U\boldsymbol{\xi}\neq\boldsymbol{\eta}_0, \tag{6.2.15}$$

where $\boldsymbol{\eta}_0$ is specified and $U\colon k\times r$, $k\leq r$. The hypothesis H is rejected if

$$\frac{N-k-p+r}{nk}N\frac{(\mathbf{d}-\boldsymbol{\eta}_0)'(UEU')^{-1}(\mathbf{d}-\boldsymbol{\eta}_0)}{1+n^{-1}T_{p-r}^{2}}>F_{k,N-k-p+r,\alpha}, \tag{6.2.16}$$

where

$$\mathbf{d}=U\boldsymbol{\xi}, \qquad T_{p-r}^{2}=N\bar{\mathbf{y}}'C'(CSC')^{-1}C\bar{\mathbf{y}} \tag{6.2.17}$$

and

$$E=(B'S^{-1}B)^{-1}. \tag{6.2.18}$$

Simultaneous $(1-\alpha)100\%$ confidence limits for parameters of the form $\mathbf{u}'\boldsymbol{\xi}$, where $\mathbf{u}$ is in the subspace spanned by the vectors of U, are

$$\mathbf{u}'\hat{\boldsymbol{\xi}}\pm N^{-1/2}T_{\alpha}\left(1+n^{-1}T_{p-r}^{2}\right)^{1/2}(\mathbf{u}'E\mathbf{u})^{1/2}, \tag{6.2.19}$$

where

$$T_{\alpha}^{2}=\frac{nk}{n-p+(r-k+1)}F_{k,n-p+(r-k+1),\alpha}. \tag{6.2.20}$$

Table 6.5[a]

Subject	Age (years) 8	10	12	14
1	21.0	20.0	21.5	23.0
2	21.0	21.5	24.0	25.5
3	20.5	24.0	24.5	26.0
4	23.5	24.5	25.0	26.5
5	21.5	23.0	22.5	23.5
6	20.0	21.0	21.0	22.5
7	21.5	22.5	23.0	25.0
8	23.0	23.0	23.5	24.0
9	20.0	21.0	22.0	21.5
10	16.5	19.0	19.0	19.5
11	24.5	25.0	28.0	28.0
means ($\bar{\mathbf{y}}'$)	21.2	22.2	23.1	24.1
population means	μ_1	μ_2	μ_3	μ_4

[a]Data from Pothoff and Roy (1964).

Example 6.2.1 The data in Table 6.5 consist of dental measurements for girls 8–14 years old (see Problem 6.4). The sample covariance matrix is given by

$$S=\begin{pmatrix} 4.51 & 3.36 & 4.33 & 4.36 \\ 3.36 & 3.62 & 4.03 & 4.08 \\ 4.33 & 4.03 & 5.59 & 5.47 \\ 4.36 & 4.08 & 5.47 & 5.94 \end{pmatrix}$$

on 10 df. Suppose we fit a linear model of the form $E(\mathbf{y})=\xi_0+\xi_1 t$ where $t+11$ denotes age (in years). Then to test for the fit of the model we test

$$H\colon \boldsymbol{\mu}=B\boldsymbol{\xi} \qquad \text{vs} \qquad A\colon \boldsymbol{\mu}\neq B\boldsymbol{\xi},$$

where $\boldsymbol{\xi}'=(\xi_0,\xi_1)$, $\boldsymbol{\mu}'=(\mu_1,\mu_2,\mu_3,\mu_4)$, and

$$B'=\begin{pmatrix} 1 & 1 & 1 & 1 \\ -3 & -1 & 1 & 3 \end{pmatrix}.$$

To find a C such that $CB=0$ and C is of rank $p-2=2$, we can go to orthogonal polynomials to find

$$C=\begin{pmatrix} 1 & -1 & -1 & 1 \\ -1 & 3 & -3 & 1 \end{pmatrix}.$$

Another choice of C would be

$$C_1=D'\left(I-B(B'B)^{-1}B'\right)$$

where

$$D=\begin{pmatrix} t_1^2 & t_2^2 & t_3^2 & t_4^2 \\ t_1^3 & t_2^3 & t_3^3 & t_4^3 \end{pmatrix}=\begin{pmatrix} 9 & 1 & 1 & 9 \\ -27 & -1 & 1 & 27 \end{pmatrix}$$

and

$$B'=\begin{pmatrix} 1 & 1 & 1 & 1 \\ -3 & -1 & 1 & 3 \end{pmatrix}.$$

Such a C_1 can tend to be unmanageable. It may be normalized to

$$C=(C_1C_1')^{-1/2}C_1$$

where $(C_1C_1')^{-1/2}$ is any square root of $(C_1C_1')^{-1}$. The SAS function HALF in PROC MATRIX produces such a square root. This choice of C yields

$$C=\begin{pmatrix} \frac{1}{2} & -\frac{1}{2} & -\frac{1}{2} & \frac{1}{2} \\ -\frac{1}{20} & \frac{3}{20} & -\frac{3}{20} & \frac{1}{20} \end{pmatrix},$$

which is identical to the matrix of orthogonal polynomial coefficients with rows normalized such that the sum of squares across is 1. Testing H: $C\boldsymbol{\mu}=0$ yields

$$F_{2,9}=0.05\leq 4.26=F_{2,9,0.05}.$$

The T^2-value was calculated to be 0.11. Hence we claim that the quadratic and cubic terms are zero; that is, a linear trend in time is adequate. Estimates of the parameters are given by

$$\hat{\boldsymbol{\xi}}=(B'S^{-1}B)^{-1}B'S^{-1}\bar{\mathbf{y}}=\begin{pmatrix} 22.70 \\ 0.491 \end{pmatrix}.$$

Hence a model for this dental measurement y is given by

$$E(y)=22.73+0.482t.$$

In terms of age, we have

$$E(y)=17.4+0.482\times\text{age}.$$

A simultaneous 95% confidence band for $\mathbf{a}'\boldsymbol{\xi}$ is given by

$$a'\hat{\xi}+N^{-1/2}T_{0.05}\left(1+n^{-1}T_{p-r}^2\right)^{1/2}\left[\mathbf{a}'(B'S^{-1}B)^{-1}\mathbf{a}\right]^{1/2}.$$

If we choose $\mathbf{a}'=(1,t)$, then we obtain a simultaneous confidence band for the line given by $\xi_0+\xi_1 t$. To obtain this confidence band for our example we use the following values: $N=11$, $n=10$, $r=2$, $p=4$, $T_{p-r}^2=0.11$,

$$T_{0.05}^2=[(10\times 2)/7]F_{2,7,0.05}=3.68,$$

$$(B'S^{-1}B)^{-1}=\begin{pmatrix} 3.807 & 0.160 \\ 0.160 & 0.045 \end{pmatrix}.$$

Hence a confidence band for $\xi_0+\xi_1 t$ is given by

$$(22.70+0.491t)\pm(11)^{-1/2}3.68[1+(0.1)(0.111)]^{1/2}$$
$$\times\left[\begin{pmatrix}1\\t\end{pmatrix}\begin{pmatrix}3.807 & 0.160\\0.160 & 0.045\end{pmatrix}\begin{pmatrix}1\\t\end{pmatrix}\right]^{1/2}$$

or

$$(22.73+0.482t)\pm1.005(3.807+0.320t+0.045t^2)^{1/2}.$$

Assuming a linear trend in time, we may now wish to test that this trend is significant. So far we know that the quadratic and cubic trends are zero, but we do not know that the linear time is not also zero. Hence we test the hypothesis H: $\mathbf{U}\boldsymbol{\xi}=\zeta_0$ where $\mathbf{U}=(0,1)$ and $\zeta_0=0$, vs A: $\mathbf{U}\boldsymbol{\xi}\neq\zeta_0$:

$$H: \xi_1=0 \qquad \text{vs} \qquad A: \xi_1\neq 0.$$

From (6.2.17) we have—with $k=1$, $r=2$, $p=4$, $T_2^2=0.11$, $n=10$, $N=11$, $d=0.482$, $\xi_0=0$—

$$E=\begin{pmatrix}3.807 & 0.160\\0.160 & 0.160\end{pmatrix} \qquad \text{and} \qquad \mathbf{U}'E\mathbf{U}=0.045.$$

Hence our test statistic has the value

$$F=8.8(0.482)^2(0.045)^{-1}(1.011)^{-1}=45.94>F_{1,8,0.05}=5.32,$$

and $\xi_1\neq 0$.

A 95% confidence interval for ξ_1 may now be obtained from (6.2.20) with $\mathbf{u}=(0,1)'$, and $T_{0.05}^2=[(10\times1)/8]F_{1,8,0.05}=6.65$. Substituting in (6.2.20) we obtain the interval

$$0.482\pm(11)^{-1/2}(6.65)^{1/2}(1.011)^{1/2}(0.045)^{1/2}$$
$$=0.482\pm0.166=(0.316,0.648).$$

6.3 Generalized MANOVA

In the previous sections a polynomial curve was fitted to the growth data of one population. If the growth curves of several populations are to be compared, one sets up the general model

$$Y_{p\times N}=B_{p\times q}\xi_{q\times m}A_{m\times N}+\varepsilon_{p\times N}, \tag{6.3.1}$$

where B and ξ are as defined in Section 6.2 and A is now the design matrix of the experiment. ε is a matrix of random variables whose columns are independently distributed as $N_p(\mathbf{0},\Sigma)$.

For example, if u_t and v_t represent the response at times $t=0,1,2,3$ from two populations and a second-degree polynomial is fitted, then

$$E(u_t)=\beta_0+\beta_1 t+\beta_2 t^2,$$
$$E(v_t)=\gamma_0+\gamma_1 t+\gamma_2 t^2,$$

$t=0,1,2,3$. Suppose N_1 and N_2 observations are taken from the two populations; that is, let

$$U=\begin{pmatrix} u_{10} & u_{20} & \cdots & u_{N_1 0} \\ u_{11} & u_{21} & \cdots & u_{N_1 1} \\ u_{12} & u_{22} & \cdots & u_{N_1 2} \\ u_{13} & u_{23} & \cdots & u_{N_1 3} \end{pmatrix},$$

$$V=\begin{pmatrix} v_{10} & v_{20} & \cdots & v_{N_2 0} \\ v_{11} & v_{21} & \cdots & v_{N_2 1} \\ v_{12} & v_{22} & \cdots & v_{N_2 2} \\ v_{13} & v_{23} & \cdots & v_{N_2 3} \end{pmatrix}$$

represent the observations matrices. Then

$$E(U)=(\boldsymbol{\mu},\boldsymbol{\mu},\ldots,\boldsymbol{\mu})=\boldsymbol{\mu}\mathbf{e}'_{N_1}=B\boldsymbol{\xi}_1\mathbf{e}'_{N_1},$$
$$E(V)=(\boldsymbol{\nu},\boldsymbol{\nu},\ldots,\boldsymbol{\nu})=\boldsymbol{\nu}\mathbf{e}'_{N_2}=B\boldsymbol{\xi}_2\mathbf{e}'_{N_2},$$

where

$$B=\begin{pmatrix} 1 & 0 & 0 \\ 1 & 1 & 1 \\ 1 & 2 & 4 \\ 1 & 3 & 9 \end{pmatrix}, \qquad \xi_1=\begin{pmatrix} \beta_0 \\ \beta_1 \\ \beta_2 \end{pmatrix}, \qquad \xi_2=\begin{pmatrix} \gamma_0 \\ \gamma_1 \\ \gamma_2 \end{pmatrix},$$

and $\mathbf{e}_r$ is an r-vector of 1s. Letting

$$Y=(U,V), \qquad \xi=(\xi_1,\xi_2), \qquad \text{and} \qquad A=\begin{pmatrix} \mathbf{e}'_{N_1} & \mathbf{0} \\ \mathbf{0} & \mathbf{e}'_{N_2} \end{pmatrix},$$

we can write the model

$$E(Y)=B\xi A.$$

For another example, suppose eight subjects are assigned to two treatments, four to each, with measurements taken at times $t=1,2,3$. Then if a linear polynomial is to be fitted to the data, the model will be

$$Y=B\xi A+\varepsilon,$$

where

$$B=\begin{pmatrix} 1 & 1 \\ 1 & 2 \\ 1 & 3 \end{pmatrix}, \qquad \xi=\begin{pmatrix} \beta_{01} & \beta_{02} \\ \beta_{11} & \beta_{12} \end{pmatrix},$$

$$A=\begin{pmatrix} 1 & 1 & 1 & 1 & 0 & 0 & 0 & 0 \\ 0 & 0 & 0 & 0 & 1 & 1 & 1 & 1 \end{pmatrix},$$

$$Y=\begin{pmatrix} y_{111} & y_{211} & y_{311} & y_{411} & y_{121} & y_{221} & y_{321} & y_{421} \\ y_{112} & y_{212} & y_{312} & y_{412} & y_{122} & y_{222} & y_{322} & y_{422} \\ y_{113} & y_{213} & y_{313} & y_{413} & y_{123} & y_{223} & y_{323} & y_{423} \end{pmatrix}.$$

That is,

$$E(y_{ijt}) = \beta_{0j} + t\beta_{1j}, \qquad i=1,2,3,4, \quad j=1,2, \quad t=1,2,3.$$

This model is generalized in the following example.

Example 6.3.1 Let us suppose that the N individuals can be divided into k homogeneous groups having $N_1, N_2, \ldots, N_k$ individuals, respectively, according to their growth pattern. For example, the growth patterns of boys and girls are different, and hence they form two different groups. We shall assume that the growth of an individual is a polynomial in time t, measured at $t=1,2,\ldots,T$. If y_{ijt} denotes the observed growth of the ith individual in the jth group at time t, then

$$y_{ijt} = \beta_{0j} + \beta_{1j}t + \cdots + \beta_{qj}t^q + \varepsilon_{ijt}$$

for $i=1,2,\ldots,N_j$, $j=1,2,\ldots,k$, and $t=1,2,\ldots,T$, where

$$E(\varepsilon_{ijt}) = 0, \qquad E(\varepsilon_{ijt}\varepsilon_{kls}) = \begin{cases} \sigma_{ts} & \text{if } i=k, \; j=l, \\ 0 & \text{otherwise.} \end{cases}$$

Let us set

$$\mathbf{y}_t' = \left(y_{11t}, \ldots, y_{N_1 1t}, y_{12t}, \ldots, y_{N_2 2t}, \ldots, y_{N_k kt}\right),$$

$$\boldsymbol{\varepsilon}_t' = \left(\varepsilon_{11t}, \ldots, \varepsilon_{N_1 1t}, \varepsilon_{12t}, \ldots, \varepsilon_{N_2 2t}, \ldots, \varepsilon_{N_k kt}\right),$$

$$\xi' = (\beta_{ij}), \qquad i=0,1,\ldots,q \quad \text{and} \quad j=1,2,\ldots,k.$$

Then

$$\mathbf{y}_t' = (1, t, \ldots, t^q)\xi A + \boldsymbol{\varepsilon}_t',$$

where $A = \text{diag}(\mathbf{e}_{N_1}', \mathbf{e}_{N_2}', \ldots, \mathbf{e}_{N_k}')$ with $\mathbf{e}_j' = (1,1,\ldots,1) : i \times j$, and $E(\mathbf{e}_t\mathbf{e}_s') = (I_N)\sigma_{t'}$, $E(\boldsymbol{\varepsilon}_t) = 0$. Hence, if $\varepsilon' = (\boldsymbol{\varepsilon}_1, \ldots, \boldsymbol{\varepsilon}_T)$, $Y' = (\mathbf{y}_1, \mathbf{y}_2, \ldots, \mathbf{y}_T)$, and

$$B = \begin{pmatrix} 1 & 1 & \cdots & 1 \\ 1 & 2 & \cdots & 2^q \\ \vdots & \vdots & & \vdots \\ 1 & T & \cdots & T^q \end{pmatrix},$$

then the above model is equivalent to (6.3.1); that is,

$$Y = B\xi A + \varepsilon$$

with $E(\varepsilon) = 0$ and $E(\boldsymbol{\varepsilon}_t\boldsymbol{\varepsilon}_s') = \sigma_{ts}I_N$. For more details, see Rao (1965, 1966, 1967), Potthoff and Roy (1964), Grizzle and Allen (1969), and Khatri (1966). Morrison (1970) discusses the selection of time intervals.

6.3.1 *Test of the Adequacy of the Model*

The full regression model of Chapter 5 was given by

$$Y = \beta A + \varepsilon, \tag{6.3.2}$$

where Y was a $p \times N$ matrix, β a $p \times m$ matrix, A an $m \times N$ matrix, and ε a $p \times N$ error matrix. For (6.3.2) to be written in the form

$$Y = B\xi A + \varepsilon, \tag{6.3.3}$$

where ξ is a $q \times m$ matrix of unknown parameter, β must be written $B\xi$. If we let C be a $(p-q)\times p$ matrix of full rank such that $CB=0$, the condition that $\beta = B\xi$ is equivalent to the condition $C\beta = 0$. Hence, to test for the adequacy of the model in (6.3.3), we test the hypothesis H: $C\beta = 0$ vs A: $C\beta \neq 0$. The test statistic is given in (5.3.3)–(5.3.7). That is, we let

$$V = Y\left(I - A'(AA')^{-1}A\right)Y' \tag{6.3.4}$$

and

$$W = YA'(AA')^{-1}AY'. \tag{6.3.5}$$

The sample covariance matrix is then $(N-m)^{-1}V$. H is rejected if

$$U_{(p-q),m,n} = |CVC'||C(V+W)C'|^{-1} \geq \text{const}, \tag{6.3.6}$$

where $n = N - m$ is the degrees of freedom for estimating the covariance matrix Σ. Asymptotically, we reject H: $C\beta = 0$ if

$$-\left[n - \tfrac{1}{2}(p - q - m + 1)\right]\ln U \geq \chi^2_{m(p-q),\alpha}. \tag{6.3.7}$$

Greater accuracy can be obtained by using (5.3.7) to find the p-value.

REMARK 6.3.1: The matrix C such that $CB = 0$ theoretically exists. However, it is not always easy to find such a C. An alternative expression is

$$U_{(p-q),m,n} = \left(|V|\,|B'V^{-1}B|\right) \big/ \left[|V+W|\,|B'(V+W^{-1})B|\right].$$

6.3.2 *Tests and Confidence Intervals*

Khatri (1966) showed that the maximum likelihood estimate of ξ is given by

$$\hat{\xi} = (B'S^{-1}B)^{-1}B'S^{-1}YA'(AA')^{-1}, \tag{6.3.8}$$

where

$$S = (N-m)^{-1}Y\left[I - A'(AA')^{-1}A\right]Y' \equiv (N-m)^{-1}v. \tag{6.3.9}$$

Note that in (6.3.8) we could use V in place of S.

For testing the hypothesis H: $F\xi G = 0$ vs A: $F\xi G \neq 0$, where $F: r \times q$ and G: $m \times t$, $\rho(F) = r \leq q$, and $\rho(G) = t \leq m$, the likelihood ratio test rejects

$H: F\xi G=0$ at the $\alpha 100\%$ level if

$$U_{r,t,n}=|V_1|/|W+V_1|\le U_{r,t,n,\alpha}, \qquad n=N-p+q-m, \quad (6.3.10)$$

where $U_{r,t,n}$ is the U-statistic defined in Chapter 4 and

$$V_1=FZ\left[I_N-X'(XX')^{-1}X\right]Z'F', \quad (6.3.11)$$

$$W=FZX'(XX')^{-1}D\left[D'(XX')D^{-1}\right]^{-1}D'(XX')^{-1}XZ'F', \quad (6.3.12)$$

where

$$Z=(B'B)^{-1}B'Y, \quad D=\binom{G}{0}, \quad X'=(A',Z_2'), \quad Z_2=CY,$$

for $Z: q\times N$, $B: p\times q$, $A: m\times N$, $C:(p-q)\times p$, $X:(p-q+\text{m})\times N$, and $D:(m+p-q)\times t$.

For large n,

$$-\left[n-\tfrac{1}{2}(r-t+1)\right]\ln U_{r,t,n}\simeq \chi^2_{rt}, \quad (6.3.13)$$

or one can use (5.3.7) to find p-values. Hence H is rejected if

$$-\left[n-\tfrac{1}{2}(r-t+1)\right]\ln U_{r,t,n}\ge \chi^2_{rt,\alpha}. \quad (6.3.14)$$

In terms of the original variables, the formulas for W and V_1 are given by

$$W=(F\hat{\xi}G)(G'RG)^{-1}(F\hat{\xi}G)', \quad (6.3.15)$$

and

$$V_1=F(B'V^{-1}B)^{-1}F', \quad (6.3.16)$$

respectively, where

$$R=(AA')^{-1}+(AA')^{-1}AY'\left[V^{-1}-V^{-1}B(B'V^{-1}B)^{-1}B'V^{-1}\right]YA'(AA')^{-1},$$

$$\hat{\xi}=(B'V^{-1}B)^{-1}B'V^{-1}YA'(AA')^{-1},$$

$$V=Y\left[I_N-A'(AA')^{-1}A\right]Y', \quad (6.3.17)$$

as defined earlier. Computations of the expressions in (6.3.8) and (6.3.15)–(6.3.17) can be made using APL programming. Some programs are given in Appendix II that seem simpler and more direct than the manipulations necessary to use statistical packages.

The $(1-\alpha)100\%$ simultaneous confidence limits of $\mathbf{a}'F\xi G\mathbf{b}$ are given by

$$\mathbf{a}'F\hat{\xi}G\mathbf{b}\pm\left[\frac{x_\alpha}{1-x_\alpha}(\mathbf{a}'V_1\mathbf{a})(\mathbf{b}'G'RG\mathbf{b})\right]^{1/2},$$

where x_α is given in Table A.5, with

$$p_0=r, \qquad m_0=t, \qquad \text{and} \qquad n=N-m-p+q.$$

Note that if $p_0=1$, $x_\alpha/(1-x_\alpha)$ should be replaced by $(m_0/n)F_{m_0,n,\alpha}$.

Table 6.6[a]

Concentration (μg/l)	Experiment 1					Experiment 2				Experiment 3				
Time (hr)	10	21	30	100	150	40	75	148	256	27	21	30	90	150
24	0	0	0	7	6	0	0	0	1	0	0	1	9	10
48	0	0	0	9	10	0	1	1	8	3	1	1	10	10
72	0	1	3	9	10	3	6	3	10	5	2	3	10	10
96	1	3	5	9	10	7	7	7	10	5	2	3	10	10
120	1	3	5	9	10	7	8	9	10	5	3	3	10	10

Concentration (μg/l)	Experiment 4					Experiment 5					Experiment 6				
Time (hr)	16	43	49	140	290	21	27	30	30	100	20	21	48	80	145
24	0	0	0	0	9	0	0	0	1	4	0	0	0	7	10
48	0	0	0	7	10	1	4	5	5	10	0	2	1	10	10
72	0	0	0	8	10	2	4	7	7	10	5	6	7	10	10
96	0	0	3	8	10	4	6	8	7	10	6	7	10	10	10
120	0	0	4	8	10	7	8	10	7	10	6	7	10	10	10

[a] Data from Hubert (1980), courtesy of J. Sprague.

Example 6.3.2 For several concentration levels of copper [(μg Cu^{++}/l H_2O)] the mortalities out of ten fish were recorded at specified time periods (in units of hours). The copper was added as sulfate, the water hardness was 355 mg/l, and the temperature was 15°C. The experiment was repeated six times (Table 6.6).

The first step was to convert the doses into $\log_{10}$ doses. Then an arcsine transformation was performed to standardize the response data:

$$y_i = \arcsin\sqrt{p_i/n_i}, \qquad i=1,2,\ldots,k,$$

where p_i is the number of fish that have died out of n_i subjects, n_i is the total number of fish per drug dose, and k is the number of dose levels. The following covariance matrix was obtained for the responses y_i, $i=1,\ldots,p$:

$$\begin{pmatrix} 576.02 & & & & \\ 441.31 & 602.60 & & & \\ 280.82 & 423.57 & 438.48 & & \\ 180.15 & 282.64 & 343.01 & 350.69 & \\ 134.08 & 274.72 & 327.69 & 350.43 & 394.67 \end{pmatrix}.$$

We let y_{ij} be transformed mortality rate for the jth dose of the ith time period, $i=1,\ldots,5$, $j=1,2,\ldots,29$. The $\log_{10}$ dose will be denoted x_j, $j=1,2,\ldots,29$. Suppose we try to fit a model of the form

$$E(y_{ij}) = \xi_{11} + \xi_{12}x_j + \xi_{21}t_i + \xi_{22}x_jt_i, \qquad i=1,\ldots,5, \quad j=1,2,\ldots,29. \tag{6.3.18}$$

That is, we fit

$$Y = B\xi A + \boldsymbol{\varepsilon},$$

where

$$Y = (y_{ij}), \qquad \xi = \begin{pmatrix} \xi_{11} & \xi_{12} \\ \xi_{21} & \xi_{22} \end{pmatrix},$$

$$A = \begin{pmatrix} 1 & 1 & \cdots & 1 \\ x_1 & x_2 & \cdots & x_{27} \end{pmatrix}, \qquad B = \begin{pmatrix} 1 & 1 \\ 1 & 2 \\ 1 & 3 \\ 1 & 4 \\ 1 & 5 \end{pmatrix}, \tag{6.3.19}$$

$\varepsilon = (\boldsymbol{\varepsilon}_1, \ldots, \boldsymbol{\varepsilon}_{29})$, and the ε_j are independent $N_5(0, \Sigma)$, $j = 1, 2, \ldots, 29$. In order to test for the adequacy of the model in time, we test the hypothesis

$$CE(Y) = 0,$$

where $CB = 0$. Once choice of C is given in (6.2.7). The value in this case is

$$C = \begin{pmatrix} 16.8 & -6.6 & -18 & -11.4 & 19.2 \\ 102.8 & -31 & -114.8 & -88.5 & 131.6 \\ 564 & -129 & -642 & -585 & 792 \end{pmatrix}. \tag{6.3.20}$$

Orthogonal polynomials could be used; but if we fitted a model, say, in $\log t$, then orthogonal polynomials are not available. The error sum of squares is given by $V = Y[\mathrm{I} - A(A'A)^{-1}A']Y'$ and the regression sum of squares by $YA(A'A)^{-1}AY'$. To test $CE(Y) = 0$, we use the formulas in (6.3.4)–(6.3.6). That is, we let

$$U_{r,m,n} = |CVC'||C(V+W)C'| = 0.3437, \tag{6.3.21}$$

with $r = 3$, $m = 2$, and $n = 27$. Asymptotically, $-[n - \frac{1}{2}(r - m + 1)]\ln U_{r,m,n} \simeq \chi^2_{rm}$. In this case, $-[n - \frac{1}{2}(r - m + 1)] \ln U = 26(0.3437) = 8.94 \leq 12.6 = \chi^2_{6,0.05}$.

Although this appears acceptable, it was felt that a quadratic model might fit a little better. Fitting a quadratic model in time yields

$$E(Y) = B\xi A$$

with

$$B = \begin{pmatrix} 1 & 1 & 1 \\ 1 & 2 & 4 \\ 1 & 3 & 9 \\ 1 & 4 & 16 \\ 1 & 5 & 25 \end{pmatrix} \quad \text{and} \quad \xi = \begin{pmatrix} \xi_{11} & \xi_{12} \\ \xi_{21} & \xi_{22} \\ \xi_{31} & \xi_{32} \end{pmatrix}. \tag{6.3.22}$$

To test the adequacy of the model, we test the hypothesis

$$H: CE(Y) = 0, \quad CB = 0, \qquad \text{vs} \qquad A: CE(Y) \neq 0.$$

In this case,

$$C = \begin{pmatrix} -1.2 & 2.4 & 0 & -2.4 & 1.2 \\ -14.06 & 27.43 & 2.06 & -30.17 & 14.74 \end{pmatrix}. \tag{6.3.23}$$

For this model, with $r=2$, $m=2$, and $n=27$,

$$-\left[n-\tfrac{1}{2}(r-m-1)\right]\ln U_{r,m,n}=\left[27-\tfrac{1}{2}(2-2+1)\right](0.1301)$$
$$=26.5(0.1301)=3.45\le 9.49=\chi^2_{4,0.05}.$$

Hence the model is not rejected. Estimates of the coefficients ξ are then

$$\hat{\xi}=(B'S^{-1}B)^{-1}B'S^{-1}YA'(AA')^{-1}=\begin{pmatrix}-69.20 & 38.95\\ 9.16 & 8.98\\ 0.22 & -1.53\end{pmatrix}, \quad (6.3.24)$$

and the estimated model is

$$\hat{y}=-69.20+38.95x+9.16t+8.98xt+0.22t^2-1.53xt^2. \quad (6.3.25)$$

Using this model, an estimate of the lethal dose required to kill 50% of the fish in a given time period (LD50) can be obtained by solving (6.3.25) for x at $Y=45$ ($\arcsin\sqrt{0.5}=45°$) for any time period t. The quadratic model yielded results similar to univariate probit analysis. See Carter and Hubert (1981).

Other models may fit the data and theory better. In Problem 6.6 the reader is asked to fit a polynomial in reciprocal time. Theoretically, it would seem to be a better representation of the process.

Example 6.3.3 Consider the data from Example 5.3.3, in which two diets were offered to rats: a control diet and a diet containing phosphorous. The 12 response variables were the amounts of each of the diets eaten by a rat over six days. It was found in the analysis that a model quadratic in the amount of phosphorous added fit the data well.

We might now wish to model the response in time as a function of time. That is, we might try to fit a third-degree polynomial to both diets. The model would then take the form

$$E(Y)=B\xi A,$$

where

$$B=\begin{bmatrix}1&1&1&1&0&0&0&0\\1&2&4&8&0&0&0&0\\1&3&9&27&0&0&0&0\\1&4&16&64&0&0&0&0\\1&5&25&125&0&0&0&0\\1&6&36&216&0&0&0&0\\0&0&0&0&1&1&1&1\\0&0&0&0&1&2&4&8\\0&0&0&0&1&3&9&27\\0&0&0&0&1&4&16&64\\0&0&0&0&1&5&25&125\\0&0&0&0&1&6&36&216\end{bmatrix}$$

and Y and A are as defined before. In symbols, the response for the control portion of the diet may be written

$$E(y)=\xi_{11}+\xi_{12}x+\xi_{13}x^2+\xi_{14}w+\xi_{21}t+\xi_{22}tx+\xi_{23}tx^2+\xi_{24}tw$$
$$+\xi_{31}t^2+\xi_{32}t^2x+\xi_{33}t^2x^2+\xi_{34}t^2w+\xi_{41}t^3$$
$$+\xi_{42}t^3x+\xi_{43}t^3x^2+\xi_{44}t^3w.$$

For the treatment diet, the response may be modeled

$$E(y)=\xi_{51}+\xi_{52}x+\xi_{53}x^2+\xi_{54}w+\xi_{61}t+\xi_{62}tx+\xi_{63}tx^2+\xi_{64}tw$$
$$+\xi_{71}t^2+\xi_{72}t^2x+\xi_{73}t^2x^2+\xi_{74}t^2w+\xi_{81}t^3$$
$$+\xi_{82}t^3x+\xi_{83}t^3x^2+\xi_{84}t^3w.$$

To test for the fit of the model, we follow the procedure in (6.2.7). That is, we complete the columns of B (for both diets) with fourth- and fifth-order terms in time to obtain the matrix D (12×4). The contrast matrix C (4×12) is then obtained:

$$C'=\left[I-B(B'B)^{-1}B'\right]D.$$

In this case

$$C'=\begin{bmatrix} 1.714285714 & 29.52380952 & 0 & 0\\ -5.142857143 & -87.61904762 & 0 & 0\\ 3.428571429 & 55.23809524 & 0 & 0\\ 3.428571429 & 64.76190476 & 0 & 0\\ -5.142857143 & -92.38095238 & 0 & 0\\ 1.714285714 & 30.47619048 & 0 & 0\\ 0 & 0 & 1.714285714 & 29.52380952\\ 0 & 0 & -5.142857143 & -87.61904762\\ 0 & 0 & 3.428571429 & 55.23809524\\ 0 & 0 & 3.428571429 & 64.76190476\\ 0 & 0 & -5.142857143 & -92.38095238\\ 0 & 0 & 1.714285714 & 30.47619048 \end{bmatrix}.$$

The U-statistic for testing

$$H\colon CE(Y)=0 \qquad \text{vs} \qquad A\colon CE(Y)\neq 0$$

is

$$U_{4,4,28}=0.3504.$$

Asymptotically,

$$-\left[28-\tfrac{1}{2}(4-4+1)\right]\ln 0.3504=28.83>\chi^2_{16,0.05}=26.29.$$

The p-value using a more accurate approximation is given by

$$\begin{aligned}&p(-27.5\ln U_{4,4,28}>28.83)\\ &\quad=p(\chi^2_{16}>28.83)+(28)^{-2}5\left[p(\chi^2_{20}>28.83)-p(\chi^2_{16}>28.83)\right]\\ &\quad=0.032+0.0003=0.032.\end{aligned}$$

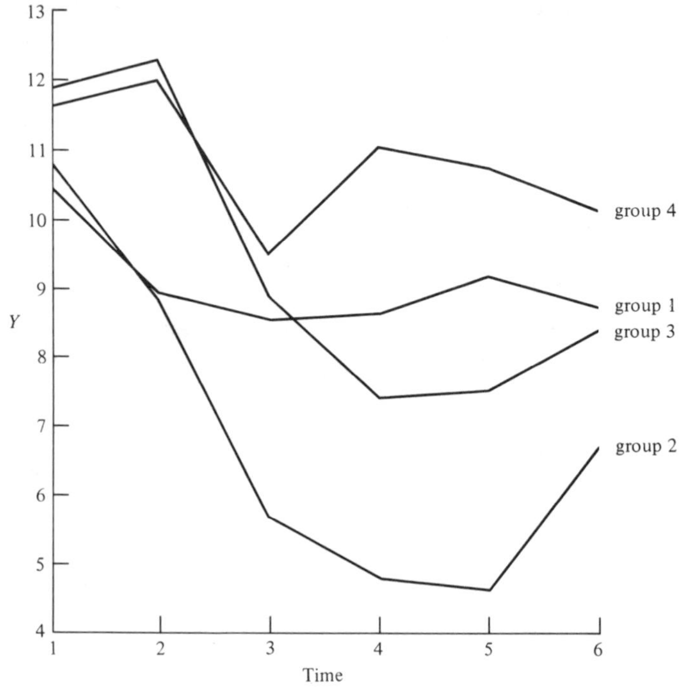

Figure 6.1 Amount Y of control diet vs time.

Hence H is rejected at the 5% level (with an observed significance level of 3.2%).

Therefore, there does not appear to be a way to reduce the dimension of the model: The control and treatment diets cannot be modeled as a polynomial in time. Perhaps other functions in time can be used to model this behavior over time properly. As it stands now, there is a quadratic response in phosphorus for each time period. The 0.25% P diet does not differ much from the control (0.1%); both are phosphorus deficient. Hence the rats chose either one arbitrarily. As the amount of phosphorus increases in the diet to 0.65%, the rats eat more of the treatment diets to consume the phosphorus. However, with ample phosphorus in the diet (1.3% and 1.71%) the rats no longer need much of the treatment diet to consume the necessary phosphorus, and they tend to eat the control diet. Looking at the plots of the mean value for each treatment group over time in Figures 6.1 and 6.2, we can see that the trend in time is not a simple polynomial. In fact, there may be a cyclic trend; that is, as the rats consume more phosphorus at one time interval, they consume less in the next. However, this behavior depends completely on the amount of phosphorus in the diet. It should be noted that a plot of the data is always useful. In this instance, Figures 6.1 and 6.2

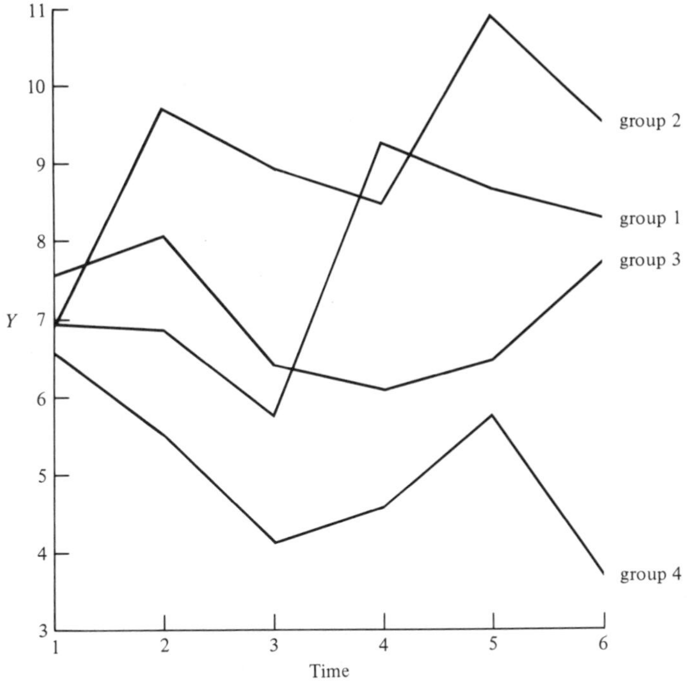

Figure 6.2 Amount Y of treatment diet vs time.

clearly indicate the lack of a polynomial trend in time. The reader is urged to plot his data before the analysis whenever possible.

6.4. Computational Procedures

To test for the degree of the polynomial in a simple case such as Example 6.1.3, one uses a Hotelling's T^2-test. The BMDP3D program performs this procedure. For this example there are four characteristics with population means μ_1, μ_2, μ_3, and μ_4. The hypothesis of interest is

$$H: \mu_1 = \mu_2 = \mu_3 = \mu_4 \qquad \text{vs} \qquad A: \mu_i \neq \mu_j$$

for some $i \neq j$, $i, j = 1,2,3,4$. Equivalently, if we define

$$\gamma_1 = \mu_1 - \mu_2, \qquad \gamma_2 = \mu_1 - \mu_3, \qquad \gamma_3 = \mu_1 - \mu_4,$$

we have

$$H: \gamma_1 = \gamma_2 = \gamma_3 = 0 \qquad \text{vs} \qquad A: \gamma_i \neq 0$$

for some $i = 1,2,3,4$. The following program cards were used:

```
/PROBLEM   TITLE IS 'KETAMIN'.
/INPUT   VARIABLES ARE 4.
    FORMAT IS '(4F2.0)'.
```

```
PROBLEM TITLE ..............................................KETAMIN
NUMBER OF VARIABLES TO READ IN ..................................      4
NUMBER OF VARIABLES ADDED BY TRANSFORMATIONS ....................      3
TOTAL NUMBER OF VARIABLES .......................................      7
NUMBER OF CASES TO READ IN ......................................1000000
CASE LABELING VARIABLES .........................................
LIMITS AND MISSING VALUE CHECKED BEFORE TRANSFORMATIONS
BLANKS ARE ........................................................ZEROS
INPUT UNIT NUMBER ...............................................      5
REWIND INPUT UNIT PRIOR TO READING..DATA .......................     NO
INPUT FORMAT
         (4F2.0)
*** CONTROL LANGUAGE TRANSFORMATIONS ARE PERFORMED ***

MAHALANOBIS D SQUARE        1.9096
HOTELLING T SQUARE         15.2769
F VALUE                     3.6374   P VALUE   0.099
  DEGREES OF FREEDOM    3,    5
```

DIFFERENCES ON SINGLE VARIABLES

```
* * * * * * * * *
*  E          *     VARIABLE NUMBER   5
* * * * * * * * *
                                          MEAN              -6.2500
  T STATISTIC    P VALUE       D.F.       STD DEV            8.0312
                                          S.E.M.             2.8395
      -2.20        0.064          7       SAMPLE SIZE             8
                                          MAXIMUM            8.0000        H                H
                                          MINIMUM          -16.0000        H        HH     HH         H
                                                                   MIN- - - - - - - - - - - - - - - - -MAX
                                                                              AN  H=          1 CASES

* * * * * * * * *
*  F          *     VARIABLE NUMBER   6
* * * * * * * * *
                                          MEAN              -5.0000
  T STATISTIC    P VALUE       D. F.      STD DEV           10.3095
                                          S.E.M.             3.6450
      -1.37        0.212          7       SAMPLE SIZE             8
                                          MAXIMUM            6.0000                          H
                                          MINIMUM          -24.0000        H         H      H  HH H H
                                                                   MIN- - - - - - - - - - - - - - - - -MAX
                                                                              AN  H=          1 CASES

* * * * * * * * *
*  G          *     VARIABLE NUMBER   7
* * * * * * * * *
                                          MEAN              -2.2500
  T STATISTIC    P VALUE       D.F.       STD DEV            8.5147
                                          S.E.M.             3.0104
      -0.75        0.479          7       SAMPLE SIZE             8
                                          MAXIMUM            8.0000                            H     H
                                          MINIMUM          -16.0000        H       H       HH  H     H
                                                                   MIN- - - - - - - - - - - - - - - - -MAX
                                                                              AN  H=          1 CASES

* * * * * * * * * * * * * * * * * * * * * * * * * * * * * * * * * * * * * * * * * * * * * * * * * * * * * * * * * * * * * * * * * * * * * * * * * * * * * * * * * * * * * *
```

```
/VARIABLE   NAMES ARE A,B,C,D,E,F,G.
    ADD=3.
/TRANSFORM   E=A-B.
    F=A-C.
    G=A-D.
/TEST   VARIABLES ARE 5 TO 7.
    HOTELLING.
/END
```

The program control cards were followed by the eight data cards. The program control cards indicate that four variables are to be read in a format 4F2.0. These are labeled A, B, C, and D. Also, three more variables are to be added. They are $E=A-B$, $F=A-C$, and $G=A-D$. The T^2-test is to be performed on the three new transformed variables. Some of the output from the program is listed below. The printout gives the T^2-value of 15.27 with a corresponding F-value of 3.634. Hence this is only significant at the 10% level. Individual t-tests are performed, and histograms are given for each variable E, F, and G.

General tests for Section 6.3 can be easily programmed using the formulas given. Copies of these programs are given in Appendix II.

Problems

6.1 The following data set, from Kovacic (1965), is analyzed by Jensen (1972). Two measurements, impact force (foot pounds) and acceleration (G units), were taken on four sides of 19 helmets, thus producing a multivariate situation of $N=19$ observations on each of $p=8$ characteristics. The experimenter wished to study the differences in the impact force and acceleration among the four sides. Jensen analyzed the data by performing two T^2-tests at the $\alpha=0.01$ level for acceleration and impact force, producing an overall level of significance of $\alpha=0.02$. The means for the four sides of the helmet for impact force are

location	front	back	right	left
mean	140.00	155.89	129.68	127.16

and the covariance is

$$\begin{pmatrix} 545.78 & 213.78 & 316.44 & 295.11 \\ 213.78 & 330.88 & 113.96 & 120.80 \\ 316.44 & 113.96 & 1049.45 & 698.39 \\ 295.11 & 120.80 & 698.39 & 753.03 \end{pmatrix}.$$

For the acceleration, the means are

location	front	back	right	left
mean	410.79	412.11	406.84	406.84

Table 6.7[a]

Rat	Week 1	2	3	4
1	29	28	25	33
2	33	30	23	31
3	25	34	33	41
4	18	33	29	35
5	25	23	17	30
6	24	32	29	22
7	20	23	16	31
8	28	21	18	24
9	18	23	22	28
10	25	28	29	30

[a]Data from Box (1950).

and the covariance is

$$\begin{pmatrix} 478.51 & 249.32 & -12.65 & 90.13 \\ 249.36 & 528.65 & 145.91 & 159.80 \\ -12.65 & 145.91 & 414.47 & 93.64 \\ 90.13 & 159.80 & 93.64 & 189.47 \end{pmatrix}.$$

Is there a difference in the impact force or acceleration among the four sides?

Table 6.8[a]

Individual	Age (years) 8	$8\frac{1}{2}$	9	$9\frac{1}{2}$
1	47.8	48.8	49.0	49.7
2	46.4	47.3	47.7	48.4
3	46.3	46.8	47.8	48.5
4	45.1	45.3	46.1	47.2
5	47.6	48.5	48.9	49.3
6	52.5	53.2	53.3	53.7
7	51.2	53.0	54.3	54.5
8	49.8	50.0	50.3	52.7
9	48.1	50.8	52.3	54.4
10	45.0	47.0	47.3	48.3
11	51.2	51.4	51.6	51.9
12	48.5	49.2	53.0	55.5
13	52.1	52.8	53.7	55.0
14	48.2	48.9	49.3	49.8
15	49.6	50.4	51.2	51.8
16	50.7	51.7	52.7	53.3
17	47.2	47.7	48.4	49.5
18	53.3	54.6	55.1	55.3
19	46.2	47.5	48.1	48.4
20	46.3	47.6	51.3	51.8
mean	48.66	49.62	50.57	51.45
s.d.	2.52	2.54	2.63	2.73

[a]Data from Elston and Grizzle (1962). Reproduced with the permission of the editors of *Biometrics*.

6.2 The data in Table 6.7 represent the weight gain in rats over a 4-week period under a standard diet. Is there a difference in the weight gain of the rats from week to week?

6.3 Consider the data in Table 6.8 on the growth of the ramus bone in boys.
(a) Test that a linear growth in time is an adequate model.
(b) Estimate the regression coefficients for the linear model and give a 95% simultaneous confidence band for the regression line.

6.4 Table 6.9 displays data collected by the investigators on 11 girls and 16 boys at four ages at the University of North Carolina Dental School. Each measurement is the distance, in millimeters, from the center of the pituitary to the pterygomaxillary fissure.
(a) Show that the growth curves due to boys and girls are linear at an appropriate significant level.
(b) Test whether the linear growth rats of boys and girls are equal at the 5% level. Give a 95% confidence interval for the difference between the two growth rates.
(c) Obtain 95% confidence intervals for boys' and girls' growth curves.

6.5 The plasma inorganic phosphate levels (mg/dl) were measured at intervals after a standard glucose challenge (Table 6.10).
(a) Fit a polynomial in time to this data. Zerbe suggests a fourth-degree polynomial. Test that the higher-order terms are insignificant.
(b) Test, using the generalized model of Section 6.3, whether the two groups have the same growth curves.

Table 6.9[a]

Girls					Boys				
	Age (years)					Age (years)			
Individual	8	10	12	14	Individual	8	10	12	14
1	21	20	21.5	23	1	26	25	29	31
2	21	21.5	24	25.5	2	21.5	22.5	23	26.5
3	20.5	24	24.5	26	3	23	22.5	24	27.5
4	23.5	24.5	25	26.5	4	25.5	27.5	26.5	27
5	21.5	23	22.5	23.5	5	20	23.5	22.5	26
6	20	21	21	22.5	6	24.5	25.5	27	28.5
7	21.5	22.5	23	25	7	22	22	24.5	26.5
8	23	23	23.5	24	8	24	21.5	24.5	25.5
9	20	21	22	21.5	9	23	20.5	31	26
10	16.5	19	19	19.5	10	27.5	28	31	31.5
11	24.5	25	28	28	11	23	23	23.5	25
					12	21.5	23.5	24	28
					13	17	24.5	26	29.5
					14	22.5	25.5	25.5	26
					15	23	24.5	26	30
					16	22	21.5	23.5	25

[a]Data from Potthoff and Roy (1964). Reproduced with the permission of the editor of *Biometrika*.

Table 6.10[a]

Patient	0	$\frac{1}{2}$	1	$1\frac{1}{2}$	2	3	4	5
				Control				
1	4.3	3.3	3.0	2.6	2.2	2.5	3.4	4.4
2	3.7	2.6	2.6	1.9	2.9	3.2	3.1	3.9
3	4.0	4.1	3.1	2.3	2.9	3.1	3.9	4.0
4	3.6	3.0	2.2	2.8	2.9	3.9	3.8	4.0
5	4.1	3.8	2.1	3.0	3.6	3.4	3.6	3.7
6	3.8	2.2	2.0	2.6	3.8	3.6	3.0	3.5
7	3.8	3.0	2.4	2.5	3.1	3.4	3.5	3.7
8	4.4	3.9	2.8	2.1	3.6	3.8	3.0	3.9
9	5.0	4.0	3.4	3.4	3.3	3.6	3.0	4.3
10	3.7	3.1	2.9	2.2	1.5	2.3	2.7	2.8
11	3.7	2.6	2.6	2.3	2.9	2.2	3.1	3.9
12	4.4	3.7	3.1	3.2	3.7	4.3	3.9	4.8
13	4.7	3.1	3.2	3.3	3.2	4.2	3.7	4.3
				Obese				
1	4.3	3.3	3.0	2.6	2.2	2.5	2.4	3.4
2	5.0	4.9	4.1	3.7	3.7	4.1	4.7	4.9
3	4.6	4.4	3.9	3.9	3.7	4.2	4.8	5.0
4	4.3	3.9	3.1	3.1	3.1	3.1	3.6	4.0
5	3.1	3.1	3.3	2.6	2.6	1.9	2.3	2.7
6	4.8	5.0	2.9	2.8	2.2	3.1	3.5	3.6
7	3.7	3.1	3.3	2.8	2.9	3.6	4.3	4.4
8	5.4	4.7	3.9	4.1	2.8	3.7	3.5	3.7
9	3.0	2.5	2.3	2.2	2.1	2.6	3.2	3.5
10	4.9	5.0	4.1	3.7	3.7	4.1	4.7	4.9
11	4.8	4.3	4.7	4.6	4.7	3.7	3.6	3.9
12	4.4	4.2	4.2	3.4	3.5	3.4	3.9	4.0
13	4.9	4.3	4.0	4.0	3.3	4.1	4.2	4.3
14	5.1	4.1	4.6	4.1	3.4	4.2	4.4	4.9
15	4.8	4.6	4.6	4.4	4.1	4.0	3.8	3.8
16	4.2	3.5	3.8	3.6	3.3	3.1	3.5	3.9
17	6.6	6.1	5.2	4.1	4.3	3.8	4.2	4.8
18	3.6	3.4	3.1	2.8	2.1	2.4	2.5	3.5
19	4.5	4.0	3.7	3.3	2.4	2.3	3.1	3.3
20	4.6	4.4	3.8	3.8	3.8	3.6	3.8	3.8

Column heading: Hours after glucose challenge.

[a]Data from Zerbe (1979). Reproduced with the permission of the American Statistical Association.

6.6 **(a)** For the data of Example 5.6.2 fit a polynomial function that is quadratic in reciprocal time and linear in $\log_{10}$ dose. Estimate the regression coefficients.

(b) Does the reduced model fit?

(c) Find the dose that kills 50% of the fish for a given time t. That is, for a given time t what is the dose (LD50) that produces a Y-value of 45? Plot LD50 vs time.

Chapter 7
Repeated Measures and Profile Analysis

7.1 Introduction

Experiments in which several treatments are applied to the same experimental subjects are called *repeated measures* experiments. For example, N subjects could be scored on their memory ability under p conditions or treatments. Such an experiment would have a model of the form

$$\mathbf{x}_j = \boldsymbol{\mu} + \boldsymbol{\varepsilon}_j, \tag{7.1.1}$$

where $\boldsymbol{\mu}' = (\mu_1, \ldots, \mu_p)$ and $\mathbf{x}_j$ is the p-vector of scores of the jth subject under each of the p treatments. The error term $\boldsymbol{\varepsilon}_j$ is distributed as $N_p(\mathbf{0}, \Sigma)$. If the p conditions are assigned to each subject in a random order, then any carryover effects from one treatment to another should be equal. Hence Σ may be assumed to be of the form

$$\Sigma = \sigma^2[I(1-\rho) + \rho\mathbf{ee}'], \tag{7.1.2}$$

where $\mathbf{e}' = (1, \ldots, 1)$, σ^2 is the variance of the responses under any treatment, and ρ is the correlation between responses under any two treatments. In this chapter we consider these problems and discuss what is known as *profile analysis*.

7.2 Repeated Measures for One Population

7.2.1 Equality of Means of Exchangeable Random Variables

In this section we consider the problem of Section 6.1.1 except that we assume a covariance matrix of the special form

$$\Sigma_I = \sigma^2 \begin{pmatrix} 1 & \rho & \cdots & \rho \\ \rho & 1 & \cdots & \rho \\ \vdots & \vdots & & \vdots \\ \rho & \rho & \cdots & 1 \end{pmatrix}$$

$$= \sigma^2[(1-\rho)I + \rho\mathbf{ee}'], \tag{7.2.1}$$

known as the *intraclass correlation* model. In this model the variables $x_1,\ldots,x_p$ are interchangeable with respect to variances and covariances. Thus, on the basis of a random sample $\mathbf{x}_1,\ldots,\mathbf{x}_N$ of size N from $N_p(\boldsymbol{\mu},\Sigma_I)$, we wish to test the hypothesis that $\mu_1=\mu_2=\cdots=\mu_p$, where $\boldsymbol{\mu}=(\mu_1,\ldots,\mu_p)'$. That is, we wish to test

$$H\colon \boldsymbol{\mu}=\gamma\mathbf{e}, \qquad \mathbf{e}=(1,1,\ldots,1)', \quad \gamma \text{ unknown},$$

vs

$$A\colon \mu_i\neq\mu_j, \qquad \text{for at least one pair} \quad (i,j), \qquad (7.2.2)$$
$$i\neq j, \quad i,j=1,2,\ldots,p.$$

Note that the matrix of observations is

$$X=(\mathbf{x}_1,\ldots,\mathbf{x}_N)=\begin{pmatrix} x_{11} & x_{12} & \cdots & x_{1N} \\ x_{21} & x_{22} & \cdots & x_{2N} \\ \vdots & \vdots & & \vdots \\ x_{p1} & x_{p1} & \cdots & x_{pN} \end{pmatrix}. \qquad (7.2.3)$$

Let

$$\bar{x}_{i\cdot}=N^{-1}\sum_{j=1}^{N}x_{ij}, \qquad \bar{x}_{\cdot j}=p^{-1}\sum_{i=1}^{p}x_{ij}, \qquad \bar{x}_{\cdot\cdot}=(Np)^{-1}\sum_{i=1}^{p}\sum_{j=1}^{N}x_{ij}, \qquad (7.2.4)$$

$$v=\sum_{j=1}^{p}\sum_{j=1}^{N}(x_{ij}-\bar{x}_{i\cdot}-\bar{x}_{\cdot j}+\bar{x}_{\cdot\cdot})^2, \qquad (7.2.5)$$

and

$$v_1=N\sum_{i=1}^{p}(\bar{x}_{i\cdot}-\bar{x}_{\cdot\cdot})^2. \qquad (7.2.6)$$

Then the likelihood ratio test rejects H if

$$F=(N-1)v_1/v>c\equiv F_{p-1,(N-1)(p-1),\alpha}, \qquad (7.2.7)$$

where $P\{F\geq c\}=\alpha$ and F has an F-distribution with $p-1$ and $(N-1)(p-1)$ df. F has a noncentral F-distribution under the alternative A with the noncentrality parameter

$$\delta^2=N\sum_{i=1}^{p}\frac{(\mu_i-\bar{\mu})^2}{\sigma^2(1-\rho)}, \qquad \text{where} \quad \bar{\mu}=p^{-1}\sum_{i=1}^{p}\mu_i. \qquad (7.2.8)$$

This extension of the one-way classification in analysis of variance was considered by Wilks (1946). That is, the data can be considered as arising from a RCBD, in which the p treatments could be on the same individual such that they are equally correlated and the N observations are independently taken over the N blocks (or individuals). The analysis of variance is

Table 7.1 ANOVA for Equally Correlated Variables

Source	df	Sum of squares
between p populations	$p-1$	$N\Sigma_{i=1}^{p}(\bar{x}_{i.}-\bar{x}_{..})^2 \equiv v_1$
between N individuals (blocks)	$N-1$	$p\Sigma_{j=1}^{N}(\bar{x}_{.j}-\bar{x}_{..})^2 \equiv v_2$
within	$(N-1)(p-1)$	$\Sigma_{i=1}^{p}\Sigma_{j=1}^{N}(x_{ij}-\bar{x}_{i.}-\bar{x}_{.j}+\bar{x}_{..})^2 \equiv v$
total	$Np-p$	$\Sigma_{i=1}^{p}\Sigma_{j=1}^{N}(x_{ij}-\bar{x}_{..})^2$

given in Table 7.1. Thus in the two-way analysis of variance the usual assumption that the p populations be independent in comparing their means is *not* required. What is required is that the p populations be equally correlated.

7.2.2 *Confidence Intervals*

The *Scheffé-type confidence interval* for $\mathbf{a}'\boldsymbol{\mu}$ for all nonnull vectors $\mathbf{a}$ such that $\mathbf{a}'\mathbf{e}=0$ is given by

$$\mathbf{a}'\bar{\mathbf{x}}-(c\mathbf{a}'\mathbf{a}v/nN)^{1/2} \leq \mathbf{a}'\boldsymbol{\mu} \leq \mathbf{a}'\bar{\mathbf{x}}+(c\mathbf{a}'\mathbf{a}v/nN)^{1/2}, \qquad (7.2.9)$$

where c is as defined in (7.2.7). The *Tukey-type confidence interval* for $\mathbf{a}'\boldsymbol{\mu}$, $\mathbf{a}\neq\mathbf{0}$ and $\mathbf{a}'\mathbf{e}=0$, is given by

$$\mathbf{a}'\bar{\mathbf{x}}-c_1(4N)^{-1/2}v\sum_{i=1}^{p}|a_i| \leq \mathbf{a}'\boldsymbol{\mu} \leq \mathbf{a}'\bar{\mathbf{x}}+c_1(4N)^{-1/2}v\sum_{i=1}^{p}|a_i|, \qquad (7.2.10)$$

where c_1 is the upper $\alpha 100\%$ point of the Studentized range on p and $(N-1)(p-1)$ df.

These two confidence intervals were given and compared by Bhargava and Srivastava (1973). It is clear that the Tukey-type confidence interval for $\boldsymbol{\alpha}'\boldsymbol{\mu}$ is shorter than the Scheffé type if and only if

$$\tfrac{1}{4}(c_1^2/c_0)\left(\sum_{i=1}^{p}|a_i|\right)^2/a_i^2<(p-1), \qquad (7.2.11)$$

where $c_0=n^{-1}c$. Thus it may be advisable to use the Tukey-type or Scheffé-type confidence intervals according to whether most of the contrasts of interest to experimenters fulfill or do not fulfill, respectively, (7.2.11). The Tukey-type confidence interval tends to be shorter than the Scheffé type if the contrast involves a small subset of $(\mu_1,\ldots,\mu_p)$; otherwise the Scheffé type tends to be shorter.

Table 7.2

	Trial							
Subject	1	2	3	Source	df	SS	MS	F
1	19	18	18	between trials	2	0.067	0.034	0.031
2	15	14	14	between subjects	9	223.0	24.78	22.3
3	13	14	15	error	18	19.9	1.11	
4	12	11	12	*total*	29			
5	16	15	15					
6	17	18	19					
7	12	11	11					
8	13	15	12					
9	19	20	20					
10	18	18	17					
means	15.4	15.4	15.3					
population means	μ_1	μ_2	μ_3					

If the hypothesis H is accepted, we may be interested in the estimate and confidence interval of γ. The maximum likelihood estimate of γ is given by

$$\hat{\gamma} = p^{-1}(\mathbf{e}'\bar{\mathbf{x}}). \tag{7.2.12}$$

A $(1-\alpha)100\%$ confidence limit for γ is given by

$$p^{-1}(\mathbf{e}'\bar{\mathbf{x}}) \pm c_2 v_2^{1/2}, \tag{7.2.13}$$

where

$$c_2^2 = (n^2 p)^{-1} F_{1,n,\alpha} \quad \text{and} \quad v_2 = p \sum_{j=1}^{N} (\bar{x}_{.j} - \bar{x}_{..})^2. \tag{7.2.14}$$

Example 7.2.1 Given data representing the scores out of 20 on a memory test, we wish to test for a significant difference in means for three trials. That is, we wish to see if memory test scores change with successive trials. The data and ANOVA table are given in Table 7.2.

As $0.031 < F_{2,18,0.05} = 3.55$, we do not reject H: $\mu_1 = \mu_2 = \mu_3$ at the 5% level. If we now assume, as is reasonable, $\mu_1 = \mu_2 = \mu_3 \equiv \gamma$, then a 95% confidence interval for γ is given from (7.2.13) and (7.2.14):

$$\tfrac{1}{3}(15.4 + 15.4 + 15.1) + \left[(9 \times 10 \times 3)^{-1} \times 5.12 \times 223\right]^{1/2}$$
$$= 15.37 \pm 2.06 = (13.3, 17.43).$$

Note that this problem can be considered as a univariate analysis of variance problem with two factors, subjects and trials. Because the subjects are randomly selected, the factor is considered random. The final confidence interval for γ is then just a confidence interval based on a subject's average score.

7.2.3 The Case of an Arbitrary Covariance Matrix

If the treatments are not assigned in a random order to the subjects, then no assumptions on the covariance matrix Σ can be made. For example, three tests could be administered to students at the beginning, middle, and end of a course. Correlations between the three tests would not be equal. Then to test H: $\mu_1 = \cdots = \mu_p$ vs A: $\mu_i \neq \mu_j$ for some $i \neq j$, $1 \leq i < j \leq p$, we apply Hotelling's T^2-procedure (Chapter 6). That is, we define

$$\bar{\mathbf{x}} = N^{-1} \sum_{j=1}^{N} \mathbf{x}_j, \qquad S = \frac{1}{N-1} \sum_{j=1}^{N} (\mathbf{x}_j - \bar{\mathbf{x}})(\mathbf{x}_j - \bar{\mathbf{x}}'),$$

and

$$C = \begin{pmatrix} 1 & -1 & & & & 0 \\ & 1 & -1 & & & \\ & & & \ddots & \ddots & \\ & 0 & & & & \\ & & & & 1 & -1 \end{pmatrix},$$

a matrix of contrasts. H is rejected if

$$\begin{aligned} T^2 &= N\bar{\mathbf{y}}'C'(CSC')^{-1}C\bar{\mathbf{y}} \\ &\geq T_\alpha^2 = \frac{(N-1)(p-1)}{(N-p+1)} F_{p-1, N-p+1, \alpha}. \end{aligned} \tag{7.2.15}$$

Note that the assumptions of equal correlations lead to a much more powerful test than Hotelling's T^2-test. In order to check for the validity of this assumption one performs the test for an intraclass correlation matrix to be given in Chapter 12.

Example 7.2.2 The data in Table 7.3 represent the responses to a reasonable request under four conditions for ten "unassertive" subjects. (See Problem 4.2.) The sample covariance is given by

$$S = \begin{pmatrix} 13.6 & -1.6 & 6.7 & -3.7 \\ 1.6 & 4.0 & 1.9 & 3.0 \\ 6.7 & 1.9 & 7.9 & 4.1 \\ -3.7 & 3.0 & 4.1 & 32.4 \end{pmatrix}.$$

Testing H: $\mu_1 = \mu_2 = \mu_3 = \mu_4$, we obtain

$$F_{3,7} = 43.06 > F_{3,7,0.05} = 4.35.$$

Hence we claim that there is a difference in the mean response for the four conditions.

Example 7.2.3 The data in Table 7.4 represent psychomotor scores for three days after radiation therapy. (See Problem 7.4 for the complete data

Table 7.3

Subjects	Conditions R_1	R_2	R_3	R_4
1	11	15	21	21
2	12	11	21	17
3	16	12	21	12
4	11	11	17	11
5	6	14	15	13
6	12	11	18	14
7	8	10	18	11
8	19	10	21	8
9	11	9	15	18
10	10	9	14	12
sample means	11.6	11.2	18.1	13.7
population means	μ_1	μ_2	μ_3	μ_4

set.) The sample covariance matrix is calculated to be

$$S = \begin{pmatrix} 6220 & 5778 & 6460 \\ 5778 & 6188 & 6650 \\ 6460 & 6650 & 7366 \end{pmatrix}$$

on 8 df. The test for equal means for the three days produces $F = 10.18$ on 2 and 7 df. The critical F-value (from Table A.4) is 4.74. Hence we claim that there is a difference in the psychomotor scores for the three days after radiation therapy at the 5% level of significance.

Example 7.2.4 Consider the data of Example 7.2.1. Analyzing the data assuming an arbitrary covariance matrix, we obtain a sample mean and

Table 7.4

Subject	Day 1	2	3
1	91	154	152
2	115	133	136
3	32	97	86
4	38	37	40
5	66	131	148
6	210	221	251
7	167	172	212
8	23	18	30
9	234	260	269
sample means	108.0	135.9	147.1
population means	μ_1	μ_2	μ_3

covariance of $\bar{\mathbf{x}}'=(15.4, 15.4, 15.3)$ and

$$S=\begin{pmatrix} 7.82 & 7.93 & 7.98 \\ 7.93 & 9.38 & 8.87 \\ 7.98 & 8.87 & 9.79 \end{pmatrix}.$$

Hotelling's T^2 for testing H: $\mu_1=\mu_2=\mu_3$ vs A: $\mu_i \neq \mu_j$ for some $1 \leq i \neq j \leq 3$ has a value of 0.09. The F-value is $F=(9.2/8)\times 0.09=0.04=F_{2,8}$. Since $F_{2,8,0.05}=4.46$ we do not reject H. Hence there is no significant difference in the mean scores of the three trials. Assuming $\mu_1=\mu_2=\mu_3=\gamma$, we may obtain a confidence interval for γ from (6.1.10). That is, $\hat{\gamma}=(\mathbf{e}'S^{-1}\bar{\mathbf{x}})(\mathbf{e}'S^{-1}\mathbf{e})^{-1}$, where $\mathbf{e}'=(1,1,1)$. Thus, $\hat{\gamma}=(1.85)/0.12=15.4$. A 95% confidence interval for γ is given by

$$\begin{aligned}\hat{\gamma} &\pm t_{n-p+1,\alpha/2}(1+n^{-1}T^2)^{1/2}(N\mathbf{e}'S^{-1}\mathbf{e})^{-1/2} \\ &=15.4 \pm 2.365(1+0.01)^{1/2}(10\times 0.12)^{-1/2} \\ &=15.4 \pm 2.17=(13.2, 17.6).\end{aligned}$$

Note that this is very similar to the interval given in Example 7.2.1 using univariate techniques.

7.3 Split-Plot and MANOVA Designs

Suppose students were given a series of p aptitude tests. Then tests for differences in scores on the tests could be carried out as in Section 7.2 using (7.2.3) with equal correlations and (7.2.15) with unequal correlations. However, it is to be expected that students would have different aptitudes. Of more importance is the question of differences between groups of students. Suppose there are J groups of students with N_j students in the jth group for a total of N students. Then the model takes the form

$$\mathbf{x}_{jk}=\boldsymbol{\mu}_j+\boldsymbol{\varepsilon}_{jk}, \qquad j=1,\ldots,J, \quad k=1,\ldots,N_j, \tag{7.3.1}$$

where $\mathbf{x}_{jk}$ is the vector of aptitude scores for the jth subject, $\boldsymbol{\mu}_j$ is the population mean for the jth group, and the $\boldsymbol{\varepsilon}_{jk}$ are independently distributed as $N_p(\mathbf{0}, \Sigma)$. The hypotheses of interest are H:

i. no interaction between groups and aptitudes,
ii. no differences in groups, and
iii. no differences in aptitudes.

See, for example, Bock (1963). In terms of the model we define $\mathbf{e}_p$ and $\mathbf{e}_J$ to be vectors of ones, and C to be a $J \times J-1$ matrix of orthogonal contrasts. Then hypotheses (i)–(iii) become

i. H_1: $C(\boldsymbol{\mu}_1-\boldsymbol{\mu}_J)=\cdots=C(\boldsymbol{\mu}_{J-1}-\boldsymbol{\mu}_J)=\mathbf{0}$;
ii. H_2: $\mathbf{e}_p'(\boldsymbol{\mu}_1-\boldsymbol{\mu}_J)=\cdots=\mathbf{e}_p'(\boldsymbol{\mu}_1-\boldsymbol{\mu}_J)=\mathbf{0}$ given H_1; and
iii. H_3: $C(\boldsymbol{\mu}_1,\ldots,\boldsymbol{\mu}_J)\mathbf{e}_J=\mathbf{0}$.

The hypothesis H_1 can be written

$$H_1\colon C(\boldsymbol{\mu}_1,\ldots,\boldsymbol{\mu}_J)F=\mathbf{0}$$

where F is a $J\times(J-1)$ matrix of orthogonal contrasts. If Σ is an intraclass correlation matrix, then the analysis is that of a split plot. In order to justify this we define

$$\mathbf{z}_j=\begin{pmatrix}\mathbf{e}_p'/\sqrt{p}\\ C\end{pmatrix}\mathbf{x}_j,\qquad j=1,\ldots,N.$$

Then

$$\mathbf{z}_j\sim N_p\left[\begin{pmatrix}\mathbf{e}_p'/\sqrt{p}\,\boldsymbol{\mu}_j\\ C\mu_j\end{pmatrix},\ \begin{pmatrix}\mathbf{e}_p'/\sqrt{p}\\ C\end{pmatrix}\Sigma\begin{pmatrix}\mathbf{e}_p/\sqrt{p} & C'\end{pmatrix}\right]$$

with

$$\begin{pmatrix}\mathbf{e}_p'/\sqrt{p}\\ C\end{pmatrix}\Sigma\begin{pmatrix}\mathbf{e}_p/\sqrt{p'} & C'\end{pmatrix}=\sigma^2\begin{pmatrix}\mathbf{e}_p'/\sqrt{p}\\ C\end{pmatrix}\left[(1-\rho)I+\rho\mathbf{e}_p\mathbf{e}_p'\right]\begin{pmatrix}\mathbf{e}_p/\sqrt{p'} & C'\end{pmatrix}$$

$$=\sigma^2\begin{pmatrix}1+(p-1)\rho & \mathbf{0}'\\ \mathbf{0} & (1-\rho)I_{p-1}\end{pmatrix}.$$

As can be seen from the covariance matrix of $\mathbf{z}_j$, the among-groups variance is different from the variances of differences among aptitudes. Hence a different error term must be used in the denominator. The ANOVA table is shown in Table 7.5, where

$$\mathbf{x}_{jk}'=(x_{1jk},\ldots,x_{pjk}),\qquad j=1,\ldots,J,\quad k=1,\ldots,N_j,$$

$$x_{\cdot jk}=\sum_{i=1}^{p}x_{ijk},\qquad x_{ij\cdot}=\sum_{k=1}^{N_j}x_{ijk},$$

$$x_{i\cdot\cdot}=\sum_{j=1}^{J}\sum_{k=1}^{N_J}x_{ijk},\qquad x_{\cdot j\cdot}=\sum_{i=1}^{p}\sum_{k=1}^{N_j}x_{ijk},$$

$$x_{\cdot\cdot\cdot}=\sum_{i=1}^{p}\sum_{j=1}^{J}\sum_{k=1}^{N_j}x_{ijk},\qquad N_1+N_2+\cdots+N_J=N,$$

$$\mathrm{SST}=\sum_{i,j,k}x_{ijk}^2-\frac{x_{\cdot\cdot\cdot}^2}{Np},$$

$$\mathrm{SS(G)}=\sum_{i=1}^{J}\frac{x_{\cdot j\cdot}^2}{pN_j}-\frac{x_{\cdot\cdot\cdot}^2}{Np},$$

Table 7.5 ANOVA for p Aptitude Tests

Source	df	SS	MS	
among groups	$J-1$	SG(G)	MS(G)	MS(G)/MS(WG)
replications within groups	$N-J$	SS(WG)	MS(WG)	—
among aptitudes	$p-1$	SS(A)	MS(A)	MS(A)/MSE
aptitudes × groups	$(p-1)(J-1)$	SS(A×G)	MS(A×G)	MS(A×G)/MSE
error	$(N-J)(p-1)$	SSE	MSE	—
total	$Np-1$	SST	—	—

$$\mathrm{SS(WG)} = \sum_{j=1}^{J} \sum_{k=1}^{N_j} \frac{x_{\cdot jk}^2}{p} - \sum_{j=1}^{J} \frac{x_{\cdot j \cdot}^2}{pN_j},$$

$$\mathrm{SS(A)} = \sum_{i=1}^{p} \frac{x_{i\cdot\cdot}^2}{N} - \frac{x_{\cdot\cdot\cdot}^2}{Np},$$

$$\mathrm{SS(AxG)} = \sum_{i=1}^{p} \sum_{j=1}^{J} \frac{x_{ij\cdot}^2}{N_j} - \sum_{i=1}^{p} \frac{x_{i\cdot\cdot}^2}{N} - \sum_{j=1}^{J} \frac{x_{\cdot j\cdot}^2}{pN_j} + \frac{x_{\cdot\cdot\cdot}^2}{Np},$$

and

$$\mathrm{SSE} = \mathrm{SST} - \mathrm{SS(G)} - \mathrm{SS(WG)} - \mathrm{SS(A)} - \mathrm{SS(AxG)}.$$

In the case where the covariance matrix Σ is not an intraclass correlation matrix Σ_I, the test of the hypothesis in (6.2.2) is performed with multivariate techniques. The case of $J=2$ groups is discussed in Section 7.5 and the general case in Section 7.6. (An analysis based on the split-plot decision is given in Example 7.6.1.)

7.4 Regression Model

In this section we generalize the results of Section 7.3 to the regression model. Let

$$X = \begin{pmatrix} x_{11} & \cdots & x_{1N} \\ x_{21} & \cdots & x_{2N} \\ \vdots & & \vdots \\ x_{p1} & \cdots & x_{pN} \end{pmatrix}$$

be the observation matrix such that the columns of X are independently distributed with covariance matrix Σ of the intraclass correlation form given by (7.2.1). Let

$$E(X)=\beta Z,$$

where β is a $p\times m$ matrix of unknown parameters and Z is an $m\times N$ matrix of constants, known as design matrix. Let Γ be the $p\times p$ orthogonal matrix

$$\Gamma=\begin{pmatrix} p^{-1/2}\mathbf{e}' \\ C \end{pmatrix}, \qquad \mathbf{e}'=(1,\dots,1).$$

Since Γ is known, observing X is equivalent to observing

$$Y=\Gamma X\equiv\begin{pmatrix} \mathbf{y}_1' \\ Y_2 \end{pmatrix}.$$

Then

$$E(\mathbf{y}_1')=p^{-1/2}\mathbf{e}'\beta Z, \qquad E(Y_2)=C\beta Z,$$

and $\mathbf{y}_1$ and Y_2 are independently distributed. The covariance of $\mathbf{y}_1$ is $\sigma_1^2 I_N$, and the covariance of each column of Y_2 is $\sigma_2^2\mathrm{I}_{p-1}$; each column of Y_2 is independently distributed. Here

$$\sigma_1^2=\sigma^2(1-\rho) \qquad \text{and} \qquad \sigma_2^2=\sigma^2[1+(p-1)\rho].$$

Thus univariate regression theory can be applied to obtain tests and estimates of $\mathbf{e}'\beta$, and multivariate regression theory can be applied to obtain tests and estimates of $C\beta$ as long as $C\mathbf{e}=\mathbf{0}$.

7.5 Profile Analysis: Two-Group Case

Suppose we wish to compare the performance of students at school A with those of students at school B. A battery of four tests T_1, T_2, T_3, and T_4 is given to N_1 students from school A and N_2 students from school B. Their sample means are plotted on a profile graph in Figure 7.1. From the graph, it is clear that the differences $\mu_{11}-\mu_{21}$, $\mu_{12}-\mu_{22}$, $\mu_{13}-\mu_{23}$, and $\mu_{14}-\mu_{24}$ are all equal, where $\boldsymbol{\mu}_1=(\mu_{11},\mu_{12},\mu_{13},\mu_{14})'$ and $\boldsymbol{\mu}_2=(\mu_{21},\mu_{22},\mu_{23},\mu_{24})'$ are the population means of the two groups. That is, in the terminology of the literature, the two groups appear to have *similar profiles*, or there appears to be no *interaction* between schools and tests. However, we choose to refer to a *similarity of profiles*. In order to test for similarity of profiles (Greenhouse and Geisser, 1959), one tests the hypothesis

$$H_1:\begin{pmatrix} \mu_{11}-\mu_{12} \\ \vdots \\ \mu_{1,p-1}-\mu_{1,p} \end{pmatrix}=\begin{pmatrix} \mu_{21}-\mu_{22} \\ \vdots \\ \mu_{2,p-1}-\mu_{2,p} \end{pmatrix}. \tag{7.5.1}$$

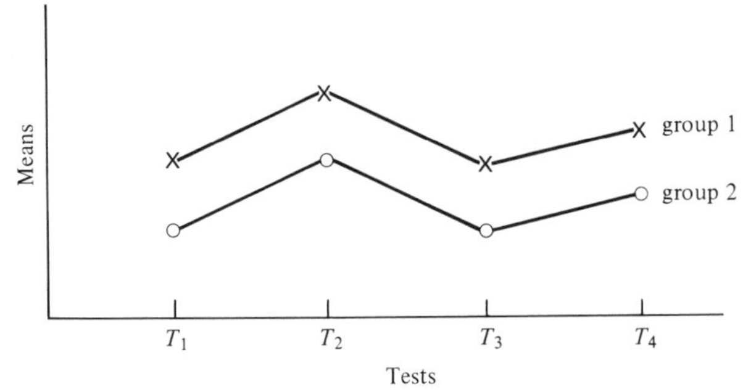

Figure 7.1 Profile graph: ×, school A; ○, school B.

Equivalently, for a $(p-1)\times p$ matrix of contrasts

$$C=\begin{pmatrix} 1 & -1 & 0 & 0 & \cdots & 0 & 0 \\ 0 & 1 & -1 & 0 & \cdots & 0 & 0 \\ 0 & 0 & 1 & -1 & \cdots & 0 & 0 \\ \vdots & \vdots & \vdots & \vdots & & \vdots & \vdots \\ 0 & 0 & 0 & 0 & \cdots & 1 & -1 \end{pmatrix}, \tag{7.5.2}$$

the hypothesis becomes

$$H_1: C(\boldsymbol{\mu}_1-\boldsymbol{\mu}_2)=\mathbf{0} \quad \text{vs} \quad A_1: C(\boldsymbol{\mu}_1-\boldsymbol{\mu}_2)\neq\mathbf{0}. \tag{7.5.3}$$

Note that the matrix C satisfies

$$C\mathbf{e}=\mathbf{0}. \tag{7.5.4}$$

As in Section 6.1.11, the test we propose does not depend on the choice of C as long as it satisfies (7.5.4). Equivalently, we could write

$$H_1: (\boldsymbol{\mu}_1-\boldsymbol{\mu}_2)=\gamma\mathbf{e} \quad \text{vs} \quad A_1: (\boldsymbol{\mu}_1-\boldsymbol{\mu}_2)\neq\gamma\mathbf{e}, \tag{7.5.5}$$

where γ represents the average difference between schools. The similarity of profiles hypothesis H_1 is also known as the *parallelism hypothesis*. When H_1 is true, there are other variations that we may like to separate out. These are represented by the following two hypotheses:

$$H_2: \gamma=0 \quad \text{vs} \quad A_2: \gamma\neq 0 \tag{7.5.6}$$

and

$$H_3: \boldsymbol{\mu}_1=\delta\mathbf{e}, \quad \boldsymbol{\mu}_2=\xi\mathbf{e}, \quad \delta, \xi \text{ unknown}$$

vs (7.5.7)

$$A_3: \boldsymbol{\mu}_1\neq\delta\mathbf{e}, \quad \boldsymbol{\mu}_2\neq\xi\mathbf{e}.$$

If an experimenter wishes to test for differences in group means, then the MANOVA methods of Chapter 4 should be used. In profile analysis an experimenter would like to test for differences in groups based on an average score. That is, in comparing two groups on the basis of four test scores, one may wish to prove that one group scored uniformly better than the other. In this case one first tests the hypothesis H_1 that the profiles are similar (parallelism) and then, assuming equal profiles, tests H_2: $\gamma = 0$ to see if the groups differ based on an average of the four tests. If the profiles are similar, then the test H_2: $\gamma = 0$ is more powerful than the more general H: $\boldsymbol{\mu}_1 = \boldsymbol{\mu}_2$.

We now turn to hypothesis H_3, known in the literature as the hypothesis of the *condition variation*. H_3 implies that the two profiles are parallel to the x axis. When H_1 is true, H_3 is equivalent to the hypothesis

$$H_3: \boldsymbol{\mu}_1 + \boldsymbol{\mu}_2 = \eta \mathbf{e}, \qquad \eta \text{ unknown}. \tag{7.5.8}$$

But for *any* C satisfying (7.5.4) this is equivalent to

$$H_3: C(\boldsymbol{\mu}_1 + \boldsymbol{\mu}_2) = \mathbf{0}. \tag{7.5.9}$$

We may, of course, take the same C as used in testing the hypothesis H_1, and indeed we shall do so since it saves computing time. These three problems will be considered in the following sections on the basis of the independent samples $\mathbf{x}_1, \ldots, \mathbf{x}_{N_1}$ and $\mathbf{y}_1, \ldots, \mathbf{y}_{N_2}$ from two normal populations $N_p(\boldsymbol{\mu}_1, \Sigma)$ and $N_p(\boldsymbol{\mu}_2, \Sigma)$, respectively. The sufficient statistics are $\bar{\mathbf{x}}$, $\bar{\mathbf{y}}$, S_1, and S_2, where

$$\bar{\mathbf{x}} = N_1^{-1} \sum_{i=1}^{N_1} \mathbf{x}_i, \qquad \bar{\mathbf{y}} = N_2^{-1} \sum_{i=1}^{N_2} \mathbf{y}_i, \qquad n_j = N_j - 1, \quad j = 1,2, \tag{7.5.10}$$

$$n_1 S_1 = \sum_{i=1}^{N_1} (\mathbf{x}_i - \bar{\mathbf{x}})(\mathbf{x}_i - \bar{\mathbf{x}})', \qquad \text{and} \qquad n_2 S_2 = \sum_{i=1}^{N_2} (\mathbf{y}_i - \bar{\mathbf{y}})(\mathbf{y}_i - \bar{\mathbf{y}})', \tag{7.5.11}$$

or equivalently,

$$\mathbf{u} \equiv \bar{\mathbf{x}} - \bar{\mathbf{y}}, \quad \mathbf{v} = \bar{\mathbf{x}} + \bar{\mathbf{y}}, \qquad S_1 \text{ and } S_2. \tag{7.5.12}$$

The pooled estimate S_p of Σ is given by

$$(n_1 + n_2) S_\mathrm{p} = n_1 S_1 + n_2 S_2. \tag{7.5.13}$$

7.5.1 *Test for Similarity of Profiles*

Using the likelihood ratio test for this problem, we reject H_1 if

$$\frac{N_1 + N_2 - p}{(N_1 + N_2 - 2)(p-1)} \frac{N_1 N_2}{N_1 + N_2} \mathbf{u}' C' (C S_\mathrm{p} C')^{-1} C \mathbf{u} \geq F_{p-1, N_1 + N_2 - p, \alpha},$$

where $\mathbf{u} = \bar{\mathbf{x}} - \bar{\mathbf{y}}$. This test does not depend on the choice of C so long as $C\mathbf{e} = \mathbf{0}$. The test was given by Greenhouse and Geisser (1959) and is a two-sample generalization of Rao's (1948) one-sample result.

7.5.2 *Test of the Level Hypothesis Given That the Profiles Are Similar*

The maximum likelihood estimate of γ is given by

$$\hat{\gamma}=\left(\mathbf{e}'S_{\mathrm{p}}^{-1}\mathbf{u}\right)/\left(\mathbf{e}'S_{\mathrm{p}}^{-1}\mathbf{e}\right).$$

A $(1-\alpha)100\%$ confidence interval for γ is given by

$$\hat{\gamma}-\left(\frac{N_1N_2}{N_1+N_2}\right)^{-1/2}t_{f-p+1,\alpha/2}\left(1+f^{-1}T_{p-1}^2\right)^{1/2}\left(\mathbf{e}'S_{\mathrm{p}}^{-1}\mathbf{e}\right)^{-1/2}$$
$$\leq\gamma\leq\hat{\gamma}+\left(\frac{N_1N_2}{N_1+N_2}\right)^{-1/2}t_{f-p+1,\alpha/2}\left(1+f^{-1}T_{p-1}^2\right)^{1/2}\left(\mathbf{e}'S_{\mathrm{p}}^{-1}\mathbf{e}\right)^{-1/2},$$

where

$$T_{p-1}^2=\frac{N_1N_2}{N_1+N_2}\mathbf{u}'C'\left(CS_{\mathrm{p}}C'\right)^{-1}C\mathbf{u}\qquad\text{and}\qquad f=n_1+n_2=N_1+N_2-2.$$

Thus the hypothesis H_2: $\gamma=0$ is accepted if zero lies in the above confidence interval.

Note that the above maximum likelihood estimate and confidence interval are different than those given in the literature and based on the idea that when Σ is known (S_{p} replaced by Σ), this estimate has smaller variance than the estimator $\mathbf{e}'\mathbf{u}/e'\mathbf{e}$ currently in use. This result is due to Srivastava (1981b).

Alternatively, H_2: $\gamma=0$ is rejected and A: $\gamma_2\neq 0$ is accepted if

$$[N_1N_2/(N_1+N_2)]\left(\mathbf{e}'S_{\mathrm{p}}^{-1}\mathbf{u}\right)^2\left(\mathbf{e}'S_{\mathrm{p}}^{-1}\mathbf{e}\right)^{-1}\left(1+f^{-1}T_{p-1}^2\right)^{-1}\geq t_{f-p+1,\alpha/2}^2$$
$$=F_{1,f-p+1,\alpha},$$

where

$$T_{p-1}^2=[N_1N_2/(N_1+N_2)]\mathbf{u}'C'\left(CS_{\mathrm{p}}C'\right)^{-1}C\mathbf{u}.$$

7.5.3 *Test of the Condition Variation*

Assuming that the profiles of the two groups are similar, one can pool the information from both to test for differences in conditions. Let $\mathbf{z}=(N_1\bar{\mathbf{x}}+N_2\bar{\mathbf{y}})/(N_1+N_2)$ be the overall mean vector for all observations and C be a $(p-1)\times p$ matrix of contrasts in (7.3.4). Then H_3 is rejected if

$$\frac{(N_1+N_2-p)(N_1+N_2)}{(N_1+N_2-2)(p-1)}\mathbf{z}'C'\left(CS_{\mathrm{p}}C'\right)^{-1}C\mathbf{z}\geq F_{p-1,N_1+N_2-p,\alpha}.$$

Note that the problem is equivalent to testing H: $C\boldsymbol{\mu}=0$ vs A: $C\boldsymbol{\mu}\neq 0$ where $E(\mathbf{z})=\boldsymbol{\mu}$.

Table 7.6[a] Example 7.5.1

	Request		
	Reasonable	Somewhat unreasonable	Highly unreasonable
unassertive people	11	16	23
	12	11	18
	16	15	22
	11	14	19
	6	12	16
	12	8	20
	8	11	14
	19	18	9
	11	9	21
	10	12	21
means	11.6	12.6	18.3
assertive people	12	13	24
	6	10	13
	12	18	22
	21	21	24
	9	10	22
	8	18	24
	7	13	24
	14	20	24
	10	20	22
	21	24	24
means	12.0	16.7	22.3

[a]Data abridged from Josefawitz (1979).

Example 7.5.1 Consider the data of Table 7.6 for two classes of people, unassertive and assertive. The data (plotted in Figure 7.2) represent the response to three types of requests (see Problem 4.2).

The F-statistic for testing for interaction or similar profiles, has the value $F = 2.88 < F_{2,17,0.05} = 3.59$. Hence one concludes that there is not a significant interaction. The result is significant at the 10% level, however. This might indicate the presence of a slight interaction. Figure 7.2 tends to support this conclusion. With the assumption of no interaction, or similar profiles, the F-value for the level hypothesis was obtained using the SAS procedure of Section 7.8:

$$F = 4.55 > F_{1,16,0.05} = 4.45.$$

For the condition hypothesis,

$$F = 22.31 > F_{2,17,0.05} = 3.59.$$

Both hypotheses are rejected, indicating a difference in the average response

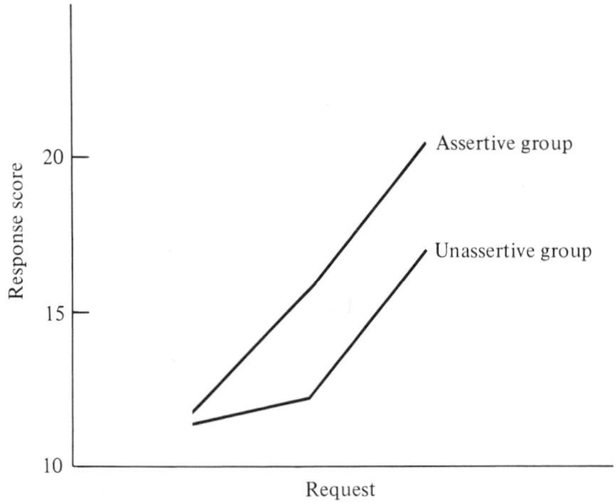

Figure 7.2 Example 7.5.1.

between the two groups and a difference in responses among the types of request. The pooled covariance matrix

$$S_p = \begin{pmatrix} 21.02 & 13.08 & 2.40 \\ 13.08 & 16.81 & 2.39 \\ 2.40 & 2.39 & 14.90 \end{pmatrix}$$

has 18 df, and the correlation matrix is

$$R = \begin{pmatrix} 1.00 & 0.70 & 0.14 \\ 0.70 & 1.00 & 0.15 \\ 0.14 & 0.15 & 1.00 \end{pmatrix}.$$

There appears to be equal correlation between reasonable requests and both somewhat and completely unreasonable requests. However there is a very high correlation of 0.70 between the two types of unreasonable requests. Hence the assumption of an intraclass correlation matrix cannot be made, and multivariate procedures such as profile analysis must be used. To obtain an estimate for the average difference in levels we use the earlier expressions

$$\hat{\gamma} = \left(\mathbf{e}'S_p^{-1}\mathbf{u}\right)\left(\mathbf{e}'S_p^{-1}\mathbf{e}\right)^{-1} = 3.47,$$

where $\mathbf{u}' = (11.6 - 12.0, 12.6 - 16.7, 18.3 - 22.3) = (-0.4, -4, 1, -4.0)$,

$$S_p^{-1} = \begin{pmatrix} 0.092 & -0.071 & -0.003 \\ -0.071 & 0.116 & -0.007 \\ -0.003 & -0.007 & 0.069 \end{pmatrix},$$

and $\mathbf{e}'=(1,1,1)$. A 95% confidence interval for γ is then given by

$$\hat{\gamma}\pm[N_1N_2/(N_1+N_2)]^{-1/2}t_{f-p+1,\alpha/2}\left(1+f^{-1}T_{p-1}^2\right)^{1/2}\left(\mathbf{e}'S_p^{-1}\mathbf{e}\right)^{-1/2},$$

where $N_1=N_2=10, f=18, p=3, \alpha=0.05$, and $t_{17,0.025}=2.11$.

The value $T_{p-1}^2=T_2^2=6.10$ is used in calculating the F-statistic for equal profiles. From a computer printout,

$$F_{r,f}=[(f-r+1)/fr]T_r^2=2.88.$$

Hence with $r=p-1=2$ and $f=18$, T_r^2 may be calculated as

$$T_2^2=T_r^2=\frac{fr}{f-r+1}F_{r,f}=\frac{18\times 2}{17}\times 2.88=6.10.$$

Substituting T_2^2 into the expression for the confidence interval for γ produces a confidence region of

$$-3.47\pm(0.447)\times 2.11\times 1.16\times 2.95=-3.47\pm 3.23.$$

Therefore we may conclude that the mean response for the unassertive group was lower than that of the assertive group.

7.6 General Case of J Groups

Suppose now that we have J schools with N_j students chosen from the jth school; then for testing for differences among schools, we use multivariate analysis of variance. That is, for observations

$$\mathbf{y}_{jk},\qquad j=1,\ldots,J,\quad k=1,\ldots,N_j,\ N_1+\cdots+N_J=N \tag{7.6.1}$$

with

$$\mathbf{y}_{jk}=\boldsymbol{\mu}_j+\boldsymbol{\varepsilon}_{jk},\qquad \boldsymbol{\varepsilon}_{jk}\sim N_p(\mathbf{0},\Sigma),$$

we define

$$\bar{\mathbf{y}}_{j\cdot}=N_j^{-1}\sum_{k=1}^{N_j}\mathbf{y}_{jk},\qquad \bar{\mathbf{y}}_{\cdot\cdot}=N^{-1}\sum_{j=1}^{J}\sum_{k=1}^{N_j}\mathbf{y}_{jk},\quad \bar{\mathbf{y}}_{\cdot\cdot}=\mathbf{y}_{\cdot\cdot}/N, \tag{7.6.2}$$

$$\mathrm{SST}=\sum_{j=1}^{J}\sum_{k=1}^{N_j}\mathbf{y}_{jk}\mathbf{y}_{jk}'-N\bar{\mathbf{y}}_{\cdot\cdot}\bar{\mathbf{y}}_{\cdot\cdot}',$$

$$\mathrm{SS(TR)}=\sum_{j=1}^{J}\mathbf{y}_{j\cdot}\mathbf{y}_{j\cdot}'/N_j-N\bar{\mathbf{y}}_{\cdot\cdot}\bar{\mathbf{y}}_{\cdot\cdot}',$$

$$\mathrm{SSE}=\mathrm{SST}-\mathrm{SS(TR)}.$$

The MANOVA table is given by Table 7.7. From Table 7.4 we can obtain the pooled estimate of Σ; namely, $S_p=\mathrm{SSE}/(N-J)$ with $N-J$ df. The

Table 7.7

Source	df	SS	$u_{p,m,n}$	p,m,n
TR	J−1	SS(TR)	|SSE|/|SSE+SS(TR)|	$p, J-1, N-J$
error	N−J	SSE		
total	N−1	SST		

test for no treatment effects is equivalent to H: $\boldsymbol{\mu}_1 = \cdots = \boldsymbol{\mu}_J$ vs A: $\boldsymbol{\mu}_{i_1} \neq \boldsymbol{\mu}_{i_2}$ for some $i_1 \neq i_2$, $i_1, i_2 = 1,\ldots,p$. If we wish to test for parallelism, for no interaction, or for similar profiles, we test the hypothesis H_1 given in (7.3.15), that is,

$$H_1\colon C(\boldsymbol{\mu}_1 - \boldsymbol{\mu}_J) = \cdots = C(\boldsymbol{\mu}_{J-1} - \boldsymbol{\mu}_J) = \mathbf{0},$$

or equivalently,

$$H_1\colon \boldsymbol{\mu}_1 - \boldsymbol{\mu}_J = \gamma_1 \mathbf{e}, \ldots, \boldsymbol{\mu}_{J-1} - \boldsymbol{\mu}_J = \gamma_{J-1}\mathbf{e},$$

where C is a $(p-1)\times p$ matrix of contrasts, as given in (7.5.2), and γ_i represents the level difference between the ith and the Jth groups. H_1 is rejected if

$$U_{p-1,J-1,N-J} = \frac{|C(\text{SSE})C'|}{|C(\text{SSE}+\text{SS(TR)})C'|} \geq U_{p-1,J-1,N-j,\alpha}. \quad (7.6.3)$$

If the parallelism hypothesis H_1 is accepted, we may desire to test the level hypothesis

$$H_2\colon \boldsymbol{\gamma} = 0 \qquad \text{where} \quad \boldsymbol{\gamma} = (\gamma_1, \ldots, \gamma_{J-1})',$$

or equivalently,

$$H_2\colon \mathbf{e}_p'(\boldsymbol{\mu}_1 - \boldsymbol{\mu}_J) = \cdots = \mathbf{e}_p'(\boldsymbol{\mu}_{J-1} - \boldsymbol{\mu}_J) = 0.$$

H_2 is rejected if

$$\frac{U_{p,J-1,N-J}}{U_{p-1,J-1,N-J}} \geq U_{1,J-1,N-J-p+1,\alpha}, \quad (7.6.4)$$

where $U_{p,J-1,N-J} = |\text{SSE}|/|\text{SSE}+\text{SS(TR)}|$, and $U_{p-1,J-1,N-J}$ is defined in (7.6.3).

Equivalently, we may use the relationship (see Theorem 4.2.4)

$$\frac{N-J-p+1}{J-1}\frac{(1-U)}{U} \sim F_{J-1,N-J-p+1}.$$

If the level hypothesis H_2 is rejected we may desire an estimate and confidence interval for $\boldsymbol{\gamma}$ or a linear combination of the components of $\boldsymbol{\gamma}$. The maximum likelihood estimate of $\boldsymbol{\gamma}$ is given by

$$\hat{\boldsymbol{\gamma}} = (Z'V^{-1}\mathbf{e})/(\mathbf{e}'V^{-1}\mathbf{e}) = (Z'S^{-1}\mathbf{e})/(\mathbf{e}'S^{-1}\mathbf{e}), \quad (7.6.5)$$

where $V = \text{SSE}$, $S = (N-J)^{-1}V$, and $Z = (\bar{\mathbf{y}}_{1\cdot} - \bar{\mathbf{y}}_{J\cdot}, \ldots, \bar{\mathbf{y}}_{(J-1)\cdot} - \bar{\mathbf{y}}_{J\cdot})$.

A $(1-\alpha)100\%$ simultaneous confidence interval for $\boldsymbol{\gamma}$ is given by

$$\mathbf{a}'\hat{\boldsymbol{\gamma}} \pm T_\alpha (\mathbf{e}'V^{-1}\mathbf{e})^{-1/2}\left\{\mathbf{a}'\left[B'B + Z'C'(CVC')^{-1}CZ\right]\mathbf{a}\right\}^{1/2}$$

where

$$T_\alpha^2 = (J-1)(f-p+1)^{-1}F_{J-1,f-p+1},$$

$f = N - J$, and

$$B = \begin{bmatrix} N_1^{-1/2} & 0 & \cdots & 0 \\ 0 & N_2^{-1/2} & \cdots & 0 \\ \vdots & \vdots & & \vdots \\ 0 & 0 & \cdots & N_{J-1}^{-1/2} \\ -N_J^{-1/2} & -N_J^{-1/2} & \cdots & -N_J^{-1/2} \end{bmatrix}.$$

Note that if our interest is a linear combination of only a small number, say, l components of $\boldsymbol{\gamma}$, and if $t_{f-p+1,\alpha/2l}^2 \le T_\alpha^2$, then use of the Bonferroni inequality will give a shorter confidence interval. It may be obtained from the preceding expression with T_α replaced by $t_{f-p+1,\alpha/2l}$.

Finally, to test for no difference among conditions or tests we reject

$$H_3\colon\ C(\boldsymbol{\mu}_1, \ldots, \boldsymbol{\mu}_J)\mathbf{e} = 0$$

if

$$\frac{N(f-p+2)}{(p-1)}\bar{\mathbf{y}}'_{\cdot\cdot}C'(CVC')^{-1}C\bar{\mathbf{y}}_{\cdot\cdot} \ge F_{p-1,f-p+2}, \qquad (7.6.6)$$

where $f = N-J$ and $V = \text{SSE}$. This generalization of profile analysis is given by Srivastava (1981b).

Example 7.6.1 The data of Table 7.8 give the blood histamine levels observed for individual dogs in various treatment groups (see Figure 7.3). Using the PROFILE program in Appendix II, we obtain the U-statistic for testing for interaction between treatments (groups) and time,

$$U_{3,3,12} = 0.1359.$$

From Theorem 4.2.1, we have that

$$\begin{aligned} &P\left\{-\left[12 - \tfrac{1}{2}(3-3+1)\right]\ln U_{3,3,12} \ge z\right\} \\ &\quad = P(-11.5\ln U_{3,3,12} \ge -11.5\ln 0.1359) \\ &\quad = P(-11.5\ln U_{3,3,12} \ge 22.95) \\ &\quad = P(\chi_9^2 \ge 22.95) + (12)^{-1}\gamma_2\left[P(\chi_{13}^2 \ge 22.95) - P(\chi_9^2 \ge 22.95)\right], \end{aligned}$$

Table 7.8[a] Example 7.6.1

	Dog	Blood histamine (μg/ml)			
		Control	1 min	3 min	5 min
group I	1	0.04	0.20	0.10	0.08
(morphine intact)	2	0.02	0.06	0.02	0.02
	3	0.07	1.40	0.48	0.24
	4	0.17	0.57	0.35	0.24
average		0.08	0.56	0.24	0.15
group II	5	0.10	0.09	0.13	0.14
(morphine depleted)	6	0.12	0.11	0.10	0.11
	7	0.07	0.07	0.07	0.07
	8	0.05	0.07	0.06	0.07
average		0.09	0.09	0.09	0.09
group III	9	0.03	0.62	0.31	0.22
(trimethaphan intact)	10	0.03	1.05	0.73	0.60
	11	0.07	0.83	1.07	0.80
	12	0.09	3.13	2.06	1.23
average		0.06	1.41	1.04	0.71
group IV	13	0.10	0.09	0.09	0.08
(trimethaphan depleted)	14	0.08	0.09	0.09	0.10
	15	0.13	0.10	0.12	0.12
	16	0.06	0.05	0.05	0.05
average		0.09	0.08	0.09	0.09

[a] Data from Cole and Grizzle (1966). Reproduced with the permission of the editors of *Biometrics*.

where

$$\gamma_2 = 9(9+3-5)/48 = 1.3125.$$

Hence the probability of observing a U-value as low as 0.1359 is

$$0.006 + 1.3125(12)^{-1}(0.028 - 0.006) = 0.009.$$

The null hypothesis of no interaction can be rejected at $\alpha = 0.01$, and we therefore reject it.

The within-groups covariance matrix is

$$\begin{pmatrix} 0.001797916667 & 0.01003958333 & 0.00776416667 & 0.00474375 \\ 0.01003958333 & 0.4280270833 & 0.23261875 & 0.1198958333 \\ 0.007760416667 & 0.23261875 & 0.15111875 & 0.08336666667 \\ 0.00474375 & 0.1198958333 & 0.08336666667 & 0.4789375 \end{pmatrix}.$$

and the correlation matrix is

$$\begin{pmatrix} 1 & 0.3619055847 & 0.4708050193 & 0.5112079424 \\ 0.3619055847 & 1 & 0.9146394229 & 0.8373924564 \\ 0.4708050193 & 0.9146394229 & 1 & 0.9799272601 \\ 0.5112079424 & 0.8373924564 & 0.9799272601 & 1 \end{pmatrix}.$$

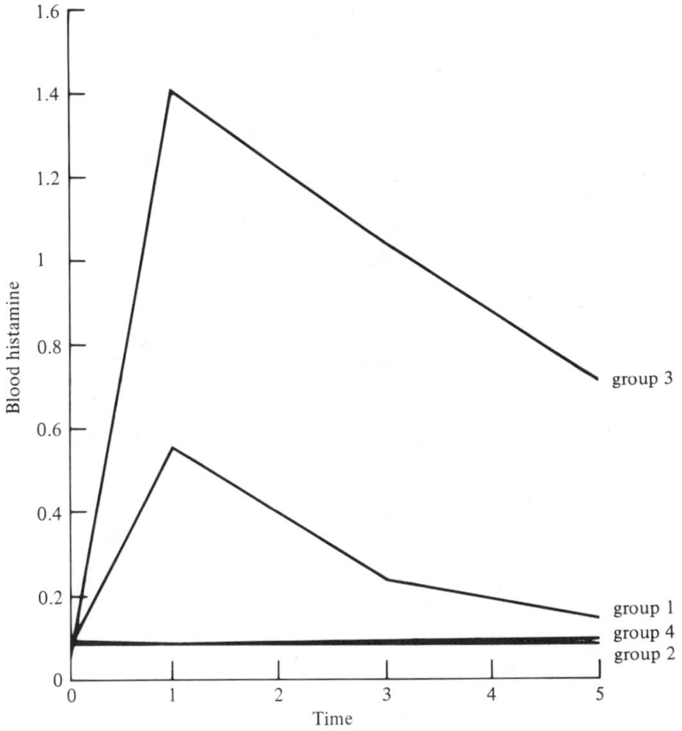

Figure 7.3 Blood histamine profiles of Example 7.6.1.

Table 7.9 ANOVA for the Split-Plot Analysis of Table 7.6

Source	df	SS	MS	F
total	63	17.584700	—	—
whole plots	15	10.169000	—	—
treatment A	3	5.531938	1.843979	4.772
error A	12	4.637063	0.386422	—
split plots	48	7.416700	—	—
treatment B	3	1.752325	0.584108	7.229
$A \times B$	9	2.755388	0.306154	3.789
error B	36	2.908988	0.080805	—

It should be noted that the assumption of equal correlations does not seem justified in this situation. If a split-plot analysis is performed, Table 7.9 is obtained.

The same conclusions would be made for this analysis. However, because of the lack of equal variances and equal correlations of the observations, the split-plot analysis is *not valid*. Since the times of 3 and 5 min give highly correlated data, knowledge of the response at 3 min gives knowledge of the response at 5 min. Hence the number of degrees of freedom in the univariate analysis is highly inflated.

Example 7.6.2 Psychomotor scores were obtained for subjects during 3 days after radiation treatment. (See Problem 7.4.) The subjects were assigned to one of four treatment groups. The data appear in Table 7.10. The sample means are given in Table 7.11, and the pooled sample covariance

Table 7.10

	Day					Day		
Subject	1	2	3		Subject	1	2	3
	Control					87.5r		
1	223	242	238		21	206	199	237
2	72	81	66		22	208	222	237
3	172	214	239		23	224	224	261
4	171	191	203		24	119	149	196
5	138	204	213		25	144	169	164
6	22	24	24		26	170	202	181
	37.5r				27	93	122	145
					28	237	243	281
7	53	102	104		29	208	235	249
8	45	50	54		30	187	199	205
9	47	45	34		31	95	102	96
10	167	188	209		32	46	67	28
11	183	206	210		33	95	137	99
12	91	154	152		34	59	76	101
13	115	133	136		35	186	198	201
14	32	97	86			187.5r		
15	38	37	40					
16	66	131	148		36	202	229	232
17	210	221	251		37	126	159	157
18	167	172	212		38	54	75	75
19	23	18	30		39	158	168	175
20	234	260	269		40	175	217	235
					41	147	183	181
					42	105	107	92
					43	213	263	260
					44	258	248	257
					45	257	269	270

Table 7.11

Group	Day 1	2	3	Sample size
control	133.0	159.3	165.5	6
37.5*r*	105.8	129.6	138.2	14
87.5*r*	151.8	169.6	178.7	15
187.5*r*	169.5	191.8	193.4	10

matrix is given by

$$S = \begin{pmatrix} 4685 & 4522 & 5096 \\ 4522 & 4779 & 5287 \\ 5096 & 5287 & 6151 \end{pmatrix}$$

on 41 df. The F-test for equal profiles has a value

$$F_{6,80} = 0.34 < F_{6,8,0.05} = 2.22.$$

Hence we do not reject the null hypothesis of equal profiles, and assuming equal profiles, we may now test for the level and condition hypotheses. For the condition hypothesis $F = 44.7$ on 2 and 40 df. Since $F_{2,40,0.05} = 3.23$, we reject the null hypothesis of equal conditions; that is, the psychomotor scores change from day to day. For the level hypothesis $F = 2.84$ on 3 and 39 df, and since $F_{3,39,0.05} = 2.84$ we reject the null hypothesis at equal levels; that is, the average psychomotor scores are significantly different for the four groups.

If we were to examine the preradiation psychomotor scores, we might notice a difference in mean among the groups. To determine whether the difference in psychomotor scores was actually due to the treatment, and not just to the differing abilities of the subjects, we would then find it necessary to adjust for the preradiation scores. Although the procedure assumes that subjects are randomly assigned to treatments, the severity of the illness or other factors might determine the treatment administered. We look at this problem of covariables in the next section.

7.7 Profile Analysis with Covariables

In Example 7.6.2 a profile analysis was performed on data from four treatment groups. The results indicated that the levels of scores of the four groups differed. For the claim that the treatments caused this difference in scores to be valid, the subjects must be assigned randomly to the four groups. This randomization is not always possible. The control group, for example, receives no treatment. One method to circumvent this problem is an analysis of the covariance. That is, we adjust the scores to account for differences in groups on other variables. In this case the preradiation

Table 7.12 Preradiation Psychomotor Scores

Control	37.5r	87.5r	187.5r
191	53	181	201
64	33	178	113
206	16	190	86
155	121	127	115
85	179	94	183
15	114	148	131
	92	99	71
	84	207	172
	30	188	224
	51	140	246
	188	109	
	137	69	
	105	69	
	205	51	
		156	
mean 119.33	100.57	137.93	154.20

psychomotor scores (Table 7.12) can be used as a covariable to adjust for differences in the subjects' psychomotor ability.

One can see that the psychomotor scores before the treatment was administered of the four treatment groups differ. Hence one should adjust for the covariable, the preradiation score. If the covariable has a different effect on score from different days, then the concept of equal profiles "adjusted" for the covariable may make no sense. If, however, one assumes that the effect of the covariable on the response is the same for all three days, then tests for equal profiles and the condition hypothesis will be unchanged. That is, one assumes that the covariable affects a subjects' average score rather than the change in score from day to day.

Suppose that we have J groups of N_j subjects $j=1,\ldots,J$ and that the observations are

$$\mathbf{y}_{jk}, \quad j=1,\ldots,J, \quad k=1,\ldots,N_j, N=\sum_{j=1}^{J} N_j, \tag{7.7.1}$$

where $\mathbf{y}_{jk}=\boldsymbol{\mu}_j+\boldsymbol{\varepsilon}_{jk}$, $\boldsymbol{\varepsilon}_{jk}\sim N_p(\mathbf{0},\Sigma)$, as in (7.6.1). In addition we observe q covariables per subject, given in vector form by [(7.7.2)] $\mathbf{x}_{jk}$, $j=1,\ldots,J$, $k=1,\ldots,N_j$. Then tests of the parallelism and condition hypotheses are given by (7.6.3) and (7.6.6). To test for differences in levels we make the following transformations. First write

$$\mathbf{y}'_{jk}=(y_{1jk},\ldots,y_{pjk}), \quad j=1,\ldots,J, \quad k=1,\ldots,N_j.$$

Then let

$$z_{jk}=p^{-1}\sum_{i=1}^{p} y_{ijk}, \qquad m_{ijk}=y_{i+1,jk}-y_{ijk}, \quad i=1,\ldots,p-1. \quad (7.7.2)$$

Then z_{jk} measures the average level of the jkth subject and m_{ijk} measures the change in the jkth subject's score from condition i to $i+1$. Under the assumption of parallelism the mean value of m_{ijk} for a given i is the same for all J groups. In addition, m_{ijk} is correlated with z_{jk}, and hence the m_{ijk}s can be used as covariables.

If we denote $E(z_{jk})=\alpha_j$, then the test of the level hypothesis is equivalent to testing

$$H: \alpha_1=\cdots=\alpha_J \qquad \text{vs} \qquad A: \alpha_i \neq \alpha \quad (7.7.3)$$

for some $i \neq j$. Note that if one ignores the variables m_{ijk}, then one can test (7.7.3) with an F-test based on a completely randomized design. However, with the information provided by the m_{ijk}s we obtain the test given in (7.6.4). In this section we consider the case where further information is provided by the covariables $\mathbf{x}_{jk}$. We have $p-1$ covariables obtained from the parallelism hypothesis $(m_{ijk},\ldots,m_{p-1,jk})$ and q additional covariables $x_{1jk},\ldots,x_{qjk}$, where $\mathbf{x}'_{jk}=(x_{1jk},\ldots,x_{qjk})$. We therefore define for uniformity $m_{p-1+i,jk}=x_{ijk}$ for $i=1,\ldots,q, j=1,\ldots,J, k=1,\ldots,N_j$. Hence we obtain the following analysis of covariance model:

$$z_{jk}=\mu+\alpha_j+\sum_{i=1}^{p+q-1}\gamma_i(m_{ijk}-\bar{m}_{i..})+\varepsilon_{jk}, \quad (7.7.4)$$

where

$$\bar{m}_{i..}=N^{-1}\sum_{i=1}^{J}\sum_{k=1}^{N_j} m_{ijk}$$

and $\varepsilon_{jk}\sim N(\mathbf{0},\sigma^2)$. Tests and confidence intervals for the differences in levels of the groups can be obtained from Section 5.5, as $\alpha_s-\alpha_t$ represents the difference in levels of groups s and t "adjusted" for the covariables.

Example 7.7.1 Consider the data from Example 7.6.2 of psychomotor scores for four groups during three days after radiation treatment, along with the initial preradiation psychomotor scores in Table 7.12. In Example 7.6.2 the hypothesis of parallelism was accepted. Then the condition and level hypotheses were rejected at $\alpha=0.05$. If we now wish to adjust the scores for the preradiation score we test the level hypothesis using the methods of this section. We have

$$\mathbf{y}'_{jk}=(y_{1jk},y_{2jk},y_{3jk}), \qquad j=1,\ldots,4, \quad k=1,\ldots,N_j,$$

$N_1=6$, $N_2=14$, $N_3=15$, $N_3=10$, and $N=45$. The number of days was $p=3$

and the number of covariables was $q=1$. The values of the one covariable are $x_{jk}, j=1,\ldots,4, k=1,\ldots,N_j$. Define

$$z_{jk}=\tfrac{1}{3}(y_{1jk}+y_{2jk}+y_{3jk}), \qquad m_{1jk}=(y_{2jk}-y_{1jk}),$$
$$m_{2jk}=(y_{3jk}-y_{2jk}), \qquad m_{3jk}=x_{jk}.$$

Then the model in (7.7.4) becomes

$$z_{jk}=\mu+\alpha_j+\sum_{i=1}^{3}\gamma_i(m_{ijk}-\bar{m}_{i..})+\varepsilon_{jk}, \qquad j=1,\ldots,4, \quad k=1,\ldots,N_j.$$

The hypothesis of equal levels becomes H: $\alpha_1=\alpha_2=\alpha_3=\alpha_4$ vs A: $\alpha_i\neq\alpha_j$, some $1\le i<j\le 4$.

The F-test for this hypothesis can be obtained from the SAS GLM procedure or by using the formulas in Section 5.5. The F-value was obtained by the SAS program listed in Section 7.8. The value was $F_{3,38}=0.33\le F_{3,38,0.05}=2.90$. Hence we claim that there is no significant difference in the levels of scores among the four treatment groups. Notice that this result contrasts sharply with the conclusions of Example 7.6.2.

Example 7.7.2 Consider the data of Example 7.6.1, in which a profile analysis was carried out on four groups, each consisting of four dogs. Blood histamine was measured before treatment and at 1, 3, and 5 min after treatment. Several points are worth noting. At the control stage the histamine levels should be the same. Hence one should test for equal profiles based on times 1, 3, and 5 min, with the control response used as a covariable. The test of parallelism based on the three measurements $y_{1jk}, y_{2jk}, y_{3jk}, j=1,\ldots,4, k=1,\ldots,4$, corresponding respectively to times 1, 3, and 5 min yields $U_{2,3,12}=0.4487$.

Hence from Theorem 4.2.3 we have

$$11(1-U_{2,3,12}^{1/2})/3U_{2,3,12}^{1/2}=F_{6,22}=1.81\le 2.55=F_{6,22,0.05}.$$

We do not reject the hypothesis of parallelism. The test of equal conditions averaged over groups yields

$$F_{2,12}=1.97\le F_{2,14,0.05}=3.74,$$

so there does not appear to be a significant change in histamine levels at times 1, 3, and 5 min. The test of equal levels among the four groups adjusted for the control histamine measurement yields

$$F_{3,9}=4.47\ge F_{3,9,0.05}=3.86.$$

Hence we claim that there is a significant difference in the blood histamine, averaged over the three times, among the four treatment groups after adjustment for the *pre*-treatment histamine levels.

In Chapters 6 and 7 we have discussed the analysis of longitudinal data using multivariate growth curve and profile analysis. A survey of these and other techniques is given by Kowalski and Guire (1974).

7.8 Computational Procedures

All the statistics given in this section can be computed from a general linear models program such as PROC GLM in SAS. For Example 7.2.2 the program statements are

```
DATA
INPUT R1-R4
Z=(R1+R2+R3+R4)/4;
Z1=R1-R2;
Z2=R2-R3;
Z3=R3-R4;
A=1.0;
CARDS;
11 15 21 21
:
.
10  9 14 12
PROC GLM;
   MODEL Z1-Z3=A/NOINT;
   MANOVA H=A/PRINTE;
   PROC GLM;
   MODEL Z=Z1 Z2 Z3/XPX I;
```

This file tests that the three differences in the means are zero. The MANOVA statement indicates that a multivariate test is to be performed. If there is no difference in the means of the four response variables R_1-R_4, then the three differences Z_1-Z_3 have mean zero and are correlated with Z, the average of the four response variables. Hence the estimate of the overall mean γ ($\mu_1=\cdots=\mu_4=\gamma$) can be obtained from the intercept of the model $Z=\beta_0+\beta_1 Z_1+\beta_2 Z_2+\beta_3 Z_3+\varepsilon$ as $E(Z_1)=E(Z_2)=E(Z_3)=0$. Alternatively, the last two lines can be omitted and an estimate and confidence interval for γ can be calculated for the sample mean and covariance matrix as in Chapter 6. These statistics can be obtained with statements such as

```
PROC GLM;
CLASS A;
MODEL R1-R4=A/NOINT;
MANOVA H=A/PRINTE;
MEANS A;
```

For Example 7.2.3 the following cards were used:

```
DATA;
INPUT Y1 Y2 Y3:
A=1;
U1=Y1-Y2;
U2=Y2-Y3;
Z=Y1+Y2+Y3;
CARDS;
 91 154 152
115 133 136
 32  97  86
 38  37  40
 66 131 148
210 221 251
```

```
167 172 212
 23  18  30
234 260 269
PROC GLM;
MODEL Y1-Y3=A/NOUNI;
MANOVA H=A/PRINTE;
PROC GLM;
MODEL U1 U2=A/NOINT NOUNI;
MANOVA H=A;
PROC GLM;
MODEL Z=U1 U2;
```

For Example 7.5.1 the following cards were used:

```
DATA REQUEST;
INPUT Y1-Y3 A;
Z=(Y1+Y2+Y3)/3;
U1=Y1-Y2;
U2=Y2-Y3;
B=1;
CARDS;
11 16 23 1


21 24 24 2
PROC GLM;
CLASS A;
MODEL Y1-Y3=A;
MANOVA H=A/PRINTE;
MEANS A;

PROC GLM;
CLASS A;
MODEL U1 U2=A;
MANOVA H=A;
PROC GLM;
CLASS A;
MODEL Z=A U1 U2;
LS MEANS A/STDERR PDIFF;
PROC GLM;
CLASS B;
MODEL U1 U2=B/NOINT;
MANOVA H=B;
```

The file gives the means of the two groups, the pooled covariance matrix and a test for equality of means for the two groups. The output for the model U1 U2 = A tests for equal profiles. The output for the model Z = A U1 U2 tests for equal levels assuming equal profiles. Finally the model U1 U2 = B can be used to test for different conditions. If the profiles are not equal, the latter two tests should not be performed. If the profiles are equal, the statement LSMEANS A/STDERR PDIFF produces estimates of the levels with the appropriate standard error and the *p*-value for a difference between two levels.

For Examples 7.6.1 and 7.6.2 the statements follow the same pattern. We give the cards for Example 7.6.2:

```
DATA;
INPUT Y1 Y2 Y3 A;
 U1=Y1-Y2;
 U2=Y2-Y3;
 Z=(Y1+Y2+Y3)/3;
 B=1;
CARDS;
      223 242 248 1

      257 269 270 4
PROC GLM;
 CLASS A;
 MODEL Y1 Y2 Y3=A/NOUNI;
 MEANS A;
MANOVA H=A/PRINTE;
PROC GLM;
CLASS A;
MODEL Z=U1 U2 A;
LSMEANS A/STDERR PDIFF;
PROC GLM;
CLASS B;
MODEL U1 U2=B/NOINT;
MANOVA H=B;
```

The file for Example 7.7.1 is a modified version of the previous computer file. The extra covariable Y_0 is inputted and is used as a covariable in the test of the level hypothesis. The file takes the following form:

```
  DATA;
INPUT Y0 Y1 Y2 Y3 A;
 U1=Y1-Y2;
 U2=Y2-Y3;
 Z=(Y1+Y2+Y3)/3;
 B=1;
CARDS;
191 223 242 248 1

246 257 269 270 4
PROC GLM;
 CLASS A;
 MODEL Y1 Y2 Y3=A/NOUNI;
 MEANS A;
MANOVA H=A/PRINTE;
PROC GLM;
CLASS A;
MODEL U1 U2=A/NOUNI;
 MANOVA H=A/PRINTE;
PROC GLM;
CLASS A;
MODEL Z=Y1 U2 Y0 A;
LSMEANS A/STDERR PDIFF;
PROC GLM;
CLASS B;
MODEL U1 U2=B/NOINT;
MANOVA H=B;
```

Problems

7.1 An experiment was conducted to examine the effect of air pollution on interpersonal attraction. Twenty-four subjects were placed in four varying environments with a stranger for a 15 min period. At the end of the encounter, each subject was asked to assess his degree of attraction to the stranger on a Likert scale of 1–10, with 10 indicating strong attraction (Table 7.13). The four experimental experiments were

i. T_1, odor-free room, culturally similar stranger,
ii. T_2, room contaminated by ammonium sulfide, culturally similar stranger,
iii. T_3, odor-free room, dissimilar stranger, and
iv. T_4, room contaminated by ammonium sulfide, dissimilar stranger.

To remove the effect of order, each subject was introduced to the four environments in one of the 24 possible orders.

(a) Analyze the data using ANOVA techniques.
(b) Analyze the data using multivariate methods (T^2).

7.2 A market researcher is interested in the effect of brand name on the consumer's perception of quality. Subjects are asked to estimate the price of a bar of

Table 7.13

	Environment			
Subject	T_1	T_2	T_3	T_4
1	8	7	4	1
2	6	6	5	3
3	5	4	3	3
4	9	4	8	2
5	7	6	7	2
6	6	3	7	3
7	9	8	8	8
8	10	7	10	8
9	4	3	5	6
10	7	5	6	4
11	8	8	9	6
12	9	2	4	4
13	7	6	6	8
14	6	9	5	4
15	6	7	3	1
16	8	8	6	6
17	6	7	6	4
18	5	3	4	2
19	2	1	2	2
20	2	1	2	2
21	4	2	3	1
22	4	3	4	6
23	8	5	6	2
24	7	3	4	2

Table 7.14 Experimental Results

Subject	Packaging P_1	P_2	P_3	P_4
labeled				
1	0.30	0.40	0.55	0.65
2	0.20	0.65	0.30	0.80
3	0.30	0.50	0.50	0.70
4	0.25	0.35	0.45	0.65
5	0.35	0.35	0.55	0.55
6	0.50	0.50	0.50	0.50
unlabeled				
1	0.40	0.40	0.60	0.60
2	0.45	0.50	0.55	0.85
3	0.90	0.95	1.10	1.10
4	0.60	0.70	0.85	0.95
5	0.55	0.75	1.00	1.20
6	0.70	0.70	1.00	1.10

ordinary hand soap that has been packaged in one of four ways:

i. P_1 plain wrapped, unboxed,
ii. P_2 plain wrapped, boxed,
iii. P_3 foil wrapped, unboxed, and
iv. P_4 foil wrapped, boxed.

Twelve subjects are each asked to estimate the price of the form packaged soaps (Table 7.14). For six of the subjects, the packages have been labeled with a well-known brand name. For the remaining six subjects, no label is used.

(a) Analyze the data using a split-plot technique.
(b) Use multivariate methods to analyze the data.
(c) Estimate the error covariance matrix. Is the assumption of an intraclass correlation matrix valid? An exact test is given in Section 12.4.

7.3 Table 7.15 gives the litter weight gains for pregnant and nonpregnant rats.
(a) Perform a profile analysis of this data using univariate split-plot tests.
(b) Perform this analysis using multivariate procedures.

7.4 The data of Table 7.16 were analyzed by Danford et al. (1960). The average daily score of four trials on a psychomotor testing device was measured on 45 subjects suffering from cancerous lesions 10 days after radiation treatment. The subjects were assigned to four groups receiving one of four dosages of radiation.
(a) Analyze the data using multivariate techniques.
(b) Test for interaction between time and dosage.
(c) If there is no significant interaction, test for the effects of time and dosage of radiation.

7.5 Carry out the profile analysis of the data given in Problem 6.1.

Table 7.15[a] Litter Weight Gains (g) for Pregnant and Nonpregnant Lactating Rats

Treatment group	Rat number	Period of lactation (days) 8–12	12–16	16–20	20–24	Average
pregnant	1	7.5	8.6	6.9	0.8	5.95
	2	10.6	11.7	8.8	1.6	8.18
	3	12.4	13.0	11.0	5.6	10.50
	4	11.5	12.6	11.1	7.5	10.68
	5	8.3	8.9	6.8	0.5	6.12
	6	9.2	10.1	8.6	3.8	7.92
average		9.92	10.82	8.87	3.30	8.23
nonpregnant	1	13.3	13.3	12.9	11.1	12.65
	2	10.7	10.8	10.7	9.3	10.38
	3	12.5	12.7	12.0	10.1	11.82
	4	8.4	8.7	8.1	5.7	7.72
	5	9.4	9.6	8.0	3.8	7.70
	6	11.3	11.7	10.0	8.5	10.38
average		10.93	11.13	10.28	8.08	10.11
period and overall average		10.42	10.98	9.58	5.69	9.17

[a]Data from Gill and Hafs (1971).

7.6 A study was made of open schools and their relation to student achievement. In order to assess the intellectual functioning of students, an intelligence test was given out. Raven's "progressive matrices," which consist of five tests measuring the different aspects of the cognitive ability, were used in the study. The subjects in the study were from two groups: one of 428 subjects aged 8 and 9 and another of 415 subjects aged 11 and 12. The researcher wants to know whether these two groups differ in terms of their level and pattern of intellectual functioning. Carry out a profile analysis of the following data. The correlation matrix, means and standard deviations for group 1 were

	1	2	3	4	5
1	1.000	0.504	0.477	0.427	0.204
2	0.504	1.000	0.581	0.629	0.377
3	0.477	0.581	1.000	0.563	0.359
4	0.427	0.629	0.563	1.000	0.448
5	0.204	0.377	0.359	0.448	1.000

$$\bar{X}_1=9.33,\quad \bar{X}_2=6.03,\quad \bar{X}_3=4.80,\quad \bar{X}_4=4.41,\quad \bar{X}_5=1.20,$$

$$S_1=1.92,\quad S_2=2.86,\quad S_3=2.68,\quad S_4=3.10,\quad S_5=1.35.$$

Table 7.16[a]

	Before radiation	Time (days after irradiation) 1	2	3	4	5	6	7	8	9	10
					Controls						
1	191	223	242	248	266	274	272	279	286	287	286
2	64	72	81	66	92	114	126	123	134	148	140
3	206	172	214	239	265	265	262	274	258	288	289
4	155	171	191	203	219	237	237	220	252	260	245
5	85	138	204	213	224	247	246	259	255	374	284
6	15	22	24	24	38	41	46	62	62	79	74
					25–50r						
7	53	53	102	104	105	125	122	150	93	127	132
8	33	45	50	54	44	47	45	61	50	60	52
9	16	47	45	34	37	61	51	28	43	40	45
10	121	167	188	209	224	229	230	269	264	249	268
11	179	193	206	210	221	234	224	255	246	225	229
12	114	91	154	152	155	174	196	207	208	229	173
13	92	115	133	136	148	159	146	180	148	168	169
14	84	32	97	86	47	87	103	124	110	162	187
15	30	38	37	40	48	61	64	65	83	91	90
16	51	66	131	148	181	172	195	170	158	203	215
17	188	210	221	251	256	268	260	281	286	290	296
18	137	167	172	212	168	213	190	196	211	213	224
19	108	23	18	30	29	40	57	37	47	56	55
20	205	234	260	269	274	282	282	290	298	304	308
					75–100r						
21	181	206	199	237	219	237	232	251	247	254	250
22	178	208	222	237	255	253	254	276	254	267	275
23	190	224	224	261	249	291	293	294	295	299	305
24	127	119	149	196	203	211	207	241	220	188	219
25	94	144	169	164	182	189	188	164	181	142	152
26	148	170	202	181	184	186	207	184	195	168	163
27	99	93	122	145	130	167	153	165	144	156	167
28	207	237	243	281	273	281	279	294	307	305	305
29	188	208	235	249	265	271	263	272	285	283	290
30	140	187	199	205	231	227	228	246	245	263	262
31	109	95	102	96	135	135	111	146	131	162	171
32	69	46	67	28	43	55	55	77	73	76	76
33	69	95	137	99	95	108	129	134	133	131	91
34	51	59	76	101	72	72	107	91	128	120	133
35	156	186	198	201	205	210	217	217	219	223	229

Table 7.16 (*continued*)

	Before radiation	Time (days after irradiation) 1	2	3	4	5	6	7	8	9	10
					125–250r						
36	201	202	229	232	224	237	217	268	244	275	246
37	113	126	159	157	137	160	162	171	167	165	185
38	86	54	75	75	71	130	157	142	173	174	156
39	115	158	168	175	188	164	184	195	194	206	212
40	183	175	217	235	241	251	229	241	233	233	275
41	131	147	183	181	206	215	197	207	226	244	240
42	71	105	107	92	101	103	78	87	57	70	71
43	172	213	263	260	276	273	267	286	283	290	298
44	224	258	248	257	257	267	260	279	299	289	300
45	246	257	269	270	289	291	306	301	295	312	311

[a]Reproduced with permission of the editors of *Biometrics*.

For group 2 the correlation matrix, means and standard deviations were

	1	2	3	4	5
1	1.000	0.437	0.413	0.456	0.304
2	0.437	1.000	0.542	0.596	0.380
3	0.413	0.542	1.000	0.604	0.474
4	0.456	0.596	0.604	1.000	0.455
5	0.304	0.380	0.474	0.455	1.000

$$\bar{x}_1 = 10.66,\quad \bar{x}_2 = 8.62,\quad \bar{x}_3 = 7.32,\quad \bar{x}_4 = 7.59,\quad \bar{x}_5 = 3.02,$$
$$s_1 = 1.60,\quad s_2 = 2.45,\quad s_3 = 2.38,\quad s_4 = 2.57,\quad s_5 = 2.29.$$

The data were collected by Dr. Ross E. Traub and associates at the Ontario Institute for Studies in Education (Lam, 1980).

7.7 Test for the normality of the four characteristics in Example 7.6.1. Transform the data by a power transformation. If possible, choose the same value for λ for all four characteristics measured. Reanalyze the data with the transformed variables.

Chapter 8
Classification and Discrimination

8.1. Introduction

A recent study focused on variables that lead to the solution of burglaries, including time until the arrival of a detective and the presence of fingerprints or eye witnesses. Based on the data, criminologists were able to derive a weighting of the variables that predicted with 95% accuracy which crimes could be solved. A procedure for obtaining weightings of variables to discriminate between populations (here solved and unsolved burglaries) is called *discriminant analysis*. New cases can then be appropriately *classified*, here as solvable or unsolvable.

In insect systematics, classification can sometimes be made only by experts. Perhaps the sole way to classify some ants is according to the internal sexual organs. Procedures for examining an ant internally often damage the specimen and can be costly. Other means of classification must then be found based on exterior measurements.

The first clear statement of the discrimination problem was given by Fisher, who was consulted by Barnard (1935) to classify skeletal remains. Fisher (1936) introduced the discriminant function for distinguishing between two multivariate normal distributions with a common covariance matrix. This will be the subject of Chapter 8.

8.2 Classifying into Two Known Normals with Common Covariance—Fisher's Discriminant Function

The problem of classifying an individual into one of two categories may be described as follows. Let π_0, π_1, and π_2 denote three populations. It is known that $\pi_0 = \pi_i$ for exactly one $i \in \{1,2\}$. The problem is to find for which i this is true. We shall assume that all populations are normally distributed with a common, known covariance matrix Σ. We shall now derive the Fisher's discriminant function.

Let $\mathbf{x}_0 \sim N_p(\boldsymbol{\mu}_0, \Sigma)$. Then $\mathbf{a}'\mathbf{x}_0 \sim N(\mathbf{a}'\boldsymbol{\mu}_0, \mathbf{a}'\Sigma\mathbf{a})$ for any nonnull vector $\mathbf{a}$. Fisher suggested that the vector $\mathbf{a}$ be chosen so that the *distance* between the two populations π_1 and π_2 is maximum. That is, $\mathbf{a}$ is chosen such that

$$(\mathbf{a}'\boldsymbol{\mu}_1 - \mathbf{a}'\boldsymbol{\mu}_2)^2/\mathbf{a}'\Sigma\mathbf{a} \equiv (\mathbf{a}'\boldsymbol{\delta})^2/\mathbf{a}'\Sigma\mathbf{a} \tag{8.2.1}$$

is a maximum, where

$$\boldsymbol{\delta} = \boldsymbol{\mu}_1 - \boldsymbol{\mu}_2, \tag{8.2.2}$$

$\boldsymbol{\mu}_1$ and $\boldsymbol{\mu}_2$ are the means of populations π_1 and π_2, respectively, and are assumed known. By the Cauchy–Schwarz inequality (Corollary 1.10.3),

$$(\mathbf{a}'\boldsymbol{\delta})^2/\mathbf{a}'\Sigma\mathbf{a} \leq \Delta^2 \equiv \boldsymbol{\delta}'\Sigma^{-1}\boldsymbol{\delta}, \tag{8.2.3}$$

where

$$\Delta^2 = (\boldsymbol{\mu}_1 - \boldsymbol{\mu}_2)'\Sigma^{-1}(\boldsymbol{\mu}_1 - \boldsymbol{\mu}_2) \tag{8.2.4}$$

is known as the *Mahalanobis squared distance*. Equality holds in (8.2.3) if and only if $\mathbf{a} = \lambda\Sigma^{-1}\boldsymbol{\delta}$, where λ is a constant of proportionality. Without loss of generality, one can take $\lambda = 1$. Thus

$$\mathbf{a}'\mathbf{x}_0 = \boldsymbol{\delta}'\Sigma^{-1}\mathbf{x}_0 = (\boldsymbol{\mu}_1 - \boldsymbol{\mu}_2)'\Sigma^{-1}\mathbf{x}_0 \equiv L(\mathbf{x}_0) \tag{8.2.5}$$

and is known as *Fisher's linear discriminant function*. Hence $\mathbf{x}_0$ is classified in π_1 if

$$L(\mathbf{x}_0) > K \tag{8.2.6}$$

and in π_2 otherwise, where K is chosen to have some desirable property. If we choose

$$K = \tfrac{1}{2}(\boldsymbol{\mu}_1 - \boldsymbol{\mu}_2)'\Sigma^{-1}(\boldsymbol{\mu}_1 + \boldsymbol{\mu}_2), \tag{8.2.7}$$

then the two errors of misclassification are the same. That is, the probability of misclassifying an individual from π_1 into π_2, denoted e_1, and the probability of misclassifying an individual from π_2 into π_1, denoted e_2, are the same. The procedure (8.2.6) becomes as follows: Classify $\mathbf{x}_0$ in π_1 if

$$(\boldsymbol{\mu}_1 - \boldsymbol{\mu}_2)'\Sigma^{-1}\mathbf{x}_0 - \tfrac{1}{2}(\boldsymbol{\mu}_1 - \boldsymbol{\mu}_2)'\Sigma^{-1}(\boldsymbol{\mu}_1 + \boldsymbol{\mu}_2) > 0, \tag{8.2.8}$$

and in π_2 otherwise. The errors of misclassification

$$e_1 = P\left[(\boldsymbol{\mu}_1 - \boldsymbol{\mu}_2)'\Sigma^{-1}\mathbf{x}_0 - \tfrac{1}{2}(\boldsymbol{\mu}_1 - \boldsymbol{\mu}_2)'\Sigma^{-1}(\boldsymbol{\mu}_1 + \boldsymbol{\mu}_2) < 0 \mid \mathbf{x}_0 \in \pi_1\right]$$

and

$$e_2 = P\left[(\boldsymbol{\mu}_1 - \boldsymbol{\mu}_2)'\Sigma^{-1}\mathbf{x}_0 - \tfrac{1}{2}(\boldsymbol{\mu}_1 - \boldsymbol{\mu}_2)'\Sigma^{-1}(\boldsymbol{\mu}_1 + \boldsymbol{\mu}_2) > 0 \mid \mathbf{x}_0 \in \pi_2\right]$$

are given by

$$e_1 = e_2 = \Phi(-\tfrac{1}{2}\Delta), \qquad \Delta^2 = (\boldsymbol{\mu}_1 - \boldsymbol{\mu}_2)'\Sigma^{-1}(\boldsymbol{\mu}_1 - \boldsymbol{\mu}_2), \tag{8.2.9}$$

where Φ denotes the standard normal cumulative distribution function,

$$\Phi(x)=\int_{-\infty}^{x}\frac{e^{-t^2/2}}{(2\pi)^{1/2}}dt. \tag{8.2.10}$$

Note that e_1 and e_2 are known since $\boldsymbol{\mu}_1$, $\boldsymbol{\mu}_2$, and Σ are assumed known. Welch (1939) has shown that (8.2.8) is also a likelihood procedure.

We may also modify the decision rule to incorporate any prior knowledge of the cost of misclassification and the probability of an observation belonging to a given population. Let c_1 be the cost of misclassifying an observation from population 1 into population 2, and let c_2 be the cost of misclassifying an observation from population 2 into population 1. Also let q_1 be the prior probability that an observation comes from population 1, and let q_2 be the prior probability that an observation comes from population 2. The value of K then becomes

$$K=\tfrac{1}{2}(\boldsymbol{\mu}_1-\boldsymbol{\mu}_2)'\Sigma^{-1}(\boldsymbol{\mu}_1+\boldsymbol{\mu}_2)+\ln(c_2q_2/c_1q_1). \tag{8.2.11}$$

The errors of misclassification are then given by

$$e_1=\Phi(-\tfrac{1}{2}\Delta+d\Delta^{-1}) \tag{8.2.12}$$

and

$$e_2=\Phi(-\tfrac{1}{2}\Delta-d\Delta^{-1}), \tag{8.2.13}$$

where $d=\ln(c_2q_2/c_1q_1)$.

8.2.1 *An Alternative Expression for the Likelihood Rule*

In this section we give an alternative expression for the classification rule (8.2.8). Since

$$\begin{aligned}
&\mathbf{x}_0'\Sigma^{-1}(\boldsymbol{\mu}_1-\boldsymbol{\mu}_2)-\tfrac{1}{2}(\boldsymbol{\mu}_1+\boldsymbol{\mu}_2)'\Sigma^{-1}(\boldsymbol{\mu}_1-\boldsymbol{\mu}_2)\\
&\quad=\tfrac{1}{2}\big(\mathbf{x}_0'\Sigma^{-1}\mathbf{x}_0-\mathbf{x}_0'\Sigma^{-1}\mathbf{x}_0+2\mathbf{x}_0'\Sigma^{-1}\boldsymbol{\mu}_1\\
&\qquad-2\mathbf{x}_0'\Sigma^{-1}\boldsymbol{\mu}_2-\boldsymbol{\mu}_1'\Sigma^{-1}\boldsymbol{\mu}_1+\boldsymbol{\mu}_2'\Sigma^{-1}\boldsymbol{\mu}_2\big)\\
&\quad=\tfrac{1}{2}\big[(\mathbf{x}_0-\boldsymbol{\mu}_2)'\Sigma^{-1}(\mathbf{x}_0-\boldsymbol{\mu}_2)-(\mathbf{x}_0-\boldsymbol{\mu}_1)'\Sigma^{-1}(\mathbf{x}_0-\boldsymbol{\mu}_1)\big]\\
&\quad\equiv\tfrac{1}{2}(T_2-T_1),
\end{aligned} \tag{8.2.14}$$

where

$$T_i=(\mathbf{x}_0-\boldsymbol{\mu}_i)'\Sigma^{-1}(\mathbf{x}_0-\boldsymbol{\mu}_i), \qquad i=1,2. \tag{8.2.15}$$

We classify $\mathbf{x}_0$ in π_1 if

$$T_1<T_2, \tag{8.2.16}$$

and in π_2 if

$$T_2<T_1. \tag{8.2.17}$$

That is, $\mathbf{x}_0$ is classified into π_i if

$$T_i = \min_{1 \le j \le 2} T_j. \tag{8.2.18}$$

8.3 Classifying into Two Normals with Known Common Covariance

Let $\mathbf{x}_i \sim N_p(\boldsymbol{\mu}_i, \Sigma)$, $i = 0, 1, 2$. We shall assume that Σ is known but $\boldsymbol{\mu}_0$, $\boldsymbol{\mu}_1$, and $\boldsymbol{\mu}_2$ unknown. On the basis of independent observations $\mathbf{x}_1, \ldots, \mathbf{x}_{N_1}$, $\mathbf{x}_{N_1+1}, \ldots, \mathbf{x}_{N_1+N_2}$, where the first N_1 observations are from π_1 and the last N_2 observations are from π_2, we wish to classify $\mathbf{x}_0$ (a subject with measurements $\mathbf{x}_0$) into π_i, $i = 1, 2$. Let

$$\bar{\mathbf{x}}_1 = N_1^{-1} \sum_{i=1}^{N_1} \mathbf{x}_i \qquad \text{and} \qquad \bar{\mathbf{x}}_2 = N_2^{-1} \sum_{i=1}^{N_2} \mathbf{x}_i.$$

The linear discriminant rule classifies x_0 in π_1 if

$$(\mathbf{x}_0 - \bar{\mathbf{x}}_1)'\Sigma^{-1}(\mathbf{x}_0 - \bar{\mathbf{x}}_1) < (\mathbf{x}_0 - \bar{\mathbf{x}}_2)'\Sigma^{-1}(\mathbf{x}_0 - \bar{\mathbf{x}}_2) \tag{8.3.1}$$

and in π_2 otherwise. Equivalently, (8.3.1) may be written

$$(\bar{\mathbf{x}}_1 - \bar{\mathbf{x}}_2)'\Sigma^{-1}\mathbf{x}_0 > \tfrac{1}{2}(\bar{\mathbf{x}}_1 - \bar{\mathbf{x}}_2)'\Sigma^{-1}(\bar{\mathbf{x}}_1 + \bar{\mathbf{x}}_2). \tag{8.3.2}$$

Letting $\mathbf{l}' = (\bar{\mathbf{x}}_1 - \bar{\mathbf{x}}_2)'\Sigma^{-1}$ and $c = \frac{1}{2}(\bar{\mathbf{x}}_1 - \bar{\mathbf{x}}_2)'\Sigma^{-1}(\bar{\mathbf{x}}_1 + \bar{\mathbf{x}}_2)$, we obtain the following linear discriminant rule: Classify x_0 in π_1 if

$$\mathbf{l}'\mathbf{x}_0 > c \tag{8.3.3}$$

and in π_2 otherwise. Asymptotic errors of misclassification are given by Okamoto (1963). For the error of classifying an observation from π_1 into π_2 we have

$$e_1 = \Phi(-\tfrac{1}{2}\Delta) + a_1 N_1^{-1} + a_2 N_2^{-1} \tag{8.3.4}$$

where

$$a_1 = (2\Delta^2)^{-1}(d_4 + 3pd_2), \tag{8.3.5}$$

and

$$a_2 = (2\Delta^2)^{-1}[d_4 - (p-4)d_2], \tag{8.3.6}$$

with

$$d_2 = (2\pi)^{-1/2}\tfrac{1}{2}\Delta \exp(-\tfrac{1}{2}\Delta^2), \tag{8.3.7}$$

and

$$d_4 = (2\pi)^{-1/2}(\tfrac{1}{8}\Delta^3 - \tfrac{3}{2}\Delta)\exp(-\Delta^2/2). \tag{8.3.8}$$

The probability of classifying an observation from π_2 in π_1 is given by

$$e_2 = \Phi(-\tfrac{1}{2}\Delta) + a_1 N_2^{-1} + a_2 N_1^{-1}. \tag{8.3.9}$$

The linear rule is used for its simplicity. Another rule, the *likelihood ratio rule*, adjusts the discriminant function for the difference in sample sizes. However this rule is quadratic and errors of misclassification are difficult to calculate. The maximum likelihood rule is to classify $\mathbf{x}_0$ in π_1 if

$$\left(1+N_1^{-1}\right)^{-1}(\mathbf{x}_0-\bar{\mathbf{x}}_1)'\Sigma^{-1}(\mathbf{x}_0-\bar{\mathbf{x}}_1)<\left(1+N_2^{-1}\right)^{-1}(\mathbf{x}_0-\bar{\mathbf{x}}_2)'\Sigma^{-1}(\mathbf{x}_0-\bar{\mathbf{x}}_2) \tag{8.3.10}$$

and in π_2 otherwise. The error of misclassification e_1 is given (John, 1961) by

$$\begin{aligned} e_1 &= P\Big[\left(1+N_1^{-1}\right)^{-1}(\mathbf{x}_0-\bar{\mathbf{x}}_1)'\Sigma^{-1}(\mathbf{x}_0-\bar{\mathbf{x}}_1) \\ &\quad > \left(1+N_2^{-1}\right)^{-1}(\mathbf{x}_0-\bar{\mathbf{x}}_2)'\Sigma^{-1}(\mathbf{x}_0-\bar{\mathbf{x}}_1)\,|\,\mathbf{x}_0\in\pi_1\Big] \\ &= P\left[F_{p,p}(\lambda_1,\lambda_2)>1\right], \end{aligned} \tag{8.3.11}$$

where $F_{p,p}(\lambda_1, \lambda_2)$ has a doubly noncentral F-distribution with p, p degrees of freedom and noncentrality parameters λ_1 and λ_2, where

$$\lambda_1=\Delta^2(1-c^{1/2})/2c\left(1+N_2^{-1}\right) \qquad \text{and} \qquad \lambda_2=\Delta^2(1+c^{1/2})/2c\left(1+N_2^{-1}\right), \tag{8.3.12}$$

$$c=1-\left[\left(1+N_1^{-1}\right)\left(1+N_2^{-1}\right)\right]^{-1},$$

and

$$\Delta^2=(\boldsymbol{\mu}_1-\boldsymbol{\mu}_2)'\Sigma^{-1}(\boldsymbol{\mu}_1-\boldsymbol{\mu}_2). \tag{8.3.13}$$

For some tabulated values of e_1 see Srivastava (1973b). A table of doubly noncentral F-distributions has been given by Tiku (1975).

The error of misclassification e_2 is given by

$$\begin{aligned} e_2 &= P\Big[\left(1+N_1^{-1}\right)^{-1}(\mathbf{x}_0-\bar{\mathbf{x}}_1)'\Sigma^{-1}(\mathbf{x}_0-\bar{\mathbf{x}}_1) \\ &\quad < \left(1+N_2^{-1}\right)^{-1}(\mathbf{x}_0-\bar{\mathbf{x}}_2)'\Sigma^{-1}(\mathbf{x}_0-\bar{\mathbf{x}}_2)\,|\,\mathbf{x}_0\in\pi_2\Big] \\ &= P\left[F_{p,p}(\lambda_1^*,\lambda_2^*)>1\right], \end{aligned} \tag{8.3.14}$$

where

$$\lambda_1^*=b\lambda_1 \quad \lambda_2^*=b\lambda_2, \qquad b=\left(1+N_1^{-1}\right)^{-1}\left(1+N_2^{-1}\right). \tag{8.3.15}$$

REMARK 8.3.1: If, instead of just one observation $\mathbf{x}_0$ from π_0, we have N_0 observations, $\mathbf{x}_0$ should be replaced by $\bar{\mathbf{x}}_0$, the sample mean vector of these N_0 observations, $(1+N_1^{-1})$ by $(N_0^{-1}+N_1^{-1})$, and $(1+N_2^{-1})$ by $(N_0^{-1}+N_2^{-1})$ in the likelihood classification rule (8.3.10)

REMARK 8.3.2: For classification rule (8.3.10), which is a maximum likelihood rule, both errors of misclassification e_1 and e_2 are monotone decreasing functions of Δ and are $\leq \frac{1}{2}$ (Das Gupta, 1974; Srivastava and Khatri, 1979; Broffit and Williams, 1973).

8.3.1 *Estimates of e_1 and e_2*

For classification rule (8.3.1) the errors of misclassification e_1 and e_2, given in (8.3.4)–(8.3.9) cannot be computed since they involve unknown parameters $\boldsymbol{\mu}_1$ and $\boldsymbol{\mu}_2$ (actually $\boldsymbol{\delta}=\boldsymbol{\mu}_1-\boldsymbol{\mu}_2$). To get some idea of the errors committed it would be desirable to estimate these quantities. There are several techniques available, and except for some Monte Carlo results there are no clear-cut ways to choose among them. We shall describe two such procedures in order of complexity.

The simplest estimate of e_1 and e_2 can be obtained by a substitution method. Substituting

$$\hat{\Delta}^2=(\bar{\mathbf{x}}_1-\bar{\mathbf{x}}_2)'\Sigma^{-1}(\bar{\mathbf{x}}_1-\bar{\mathbf{x}}_2)$$

for Δ^2 in the formulas (8.3.4)–(8.3.9). These estimates are, however, biased. But for large samples the bias will be negligible.

We may also use a proportion method. It is known that the observations $\mathbf{x}_1,\ldots,\mathbf{x}_{N_1}$ have come from π_1 and the observations $\mathbf{x}_{N_1+1},\ldots,\mathbf{x}_{N_1+N_2}$ from π_2. We can use these to assess the performance of classification rule (8.3.1). If the rule is a reasonable one, only a few should be misclassified. The ratio of the misclassified observations from π_1 to the total N_1 will thus give an unbiased estimate of e_1. Similarly, the ratio of the misclassified observations from π_2 to the total N_2 will give an unbiased estimate of e_2. The procedure can be described formally as follows. Let $\bar{\mathbf{x}}^{(i)}$, $i=1,2,\ldots,N_1$, be the mean of only N_1-1 observations of $\mathbf{x}_1,\ldots,\mathbf{x}_{N_1}$ for which the ith observation has been left out; then

$$T_i^{(1)}=\left(\mathbf{x}_i-\bar{\mathbf{x}}_1^{(i)}\right)'\Sigma^{-1}\left(\mathbf{x}_i-\bar{\mathbf{x}}_1^{(i)}\right),$$

$$Q_i^{(1)}=(\mathbf{x}_i-\bar{\mathbf{x}}_2)'\Sigma^{-1}(\mathbf{x}_i-\bar{\mathbf{x}}_2).$$

Let m_1 be the number of $T_i^{(1)}$s for which $T_i^{(1)}>Q_i^{(1)}$. Then an unbiased estimate of e_1 is given by m_1/N_1. Similarly, for $i=1,2,\ldots,N_2$ let

$$T_i^{(2)}=\left(\mathbf{x}_{N_1+i}-\bar{\mathbf{x}}_1\right)'\Sigma^{-1}\left(\mathbf{x}_{N_1+i}-\bar{\mathbf{x}}_1\right),$$

$$Q_i^{(2)}=\left(\mathbf{x}_{N_1+i}-\bar{\mathbf{x}}_2^{(i)}\right)'\Sigma^{-1}\left(\mathbf{x}_{N_1+i}-\bar{\mathbf{x}}_2^{(i)}\right),$$

where $\bar{\mathbf{x}}_2^{(i)}$ is the mean of only N_2-1 observations of $\mathbf{x}_{N_1+1},\ldots,\mathbf{x}_{N_1+N_2}$ for which the (N_1+i)th observation has been left out. Let m_2 be the number of $T_i^{(2)}$s for which $T_i^{(2)}<Q_i^{(2)}$. Then an unbiased estimate of e_2 is given by m_2/N_2.

Example 8.3.1 Table 8.1 gives the means for three psychological tests administered to patients suffering from anxiety and hysteria (Rao, 1973). The covariance matrix

$$\begin{pmatrix} 2.3008 & 0.2516 & 0.4742 \\ 0.2516 & 0.6075 & 0.0358 \\ 0.4742 & 0.0358 & 0.5951 \end{pmatrix}$$

Table 8.1

Illness	Test x_1	x_2	x_3	Sample size
anxiety	2.9298	1.667	0.7281	114
hysteria	3.0303	1.2424	0.5455	33

was obtained from these and other measurements to give 250 df. Hence we may assume that Σ is known. The linear rule classifies $\mathbf{x}_0$ in π_1 if $\mathbf{l}'\mathbf{x}_0 > c$, where $\mathbf{l}' = (-0.217, 0.764, 0.069)$ and $c = 0.512$. Hence a patient would be diagnosed as suffering from anxiety, as opposed to hysteria, based on these three psychological tests if $-0.217x_1 + 0.764x_2 + 0.069x_3 > 0.512$. In order to assess the accuracy of this procedure we must now calculate the error of misclassification. The estimate of Δ^2 is given by $\hat{\Delta}^2 = 0.359$, and $d_2 = 0.10$, $d_4 = -0.29$, $a_1 = 0.85$, and $a_2 = -0.26$. With $N_1 = 114$ and $N_2 = 33$ we obtain

$$\begin{aligned} e_1 &= \Phi(-\tfrac{1}{2}\hat{\Delta}) + a_1 N_1^{-1} + a_2 N_2^{-1} \\ &= 0.4032 + 0.0074 - 0.0079 = 0.4027, \\ e_2 &= \Phi(-\tfrac{1}{2}\hat{\Delta}) + a_2 N_1^{-1} + a_1 N_2^{-1} \\ &= 0.4032 - 0.0022 + 0.0257 = 0.4267. \end{aligned}$$

Hence we find that there is a high probability of misclassifying patients suffering from anxiety or hysteria solely on the results of the three psychological tests. The proportion method of estimating the errors of misclassification could not be used in this situation as the original data were not available.

8.4 Classifying into Two Completely Unknown Normals

Let $\mathbf{x}_1, \ldots, \mathbf{x}_{N_1}, \mathbf{x}_{N_1+1}, \ldots, \mathbf{x}_{N_1+N_2}$ denote independent observations with the first N_1 observations from π_1 and the last N_2 observations from π_2, where $\pi_i \sim N_p(\boldsymbol{\mu}_i, \Sigma)$, $i = 0, 1, 2$. Let

$$\bar{\mathbf{x}}_1 = N_1^{-1} \sum_{i=1}^{N_1} \mathbf{x}_i, \qquad \bar{\mathbf{x}}_2 = N_2^{-1} \sum_{i=N_1+1}^{N_1+N_2} \mathbf{x}_i, \qquad f = N_1 + N_2 - 2, \quad (8.4.1)$$

and

$$fS_p = \sum_{i=1}^{N_1} (\mathbf{x}_i - \bar{\mathbf{x}}_1)(\mathbf{x}_i - \bar{\mathbf{x}}_1)' + \sum_{i=N_1+1}^{N_1+N_2} (\mathbf{x}_i - \bar{\mathbf{x}}_2)(\mathbf{x}_i - \bar{\mathbf{x}}_2)'. \quad (8.4.2)$$

Then an individual with measurements $\mathbf{x}_0$ (from π_0) is classified in π_1 if

$$(\mathbf{x}_0 - \bar{\mathbf{x}}_1)' S_p^{-1} (\mathbf{x}_0 - \bar{\mathbf{x}}_1) < (\mathbf{x}_0 - \bar{\mathbf{x}}_2)' S_p^{-1} (\mathbf{x}_0 - \bar{\mathbf{x}}_2) \qquad (8.4.3)$$

and in π_2 otherwise. This is a linear discriminant rule, called *Anderson's rule* (Anderson, 1951). Letting $\mathbf{l}'=(\bar{\mathbf{x}}_1-\bar{\mathbf{x}}_2)'S_p^{-1}$ and $c=\frac{1}{2}(\bar{\mathbf{x}}_1-\bar{\mathbf{x}}_2)'S_p^{-1}(\bar{\mathbf{x}}_1+\bar{\mathbf{x}}_2)$, we classify $\mathbf{x}_0$ in π_1 if

$$\mathbf{l}'\mathbf{x}_0>c, \tag{8.4.4}$$

and in π_2 otherwise. The distribution of the errors of misclassification are given by Anderson (1951), Bowker and Sitgreaves (1961), and Sitgreaves (1961). We shall use the asymptotic expansion of Okamoto (1963) for errors of misclassification. That is, the probability of misclassifying an observation from π_1 in π_2 is given by

$$e_1=\Phi(-\tfrac{1}{2}\Delta)+a_1N_1^{-1}+a_2N_2^{-1}+a_3f^{-1}, \tag{8.4.5}$$

where $a_3=\frac{1}{2}(p-1)d_2$, and a_1, a_2, and d_2 are defined by (8.3.5)–(8.3.7), respectively. The probability of misclassifying an observation from π_2 in π_1 is given by

$$e_2=\Phi(-\tfrac{1}{2}\Delta)+a_2N_1^{-1}+a_1N_2^{-1}+a_3f^{-1}. \tag{8.4.6}$$

Estimates of the errors of misclassification may be obtained by substituting the consistent estimator

$$\hat{\Delta}^2=(\bar{\mathbf{x}}_1-\bar{\mathbf{x}}_2)'S_p^{-1}(\bar{\mathbf{x}}_1-\bar{\mathbf{x}}_2)(f-p+3)/(f-2) \tag{8.4.7}$$

for Δ^2 in (8.4.5) and (8.4.6). Note that this method assumes normality.

An alternative method for estimating the errors of misclassification that does not require the normality assumption is given by the *proportion* or *jackknife method*. Let

$$T_i^{(1)}=(\mathbf{x}_i-\bar{\mathbf{x}}_1^{(i)})'S_p^{-1}(\mathbf{x}_i-\bar{\mathbf{x}}_1^{(i)}),$$

and

$$Q_i^{(1)}=(\mathbf{x}_i-\bar{\mathbf{x}}_2)'S_p^{-1}(\mathbf{x}_i-\bar{\mathbf{x}}_2),$$

where $\bar{\mathbf{x}}^{(i)}$, $i=1,2,\ldots,N_1$, is the mean of only N_1-1 observations of $\mathbf{x}_1,\ldots,\mathbf{x}_N$ for which the ith observation has been left out and S_p has been defined in (8.4.2). Let m_1 be the number of $T_i^{(1)}$s for which $T_i^{(1)}>Q_i^{(1)}$. Then an estimate of e_1 is m_1/N_1. An estimate of e_2 can be made similarly. A comparison of these and other estimates of misclassification is given by Lachenbruch and Mickey (1968). The method using Okamoto's expansion in (8.4.5) and (8.4.6) is preferable to the jackknife method when the data are normally distributed. When the data are not normal, the linear rule may still be used for discrimination, but the jackknife method should be used to estimate the errors of misclassification.

An alternative procedure that adjusts the discriminary function for the sample sizes used in estimating the population means is the likelihood ratio

Table 8.2[a] Example 8.4.1

Ch. concinna			*Ch. heikertlingeri*		
x_1	x_2	x_3	x_1	x_2	x_3
191	131	53	186	107	49
185	134	50	211	122	49
200	137	52	201	144	47
173	127	50	242	131	54
171	128	49	184	108	43
160	118	47	211	118	51
188	134	54	217	122	49
186	129	51	223	127	51
174	131	52	208	125	50
163	115	47	199	124	46

[a]Partial data from Lubischew (1962).

rule: $\mathbf{x}_0$ is classified in π_1 if

$$\left(1+N_1^{-1}\right)^{-1}(\mathbf{x}_0-\bar{\mathbf{x}}_1)'S_p^{-1}(\mathbf{x}_0-\bar{\mathbf{x}}_1)<\left(1+N_2^{-1}\right)^{-1}(\mathbf{x}_0-\bar{\mathbf{x}}_2)'S_p^{-1}(\mathbf{x}_0-\bar{\mathbf{x}}_2) \tag{8.4.8}$$

and in π_2 otherwise.

Example 8.4.1 Measurements made on two species of *Chaetocnema* were x_1, the width of the first joint of the first tarsus in microns, x_2, the width for second joint, and x_3, the maximal width of aedeagus in microns (Table 8.2; see Problem 8.4). The means are $\bar{\mathbf{x}}_1'=(179.1, 127.4, 50.5)$ and $\bar{\mathbf{x}}_2'=(208.2, 119.8, 48.9)$. The pooled covariance matrix is calculated to be

$$S_p=\begin{pmatrix} 231.25 & 103.83 & 34.15 \\ 103.83 & 61.67 & 15.66 \\ 34.15 & 15.66 & 7.41 \end{pmatrix}.$$

The likelihood discriminant function and the linear discriminant function are identical in this case. That is, we classify $\mathbf{x}_0$ in π_1 if

$$(\mathbf{x}_0-\bar{\mathbf{x}}_1)'S_p^{-1}(\mathbf{x}_0-\bar{\mathbf{x}}_1)<(\mathbf{x}_0-\bar{\mathbf{x}}_2)'S_p^{-1}(\mathbf{x}_0-\bar{\mathbf{x}}_2)$$

or, equivalently, if

$$(\bar{\mathbf{x}}_1-\bar{\mathbf{x}}_2)'S_p^{-1}\mathbf{x}_0>\tfrac{1}{2}\left(\bar{\mathbf{x}}_1'S_p^{-1}\bar{\mathbf{x}}_1-\bar{\mathbf{x}}_2'S_p^{-1}\bar{\mathbf{x}}_2\right),$$

and in π_2 otherwise.

In terms of numerical values $\mathbf{x}_0$ is classified in π_1 if

$$(-1.06, 1.34, 2.31)\mathbf{x}_0>72.42.$$

The sample Mahalanobis squared distance is given by

$$\hat{\Delta}^2=(\bar{\mathbf{x}}_1-\bar{\mathbf{x}}_2)'S_p^{-1}(\bar{\mathbf{x}}_1-\bar{\mathbf{x}}_2)(f-p+1)/(f-2)=50.23$$

with $\hat{\Delta}=7.1$. The larger the value of $\hat{\Delta}^2$, the greater is the distance between the two groups. Estimates of the errors of misclassification by the proportion method are zero because any observations can be correctly classified with the discriminant rule based on the remaining observations. In this example the two species can be correctly classified with almost no error. Using (8.4.5) and (8.4.6) we estimate the errors of misclassification as $e_1=e_2=0.0003$.

8.5 Classifying into k Normals with Common Covariance

Let $N_p(\boldsymbol{\mu}_i, \Sigma)$ be the distribution of π_i, $i=0,1,2,\dots,k$. Let $\bar{\mathbf{x}}_i$ denote the sample mean vector based on N_i observations from π_i, $i=1,2,\dots,k$, and S_p the pooled sample covariance matrix on r df where $r=\Sigma_{i=1}^{k}(N_i-1)$. Then the observation $\mathbf{x}_0$ from π_0 is classified in population π_i iff $T_i=\min_{1\le j\le k}T_j$, where

$$T_j=(\mathbf{x}_0-\bar{\mathbf{x}}_j)'S_\mathrm{p}^{-1}(\mathbf{x}_0-\bar{\mathbf{x}}_j).$$

Expressions for the errors of misclassification are not available. However, an estimate can be obtained by the proportion method described in Sections 8.3 and 8.4. If we define $\mathbf{l}_j'=\bar{\mathbf{x}}_j'S_\mathrm{p}^{-1}$ and $c_j=-\frac{1}{2}\bar{\mathbf{x}}_j'S_\mathrm{p}^{-1}\bar{\mathbf{x}}_j$, $j=1,\dots,k$, then the linear discriminant rule classifies x_0 in π_i for

$$\mathbf{l}_i'\mathbf{x}_0+c_i=\max_{1\le j\le k}(\mathbf{l}_j'\mathbf{x}_0+c_j).$$

The linear functions $\mathbf{l}_j'$ and the constants c_j are usually given by computer outputs from SAS and BMDP.

Example 8.5.1 Consider the measurements of Example 8.4.1 together with the measurements on a third group, shown in Table 8.3. The means are $\bar{\mathbf{x}}_3'=(136.8,\ 122.8,\ 50.7)$. The sample covariance matrix (pooled on 27 df)

Table 8.3 Example 8.5.1

Ch. heptapotamica		
x_1	x_2	x_3
158	141	58
146	119	51
151	130	51
122	113	45
138	121	53
132	115	49
131	127	51
135	123	50
125	119	51
130	120	48

is

$$S_p = \begin{pmatrix} 198.67 & 94.13 & 32.82 \\ 94.13 & 63.24 & 18.08 \\ 32.82 & 18.08 & 8.72 \end{pmatrix}.$$

The linear discriminant functions and constants are

$$\mathbf{l}'_1 = (-.0655, 1.496, 5.150), \qquad c_1 = -166.7,$$

$$\mathbf{l}'_2 = (0.129, 0.562, 3.962), \qquad c_2 = -144.0,$$

and

$$\mathbf{l}'_3 = (-1.426, 2.070, 6.892), \qquad c_3 = -205.1.$$

Again in this case the jackknife method gives 100% correct classification.

8.6 Classifying into Two Normals with Unequal Covariance

Let $\mathbf{x}_0$ be the measurements on a subject from π_0. It is known that $\mathbf{x}_0$ belongs to either π_1 or π_2, where $\pi_j \sim N_p(\boldsymbol{\mu}_j, \Sigma_j)$, $j = 1, 2$. When all the parameters are known, the likelihood rule classifies x_0 in π_1 if

$$(\mathbf{x}_0 - \boldsymbol{\mu}_1)'\Sigma_1^{-1}(\mathbf{x}_0 - \boldsymbol{\mu}_1) < (\mathbf{x}_0 - \boldsymbol{\mu}_2)'\Sigma_2^{-1}(\mathbf{x}_0 - \boldsymbol{\mu}_2) + C,$$

and in π_2 otherwise, where

$$C = \ln|\Sigma_2| - \ln|\Sigma_1| + k.$$

For the minimum distance rule, $C \equiv 0$. When all the parameters are unknown, the maximum likelihood rule is somewhat involved (Srivastava and Khatri, 1979), and the minimum distance rule is suggested, with the parameters replaced by their estimates.

8.7 Step-down Procedure

For convenience the notation of this section will differ from that elsewhere in this chapter. Let $\mathbf{x}$ denote the observation to be classified, and assume it to be distributed as $N_p(\boldsymbol{\mu}_0, \Sigma)$. Let $\mathbf{y}$ and $\mathbf{z}$ denote the sample means based on N observations from population π_1 and π_2, respectively. That is, $\mathbf{y} \sim N_p(\boldsymbol{\mu}_1, N_1^{-1}\Sigma)$ and $\mathbf{z} \sim N_p(\boldsymbol{\mu}_2, N_2^{-1}\Sigma)$. Let S denote the pooled estimate of Σ based on $f = (N_1 - 1) + (N_2 - 1)$ df. That is, $fS \sim W_p(\Sigma, r)$. The problem is to classify $\mathbf{x}$ into either π_1 or π_2. Suppose that the variates can be arranged in descending order of importance.

Let $\mathbf{x}' = (x_1, \ldots, x_p)$, and $\mathbf{x}'_{(i)} = (x_1, \ldots, x_i)$. We define $\mathbf{y}'_{(i)}$ and $\mathbf{z}'_{(i)}$ similarly. Further, we shall denote the top left-hand $i \times i$ submatrix of $S = (S_{ij})$ by $S_{(i)}$. Also, let

$$\hat{\boldsymbol{\beta}}_{(i)} = S_{(i)}^{-1} \begin{pmatrix} S_{1,i+1} \\ \vdots \\ S_{i,i+1} \end{pmatrix}, \qquad i = 1, \ldots, p.$$

For $i=0,1,\ldots,p-1$, $\hat{\beta}_0=0$, and $S_0=1$, define

$$\tilde{x}_{i+1}=x_{i+1}-\mathbf{x}'_{(i)}\hat{\boldsymbol{\beta}}_{(i)},$$

$$\tilde{y}_{i+1}=y_{i+1}-\mathbf{y}'_{(i)}\hat{\boldsymbol{\beta}}_{(i)},$$

$$\tilde{z}_{i+1}=z_{i+1}-\mathbf{z}'_{(i)}\hat{\boldsymbol{\beta}}_{(i)}.$$

Then $\mathbf{x}$ is classified in π_1 if, for all $i=1,2,\ldots,p$,

$$Q_i \equiv \tilde{x}_i(\tilde{y}_i-\tilde{z}_i)-\tfrac{1}{2}(\tilde{y}_i+\tilde{z}_i)(\tilde{y}_i-\tilde{z}_i)>0$$

and in π_2 if, for all $i=1,2,\ldots,p$, $Q_i<0$; otherwise, it is assigned to neither π_1 nor π_2.

Estimates of the errors of misclassification can be obtained by the proportion method. Asymptotic expansions for the errors of misclassification are given as follows (Srivastava, 1973a). Let u and v be standard normal random variables with correlation ρ. Then we define $G(a,b,\rho)=P(u<a, v<b)$. Then for large N_1 and N_2 the errors of misclassification are

$$e_1=e_2=\prod_{i=1}^{p}\{G(a_i,b_i,\rho_i)+G(-a_i,-b_i,\rho_i)\}, \tag{8.7.1}$$

where

$$a_i=[N_1N_2/(N_1+N_2+4N_1N_2)]^{1/2}(\delta_i/\sigma_i),$$

$\delta_i=E(y_i-z_i)$, $\sigma_i^2=|\Sigma_{(i)}|/|\Sigma_{(i-1)}|$, $i=2,\ldots,p$, where $\Sigma_{(i)}$ is the top left-hand $i\times i$ matrix of Σ, and

$$\rho_i=[(N_1-N_2)/(N_1+N_2)][N_1N_2/(N_1+N_2+4N_1N_2)]^{1/2}$$
$$\times[N_1N_2/(N_1+N_2)]^{1/2},$$

$$b_i=[N_1N_2/(N_1N_2)]^{1/2}(\delta_i/\sigma_i).$$

The step-down procedure assumes that a preassigned order exists for the p characteristics. In Section 8.9 a stepwise procedure is discussed for cases in which the characteristics are entered in an order determined by the data. Thus the importance of the characteristics in classification is determined in order to reduce the number of characteristics necessary for correct classification. The reduced number of characteristics will produce a simpler model for future use.

Example 8.7.1 Consider the data of Example 8.4.1. We have $\mathbf{y}'=(179.1, 127.4, 50.5)$ and $\mathbf{z}'=(208.2, 119.8, 48.9)$. The sample covariance matrix has entries

$$S=\begin{pmatrix} 231.25 & 103.83 & 34.15 \\ 103.83 & 61.67 & 15.66 \\ 34.15 & 15.66 & 7.41 \end{pmatrix}.$$

If the variables have been listed in order of importance, then we may apply the step-down procedure. In general the researcher dictates this order. We obtain values

$$S_{(1)} = 231.25 \qquad \text{and} \qquad S_{(2)} = \begin{pmatrix} 231.25 & 103.83 \\ 103.83 & 61.67 \end{pmatrix}.$$

Calculations yield $\hat{\beta}_1 = 0.449$ and $\hat{\beta}_2' = (0.14, 0.02)$. Hence $\tilde{y}_1 = 179.1$, $\tilde{z}_1 = 208.2$, $\tilde{y}_2 = 47.0$, $\tilde{z}_2 = 26.3$, $\tilde{y}_3 = 22.9$, $\tilde{z}_3 = 17.4$. The values of Q_i, $i = 1,2,3$, are $Q_1 = -29.1\tilde{x}_1 + 5635.2$, $Q_2 = 20.7\tilde{x}_2 - 758.66$, and $Q_3 = 5.5\tilde{x}_3 - 110.825$. Hence x is classified in π_1 if $\tilde{x}_1 < 193.65$, $\tilde{x}_2 > 36.65$, and $\tilde{x}_3 > 20.15$. For the first observation from *concinna*, we obtain $\mathbf{x}' = (191, 131, 53)$ and $\tilde{x}_1 = 191$, $\tilde{x}_2 = 45.2$, and $\tilde{x}_3 = 23.6$. Hence, this observation would be correctly classified in π_1.

We may use the proportion method to estimate the errors of misclassification. In this case nine of ten observations from π_1 were correctly classified, eight of ten from π_2. The three remaining observations were unclassified.

Note that the sample means and the covariance matrix should be based on all the observations except the one being classified. If the means and the covariance matrix reflect all the observations, the calculations are easier, but the estimates of the errors of misclassification will be slightly biased.

8.8 Stepwise Discriminant Analysis: A Variable Selection Procedure

8.8.1 Introduction

Researchers using discriminant functions like to use as few variables as possible for classification. Use of fewer variables makes the discriminant function easier to work with and allow better estimates of the parameters remaining in the model. The effects of dimension in discriminant analysis have been considered by Van Ness and Simpson (1976).

The relative efficiency of a subset of variables may be judged by examining

i. the reduction of the between group variances with respect to the within group variances, using an F-statistic or U-statistic; or
ii. the decrease in the correct classification rate.

Lubischew (1962) gave an excellent account of how to select variables, based on their between group and within group correlations, before using an F-criterion on his selected variables. With the computing facilities available, it is now easier to use all variables and to select a subset based on one or both of the criteria.

Ideally, it would be best to look at all possible subsets of variables. Since this is supposedly too expensive for any but a small number of variables, most computer programs use some form of stepwise procedure. Both BMDP7M and SPSS use a stepwise F-statistic criterion. At each step until a threshold value for the F-statistic is reached the variable that will produce the maximal F-value is added. Variables are then removed by deleting at each succeeding step the variable that will produce the minimal F-value, until again the threshold value is reached. Habbema and Hermans (1977) have discussed the disadvantage of this F-statistic criterion: for more than two groups it tends to choose variables that separate groups that are far apart, and it does not reduce the errors of misclassification of groups that are close together. One procedure to eliminate this drawback is first to perform a stepwise discriminant analysis on all groups. One may then perform a stepwise discriminant analysis separately on those groups that are not yet separated.

8.8.2 Procedure for Two Groups

Suppose we have q variables to choose from; that is, suppose we have the observations $\mathbf{x}_1,\ldots,\mathbf{x}_{N1}\sim N_q(\boldsymbol{\mu},\Sigma)$ and $\mathbf{y}_1,\ldots,\mathbf{y}_{N2}\sim N_q(\boldsymbol{\gamma},\Sigma)$. We first calculate the sample means and covariance matrix

$$\bar{\mathbf{x}}=N_1^{-1}\sum_{j=1}^{N_1}\mathbf{x}_j,\qquad \bar{\mathbf{y}}=N_2^{-1}\sum_{i=1}^{N_2}\mathbf{y}_j,$$

$$S=f^{-1}\left[\sum_{j=1}^{N_1}(\mathbf{x}_j-\bar{\mathbf{x}})(\mathbf{x}_j-\bar{\mathbf{x}})'+\sum_{j=1}^{N_2}(\mathbf{y}_j-\bar{\mathbf{y}})(\mathbf{y}_j-\bar{\mathbf{y}})'\right],$$

respectively, where $f=N_1+N_2-2$. To decide which variable should be entered first into the model, we calculate F-tests for differences between the two groups. There will be q values for F, corresponding to each of the q variables. For the ith variable the F-test is given by

$$F=\left(N_1^{-1}+N_2^{-1}\right)\left(\bar{x}_i-\bar{y}_i\right)^2/s_{ii}$$

where $\bar{\mathbf{x}}'=(\bar{x}_1,\ldots,\bar{x}_q)$, $\bar{\mathbf{y}}'=(\bar{y}_1,\ldots,\bar{y}_q)$, and $S=(s_{ij})$. The variable with the largest F-value is entered into the model if this F-value exceeds a critical point, which could be determined from F-tables or can be arbitrary. (The BMDP program uses a default value of 4, which is taken as a reasonable indication that the variable can separate the groups fairly well.)

When some variables have been entered in the model we again calculate F-statistics to test for differences between groups for each of the variables not in the model. These tests, which must be adjusted for the variables already in the model, are called *partial F-tests*. (See the Appendix to this chapter for details.) If some partial F-test exceeds the cutoff point, again the variable with the largest partial F-value is then entered into the model, and a partial F-test is performed on each of the other variables in the model. If any partial F-test fails to exceed the cutoff point, then the variable with the lowest partial F-value is deleted. The procedure stops when no variables can be added to or deleted from the model.

Example 8.8.1 Lubischew (1962) separated two species of the flea beatle, genus *Halticus* (emendation of *Haltica*), using a discriminant function based on external characters only. The species could be easily identified by the male copulatory organs, but each external character showed considerable measurement overlap between the populations. Twenty-one characters were originally studied, then the best four were selected for deriving a discriminant function. "Best" variables were those with the highest coefficients of discrimination K, where

$$K=(\bar{x}_1-\bar{x}_2)^2/2v.$$

Table 8.4 Measurements of Four Characters for Samples of Males of *Haltica oleracea* and *H. carduorum*[a]

H. oleracea					*H. carduorum*				
Sample	x_1	x_2	x_3	x_4	Sample	x_1	x_2	x_3	x_4
1	189	245	137	163	1	181	305	184	209
2	192	260	132	217	2	158	237	133	188
3	217	276	141	192	3	184	300	166	231
4	221	299	142	213	4	171	273	162	213
5	171	239	128	158	5	181	297	163	224
6	192	262	147	173	6	181	308	160	223
7	213	278	136	201	7	177	301	166	221
8	192	255	128	185	8	198	308	141	197
9	170	244	128	192	9	180	286	146	214
10	201	276	146	186	10	177	299	171	192
11	195	242	128	192	11	176	317	166	213
12	205	263	147	192	12	192	312	166	209
13	180	252	121	167	13	176	285	141	200
14	192	283	138	183	14	169	287	162	214
15	200	294	138	188	15	164	265	147	192
16	192	277	150	177	16	181	308	157	204
17	200	287	136	173	17	192	276	154	209
18	181	255	146	183	18	181	278	149	235
19	192	287	141	198	19	175	271	140	192
					20	197	303	170	205
means	194.5	267.1	137.4	185.9		179.6	290.8	157.2	209.2
s	13.7	18.6	8.2	15.6		10.1	19.7	12.9	13.3
K	0.83	0.80	1.68	1.30					

[a]Reproduced with the permission of the editors of *Biometrics*.

Table 8.5 Number of Cases Classified

Group	Percent correct	*H. oleracea*	*H. carduorum*
H. oleracea	94.7	18	1
H. carduorum	95.0	1	19
total	94.9		

$\bar{x}_1$ and $\bar{x}_2$ are the sample means of x for the two species, and v is the pooled estimate of the within-species variance.

The four variables chosen were thorax length x_1 in microns (THORAX), elytra length x_2 times 0.01 mm (ELYTRA), length x_3 of the second antennal joint in microns (AJ2), and length x_4 of the third antennal joint in microns (AJ3). The resulting data are shown in Table 8.4.

The data were analyzed using the BMDP7M program, as will be explained in Section 8.10. This computer program gives the linear discriminant function in (8.4.3). When the sample sizes are the same for each group, this discriminant function will be identical to the maximum likelihood rule of (8.4.8). In this example the sample sizes are 19 and 20, and the two discriminant functions will be essentially the same. The highest F-value corresponds to the variable AJ2. THORAX had the next greatest F-value conditional on AJ2 being in the model. Then came AJ3 and ELYTRA. After all four variables were entered, it was found that AJ2 became insignificant, and hence it was deleted. The final discriminant function classified an observation into population 1 (*oleracea*) if

$$-159.676+0.876x_1+0.227x_2+0.474x_3>-169.466+0.519x_1 +0.396x_2+0.624x_3,$$

or if

$$0.357x_1-0.169x_2-0.150x_3>-9.790.$$

The errors of misclassification were estimated by the proportion method. The values are summarized in Table 8.5. By this method $e_1=0.056$ and $e_2=0.050$. Using (8.4.5)–(8.4.7) we obtain $e_1=e_2=0.0329$, with $\hat{\Delta}^2=13.56$. Hence we estimate the chance of misclassification at 3.3%.

8.8.3 Procedure for k Groups

The procedure for k groups is almost identical to that for two groups. F-tests and partial F-tests are calculated using the results of the appendix. For q characteristics or variables the variable with the highest F-value that exceeds a critical value or cutoff point F_1 is entered into the model first. If variables, say $x_1,\ldots,x_m'$, are in the model, then partial F-statistics to test for differences in the k groups, adjusted for $x_1,\ldots,x_m$ in the model, are calculated for each of the remaining variables $x_{m+1},\ldots,x_q$ not in the model.

The variable with the largest F-value exceeding F_1 is then added to the model.

For example, suppose this variable is x_{m+1}, so that the model contains $x_1,\ldots,x_m, x_{m+1}$. At this point partial F-values for each of $x_1,\ldots,x_m$ are calculated for the new model. If any of the variables has a partial F-value less than a cutoff point F_2, then the variable with the lowest F-value is deleted from the model. The procedure stops when no variables can be added to or deleted from the model. BMDP uses $F_1 = 4$ and $F_2 = 3.996$ to avoid problems that arise if $F = 4$ is observed, in which case the procedure does not stop if $F_1 = F_2 = 4$.

It should be noted that, unlike in Section 8.7 (for step-down procedures), the order of the variables here is based on the *data*, and hence the significance level of the tests is not a true significance level. However, as the procedure is used for variable selection between groups that are known to be different, this drawback can be ignored.

8.9 Canonical Variates

Two-dimensional plots are often useful in separating the different populations, especially when more than two populations are involved. Suppose p variables are used in a discriminant analysis of k groups. Then we would like to find new variables that are independent and have the largest F-values for testing equality of the k means. That is, we would like to find a vector $\mathbf{a} = (a_1,\ldots,a_p)'$ such that

$$\sum_{i=1}^{k} N_i(\mathbf{a}'\bar{\mathbf{x}}_i + \mathbf{a}'\bar{\mathbf{x}})^2(\mathbf{a}'S\mathbf{a})^{-1} \tag{8.9.1}$$

is a maximum, where $\bar{\mathbf{x}}_i$ is the mean vector of the ith population, N_i is the number of observations from the ith population, $\bar{\mathbf{x}}$ is the average of all the observations, and S is the pooled covariance matrix. The maximum of (8.9.1) occurs when $\mathbf{a}$ satisfies the equation

$$(B - \lambda S)\mathbf{a} = 0, \tag{8.9.2}$$

where B is the between groups mean sum of squares $\Sigma_{i=1}^{k} N_i(\mathbf{x}_i - \bar{\mathbf{x}})(\mathbf{x}_i - \bar{\mathbf{x}}_i)'/(k-1)$, and λ is the maximum eigenvalue of $S^{-1}B$.

If we are interested in the first m ($m < \min(p, k-1)$) canonical variates, then we calculate

$$y_i = \mathbf{a}_i'\mathbf{x}, \qquad i = 1,\ldots,m,$$

where $\mathbf{a}_i$ is the solution of (8.9.2) for λ equal to the ith largest eigenvalue of $S^{-1}B$. In general, the first two canonical variables or all pairs of the first three canonical variables are plotted to show the separation of the k groups, but the standardized canonical variables

$$Z_i = \mathbf{a}_i'(\mathbf{x} - \bar{\mathbf{x}}) \tag{8.9.3}$$

may be plotted instead.

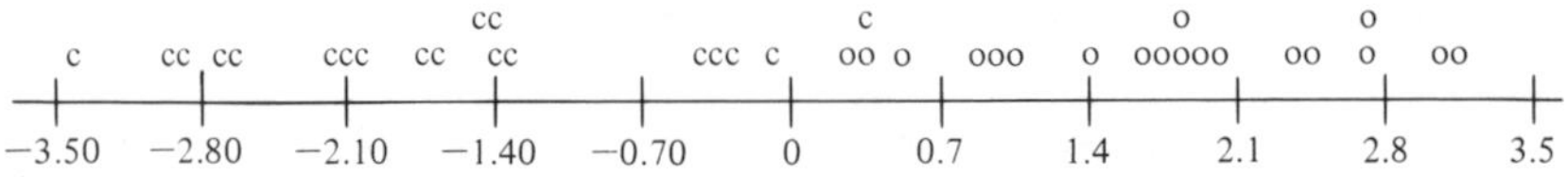

Figure 8.1 Standardized canonical variable for Example 8.9.1.

The canonical variables are approximately normally distributed with variance 1 and mean $\mathbf{a}_i'(\boldsymbol{\mu}_i - \boldsymbol{\mu})$, where $\boldsymbol{\mu}_i$ is the mean of the group from which $\mathbf{x}$ was taken, if the $\mathbf{a}_i$s are normalized such that $\mathbf{a}_i' S \mathbf{a}_i = 1$.

Example 8.9.1 Consider the data of Example 8.8.1. In this situation, with $k-1=1$, there exists only one standardized canonical variable, $y=$ 0.09983(THORAX)−0.04718(ELYTRA)−0.04171(AJ3)+2.777. A plot of this variable for all data points is shown in Figure 8.1, where c stands for *carduorum*, o for *oleracea*.

Example 8.9.2 Consider again the data of Example 8.5.1 on the *Chaetocnema*. To plot the canonical variables, we must solve (8.9.7). This can be achieved in the following manner. Let S^{-1} be written $S^{-1} = T'T$, where T is upper triangular, and let $\mathbf{a} = T'\mathbf{b}$; then $\mathbf{b}$ satisfies

$$(TBT' - \lambda I)\mathbf{b} = 0.$$

In this case $\mathbf{b}$ is the eigenvector associated with the eigenvalue λ of TBT'. Calculations were performed by the SAS MATRIX procedure. The values obtained were

$$S^{-1} = \begin{pmatrix} 0.0205 & -0.0208 & -0.0341 \\ -0.0208 & 0.0599 & -0.0459 \\ -0.0341 & -0.0459 & 0.3382 \end{pmatrix},$$

$$T = \begin{pmatrix} 0.1432 & -0.1453 & -0.2382 \\ 0 & 0.1970 & -0.4087 \\ 0 & 0 & 0.3383 \end{pmatrix},$$

$$\lambda_1 = 308, \qquad \lambda_2 = 3.0,$$

$$\mathbf{a}_1' = (0.143, -0.143, -0.261), \qquad \mathbf{a}_2' = (0, 0.191, -0.314),$$

$$\mathbf{b}_1' = (0.998, 0.012, -0.055), \qquad \text{and} \qquad \mathbf{b}_2' = (-0.002, -0.970, -0.244).$$

The two canonical variables are plotted in Figure 8.2.

Interpreting the above values gives additional information. $\lambda_1 = 308$ and $\lambda_2 = 3.0$ indicate that there is only one dominant canonical variable separating the group. This is clear from Figure 8.2. There is a definite separation of the groups on the horizontal axis but very little on the vertical axis. This situation arises when one combination of variables discriminates among *all* groups.

If we were to perform a discriminant analysis with the two canonical variables $Z_1 = \mathbf{a}_1'(\mathbf{x} - \bar{\mathbf{x}})$ and $Z_2 = \mathbf{a}_2'(\mathbf{x} - \bar{\mathbf{x}})$, we would obtain the same linear discriminant rule, since all the information for separating the three groups is

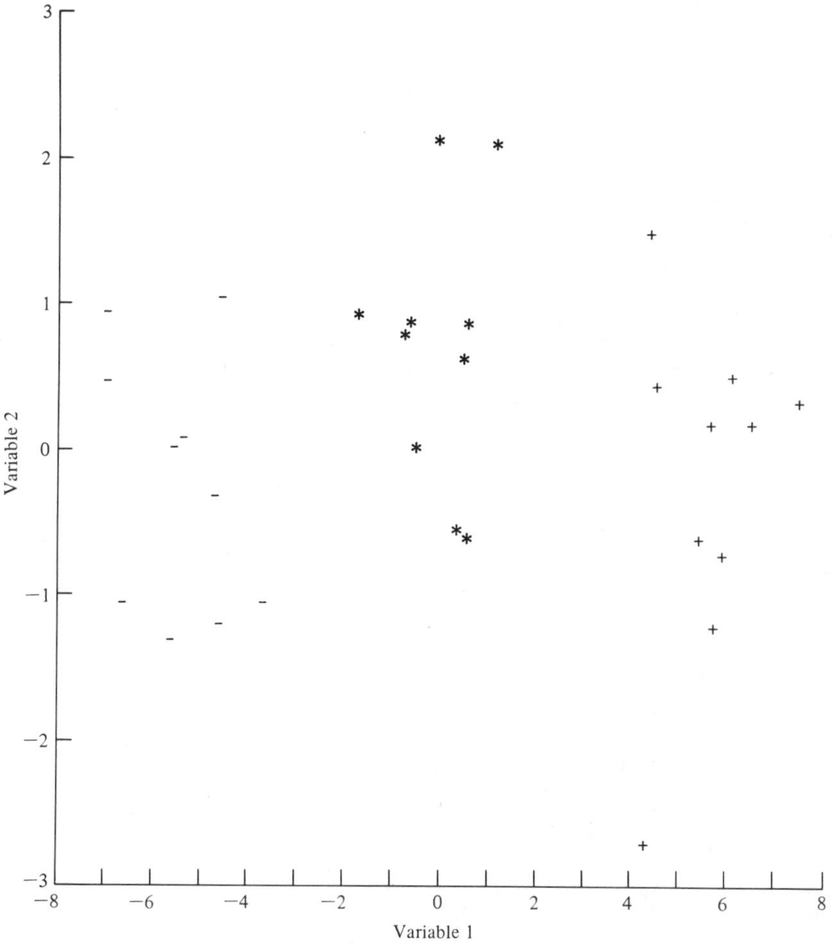

Figure 8.2 Canonical variables for Example 8.9.2: $*$, *concinna*; $+$, *heikertlinger*; —, *hepotemica*.

contained in the two canonical variables. Now Z_2 is of little use in discriminating among the three species. Hence it would be simpler to delete Z_2 and use only Z_1 for discrimination. The linear functions and constants for discriminating among the three species are then given by

$$l_1 = -5.62, \qquad c_1 = 15.78$$
$$l_2 = -0.118, \qquad c_2 = 0.665$$
$$l_3 = -10.99, \qquad c_3 = 0.483.$$

An observation Z_1 is classified in π_i if $l_i Z_1 + c_i$ is the maximum of $l_j Z_1 + c_j$, $j = 1,2,3$. Equivalently, we classify an observation Z_1 in π_i rather than π_j if

$$l_i Z_1 + c_i > l_j Z_1 + c_j.$$

Then for $l_i - l_j > 0$ we classify Z_1 in π_i rather than π_j if

$$Z_1 > (c_j - c_i)/(l_i - l_j).$$

If $l_i - l_j < 0$, we classify Z_1 into π_i and not π_j if

$$Z_1 < (c_j - c_i)/(l_i - l_j).$$

For this example we classify an observation in π_1 and not π_2 if $Z_1 < 2.75$, in π_1 and not π_3 if $Z_1 > -2.84$, and in π_2 and not π_3 if $Z_1 > -0.17$.

Combining the three regions, we classify Z_1 in

$$\begin{aligned} &\pi_2 \quad \text{if} \quad Z_1 > 2.75, \\ &\pi_1 \quad \text{if} \quad -2.84 < Z_1 < 2.75, \\ &\pi_3 \quad \text{if} \quad Z_1 < -2.84. \end{aligned}$$

This procedure is much simpler than using three discriminant functions. If we were to approach the problem from a graphical viewpoint, we would reach approximately the same decision rule based on Figure 8.2.

8.10 Computational Procedures

Discriminant analysis can be run on the BMDP 7M program. The program cards for Example 8.18.1 were

```
/PROBLEM        TITLE IS 'HALTICA DATA'.
/INPUT          VARIABLES ARE 5.
                FORMAT IS '(F1.0,4F4.0)'.
/VARIABLE       NAMES ARE SPECIES, THORAX, ELYTRA,
                AJ2,AJ3.
/GROUP          CODES (1) ARE 1,2.
                NAMES ARE OLERACEA, CARDUORUM.
/DISCRIMINANT   JACKKNIFE.
/PRINT          WITHIN
/END
```

As in the previous chapters, the PROBLEM, INPUT, and VARIABLE statements are self-explanatory. The GROUP paragraph designates variable 1 (SPECIES) as the grouping variable with values 1 for *oleracea* and 2 for *carduorum*. The JACKKNIFE statement asks for the proportion estimates of the error of misclassification as discussed in this text. The WITHIN statement asks for the pooled (within) covariance matrix.

The following partial output resulted.

WITHIN COVARIANCE MATRIX

	THORAX	ELYTRA	AJ2	AJ3
THORAX	143.559			
ELYTRA	151.803	367.788		
AJ2	42.527	121.876	118.314	
AJ3	71.993	106.245	42.064	208.073

STEP NUMBER 0

VARIABLE	F TO REMOVE DF=1 38	FORCE LEVEL	TOLERANCE	*VARIABLE	F TO ENTER *DF=1 37	FORCE LEVEL	TOLERANCE
				*2 THORAX	15.116	1	1.0
				*3 ELYTRA	14.940	1	1.0
				*4 AJ2	32.389	1	1.0
				*5 AJ3	25.428	1	1.0

At this stage variable 4(AJ2) has the highest significant F-value with 1 and 37 d. The next table produced was

VARIABLE	F TO REMOVE DF=1 37	FORCE LEVEL	TOLERANCE	*VARIABLE	F TO ENTER DF=1 36	FORCE LEVEL	TOLERANCE
4 AJ2	32.389	1	1.00	*			
				*2 THORAX	19.164	1	0.893
				*3 ELYTRA	0.230	1	0.659
				*5 AJ3	6.914	1	0.928

On the left side of the table the partial F-test to remove variables is given and on the right the partial F-test to add variables is given. Variable 2 (THORAX) was significant with 1 and 36 df. Hence it was added to the model. Variables 5(AJ3) and then 3(ELYTRA) were similarly added to the model. The table with all four variables in the model was then printed.

VARIABLE	F TO REMOVE DF=1 34	FORCE LEVEL	TOLERANCE	VARIABLE	F TO ENTER DF=1 33	FORCE LEVEL	TOLERANCE
2 THORAX	35.934	1	0.525				
3 ELYTRA	5.799	1	0.408				
4 AJ2	1.775	1	0.648				
5 AJ3	6.259	1	0.800				

With all four variables in the model, AJ2 became insignificant and was removed from the model. The final table was then given.

VARIABLE	F TO REMOVE DF=1 35	FORCE LEVEL	TOLERANCE	*VARIABLE	F TO ENTER DF=1 34	FORCE LEVEL	TOLERANCE
2 THORAX	49.907	1	0.532	*4 AJ2	1.775	1	0.6477
3 ELYTRA	18.597	1	0.549	*			
5 AJ3	10.163	1	0.605	*			

At this stage all variables in the model are significant, and all variables not in the model are insignificant.

The CLASSIFICATION FUNCTIONS were then given.

GROUP = VARIABLE	OLERACEA	CARDUORUM
2 THORAX	0.876	0.519
3 ELYTRA	0.227	0.396
5 AJ3	0.474	0.624
CONSTANT	−160.369	−170.159.

An observation is classified into the population with the highest classification. That is, an observation is classified into OLERACEA if

$$0.876\ \text{THORAX}+0.227\ \text{ELYTRA}+0.474\ \text{AJ3}-160.369$$
$$>0.519\ \text{THORAX}+0.396\ \text{ELYTRA}+0.624\ \text{AJ3}-170.159$$

or, equivalently, if

$$0.357\ \text{THORAX}-0.169\ \text{ELYTRA}-0.150\ \text{AJ3}>-9.790.$$

If all variables are to put into the model, the FORCE and LEVEL statements should be put in the program cards in the DISCRIMINANT paragraph. In the following example

```
/DISC   LEVEL = 0, 1, 1, 1, 1.
        FORCE = 1.
```

the grouping variable is assigned a value of zero and will not enter into the model. All other variables are assigned the highest level 1, and all variables with a nonzero level value with level less than or equal to 1 are forced into the model.

```
/DISC   LEVEL = 0, 1, 2, 3, 3
        FORCE = 1
```

In this example the variable THORAX is forced into the model. ELYTRA will then be put in the model if significant with a level of 2. Then AJ2 and AJ3 will be placed into the model if significant with a level of 3.

The SAS procedure for Example 8.5.1 is

```
DATA BUGS;
INPUT Y1 Y2 Y3 A;
CARDS;
191 131 53 1
185 134 50 1

125 119 51 3
130 120 48 3
PROC DISCRIM POOL = YES S WCOV WCORR PCOV PCORR LIST;
CLASS A;
VAR Y1-Y3;
```

The procedure DISCRIM calculates a linear discriminant function if there are equal covariance matrices, and also gives estimates of the errors of misclassification. It neither performs a stepwise procedure nor calculates canonical variates.

Appendix. Partial F-Tests

Suppose we have been interested in m variables $x_1,\ldots,x_m$ and wish to add another variable x_{m+1} to the model. Then we look at the sample means for the k populations for these variables, based on $N_1,\ldots,N_k$ observations. That is, we observe

$$\bar{\mathbf{x}}_1=\begin{pmatrix}\bar{x}_{11}\\ \vdots\\ \bar{x}_{m1}\\ \bar{x}_{m+1,1}\end{pmatrix},\qquad \ldots,\qquad \bar{\mathbf{x}}_k=\begin{pmatrix}\bar{x}_{1k}\\ \vdots\\ \bar{x}_{mk}\\ \bar{x}_{m+1,k}\end{pmatrix}.$$

In addition, the sample covariance matrix is given by S on $f=\Sigma_{j=1}^{k}(N_j-1)$ df. We wish to test that the k population means for variable x_{m+1} are the same, adjusting for, or given, $x_1,\ldots,x_m$. Thus we test the null hypothesis

$$H\colon E(\bar{x}_{m+1,1}\mid \bar{x}_{11},\ldots,\bar{x}_{m1})=\cdots=E(\bar{x}_{m+1,k}\mid \bar{x}_{1k},\ldots\bar{x}_{mk})$$

vs the general alternative. Following the procedure of Section 4.6 we let $V=fS$ be the error sum of squares for all variables and let

$$W=\sum_{j=1}^{k} N_j(\bar{\mathbf{x}}_j-\bar{\mathbf{x}})(\bar{\mathbf{x}}_j-\bar{\mathbf{x}})',$$

where $\bar{\mathbf{x}}=(\Sigma_{j=1}^{k}N_j\bar{\mathbf{x}}_j)/(\Sigma_{j=1}^{k}N_j)$, be the sum of squares due to differences in groups. Partition the matrices

$$V=\begin{pmatrix}V_{11} & V_{12}\\ V_{12}' & V_{22}\end{pmatrix}\qquad\text{and}\qquad W=\begin{pmatrix}W_{11} & W_{12}\\ W_{12}' & W_{22}\end{pmatrix},$$

where V_{11} and W_{11} are $m\times m$ matrices; then the U-statistic for testing for differences among the k groups, for the mean of x_{m+1} given $x_1,\ldots,x_m$, is given by

$$U_{1,k-1,f-m}=|V_{11}+W_{11}||V||V_{11}|^{-1}|V+W|^{-1}.$$

From Theorem 4.2.4 we obtain

$$F=(f-m)(1-U_{1,k-1,f-m})/\big[(k-1)U_{1,k-1,f-m}\big].$$

This is called a *partial F-statistic*. H is rejected at the $\alpha 100\%$ level if $F>F_{k-1,f-m,\alpha}$.

Problems

8.1 Show that (8.2.8) is a maximum likelihood rule.

8.2 Prove (8.2.9).

8.3 Three psychological measurements x_1, x_2, and x_3 were taken on subjects in six neurotic groups. The sample means covariance matrix are given in Table 8.6. The pooled covariance matrix is

$$S_p=\begin{pmatrix}2.3008 & 0.2516 & 0.4742\\ & 0.6075 & 0.0358\\ & & 0.5951\end{pmatrix}.$$

Obtain a linear discriminant function for classifying the data.

Table 8.6[a]

Group	Sample size	x_1	x_2	x_3
anxiety	114	2.9298	1.667	0.7281
hysteria	33	3.0303	1.2424	0.5455
psychopathy	32	3.8125	1.8438	0.8125
obsession	17	4.7059	1.5882	1.1176
personality change	5	1.4000	0.2000	0.0000
normal	55	0.6000	0.1455	0.2182

[a]Data from Rao and Slater (1949) and Rao (1973).

8.4 The data of Table 8.7 are from Lubischew (1962). Six variables were measured on males for three species of *Chaetocnema concinna*, *Ch. heikertlingeri*, and *Ch. heptapotomica*: the width x_1 of the first joint of the first tarsus in microns (the sum of the measurements for both tarsi), the width x_2 for the second joint, the maximal width x_3 of the aedeagus in the fore-part in microns, the front angle x_4 of the aedeagus (1 unit $\equiv$ 7.5°), the maximal width x_5 of the head between the external edges of the eyes in 0.01 mm, and the aedeagus width x_6 from the side in microns.

(a) Derive a classification rule for these data.

(b) Estimate the errors of misclassification.

Table 8.7[a]

Number	x_1	x_2	x_3	x_4	x_5	x_6
			Ch. concinna			
1	191	131	53	150	15	104
2	185	134	50	147	13	105
3	200	137	52	144	14	102
4	173	127	50	144	16	97
5	171	118	49	153	13	106
6	160	118	47	140	15	99
7	188	134	54	151	14	98
8	186	129	51	143	14	110
9	174	131	52	144	14	116
10	163	115	47	142	15	95
11	190	143	52	141	13	99
12	174	131	50	150	15	105
13	201	130	51	148	13	110
14	190	133	53	154	15	106
15	182	130	51	147	14	105
16	184	131	51	137	14	95
17	177	127	49	134	15	105
18	178	126	53	157	14	116
19	210	140	54	149	13	107
20	182	121	51	147	13	111
21	186	136	56	148	14	111

Table 8.7 (*continued*)

Number	x_1	x_2	x_3	x_4	x_5	x_6
			Ch. heikertlingeri			
1	186	107	49	120	14	84
2	211	122	49	123	16	95
3	201	114	47	130	14	74
4	242	131	54	131	16	90
5	184	108	43	116	16	75
6	211	118	51	122	15	90
7	217	122	49	127	15	73
8	223	127	51	132	16	84
9	208	125	50	125	14	88
10	199	124	46	119	13	78
11	211	129	49	122	13	83
12	218	126	49	120	15	85
13	203	122	49	119	14	73
14	192	116	49	123	15	90
15	195	123	47	125	15	77
16	211	122	48	125	14	73
17	187	123	47	129	14	75
18	192	109	46	130	13	90
19	223	124	53	129	13	82
20	188	114	48	122	12	74
21	216	120	50	129	15	86
22	185	114	46	124	15	92
23	178	119	47	120	13	78
24	187	111	49	119	16	66
25	187	112	49	119	14	55
26	201	130	54	133	13	84
27	187	120	47	121	15	86
28	210	119	50	128	14	68
29	196	114	51	129	14	86
30	195	110	49	124	13	89
31	187	124	49	129	14	88
			Ch. heptapotamica			
1	158	141	58	145	8	107
2	146	119	51	140	11	111
3	151	130	51	140	11	113
4	122	113	45	131	10	102
5	138	121	53	139	11	106
6	132	115	49	139	10	98
7	131	127	51	136	12	107
8	135	123	50	129	11	107
9	125	119	51	140	10	110

(*continued*)

Table 8.7 (*continued*)

Number	x_1	x_2	x_3	x_4	x_5	x_6
			Ch. heptapotamica			
10	130	120	48	137	9	106
11	130	131	51	141	11	108
12	138	127	52	138	9	101
13	130	116	52	143	9	111
14	143	123	54	142	11	95
15	154	135	56	144	10	123
16	147	132	54	138	10	102
17	141	131	51	140	10	106
18	131	116	47	130	9	102
19	144	121	53	137	11	104
20	137	146	53	137	10	113
21	143	119	53	136	9	105
22	135	127	52	140	10	108

[a]Reproduced with the permission of the editors of *Biometrics*.

(c) Use a stepwise discriminant procedure to derive a simpler discriminant function and to estimate the errors of misclassification.

(d) For the simple model derived in **(c)** plot the two canonical variables.

8.5 Check the data of Problem 8.4 for normality. Where necessary, make a transformation along the lines of Section 4.7 to obtain data that are approximately normally distributed and perform a discriminant analysis on the transformed date. Finally, estimate the errors of misclassification using the jackknife method and the approximations given in (8.4.5) and (8.4.6).

8.6 **(a)** For the data of Problem 4.5 calculate the linear discriminant functions for discriminating between the three groups of criminals.

(b) Estimate the errors of misclassification by two methods.

(c) Can ear measurements discriminate between "types" of criminals?

Griffiths (1904) gives many measurements on the heads of criminals of different types. The reader may wish to consult this paper and perform a stepwise discriminant analysis on all the measurements.

Chapter 9
Correlation

9.1 Introduction

In this chapter four forms of correlation are considered: simple, partial, multiple, and canonical. *Simple correlation* measures the degree of association between two random variables, such as yield and rainfall. However, this association can be affected by other variables, such as temperature. For example, the rainfall may occur with low temperatures, which could cause a decrease in yield and hence a negative simple correlation between rainfall and yield. *Partial correlation* measures the degree of association between two random variables, such as yield and rainfall, while adjusting for other variables, such as temperature. *Multiple correlation* is defined as the maximum correlation between one random variable and a linear combination of a set of random variables. *Canonical correlations* investigate the degree of association between two sets of random variables.

9.2 Correlation Between Two Random Variables

Let $\mathbf{x} = (x_1, x_2)'$ with $E(\mathbf{x}) = \boldsymbol{\mu}$, $\operatorname{cov}(\mathbf{x}) = \Sigma$, where

$$\Sigma = \begin{pmatrix} \sigma_{11} & \sigma_{12} \\ \sigma_{12} & \sigma_{22} \end{pmatrix}. \tag{9.2.1}$$

Then the population correlation between the two random variables x_1 and x_2 is defined by

$$\rho = \sigma_{12}/\sigma_{11}^{1/2}\sigma_{22}^{1/2}, \qquad -1 \leq \rho \leq 1. \tag{9.2.2}$$

Note that if we relabel the random variables x_1 and x_2 by $y_1 = ax_1 + b_1$, $a > 0$, and $y_2 = cx_2 + d_2$, $c > 0$, then the correlation between y_1 and y_2 is the same as the correlation between x_1 and x_2. Given a sample (x_{11}, x_{21}), $(x_{12}, x_{22}), \ldots, (x_{1N}, x_{2N})$, the correlation ρ is estimated by the sample correlation

$$r=\frac{\Sigma_{i=1}^{N}(x_{1i}-\bar{x}_1)(x_{2i}-\bar{x}_2)}{\left[\Sigma_{i=1}^{N}(x_{1i}-\bar{x}_1)^2\Sigma_{i=1}^{N}(x_{2i}-\bar{x}_2)^2\right]^{1/2}}, \qquad -1\leq r\leq 1, \quad (9.2.3)$$

where $\bar{x}_1=N^{-1}\Sigma_{i=1}^{N}x_{1i}$ and $\bar{x}_2=N^{-1}\Sigma_{i=1}^{N}x_{2i}$. If the sample is from $\mathbf{N}_2(\boldsymbol{\mu}, \Sigma)$, then r is the maximum likelihood estimate of ρ, but it is a biased estimate, where the bias goes to zero as N tends large. r is sufficient for ρ. Note that the sample covariance is given by

$$n^{-1}V\equiv S\equiv\begin{pmatrix} s_{11} & s_{12} \\ s_{12} & s_{22} \end{pmatrix}$$

$$\equiv\begin{pmatrix} \Sigma(x_{1i}-\bar{x}_1)^2 & \Sigma(x_{1i}-\bar{x}_1)(x_{2i}-\bar{x}_2) \\ \Sigma(x_{1i}-\bar{x}_1)(x_{2i}-\bar{x}_2) & \Sigma(x_{2i}-\bar{x}_2)^2 \end{pmatrix}, \quad (9.2.4)$$

and in terms of the elements of the sample covariance matrix S, r is given by

$$r=s_{12}/s_{11}^{1/2}s_{22}^{1/2}=v_{12}/v_{11}^{1/2}v_{22}^{1/2}. \quad (9.2.5)$$

The sample correlation r remains unchanged under the transformation $x_1\rightarrow ax_1+b$, $a>0$, and $x_2\rightarrow cx_2+d$, $c>0$. The distribution of r depends only on ρ and is available in any textbook [see, for example, Srivastava and Khatri (1979)]. When $\rho=0$, the distribution of

$$(N-2)^{1/2}r/(1-r^2)^{1/2}\equiv(n-1)^{1/2}r(1-r^2)^{-1/2} \quad (9.2.6)$$

is a Student's t-distribution with $n-1=N-2$ df. Thus to test hypothesis H: $\rho=0$ vs the alternative A: $\rho\neq 0$ (or one-sided), we use the statistic in (9.2.6). Hence the hypothesis H: $\rho=0$ (vs A: $\rho\neq 0$) is rejected if

$$\left|(N-2)^{1/2}r(1-r^2)^{-1/2}\right|\geq t_{N-2,\alpha/2}.$$

The problem is somewhat difficult if we wish to test the hypothesis H: $\rho=\rho_0$, ρ_0 specified, vs the alternative A: $\rho\neq\rho_0$. In this case, however, we have an approximate solution suggested by Fisher (1921). Let

$$z=\frac{1}{2}\ln\frac{1+r}{1-r} \qquad \text{and} \qquad \zeta=\frac{1}{2}\ln\frac{1+\rho}{1-\rho}. \quad (9.2.7)$$

Then, as $n\rightarrow\infty$,

$$z\rightarrow N\left(\zeta,(N-2)^{-1}\right). \quad (9.2.8)$$

Hence, under H,

$$(N-2)^{1/2}(z-\zeta_0)\rightarrow N(0,1), \quad (9.2.9)$$

and a test or confidence interval for ζ can be obtained.

9.2.1 *Estimating* ρ *When* $\sigma_{11}=\sigma_{22}=\sigma^2$

For problems in which two characteristics are measured in the same units, we can sometimes assume that the variances of the characteristics are the same. In that case the covariance matrix is of the form

$$\Sigma=\sigma^2\begin{pmatrix}1 & \rho\\ \rho & 1\end{pmatrix}. \tag{9.2.10}$$

If

$$S=\begin{pmatrix}s_{11} & s_{12}\\ s_{12} & s_{22}\end{pmatrix}$$

is the sample covariance matrix, then maximum likelihood estimators for σ^2 and ρ, respectively, are given by

$$\hat{\sigma}^2=\tfrac{1}{2}(s_{11}+s_{22}) \tag{9.2.11}$$

and

$$\hat{\rho}=2s_{12}/(s_{11}+s_{22}). \tag{9.2.12}$$

Example 9.2.1 For the first two trials of the memory experiment described in Example 7.2.1, the covariance matrix is

$$S=\begin{pmatrix}7.82 & 7.93\\ 7.93 & 9.38\end{pmatrix}.$$

The simple correlation was calculated to be $r=0.926$. If, as is reasonable, we assume that the population variances are equal, we obtain estimates of ρ and σ^2 as $\hat{\rho}=0.922$ and $\hat{\sigma}^2=8.93$, respectively. In this instance the two estimates of ρ are almost the same.

9.2.2 *Estimating* ρ *When* $\sigma_{11}=\sigma_{22}=1$ *and* $\rho>0$ *(or* <0*)*

In some practical situations it is known whether ρ is positive or negative. Suppose further we know that $\sigma_{11}=1$ and $\sigma_{22}=1$. The maximum likelihood estimation becomes difficult; in practice, the estimate r given in (9.2.3) or (9.2.5) is used and this information is ignored. In this section we give an alternative method. To fix our ideas, suppose it is known that ρ is positive. Since S is sufficient for Σ, any estimation of ρ is to be based on S. Let $\lambda_1>\lambda_2$ be the ordered characteristic roots of Σ and $l_1>l_2$ be the ordered characteristic roots S. Then the maximum likelihood estimates of λ_1 and λ_2 are l_1 and l_2, respectively. Note that

$$\lambda_1=1+\rho \quad\text{and}\quad \lambda_2=1-\rho. \tag{9.2.13}$$

Hence

$$\alpha(l_1-1)+(1-\alpha)(1-l_2),\qquad 0\le\alpha\le1, \tag{9.2.14}$$

is a class of estimators of ρ when ρ is positive. Choosing $\alpha=\frac{1}{2}$ gives l_1-l_2 as an estimator of ρ. Some tests in correlation analysis when partial information is known are given by Olkin and Sylvan (1977).

9.3 Estimating ρ in the Intraclass Correlation Model

Let $S\sim W_p(\Sigma, n)$, where Σ takes the form

$$\Sigma=\sigma^2\big[(1-\rho)I_p+\rho\mathbf{e}\mathbf{e}'\big]. \tag{9.3.1}$$

Maximum likelihood estimators are

$$\hat{\sigma}^2=(\operatorname{tr} S)/p \tag{9.3.2}$$

and

$$\rho=(\mathbf{e}'S\mathbf{e}-\operatorname{tr} S)/[(p-1)+\operatorname{tr} S]. \tag{9.3.3}$$

Snedecor and Cochran discuss this problem in the context of experimental design (1980, p. 243).

Example 9.3.1 Consider the full covariance matrix of Example 7.2.4, which concerned three memory trials administered to each of several subjects. The sample covariance matrix

$$S=\begin{pmatrix} 7.82 & 7.93 & 7.98\\ 7.93 & 9.38 & 8.87\\ 7.98 & 8.87 & 9.79\end{pmatrix}$$

and the correlation matrix was calculated to be

$$R=\begin{pmatrix} 1 & 0.926 & 0.912\\ 0.926 & 1 & 0.926\\ 0.912 & 0.926 & 1\end{pmatrix}.$$

From (9.3.2) and (9.3.3) we obtain the estimates $\hat{\sigma}^2=9.0$ and $\hat{\rho}=0.918$.

9.4 Simple and Partial Correlations

Let $\mathbf{x}_1,\ldots,\mathbf{x}_N$ be a sample of $N_p(\boldsymbol{\mu},\Sigma)$ random variables. Then the sample covariance matrix

$$S=\frac{1}{n}\sum_{i=1}^{N-1}(\mathbf{x}_i-\bar{\mathbf{x}})(\mathbf{x}_i-\bar{\mathbf{x}})',\qquad \text{with}\quad n=N-1,$$

has a Wishart distribution with scale matrix Σ and degrees of freedom n. That is, $S\sim W_p(\Sigma,n)$. The sample variance of the ith component of $\mathbf{x}$ is s_{ii}. Hence we may write the correlation between the ith and jth component

$$r_{ij}=s_{ij}\Big/\sqrt{s_{ii}s_{jj}}\,. \tag{9.4.1}$$

If we define $D_s = \mathrm{diag}(s_{11},\ldots,s_{pp})$, then the matrix

$$R = D_s^{-1/2} S D_s^{-1/2} \tag{9.4.2}$$

has entries

$$r_{ii} = 1 \qquad \text{and} \qquad r_{ij} = s_{ij} \Big/ \sqrt{s_{ii}s_{jj}}, \quad i \neq j,$$

and is called the *correlation matrix*. The *population correlation matrix* is defined similarly:

$$\rho = D_\sigma^{-1/2} \Sigma D_\sigma^{-1/2}, \tag{9.4.3}$$

where $\Sigma = (\sigma_{ij})$ and $D_\sigma = \mathrm{diag}(\sigma_{11},\ldots,\sigma_{pp})$.

To test for a given i and j

$$H: \rho_{ij} = 0 \qquad \text{vs} \qquad A: \rho_{ij} \neq 0,$$

one rejects H if

$$|t| = \left| r_{ij}\sqrt{(n-1)/\left(1-r_{ij}^2\right)} \right| \geq t_{n-1,\alpha/2}. \tag{9.4.4}$$

Confidence intervals for ρ_{ij} may be obtained from the expansion

$$\frac{1}{2}\ln\frac{1+r_{ij}}{1-r_{ij}} - \frac{1}{2}\ln\frac{1+\rho_{ij}}{1-\rho_{ij}} \sim N\left(0,(N-2)^{-1}\right). \tag{9.4.5}$$

Hence by finding the confidence interval for $\frac{1}{2}\ln(1+\rho_{ij})/(1-\rho_{ij})$ one can then find the confidence interval for ρ_{ij}.

For partial correlation one partitions $\mathbf{x}'$ as $(\mathbf{x}_1,\mathbf{x}_2)'$, where $\mathbf{x}_1$ is a p_1-vector and $\mathbf{x}_2$ is a p_2-vector. Then, from Chapter 2, the distribution of $\mathbf{x}_1$, given $\mathbf{x}_2$, is

$$N_{p_1}\left(\boldsymbol{\mu}_1 + \Sigma_{12}\Sigma_{22}^{-1}(\mathbf{x}_2 - \boldsymbol{\mu}_2), \Sigma_{1\cdot 2}\right),$$

where

$$\boldsymbol{\mu} = \begin{pmatrix} \boldsymbol{\mu}_1 \\ \boldsymbol{\mu}_2 \end{pmatrix}, \qquad \Sigma = \begin{pmatrix} \Sigma_{11} & \Sigma_{12} \\ \Sigma_{12} & \Sigma_{22} \end{pmatrix}$$

and $\Sigma_{1\cdot 2} = \Sigma_{11} - \Sigma_{12}\Sigma_{22}^{-1}\Sigma_{12}'$. Hence the partial correlation between two components of $\mathbf{x}_1$, given $\mathbf{x}_2$, may be obtained from $\Sigma_{1\cdot 2}$. That is, if we define $\Sigma_{1\cdot 2} = (a_{ij})$, then the population partial correlation between the ith and jth component of $\mathbf{x}_1$, $1 \leq i,\ j \leq p_1$, is given by

$$\rho_{ij\cdot p_1+1,\ldots,p} = a_{ij} \Big/ \sqrt{a_{ii}a_{jj}}\,. \tag{9.4.6}$$

Similarly, if

$$S = \begin{pmatrix} S_{11} & S_{12} \\ S_{12} & S_{22} \end{pmatrix}$$

with S_{11}, S_{22}, and S_{12} $p_1\times p_1$ symmetric, $p_2\times p_2$ symmetric, and $p_1\times p_2$ matrices, respectively, then one defines $S_{1\cdot 2}=S_{11}-S_{12}S_{22}^{-1}S_{12}'$. The sample partial correlation $r_{ij\cdot p_1+1,\ldots,p}$ is defined from the matrix $S_{1\cdot 2}=(b_{ij})$ by

$$r_{ij\cdot p_1+1,\ldots,p}=b_{ij}/\sqrt{b_{ii}b_{jj}}\,. \tag{9.4.7}$$

The distribution of the partial correlations given $\mathbf{x}_2$ is that defined for simple correlations in (9.4.4) and (9.4.5) with degrees of freedom $n-p_2$ instead of n.

9.5 Multiple Correlation

In regression one would like to examine a variable by a linear combination of other variables. *Multiple correlation* measures the association between a random variable x_1 and a group of random variables $\mathbf{x}_2:(p-1)\times 1$. It is defined as the maximum correlation between x_1 and $\boldsymbol{\alpha}'\mathbf{x}_2$, a linear combination of the variables $x_2,\ldots,x_p$. Let

$$\mathbf{x}=\begin{pmatrix} x_1 \\ \mathbf{x}_2 \end{pmatrix} \quad \text{and} \quad \operatorname{cov}(\mathbf{x})=\begin{pmatrix} \sigma_{11} & \boldsymbol{\sigma}_{12}' \\ \boldsymbol{\sigma}_{12} & \Sigma_{22} \end{pmatrix}. \tag{9.5.1}$$

That is, $\operatorname{var}(x_1)=\sigma_{12}$, $\operatorname{cov}(\mathbf{x}_2)=\Sigma_{22}$, and $\operatorname{cov}(x_1,\mathbf{x}_2)=\sigma_{12}$. Then the correlation between x_1 and $\boldsymbol{\alpha}'\mathbf{x}_2$ is given by

$$\rho_{x_1,\boldsymbol{\alpha}'\mathbf{x}_2}=\boldsymbol{\alpha}'\boldsymbol{\sigma}_{12}/\sigma_{11}^{1/2}(\boldsymbol{\alpha}'\Sigma_{22}\boldsymbol{\alpha})^{1/2} \tag{9.5.2}$$

since $\operatorname{cov}(x_1,\boldsymbol{\alpha}'\mathbf{x}_2)=\boldsymbol{\alpha}'\sigma_{12}$ and $\operatorname{var}(\boldsymbol{\alpha}'\mathbf{x}_2)=\boldsymbol{\alpha}'\Sigma_{22}\boldsymbol{\alpha}$. The maximum value (shown in Section 9.5.1) is given by

$$\rho_{1\cdot 23\ldots p}=(\boldsymbol{\sigma}_{12}'\Sigma_{22}^{-1}\boldsymbol{\sigma}_{12})^{1/2}/\sigma_{11}^{1/2}. \tag{9.5.3}$$

Thus the multiple correlation always lies between 0 and 1, $0\le\rho_{1\cdot 23\ldots p}\le 1$.

Next, we define the sample multiple correlation coefficient based on a random sample of size N. Let S be the sample covariance matrix partitioned as Σ; that is, let

$$S=\begin{pmatrix} s_{11} & \mathbf{s}_{12}' \\ \mathbf{s}_{12} & S_{22} \end{pmatrix}. \tag{9.5.4}$$

Then the sample multiple correlation coefficient is defined as

$$r_{1\cdot 23\ldots p}=(\mathbf{s}_{12}'S_{22}^{-1}\mathbf{s}_{12}/s_{11})^{1/2}. \tag{9.5.5}$$

The square of the multiple correlation can be defined as

$$R^2\equiv r_{1\cdot 2\ldots p}^2=\mathbf{s}_{12}'S_{22}^{-1}\mathbf{s}_{12}/s_{11}. \tag{9.5.6}$$

Under the assumption of normality, the distribution of R^2 when $\rho_{1\cdot 2\ldots p}=0$ is similar to the ordinary correlation coefficient r^2. That is, if $\mathbf{x}\sim N_p(\boldsymbol{\mu},\Sigma)$, then if $\rho_{1\cdot 2\ldots p}=0$,

$$\frac{n-p+1}{p-1}\,\frac{R^2}{1-R^2}\sim F_{p-1,\,n-p+1},\qquad n=N-1. \tag{9.5.7}$$

Thus, for testing the hypothesis that $\rho_{1\cdot 23\ldots p}=0$, we can use the tables for F.

Example 9.5.1 The following data consist of the results of 12 experiments measuring the variables X_1, the natural logarithm of the oxygen concentration; X_2, the natural logarithm of the water velocity; and X_3, the natural logarithm of the water depth, with $N=12$ and $n=11$. The covariance matrix is

$$S=\begin{pmatrix} 7.250 & 0.507 & -3.878\\ 0.507 & 0.480 & -0.056\\ -3.878 & 0.056 & 2.550 \end{pmatrix}$$

on 11 df and the correlation matrix is

$$R=\begin{pmatrix} 1 & 0.272 & -0.902\\ 0.272 & 1 & -0.051\\ -0.902 & -0.051 & 1 \end{pmatrix}.$$

To obtain a 95% confidence interval for ρ_{12} we calculate

$$Z=\tfrac{1}{2}\ln[(1+r_{12})/(1-r_{12})]=0.272.$$

Now let $Z_1=0.272-1.96(12-2)^{-1/2}=-0.341$ and $Z_2=0.272+1.96\times(12-2)^{1/2}=0.899$. Then a 95% confidence interval for $\frac{1}{2}\ln[(1+\rho_{12})/(1-\rho_{12})]$ is given by $(Z_1, Z_2)=(-0.341, 0.899)$. The confidence interval for ρ_{12} is obtained by letting $a_i=\frac{1}{2}(e^{Zi}-e^{-Zi})$, $i=1,2$. Then the 95% confidence interval for ρ_{12} is given by $(a_1, a_2)=(-0.328, 0.716)$. Note that zero lies in this interval, and thus the hypothesis H: $\rho_{12}=0$ vs A: $\rho_{12}\neq 0$ would not be rejected.

If we now wish to look at the relationship between X_1 and X_2, adjusting for X_3, we form the partial covariance matrix $S_{1\cdot 2}=S_{11}-S_{12}S_{22}^{-1}S_{21}$, where

$$S_{11}=\begin{pmatrix} 7.250 & 0.507\\ 0.507 & 0.480 \end{pmatrix},\qquad S_{12}=\begin{pmatrix} -3.878\\ -0.056 \end{pmatrix},$$

$$S_{21}=(-3.878, -0.056),\qquad \text{and}\qquad S_{22}=2.550.$$

Therefore we obtain

$$S_{1\cdot 2}=\begin{pmatrix} 1.35 & 0.42\\ 0.42 & 0.48 \end{pmatrix},$$

and hence $r_{12\cdot 3}=0.42/(1.35\times 0.48)^{1/2}=0.52$. The corresponding Z-value for $r_{12\cdot 3}$ is given by

$$Z=\tfrac{1}{2}\ln[(1+0.52)/(1-0.52)]=0.582.$$

A 95% confidence interval for $\frac{1}{2}\ln[(1+\rho_{12\cdot 3})/(1-\rho_{12\cdot 3})]$ is given by $Z\pm$

$1.96\times(12-1-2)^{-1/2}=(-0.07, 1.24)$. Note that because we are conditioning on *one* variable X_3 the variance of Z becomes $(N-1-2)^{-1}$.

Transforming back to $\rho_{12\cdot 3}$ we obtain a 95% confidence interval for $\rho_{12\cdot 3}$ of $(-0.07, 0.85)$. Note that the null hypothesis H: $\rho_{12\cdot 3}=0$ vs A: $\rho_{12\cdot 3}\neq 0$ would not be rejected at $\alpha=0.05$. Hence there does not appear to be a significant correlation between X_1 and X_2, with or without adjusting for X_3, based on the data.

If we calculate the amount of variation in X_1 accounted for by X_2 and X_3, we obtain

$$R^2=r_{1\cdot 23}^2=(7.25)^{-1}(0.507, -3.878)\begin{pmatrix} 0.48 & -0.056 \\ -0.056 & 2.55 \end{pmatrix}^{-1}\begin{pmatrix} 0.507 \\ -3.878 \end{pmatrix}$$
$$=0.864.$$

Hence 86.4% of the variation in X_1 is accounted for by X_2 and X_3. The multiple correlation is then given by $r_{1\cdot 23}=0.930$. The test of H: $\rho_{1\cdot 23}=0$ vs A: $\rho_{1\cdot 23}\neq 0$ has the F-value

$$F=\frac{n-p+1}{p-1}\,\frac{R^2}{1-R^2}=\frac{11-3+1}{3-1}\,\frac{0.864}{1-0.864}\doteq 28.6$$
$$>F_{2,9,0.05}=4.26.$$

Hence there is a significant linear relationship between X_1 and (X_2, X_3). The water velocity did not significantly affect the oxygen concentration, even after adjusting for the effect of the water depth. Water depth accounts for 81.4% of the variation in oxygen concentration ($r_{13}^2=0.814$), and water depth and water velocity together account for 86.4% of the variation in oxygen concentration. Hence the effect of using water velocity in addition to water depth to explain oxygen concentration is fairly small. We account for only an additional 5% of the variation in oxygen concentration.

9.5.1 *Maximization of* $\rho_{x_1, \boldsymbol{\alpha}'\mathbf{x}_2}$

Consider the maximization of $\rho^2_{x_1, \boldsymbol{\alpha}'\mathbf{x}_2}$. We have

$$\rho^2_{x_1, \boldsymbol{\alpha}'\mathbf{x}_2}=(\boldsymbol{\alpha}'\boldsymbol{\sigma}_{12})^2/\sigma_{11}\boldsymbol{\alpha}'\Sigma_{22}\boldsymbol{\alpha}=(\boldsymbol{\alpha}'\Sigma_{22}^{1/2}\Sigma_{22}^{-1/2}\boldsymbol{\sigma}_{12})^2/\sigma_{11}\boldsymbol{\alpha}'\Sigma_{22}\boldsymbol{\alpha}$$
$$\leq\boldsymbol{\alpha}'\Sigma_{22}\boldsymbol{\alpha}\boldsymbol{\sigma}_{12}'\Sigma_{22}^{-1}\boldsymbol{\sigma}_{12}/\sigma_{11}\boldsymbol{\alpha}'\Sigma_{22}\boldsymbol{\alpha}=\boldsymbol{\sigma}_{12}'\Sigma_{22}^{-1}\boldsymbol{\sigma}_{12}/\sigma_{11}$$

from the Cauchy–Schwarz inequality. The equality holds iff $\boldsymbol{\alpha}=\lambda\Sigma_{22}^{-1}\boldsymbol{\sigma}_{12}$, where $\lambda\neq 0$ (but could be positive as well as negative). The positive value of λ is the maximum (the negative value gives the minimum). Hence

$$\max_{\boldsymbol{\alpha}}\rho_{x_1, \boldsymbol{\alpha}'\mathbf{x}_2}=\rho_{1\cdot 23\ldots p}=(\boldsymbol{\sigma}_{12}'\Sigma_{22}^{-1}\boldsymbol{\sigma}_{12})^{1/2}/\sigma_{11}^{1/2},$$

and $0\leq\rho_{1\cdot 23\ldots p}\leq 1$.

9.6 Canonical Correlation

Let $\mathbf{x}=(\mathbf{x}_1,\mathbf{x}_2)'$, where $\mathbf{x}_1$ is a p-vector and $\mathbf{x}_2$ is a q-vector. The canonical correlation measures the association between the random vectors $\mathbf{x}_1$ and $\mathbf{x}_2$; this is a generalization of multiple correlation. Let

$$\operatorname{cov}(\mathbf{x})=\operatorname{cov}\begin{pmatrix}\mathbf{x}_1\\ \mathbf{x}_2\end{pmatrix}=\begin{pmatrix}\Sigma_{11} & \Sigma_{12}\\ \Sigma_{12} & \Sigma_{22}\end{pmatrix}. \tag{9.6.1}$$

Assume without loss of generality that $p\le q$. Then the first canonical correlation is the maximum correlation between a linear combination of $\mathbf{x}_1$, say $\boldsymbol{\alpha}_1'\mathbf{x}_1$, and a linear combination of $\mathbf{x}_2$, say $\boldsymbol{\beta}_1'\mathbf{x}_2$. That is,

$$\rho_1=\max_{\boldsymbol{\alpha}_1,\boldsymbol{\beta}_1}\boldsymbol{\alpha}_1'\Sigma_{12}\boldsymbol{\beta}_1/(\boldsymbol{\alpha}_1'\Sigma_{11}\boldsymbol{\alpha}_1)^{1/2}(\boldsymbol{\beta}_1'\Sigma_{22}\boldsymbol{\beta}_1)^{1/2}. \tag{9.6.2}$$

It can be shown that ρ_1^2 is the maximum characteristic root of $\Sigma_{11}^{-1}\Sigma_{12}\Sigma_{22}^{-1}\Sigma_{12}'$. But the matrix $\Sigma_{11}^{-1}\Sigma_{12}\Sigma_{22}^{-1}\Sigma_{12}'$ can have at most p nonzero roots. Let $\rho_1^2\ge\rho_2^2\ge\cdots\ge\rho_p^2$ be the roots of $\Sigma_{11}^{-1}\Sigma_{12}\Sigma_{22}^{-1}\Sigma_{12}'$ (some of these could be zero). Then $\rho_1,\rho_2,\ldots,\rho_p$ are called the *population canonical correlations.* Note that the characteristic roots of $\Sigma_{11}^{-1}\Sigma_{12}\Sigma_{22}^{-1}\Sigma_{12}'$ and of the matrix $\Sigma_{11}^{-1/2}\Sigma_{12}\Sigma_{22}^{-1}\Sigma_{12}'\Sigma_{11}^{-1/2}$ are the same. Let

$$A=\Sigma_{11}^{-1/2}\Sigma_{12}\Sigma_{22}^{-1/2}.$$

Then $\rho_1^2,\ldots,\rho_p^2$ are ordered characteristic roots of AA' and of $A'A$. Let $\boldsymbol{\alpha}_1,\ldots,\boldsymbol{\alpha}_p$ be the characteristic vectors corresponding to the ordered roots $\rho_1^2,\ldots,\rho_p^2$ of AA', and let $\boldsymbol{\beta}_1,\ldots,\boldsymbol{\beta}_p$ be the characteristic vectors corresponding to the ordered roots $\rho_1^2,\ldots,\rho_p^2$ of $A'A$. Note that the $\boldsymbol{\alpha}_i$s are p-vectors and the $\boldsymbol{\beta}_i$s q-vectors. Then $(\boldsymbol{\alpha}_i'\mathbf{x}_1,\boldsymbol{\beta}_i'\mathbf{x}_2)$ are called the ith canonical variables. These canonical variables have the property that the random variables $(\boldsymbol{\alpha}_1'\mathbf{x}_1,\ldots,\boldsymbol{\alpha}_p'\mathbf{x}_1)$ are uncorrelated among themselves (and independent if we assume normality). Similarly the variables $(\boldsymbol{\beta}_1'\mathbf{x}_2,\ldots,\boldsymbol{\beta}_p'\mathbf{x}_2)$ are uncorrelated among themselves. Also, $\operatorname{var}(\boldsymbol{\alpha}_i'\mathbf{x}_1)=1$, $i=1,2,\ldots,p$, $\operatorname{var}(\boldsymbol{\beta}_i'\mathbf{x}_2)=1$, $i=1,2,\ldots,p$, and $\operatorname{cov}(\boldsymbol{\alpha}_i'\mathbf{x}_1,\boldsymbol{\beta}_i'\mathbf{x}_2)=\rho_i$.

9.6.1 Sample Canonical Correlation

Suppose a sample of size $N>p+q$ is taken from the distribution of $\mathbf{x}$, which is assumed continuous. Let S denote the sample covariance matrix. Let it be partitioned as in Σ;

$$S=\begin{pmatrix}S_{11} & S_{12}\\ S_{12}' & S_{22}\end{pmatrix}. \tag{9.6.3}$$

Let $r_1^2>r_2^2>\cdots>r_p^2$ be the ordered characteristic roots of $S_{11}^{-1}S_{12}S_{22}^{-1}S_{12}'$. Then $1>r_1>r_2>\cdots>r_p>0$ are called the sample canonical correlations. We now propose some tests.

9.6.2 *Some Tests*

Let $\mathbf{x} \sim N_{p+q}(\boldsymbol{\mu}, \Sigma)$, where Σ is positive definite. Let S be the sample covariance based on a sample of size $N > p+q$. Under this assumption the p canonical variables $(\boldsymbol{\alpha}_i\mathbf{x}_1, \boldsymbol{\beta}_i\mathbf{x}_2)$, $i=1,2,\ldots,p$, are independently distributed, but the correlations $\rho_1,\ldots,\rho_p$ are ordered. We may be interested to test whether the last $p-k$ correlations are zero, for in this case the last $p-k$ canonical variables $(\boldsymbol{\alpha}_i\mathbf{x}_1, \boldsymbol{\beta}_i\mathbf{x}_2)$, $i=k+1,\ldots,p$, are of no predictive value, and the relationship between $\mathbf{x}_1$ and $\mathbf{x}_2$ can be measured by the first k canonical variables. When there are many variables in each set, this can often lead to a substantial reduction in dimensionality. Formally, we wish to test the hypothesis

$$H: \rho_{k+1} = \cdots = \rho_p = 0, \qquad \rho_k > 0,$$

vs

$$A: \rho_i \neq 0 \qquad \text{at least one } i = k+1,\ldots,p.$$

The likelihood ratio test is based on the statistic

$$Q_{p-k} = \prod_{i=k+1}^{p} (1 - r_i^2). \tag{9.6.4}$$

Bartlett (1937, 1947) has shown that as $N \to \infty$, $-N \ln Q_{p-k}$ is asymptotically distributed as chi-square with $f = (p-k)(q-k)$ df. Further improvement in describing this distribution is currently under investigation. For example, it has been shown by Fujikoshi (1977) that as $N \to \infty$,

$$\left[n - k - \tfrac{1}{2}(p+q+1) + \sum_{i=1}^{k} r_i^{-2} \right] \ln Q_{p-k} \to \chi_f^2.$$

9.6.3 *Likelihood Test for Independence*

At the outset we may be interested in testing that the two subvectors are independently distributed. That is, we may wish to test the hypothesis

$$H: \rho_1 = \cdots = \rho_p = 0$$

vs

$$A: \rho_i \neq 0 \qquad \text{at least one } i = 1,2,\ldots,p.$$

The likelihood ratio test is given by

$$\lambda = \frac{|S|}{|S_{11}||S_{22}|} = \prod_{i=1}^{p} (1 - r_i^2),$$

and $-m \ln \lambda$ is asymptotically χ_f^2 with $f = pq$, where

$$m = N - 2\alpha, \qquad \alpha = \tfrac{1}{4}(p+q+6).$$

To obtain p-values, one can use the expression

$$P(-m\ln\lambda \geq z) = P(\chi_f^2 \geq z) + (\gamma/m^2)\left[P(\chi_{f+4}^2 \geq z) - P(\chi_f^2 \geq z)\right],$$

where

$$\gamma = \tfrac{1}{48}pq(p^2 + q^2 - 5).$$

9.7 Computational Procedures

Partial correlations and simple correlations can be calculated on most calculators and computers. Canonical correlation, however, is best handled by a package such as SPSS, BMDP, or SAS.

Example 9.7.1 Grade point averages and SAT scores (Hosseini, 1978) were analyzed using a CMS file editing system to run the BMDP6M program canonical correlation analysis ($N = 406$), using the data in Table 9.1. (Variables 5 and 6 are university grade point averages.) The program statements required to run the program were

```
/PROBLEM TITLE IS 'TESTS'.
/INPUT VARIABLES ARE 6.
TYPE = COVA.
SHAPE = LOWER.
FORMAT IS '(6F8.3)'.
/VARIABLES NAMES ARE A, B, C, D, E, F.
/CANONICAL FIRST ARE A, B, C, D.
SECOND ARE E, F.
/END.
```

The covariance matrix was entered in the shape of a lower triangular matrix with the format of the last line as 6F8.3 with variables 1–6 assigned the letters $A-F$, respectively. The ensuing correlations are displayed in Table 9.2. The squared multiple correlations of each variable in the set $A-D$ with all other variables in the set and of E and F with each other appear in Table 9.3.

Table 9.1[a]

	1	2	3	4	5	6	Mean
1. grade 10	1.488						17.48
2. grade 11	0.988	1.166					17.85
3. grade 12	0.409	0.381	0.740				16.95
4. SAT score	2.081	2.763	13.2	7270.973			574.78
5. year 1	0.121	0.124	0.188	10.198	0.27		2.58
6. year 2	0.104	0.097	0.151	6.907	0.190	0.203	2.61

[a]Reproduced with the permission of the editors of *Educational and Psychological Measurement*.

Table 9.2

	$E(5)$	$F(6)$	$A(1)$	$B(2)$	$C(3)$	$D(4)$
E	1.000					
F	0.812	1.000				
A	0.191	0.189	1.000			
B	0.221	0.199	0.750	1.000		
C	0.421	0.390	0.390	0.410	1.000	
D	0.230	0.180	0.020	0.030	0.180	1.000

Table 9.3

Variable	r^2
A	0.57110
B	0.57906
C	0.21194
D	0.03554
E	0.65864
F	0.65864

Table 9.4

Variable	CNVRF1	CNVRF2
E	0.783	−1.522
F	0.254	1.693
	CNVRS1	CNVRS2
A	−0.002	0.898
B	0.146	−0.797
C	0.814	0.395
D	0.342	−0.745

The canonical correlations r_1 and r_2 were $r_1=0.45808$ and $r_2=0.06772$. Their squares were $r_1^2=0.20984$ and $r_2^2=0.00459$. These values indicate a relationship between the first linear combination of variables in each group but not in the second. If we let $x_1,\ldots,x_6$ denote the six variables and $s_1,\ldots,s_6$ their standard deviations, then the groups of canonical variates are given by the standardized coefficients of Table 9.4. That is,

$$y_1=-0.002x_1/s_1+0.146x_2/s_2+0.814x_3/s_3+0.342x_4/s_4,$$
$$y_2=0.783x_5/s_5+0.254x_6/s_6$$

have a correlation of 0.458. In terms of the square of the canonical correlation, y_1 accounts for 21% of the variation in y_2.

It is quite clear from the second eigenvalue that $0.898x_1/s_1-0.797x_2/s_2+0.395x_3/s_3-0.745x_4/s_4$ accounts for only 0.4% of the variation in $-1.522x_5/s_5+1.693x_6/s_6$.

In conclusion, y_1 represents an average of the high-school averages (suitably standardized), with more weight given to grade 12 marks. This average is a reasonable predictor of y_2, the average GPA over the first two years at a university. The second set of canonical variates measure change in high-school marks and change in first- and second-year university marks. These changes do not appear correlated at all.

To test for independence between the two groups, one looks at the test given in Section 9.6.3. That is,

$$\lambda = \frac{|S|}{|S_{11}||S_{21}|} = \prod_{i=1}^{2} (1 - r_i^2) = (1 - 0.21)(1 - 0.004) = 0.7868$$

for $\alpha = 3$, $m = 406 - 6 = 400$,

$$-m \ln \lambda = 400(0.240) = 95.9 \geq \chi^2_{8,0.05} = 15.50.$$

Hence there is a significant interdependence between the two groups of variables.

If we are interested in testing that there is only one significant correlation, then the test statistic for the hypothesis

$$H: \rho_2 = 0 \qquad \text{vs} \qquad A: \rho_2 \neq 0$$

is given in this case, from (9.6.4), by

$$Q = 1 - r_2^2 = 1 - 0.004.$$

Asymptotically,

$$(400.5 + 1/0.2) \ln Q = 1.624 \leq \chi^2_{3,0.05} = 7.8741.$$

Hence we conclude that $\rho_1^2 \neq 0$ and $\rho_2^2 = 0$, and we may reasonably conclude that a student's performance in high school during the last two years is correlated with his or her university performance. However, the fluctuations of a student's high-school grades in no way influence fluctuations in university grades.

Problems

9.1 Show that r is the maximum likelihood estimator of ρ.

9.2 The matrix of Table 9.5 (Batlis, 1978) gives the correlations of measures of job involvement, age, SAT scores, grade point average (GPA), satisfaction, and performance.

- **(a)** Owing to incomplete data, some of the correlations were based on different sample sizes. Assuming a minimum sample size of 34, estimate the partial correlation between performance and job involvement, adjusting for the other variables. Give a 95% confidence interval for this partial correlation.
- **(b)** Compute the multiple correlation between job performance and the other five variables.

Table 9.5[a]

Variable	1	2	3	4	5	6	Mean	s.d.
1. job involvement	1.00						15.82	3.73
2. age	0.13	1.00					21.50	6.03
3. SAT score	0.12	0.02	1.00				1071.98	134.52
4. GPA	0.37	0.29	0.26	1.00			2.78	0.63
5. satisfaction	−0.01	0.03	0.27	0.08	1.00		6.21	1.77
6. performance	0.43	0.33	0.37	0.64	0.22	1.00	3.54	1.24

[a]Reproduced with the permission of the editors of *Educational and Psychological Measurement*.

9.3 The correlation matrix of Table 9.6 is given by Carpenter and Carpenter (1978), who investigated the relationship between California achievement test (CAT) scores and scores on the Larsen–Hammill test of written spelling (TWS), for predictable and unpredictable words ($N = 45$). A predictable word, for example, would be "up." An unpredictable word would be "enough."

(a) Are the criterion variables (3–6) and predictor variables (1 and 2) independent?

(b) Perform a canonical correlation analysis on the data. Note that as only the correlation matrix is given, the variables can be considered to have been standardized with mean 0 and variance 1.

9.4 Table 9.7 presents data from Hosseini (1978) on grade point averages and SAT scores ($N = 44$). (The variables are similar to those of Example 9.7.1 except that only courses in mathematics and the natural sciences are considered.)

(a) Estimate the canonical variates.

(b) Test the hypothesis of independence between the set of variables 1–4 and the set of variables 5 and 6.

Table 9.8[a]

	1	2	3	4	5	6	7	8
1	1.00							
2	−0.08	1.00						
3	0.04	0.16	1.00					
4	−0.18	−0.29	−0.53	1.00				
5	0.04	−0.17	0.25	−0.25	1.00			
6	−0.30	−0.48	−0.48	0.18	−0.31	1.00		
7	0.10	−0.03	−0.47	0.15	−0.17	0.23	1.00	
8	0.42	−0.33	−0.15	0.12	0.13	0.02	0.11	1.00
9	−0.13	0.33	0.07	−0.04	−0.00	−0.36	−0.06	−0.30
10	0.06	−0.38	−0.03	0.19	0.13	0.18	0.09	0.12
11	0.24	−0.11	−0.25	−0.11	−0.09	0.36	0.33	0.10
12	0.18	−0.24	−0.35	0.18	−0.26	0.37	0.25	0.13
13	−0.08	−0.11	−0.26	0.41	0.14	−0.10	0.13	0.05
14	0.04	−0.10	0.02	0.00	0.38	−0.09	−0.05	−0.02
15	0.02	−0.01	−0.00	−0.09	−0.37	0.26	−0.07	−0.13
16	0.22	−0.01	−0.11	0.29	0.10	−0.24	0.03	0.08

[a]Reproduced with the permission of the editors of *Multivariate Behavioral Research*.

Table 9.6[a]

Variable	1	2	3	4	5	6
1. TWS predictable	1.00	0.82	0.32	0.21	0.38	0.31
2. TWS unpredictable	0.82	1.00	0.26	0.20	0.33	0.28
3. CAT reading	0.32	0.26	1.00	0.90	0.90	0.82
4. CAT mathematics	0.21	0.20	0.90	1.00	0.87	0.86
5. CAT language	0.38	0.33	0.90	0.87	1.00	0.85
6. CAT spelling	0.31	0.28	0.82	0.86	0.85	1.00

[a]Reproduced with the permission of the editors of *Educational and Psychological Measurement*.

Table 9.7[a]

Variable	2	3	4	5	7	Mean	s.d.
1. grade 10	0.71	0.46	−0.24	0.05	−0.06	17.14	1.19
2. grade 11		0.33	−0.23	0.11	−0.01	17.52	1.11
3. grade 12			0.15	0.35	0.24	16.23	0.78
4. SAT				0.36	0.26	534.29	80.12
5. year 1					0.75	2.75	0.51
6. year 2						2.64	0.42

[a]Reproduced with the permission of the editors of *Educational and Psychological Measurement*.

9.5 The correlation matrix of Table 9.8, from Fornell (1979), considers the following variables:

1. foam,
2. pill,
3. I.U.D.,
4. prophylactics,

Table 9.8 (*continued*)

9	10	11	12	13	14	15	16
1.00							
−0.16	1.00						
−0.22	0.26	1.00					
−0.10	0.03	0.21	1.00				
0.23	−0.16	−0.14	0.13	1.00			
−0.08	0.14	0.10	−0.25	0.14	1.00		
0.14	0.00	0.24	0.18	−0.22	−0.21	1.00	
0.00	−0.01	−0.02	0.08	0.28	0.13	−0.21	1.00

5. diaphragm,
6. rhythm method,
7. "I enjoy getting dressed up for parties,"
8. "Liquor adds to the party,"
9. "I study more than most students,"
10. "Varsity athletes earn their scholarships,"
11. "I am influential in my living group,"
12. "Good grooming is a sign of self-respect,"
13. "My days seem to follow a definite routine,"
14. "Companies generally offer what consumers want,"
15. "Advertising results in higher prices," and
16. "There should be a gun in every home."

Perform a canonical correlation analysis between birth control preference (variables 1–6) scores and social attitude scores (variables 7–16). (Fornell suggests an alternative method of analysis for this data using external single-set component analysis. The reader is referred to his paper for details.)

9.6 Table 9.9 is the correlation matrix of characteristics of a consumer affairs department (variables 1–14) and involvement in management decision making for 128 executives (variables 15–22), taken from Fornell (1979). The explanatory variables were

1. middle-aged CA executive,
2. CA executive's title,
3. CA executive's previous title,

Table 9.9[a]

	1	2	3	4	5	6	7	8	9	10	11
1	1.00										
2	−0.05	1.00									
3	−0.08	0.49	1.00								
4	0.10	−0.15	0.16	1.00							
5	0.12	0.04	0.15	0.02	1.00						
6	0.20	0.19	0.30	0.21	0.53	1.00					
7	0.10	0.18	0.03	0.03	0.28	0.43	1.00				
8	0.05	0.19	0.10	−0.02	0.17	0.32	0.59	1.00			
9	0.01	0.19	0.15	−0.12	0.21	0.27	0.38	0.26	1.00		
10	−0.10	0.09	0.19	0.11	0.12	0.30	0.15	0.30	0.24	1.00	
11	−0.00	0.07	0.08	0.02	0.08	0.27	0.42	0.52	0.40	0.19	1.00
12	−0.01	0.04	0.01	0.12	0.19	0.20	0.40	0.53	0.20	0.08	0.45
13	0.06	0.02	−0.04	0.02	0.07	0.19	0.38	0.49	0.30	0.06	0.47
14	−0.20	−0.15	−0.10	−0.13	−0.06	−0.10	−0.07	−0.04	−0.14	0.04	0.13
15	0.18	0.17	0.12	0.09	0.37	0.48	0.20	0.14	0.17	0.19	0.09
16	0.10	0.23	0.15	0.26	0.33	0.20	0.13	0.10	−0.04	0.22	−0.04
17	0.20	0.08	0.10	0.18	0.45	0.42	0.22	0.21	0.26	0.09	0.20
18	0.30	0.22	0.20	0.05	0.54	0.44	0.40	0.38	0.22	0.15	0.26
19	0.06	0.06	0.17	0.11	0.26	0.38	0.14	0.31	0.08	0.21	0.33
20	0.28	0.04	0.15	0.16	0.30	0.30	0.34	0.25	0.25	0.13	0.27
21	0.08	−0.04	−0.01	0.19	0.22	0.30	0.56	0.48	0.35	0.13	0.51
22	−0.02	0.22	0.17	0.07	0.26	0.45	0.63	0.62	0.45	0.23	0.50

[a]Reproduced with the permission of the editors of *Multivariate Behavioral Research*.

4. CA executive's education,
5. veto power,
6. reports on committees,
7. coordinates consumer-related activities,
8. educating personnel,
9. top management support,
10. budget responsibility,
11. responsibility for long-term goal development,
12. publishing factual information,
13. research on consumer behavior, and
14. consumer complaints.

The criterion variables measured involvement in

15. advertising copy,
16. sales promotion,
17. concept generation,
18. new-product launching,
19. market research,
20. consumer service,
21. new consumer programs, and
22. influence on top management.

(a) Perform a canonical correlation analysis on the data and interpret the first two canonical variables.

(b) Test the null hypothesis that there are only two nonzero canonical correlations.

Table 9.9 (*continued*)

12	13	14	15	16	17	18	19	20	21	22
1.00										
0.38	1.00									
−0.08	0.10	1.00								
0.07	0.05	−0.07	1.00							
0.03	−0.12	−0.11	0.54	1.00						
0.11	0.08	−0.26	0.20	0.18	1.00					
0.16	0.14	−0.23	0.23	0.31	0.48	1.00				
0.27	0.40	0.04	0.13	0.07	0.26	0.18	1.00			
0.12	0.18	−0.03	0.26	0.18	0.38	0.42	0.27	1.00		
0.43	0.47	−0.04	0.18	0.09	0.27	0.32	0.25	0.43	1.00	
0.46	0.38	−0.06	0.33	0.13	0.29	0.33	0.31	0.34	0.60	1.00

Chapter 10
Principal Component Analysis

10.1 Introduction

Suppose we are interested in finding the level of performance in mathematics of the ninth grade students of a certain school. We may then record their scores in mathematics; that is, we consider just one characteristic of each student. Now suppose we are instead interested in *overall* performance and select some p characteristics, such as mathematics, English, history, and science. These characteristics, although related to each other, may not all contain the same amount of information, and in fact some characteristics may be completely redundant. Obviously, this will result in a loss of information and waste of resources in analyzing the data. Thus we should select only those characteristics that will truly discriminate one student from another, while those least discriminatory should be discarded. Often this is not an easy task.

Alternatively, we could consider p linear functions of these p characteristics and choose the best discriminators among these. This task is simplified if the transformations are made in such a way that these p linear functions become uncorrelated (independent if the original p characteristics are normally distributed), for then we could discard from study functions reflecting less variability and consider only those functions that have higher variances.

These p uncorrelated linear functions are called *principal components*. Their study, principal component analysis, was essentially developed by Hotelling (1933) after its origination by Pearson (1901). For details and various applications the reader is referred to Rao (1964).

10.2 Analysis Based on the Covariance Matrix

In this section we shall assume that the variates x_i of the vector $\mathbf{x} = (x_1, \ldots, x_p)'$ are measured in the same or comparable units. (See Section 10.2.5 for the effects of not meeting this condition.) We shall also assume that the variances of the variables do not vary too much (see Example

10.3.1). Let $\mathbf{x}$ be a p-component vector with covariance matrix Σ. Since Σ is a symmetric and positive definite matrix, all the characteristic roots of Σ are real and positive (>0). Let $\lambda_1 \geq \lambda_2 \geq \cdots \geq \lambda_p > 0$ be the ordered characteristic roots of Σ. Then there exists an orthogonal matrix $\Gamma = (\gamma_1, \ldots, \gamma_p)$, $\Gamma\Gamma' = I_p$, such that

$$\Gamma'\Sigma\Gamma = D_\lambda, \qquad D_\lambda = \operatorname{diag}(\lambda_1, \lambda_2, \ldots, \lambda_p). \tag{10.2.1}$$

Hence, if we let

$$\mathbf{y} = \Gamma'\mathbf{x} = \begin{pmatrix} \gamma_1' \\ \vdots \\ \gamma_p' \end{pmatrix} \mathbf{x} = \begin{pmatrix} \gamma_1'\mathbf{x} \\ \vdots \\ \gamma_p'\mathbf{x} \end{pmatrix}, \tag{10.2.2}$$

then $\operatorname{cov}(\mathbf{y}) = \Gamma'\Sigma\Gamma = D_\lambda$, and the components $y_1 = \gamma_1'\mathbf{x}$, $y_2 = \gamma_2'\mathbf{x}, \ldots, y_p = \gamma_p'\mathbf{x}$ are uncorrelated. These components $y_1, \ldots, y_p$ are the principal components. The variance of y_i is λ_i. Since $\lambda_1 \geq \lambda_2 \geq \cdots \geq \lambda_p$, y_1 has the largest variance λ_1, y_2 has the second-largest variance λ_2, and so on. Usually y_1 is called the first principal component, y_2 the second principal component, and so on. Since

$$\lambda_1 + \lambda_2 + \cdots + \lambda_p = \operatorname{tr} \Sigma = \sum_{i=1}^{p} \sigma_{ii}, \tag{10.2.3}$$

the sum of the variances of the p principal components is the same as the sum of the variances of the old variables $x_1, \ldots, x_p$. Thus the components with smaller variances could be ignored without significantly effecting the total variance, thereby reducing the number of variables from p to, say, $k \leq p$.

10.2.1 Some Relationships Between the Original Variables and the Principal Components

Let $\mathbf{x} \sim N_p(\mathbf{0}, \Sigma)$. Then $y_1 = \gamma_1'\mathbf{x}, \ldots, y_p = \gamma_p'\mathbf{x}$ are the p principal components, where $\Gamma = (\gamma_1, \ldots, \gamma_p)$ and $\Gamma'\Sigma\Gamma = D_\lambda = \operatorname{diag}(\lambda_1, \lambda_2, \ldots, \lambda_p)$. Let us now calculate the correlations between x_i and y_j. The correlation between x_i and y_j is given by

$$\frac{E(x_i y_j)}{\left[\operatorname{var}(x_i)\operatorname{var}(y_j)\right]^{1/2}} = \frac{E(x_i \mathbf{x}'\gamma_j)}{\sigma_{ii}^{1/2}\lambda_j^{1/2}} = \frac{(\sigma_{i1}, \sigma_{i2}, \ldots, \sigma_{ip})\gamma_j}{\sigma_{ii}^{1/2}\lambda_j^{1/2}}. \tag{10.2.4}$$

Since

$$\Sigma\gamma_j = \lambda_j \gamma_j, \tag{10.2.5}$$

we get

$$(\sigma_{i1}, \sigma_{i2}, \ldots, \sigma_{ip})\gamma_j = \lambda_j \gamma_{ij} = E(x_i y_j). \tag{10.2.6}$$

Table 10.1 Correlations Between Original Variables and Principal Components

Original variables	Principal components			
	y_1	y_2	$\cdots$	y_p
x_1	$\lambda_1^{1/2}\gamma_{11}/\sigma_{11}^{1/2}$	$\lambda_2^{1/2}\gamma_{12}/\sigma_{11}^{1/2}$	$\cdots$	$\lambda_p^{1/2}\gamma_{1p}/\sigma_{11}^{1/2}$
x_2	$\lambda_1^{1/2}\gamma_{21}/\sigma_{22}^{1/2}$	$\lambda_2^{1/2}\gamma_{22}/\sigma_{22}^{1/2}$	$\cdots$	$\lambda_p^{1/2}\gamma_{2p}/\sigma_{22}^{1/2}$
$\vdots$	$\vdots$	$\vdots$		$\vdots$
x_p	$\lambda_1^{1/2}\gamma_{p1}/\sigma_{pp}^{1/2}$	$\lambda_2^{1/2}\gamma_{p2}/\sigma_{pp}^{1/2}$	$\cdots$	$\lambda_p^{1/2}\gamma_{pp}/\sigma_{pp}^{1/2}$

where

$$\gamma_j' = (\gamma_{1j},\ldots,\gamma_{pj}), \quad j=1,2,\ldots,p, \qquad \Gamma = (\gamma_{ij}). \tag{10.2.7}$$

Hence the correlation between x_i and y_j is given by $\lambda_j^{1/2}\gamma_{ij}/\sigma_{ii}^{1/2}$. These correlations are shown in Table 10.1.

Thus the magnitudes and the signs of the correlations help determine the relationship between the original variables and the principal components. In practice we may be interested in only, say, the first k components and would have only to complete a $p \times k$ table of correlations.

In order to judge how good the first k $(k < p)$ principal components are in approximating the original p variables $x_1, x_2, \ldots, x_p$, we find the multiple correlations between x_1 and $y_1,\ldots,y_k$; between x_2 and $y_1,\ldots,y_k$; and between x_p and $y_1,\ldots,y_k$. The squared multiple correlation between x_i and $y_1,\ldots,y_k$ is given by

$$\sigma_{ii}^{-1}\tau_i' \begin{pmatrix} \lambda_1 & & \\ & \ddots & \\ & & \lambda_k \end{pmatrix}^{-1} \tau_i, \tag{10.2.8}$$

where

$$\tau_i' = E(x_i y_1, x_i y_2, \ldots, x_i y_k). \tag{10.2.9}$$

From (10.2.6) we have

$$\tau_i' = (\lambda_1\gamma_{i1},\ldots,\lambda_k\gamma_{ik}). \tag{10.2.10}$$

Hence the squared multiple correlation between x_i and $y_1,\ldots,y_k$ is given by

$$\rho^2_{x_i(y_1,\ldots,y_k)} = \sum_{j=1}^{k} \lambda_j\gamma_{ij}^2/\sigma_{ii}. \tag{10.2.11}$$

Table 10.2 Multiple Correlation

$\rho^2_{x_1(y_1,\ldots,y_k)}$	$\rho^2_{x_2(y_1,\ldots,y_k)}$	$\cdots$	$\rho^2_{x_p(y_1,\ldots,y_k)}$
$\Sigma_{j=1}^k\lambda_j\gamma_{1j}^2/\sigma_{11}$	$\Sigma_{j=1}^k\lambda_j\gamma_{2j}^2/\sigma_{22}$	$\cdots$	$\Sigma_{j=1}^k\lambda_j\gamma_{pj}^2/\sigma_{kk}$

The multiple correlations obtained are shown in Table 10.2. The residual variances are given by

$$\sigma_{11}-\sum_{j=1}^{k}\lambda_j\gamma_{1j}^2,\qquad \sigma_{22}-\sum_{j=1}^{k}\lambda_j\gamma_{2j}^2,\qquad \ldots,\qquad \sigma_{pp}-\sum_{j=1}^{k}\lambda_j\gamma_{pj}^2. \tag{10.2.12}$$

The multiple correlations and the residual variances enable us to see how well the first k principal components approximate the original variables.

10.2.2 Principal Component Analysis Based on the Sample Covariance Matrix

Usually, the covariance matrix Σ is not known, and we may consider the principal components based on the sample covariance matrix,

$$S=n^{-1}\sum_{i=1}^{N}(\mathbf{x}_i-\bar{\mathbf{x}})(\mathbf{x}_i-\bar{\mathbf{x}})', \tag{10.2.13}$$

where $\mathbf{x}_1,\ldots,\mathbf{x}_N$ are independent samples from $N_p(\boldsymbol{\mu},\Sigma)$, $\bar{\mathbf{x}}=N^{-1}\Sigma_{i=1}^{N}\mathbf{x}_i$, and $n=N-1$. Let $C=(\mathbf{c}_1,\ldots,\mathbf{c}_p)$ be an orthogonal matrix with positive diagonal elements such that

$$C'SC=D_l, \tag{10.2.14}$$

where $l_1>l_2>\cdots>l_p$ are the ordered roots of S. Note that C is unique. If the λ_is are distinct ($\lambda_1>\lambda_2>\cdots>\lambda_p$), then C is the maximum likelihood estimate of Γ, and $(n/N)l_i$ are the maximum likelihood estimates of λ_i. However, if the λ_is are not distinct, some adjustments are needed to obtain maximum likelihood estimates. For example, if, say, λ_1 is repeated r times, then the maximum likelihood estimate of λ_1 is given by

$$n(l_1+\cdots+l_r)/Nr, \tag{10.2.15}$$

and when λ_1 is repeated r times, the characteristic vectors $\gamma_1,\ldots,\gamma_r$ are not unique but can be chosen so that $\gamma_1\neq\gamma_2\neq\cdots\neq\gamma_r$. However, $(\gamma_1,\ldots,\gamma_r)O$ will also be a characteristic matrix, where O is any arbitrary $r\times r$ orthogonal matrix. Thus the maximum likelihood estimate of $(\gamma_1,\ldots,\gamma_r)$ is $(\mathbf{c}_1,\ldots,\mathbf{c}_r)G$, where G is any $q\times q$ orthogonal matrix. For details, see Anderson (1963) or Srivastava and Khatri (1979).

The sample principal components are given by $\mathbf{c}_1'\mathbf{x},\ldots,\mathbf{c}_p'\mathbf{x}$. These sample principal components are only *asymptotically* independently distributed. The variances of these components are estimated by $l_1,\ldots,l_p$, respectively. Thus we consider only those sample principal components whose variances l_i are large and ignore the ones with smaller l_i. Since $\operatorname{tr}S=\Sigma_{i=1}^{p}s_{ii}=\Sigma_{i=1}^{p}l_i$, one need not obtain all the characteristic roots and vectors of S. For example, if for some k

$$(l_1+\cdots+l_k)/(s_{11}+\cdots+s_{pp}) \tag{10.2.16}$$

Table 10.3 Sample Correlations Between Original Variables and k Principal Components

Original variables	Principal components			
	y_1	y_2	$\cdots$	y_k
x_1	$l_1^{1/2}c_{11}/s_{11}^{1/2}$	$l_2^{1/2}c_{12}/s_{11}^{1/2}$	$\cdots$	$l_k^{1/2}c_{1k}/s_{11}^{1/2}$
x_2	$l_1^{1/2}c_{21}/s_{22}^{1/2}$	$l_2^{1/2}c_{22}/s_{22}^{1/2}$	$\cdots$	$l_k^{1/2}c_{2k}/s_{22}^{1/2}$
$\vdots$	$\vdots$	$\vdots$		$\vdots$
x_p	$l_1^{1/2}c_{p1}/s_{pp}^{1/2}$	$l_2^{1/2}c_{p2}/s_{pp}^{1/2}$	$\cdots$	$l_k^{1/2}c_{pk}/s_{pp}^{1/2}$

Table 10.4 Sample Multiple Correlations

$r^2_{x_1(y_1,\ldots,y_k)}$	$r^2_{x_2(y_1,\ldots,y_k)}$	$\cdots$	$r^2_{x_p(y_1,\ldots,y_k)}$
$\Sigma_{j=1}^k l_j c_{1j}^2/s_{11}$	$\Sigma_{j=1}^k l_j c_{2j}^2/s_{22}$	$\cdots$	$\Sigma_{j=1}^k l_j c_{pj}^2/s_{kk}$

accounts for large percentage, say, 80%, then one needs to obtain $l_1,\ldots,l_k$ only.

After we have decided upon the number of principal components, say, k, $\mathbf{c}_1'\mathbf{x},\ldots,\mathbf{c}_k'\mathbf{x}$, we may wish to know about its relationships with the original variables. This transformation of Table 10.1 is shown in Table 10.3. Here $S=(s_{ij})$; that is, s_{ii} is the sample variance of the variate x_i.

The sample multiple correlations corresponding to Table 10.2 are given in Table 10.4. The residual sample variances are

$$s_{11}-\sum_{j=1}^{k} l_j c_{1j}^2, \qquad s_{22}-\sum_{j=1}^{k} l_j c_{2j}^2, \qquad \ldots, \qquad s_{pp}-\sum_{j=1}^{k} l_j c_{pj}^2. \tag{10.2.17}$$

Example 10.2.1 The appearance of a color transparency is quite sensitive to changes in the chemical composition of the processing solution. The problem of statistical quality control of processing solutions is to differentiate between observed variations that are consistent and those that are inconsistent with the inherent variability of the process (Jackson and Morris, 1957). The most common procedure for a photographic check on the process is as follows: A film strip is given a graded series of exposures to white light and is processed; optical densities of the resultant "steps" are then measured through red, green, and blue filters at the high-density portion (shoulder), at the middle-tone portion, and at the toe portion. Thus there are nine characteristics. The variance–covariance matrix based on 108 df is given in Table 10.5.

The eigenvalues, obtained by the use of BMDP4M program, are given in Table 10.6. The first three components account for 83% of the total

Table 10.5 Variance–Covariance Matrix[a]

Shoulder			Middle tone			Toe		
Red	Green	Blue	Red	Green	Blue	Red	Green	Blue
177	179	95	96	53	32	−7	−4	−3
	419	245	131	181	127	−2	1	4
		302	60	109	142	4	4	11
			158	102	42	4	3	2
				137	96	4	5	6
					128	2	2	8
						34	31	33
							39	39
								48

[a]Reproduced wtih the permission of the American Statistical Association.

Table 10.6 Eigenvalues

Component	Eigenvalue	Cumulative proportion of total variance
1	878.52	0.6092
2	196.10	0.7452
3	128.64	0.8344
4	103.43	0.9062
5	81.26	0.9625
6	37.85	0.9888
7	6.98	0.9936
8	5.71	0.9976
9	3.52	1.0000

variation, and the first two components for 75%. It appears that two or at most three components are sufficient to account for most of the variation.

The eigenvectors for the first three components are given in Table 10.7. The first component, the brightness, measures the average density at the shoulder and the middle tone. The second component, the tint, contrasts the red and blue colors in the shoulder and the middle tone. The third component, the contrast, compares the middle-tone and toe densities with the shoulder densities. Other comparisons have similar interpretations.

Table 10.7 Characteristic Vectors of the Covariance Matrix

	Shoulder			Middle tone			Toe		
Component	Red	Green	Blue	Red	Green	Blue	Red	Green	Blue
I	0.305	0.654	0.482	0.261	0.323	0.271	0.002	0.006	0.014
II	−0.485	−0.150	0.587	−0.491	−0.038	0.376	0.057	0.053	0.088
III	−0.412	−0.182	−0.235	0.457	0.495	0.268	0.256	0.266	0.282

Table 10.8 Correlations Between Components and Variables

	Shoulder			Middle tone			Toe		
Component	Red	Green	Blue	Red	Green	Blue	Red	Green	Blue
I	0.679	0.946	0.824	0.616	0.819	0.710	0.012	0.028	0.062
II	−0.511	−0.103	0.473	−0.547	−0.045	0.462	0.137	0.120	0.179
III	−0.351	−0.101	−0.154	0.412	0.480	0.269	0.497	0.483	0.462
r^2	0.845	0.916	0.928	0.849	0.903	0.791	0.266	0.249	0.249

The correlations between the original variables and the first three components are given in Table 10.8. The multiple correlation between each variable and the first two components are also given, along with the residual variance. It should be noted that the toe gives almost no contribution to the first two components and only a small contribution to the third.

Also, R^2 using three factors, except at the shoulder and middle tone, is not as high as we would like it to be. If we investigate the covariance, we see that the toe portion is uncorrelated with the shoulder and middle tone. It also has a smaller variance. Hence it might be of interest to perform principal component analysis on the correlation matrix. (See Sections 10.3 and 10.5.)

10.2.3 Likelihood Ratio Test for the Equality of the Last $p-k$ Roots of Σ

Let $\mathbf{x}_1,\ldots,\mathbf{x}_N$ be independently and identically distributed as $N_p(\boldsymbol{\mu},\Sigma)$. Let

$$nS=\sum_{i=1}^{N}(\mathbf{x}_i-\bar{\mathbf{x}})(\mathbf{x}_i-\bar{\mathbf{x}})', \qquad n=N-1,$$

and let $l_1>l_2>\cdots>l_p$ be the ordered characteristic roots of S and $\lambda_1\geq\lambda_2\geq\cdots\geq\lambda_p$ be the ordered characteristic roots of Σ. Suppose we wish to test the hypothesis that the last $p-k$ roots of Σ are equal to λ. That is, we wish to test

$$H: \lambda_{k+1}=\cdots=\lambda_p=\lambda$$

vs

$$A: \lambda_i\neq\lambda, \qquad \text{for at least one } i=k+1,\ldots,p.$$

The likelihood ratio test (Lawley, 1956) for the above hypothesis is based on the statistic

$$Q=\left(\prod_{j=k+1}^{p} l_j\right)\Big/\left(\sum_{j=k+1}^{p} l_j/q\right)^{q}, \qquad q\equiv p-k.$$

It has been shown by Fujikoshi (1978) [see also Lawley (1956) and James

(1969)] that

$$\tilde{Q} \equiv -\left\{n-k-\tfrac{1}{6}\left[2(p-k)+1+2(p-k)^{-1}\right]+\hat{\lambda}^2 \sum_{i=1}^{k} (l_i-\hat{\lambda})^{-2}\right\} \ln Q,$$

where $\hat{\lambda}=\Sigma_{r=k+1}^{p} l_i/q$, is asymptotically distributed as chi-square with $f \equiv \frac{1}{2}(p-k)(p-k+1)-1$ degrees of freedom. Thus the hypothesis H is rejected if $\tilde{Q}>\chi^2_{f,\alpha}$.

10.2.4 Some Asymptotic Distributions

Let $l_1>l_2>\cdots>l_p$ be the roots of S defined by (10.2.13), and let $\lambda_1>\lambda_2>\cdots>\lambda_p$ be the roots of Σ; that is, the covariance matrix Σ has all distinct roots. Then $n^{1/2}(l_i-\lambda_i)/(2\lambda_i^2)^{1/2}$, $i=1,2,\ldots,p$, is asymptotically independently distributed as $N(0,1)$. Even if other roots of Σ have multiplicity (but not the ith root λ_i, which has no multiplicity), the marginal distribution of $n^{1/2}(l_i-\lambda_i)/(2\lambda_i^2)^{1/2}$ is normal with mean 0 and variance 1. The above result is due to Girshick (1939) and Anderson (1963).

10.2.5 Effect of Unit of Measurements

The principal component analysis of this section suffers from the disadvantage that it is *not* invariant under linear transformations. In particular, a change in the scale of measurements of some or all of the variables results in the covariance matrix being multiplied on both sides by a diagonal matrix. Thus the role of important and unimportant variables could be reversed. The above analysis should therefore be carried out only if the variates are measured in the same units or comparable units. Alternatively, we may obtain principal components based on the sample correlation matrix, as discussed in Section 10.3.

10.3 Analysis Based on the Sample Correlation Matrix

Let $\mathbf{x}_1,\ldots,\mathbf{x}_N$ be independently distributed as $N_p(\boldsymbol{\mu},\Sigma)$. As before, let S denote the sample covariance matrix:

$$S=n^{-1}\sum_{i=1}^{N}(\mathbf{x}_i-\bar{\mathbf{x}})(\mathbf{x}_i-\bar{\mathbf{x}})', \qquad \bar{\mathbf{x}}=N^{-1}\sum x_i, \quad n=N-1. \tag{10.3.1}$$

We can write

$$S=(s_{ij})=D_{s^{1/2}}RD_{s^{1/2}}, \tag{10.3.2}$$

where $D_{s^{1/2}}=\operatorname{diag}(s_{11}^{1/2},\ldots,s_{pp}^{1/2})$ and $R=(r_{ij})$ with $r_{ii}=1$, $i=1,2,\ldots,p$, and

Table 10.9 Sample Correlations Between Original Variables and the Principal Components Based on the Correlation Matrix R

Original variables	Principal components			
	z_1	z_2	$\cdots$	z_k
x_1	$f_1^{1/2}h_{11}$	$f_2^{1/2}h_{12}$	$\cdots$	$f_k^{1/2}h_{1k}$
x_2	$f_1^{1/2}h_{21}$	$f_2^{1/2}h_{22}$	$\cdots$	$f_k^{1/2}h_{2k}$
$\vdots$	$\vdots$	$\vdots$		$\vdots$
x_p	$f_1^{1/2}h_{p1}$	$f_2^{1/2}h_{p2}$	$\cdots$	$f_k^{1/2}h_{pk}$

Table 10.10 Sample Multiple Correlations

$r^2_{x_1(z_1,\ldots,z_k)}$	$r^2_{x_2(z_1,\ldots,z_k)}$	$\cdots$	$r^2_{x_p(z_1,\ldots,z_k)}$
$\Sigma_{j=1}^k f_j h_{1j}^2$	$\Sigma_{j=1}^k f_j h_{2j}^2$	$\cdots$	$\Sigma_{j=1}^k f_j h_{pj}^2$

$r_{ij} = s_{ij}/s_{ii}^{1/2}s_{jj}^{1/2}$. Let H be any orthogonal matrix such that

$$R = HD_fH', \qquad H = (\mathbf{h}_1,\ldots,\mathbf{h}_p), \tag{10.3.3}$$

where $D_f = \text{diag}(f_1,\ldots,f_p)$ and $f_1 \geq \cdots \geq f_p$ are the roots of the correlation matrix R. Then the principal components based on the correlation matrix R are given by

$$Z_1 = \mathbf{h}_1' D_{s^{-1/2}}\mathbf{x}, \qquad \ldots, \qquad Z_p = \mathbf{h}_p' D_{s^{-1/2}}\mathbf{x}. \tag{10.3.4}$$

Note that $\operatorname{tr} R = p$. Hence, we consider only k principal components if $(f_1 + \cdots + f_k)/p$ accounts for most of the variation, say 80%.

The correlation matrix between the original variables $x_1,\ldots,x_p$ and the k principal components $z_1,\ldots,z_k$ is given in Table 10.9. Note that each column of this table is simply a scaled characteristic vector (the ith column scaled by $f_i^{1/2}$).

The multiple correlations between x_1 and $z_1,\ldots,z_k$; between x_2 and $z_1,\ldots,z_k,\ldots$; and between x_p and $z_1,\ldots,z_k$ are given in Table 10.10. The residual variances are given by

$$1 - \sum_{j=1}^{k} f_j h_{1j}^2, \qquad 1 - \sum_{j=1}^{k} f_j h_{2j}^2, \qquad \ldots, \qquad 1 - \sum_{j=1}^{k} f_j h_{pj}^2.$$

Example 10.3.1 The Forestry Commission, in cooperation with the Forest Research Laboratory of the United Kingdom, was concerned with a study of the strength of pitprops cut from home-grown timber (Jeffers, 1967). Tests were carried out on all species of commercial importance grown in that country with the object of determining whether or not pitprops cut

Table 10.11

Variable	Mean	s.d.
1	4.21	0.98
2	46.88	11.29
3	114.6	57.1
4	0.877	0.231
5	0.415	0.070
6	13.3	3.24
7	16.3	3.95
8	0.65	0.43
9	23.4	9.06
10	2.49	1.17
11	10.7	6.27
12	5.45	1.65
13	0.82	0.325

from home-grown timber are sufficiently strong for use in the mines. Load was applied at the rate of 200 $lb/in.^2$-min, calculated on the small end of the prop, until failure occurred. The maximum load and the position of the fractures were recorded. Both variables were subjected to analysis, but only maximum load was considered in the paper. Means and standard deviations are shown in Table 10.11 for the following variables.

1. the top diameter of the prop (in.),
2. the length of the prop (in.),
3. the moisture content of the prop (% of dry weight),
4. the specific gravity of the timber at time of test,
5. the oven-dry specific gravity of the timber,
6. the number of annual rings at the top of the prop,
7. the number of annual rings at the base of the prop,
8. the maximum bow (in.),
9. the distance of the point of maximum bow from the top of the prop (in.),
10. the number of knot whorls,
11. the length of clear prop from the top of the prop (in.),
12. the average number of knots per whorl, and
13. the average diameter of the knots (in.).

It is necessary to decide which components have any practical significance. Since the *units of measurements are not the same*, it is necessary to do the principal component analysis based on the correlation matrix. Table 10.13 gives the eigenvalues, the characteristic roots of the correlation matrix given in Table 10.12. The first four components account for 74% of the total variability, and the first six components account for 87% of the total variability. Thus probably the first four components, or at the most the first

Table 10.12 Correlation Matrix for Corsican Pine ($N = 180$)[a]

	1	2	3	4	5	6	7	8	9	10	11	12	13
1	1												
2	0.954	1											
3	0.364	0.297	1										
4	0.342	0.284	0.882	1									
5	−0.129	−0.118	−0.148	0.220	1								
6	0.313	0.291	0.153	0.381	0.364	1							
7	0.496	0.503	−0.029	0.174	0.296	0.813	1						
8	0.424	0.419	−0.054	−0.059	0.004	0.090	0.372	1					
9	0.592	0.648	0.125	0.137	−0.039	0.211	0.465	0.482	1				
10	0.545	0.569	−0.081	−0.014	0.037	0.274	0.679	0.557	0.526	1			
11	0.084	0.076	0.162	0.097	−0.091	−0.036	−0.113	0.061	0.085	−0.319	1		
12	−0.019	−0.036	0.220	0.169	−0.145	0.024	−0.232	−0.357	−0.127	−0.368	0.029	1	
13	0.134	0.144	0.126	0.015	−0.208	−0.329	−0.424	−0.202	−0.076	−0.291	0.007	0.184	1

[a]Reproduced with the permission of the editors of *Applied Statistics*.

Table 10.13 The First Ten Eigenvalues of the Correlation Matrix of Table 10.12

		Percentage of variability	
Component	Eigenvalue	Component	Cumulative
1	4.219	32.4	32.4
2	2.378	18.3	50.7
3	1.878	14.4	65.1
4	1.109	8.5	73.6
5	0.910	7.0	80.6
6	0.815	6.3	86.9
7	0.576	4.4	91.3
8	0.440	3.4	94.7
9	0.353	2.7	97.4
10	0.191	1.5	98.9

Table 10.14 Eigenvectors for the First Six Components

	Component					
Variable	1	2	3	4	5	6
1	0.39	0.21	−0.21	−0.07	−0.07	0.11
2	0.42	0.19	−0.23	−0.09	−0.09	0.16
3	0.12	0.52	−0.15	0.07	0.07	−0.27
4	0.17	0.45	0.35	0.04	0.04	−0.06
5	0.05	−0.17	0.48	0.04	0.04	0.63
6	0.29	−0.13	0.48	−0.05	−0.06	0.05
7	0.39	−0.19	0.25	−0.55	−0.55	0.00
8	0.29	−0.19	−0.25	0.23	0.24	0.06
9	0.37	0.17	−0.21	0.07	0.07	−0.03
10	0.37	−0.24	−0.12	−0.18	−0.18	−0.18
11	0.00	0.21	−0.06	0.67	0.67	0.18
12	−0.12	0.34	0.08	−0.25	−0.25	−0.18
13	−0.12	0.30	−0.33	−0.25	−0.25	0.63
eigenvalue	4.219	2.378	1.878	1.109	0.910	0.815

six components, need to be considered in relation to the maximum compressive strength. By inspecting the coefficients in Tables 10.14 and 10.15 we can interpret the meaning of the components. The first component is a measure of the general size of the prop. The second measures the moisture content and specific gravity of the prop (a measure of the degree of seasoning). The third compares the size of the prop with the number of rings and specific gravity. This component can be considered the strength component. Other components have similar interpretations.

The above procedure has reduced the number of characteristics from 13 to at most 6. In addition, the six components extracted can be considered as six independent factors of interest.

Table 10.15 Correlation Between Original Variables and Components

Variable	Component 1	2	3	4	5	6	r^2
1	0.80	0.32	−0.28	−0.08	0.08	0.10	0.8436
2	0.85	0.29	−0.31	−0.09	0.11	0.15	0.9454
3	0.25	0.81	−0.20	0.07	−0.33	−0.25	0.9349
4	0.35	0.69	0.48	0.05	−0.33	−0.06	0.9440
5	0.10	−0.26	0.65	0.05	−0.16	0.56	0.8418
6	0.60	−0.20	0.65	−0.06	0.30	0.04	0.9177
7	0.80	−0.29	0.34	−0.58	0.21	0.00	0.9382
8	0.60	−0.29	−0.34	0.25	−0.18	0.06	0.6582
9	0.75	0.26	−0.28	0.07	−0.10	−0.03	0.7243
10	0.75	−0.37	−0.17	−0.19	−0.14	−0.16	0.8096
11	0.00	0.32	−0.08	0.71	0.33	0.16	0.7474
12	−0.25	0.52	0.11	−0.27	0.57	−0.16	0.7684
13	−0.25	0.46	−0.46	−0.27	−0.08	0.56	0.8786

10.4 Tests for Multivariate Normality

Although many good tests and techniques exist for testing univariate normality, the same cannot be said for assessing multivariate normality. Generalizations of univariate tests of normality have been given by Mardia (1970), Malkovich and Afifi (1973), and Cox and Small (1978), but the power of these tests still needs to be shown. Even the graphical method of Healy (1968) [see also Small (1978)] as used by Andrews et al. (1973) needs further investigation. In this section we propose some multivariate generalizations of univariate tests and a graphical method.

10.4.1 Multivariate Generalizations of Some Univariate Tests

Let $\mathbf{x}_1,\ldots,\mathbf{x}_N$ be a sample from a multivariate normal population. Tests of the original variables for marginal normality do not examine multivariate normality. In particular, they ignore the correlations between these variables. However, the principal components are asymptotically independent, so we can base our test statistics on the components instead. That is, letting $\mathbf{y}_1,\ldots,\mathbf{y}_N$ be the N vectors of principal components, we can base our univariate test statistics on the values $y_{i1},\ldots,y_{iN}$ for each $i=1,\ldots,p$.

There are many reasonable tests for assessing univariate normality, notably that formulated by Shapiro and Wilk (1965) and the less known test of Vasicek (1976). Multivariate generalizations of these can easily be given in terms of the transformed variables y_{ij}. For example, a multivariate generalization of the Shapiro–Wilk statistic is given by

$$M_1=\sum_{i=1}^{p} w_i^{-1}\left[\sum_{j=1}^{N} a_j\left(y_{i(j)}-\bar{y}_i\right)\right]^2,$$

where $w_1 \geq \cdots \geq w_p$ are the eigenvalues of the sample covariance matrix, the constants a_j are tabulated by Shapiro and Wilk (1965), and $y_{i(j)}$ is the jth order statistic in the ith sample, $y_{i(1)} < y_{i(2)} < \cdots < y_{i(N)}$, $i=1,2,\ldots,p$. The observations are ordered separately in each sample. Similarly, a generalization of the Vasicek statistic is given by

$$M_2 = \sum_{i=1}^{p} \frac{N}{2mw_i} \left[\prod_{j=1}^{N} \left(y_{i(j+m)} - y_{i(j-m)} \right) \right]^{1/N},$$

where m is a positive integer less than $N/2$ and $y_{i(j)} = y_{i(1)}$ for $j<1$ and $y_{i(j)} = y_{i(N)}$ for $j>N$. Some choices of m are given by Vasicek (1976).

Percentage points and the power of these tests are currently under investigation.

10.4.2 Graphical Method

Asymptotically, under normality the y_{ij}, $i=1,2,\ldots,p$, $j=1,2,\ldots,N$, are independently normally distributed. We can therefore obtain p normal probability plots. We order the observations in the ith sample, where again $i=1,2,\ldots,p$, as $y_{i(1)} < \cdots < y_{i(N)}$ and plot the resulting set against $\Phi^{-1}[(i-\frac{3}{8})/(N+\frac{1}{4})]$, where $\Phi^{-1}(p)$ is defined by $P\{Z \leq \Phi^{-1}(p)\} = p$ and where $Z \sim N(0,1)$. If the data are normal, the p plots should result in reasonably straight lines.

Example 10.4.1 Let us reconsider Example 3.7.1 using the graphical method to assess multivariate normality for the original and the transformed data. (The multivariate tests developed in Section 10.4.1 cannot be applied because of the lack of tabulated values.) PROC MATRIX and its EIGEN function were used to find the change of basis matrix to transform the raw data into its principle components. Bout length dominates the first principle component, as expected because of its large variance.

The principal components are

$$Y_1 = 0.0648X_1 + 0.998X_2, \qquad Y_2 = -0.998X_1 + 0.0648X_2.$$

Proc univariate returns

	mean	variance	skewness	kurtosis	Shapiro–Wilk W	p-value
Y_1	94.74	9977	1.44	0.79	0.739	<0.01
Y_2	-62.44	1894	-0.824	0.47	0.944	<0.215

Hence we reject marginal normality for Y_1 at the 5% significance level. The graph of Y_1 vs its normal rank is not a straight line, while that for Y_2 is, in agreement with these results.

Performing the transformations of Example 3.7.1 and repeating our procedure yields the components

$$Y_1^* = 0.327X_1^* + 0.945X_2^*, \qquad Y_2^* = -0.945X_1^* + 0.327X_2^*,$$

and

	mean	variance	skewness	kurtosis	Shapiro–Wilk W	p-value
Y_1^*	268	3397	−0.078	−1.05	0.957	0.395
Y_2^*	−11.67	1381	0.468	0.429	0.973	0.689

We do not reject marginal normality for the principal components of the transformed data. Moreover, the plots yield straight lines.

Since the sample principal components are asymptotically independent and normal if the underlying distribution is normal, it is reasonable to conclude that the transformed data are multivariate normal.

10.5 Computational Procedures

Most of the calculations given in this chapter can be obtained from such standard factor analysis programs as BMDP4M. Example 10.2.1 was run on BMDP4M. The program cards, which are common to every system and relate to the BMDP4M program, are

```
/PROBLEM TITLE IS 'PHOTO'.
/INPUT TYPE = COVA.
        SHAPE = LOWER.
         FORMAT IS '(9F3.0)'.
          VARIABLES ARE 9.
/VARIABLE      SCORES ARE 1 TO 9.
/FACTOR         FORM = COVA.
/END
```

The problem paragraph gives the title of the data set. The input paragraph describes the form of the input. There are nine variables, and the input is in the form of a lower triangular portion of a covariance matrix. The data are entered one row at a time: one entry on the first card, two on the second, and so on. The format of the last row is 9F3.0. If the original data are to be used, then the type and shape cards are not necessary. The variable paragraph gives the names of variables, in this case codes from 1 to 9. Finally, the factor paragraph indicates that the analysis should be performed on the covariance matrix. If this card is deleted, the analysis is performed on the correlation matrix by default. Many other options are available, and the reader is referred to the BMDP manual.

Output tables 10.16 and 10.17 were used for the calculations in Example 10.2.1 for two factors only. If three components are required, an additional statement is necessary. The number of factors chosen is the number of components whose variance is greater than the average variance (160 in this case). If the third factor is to be included, then we add in the Factor paragraph NUMBER = 3, CONST = 0.

Table 10.16 Unrotated Factor Loadings (Pattern) for Principal Components

Variable	Loadings	
	Factor 1	Factor 2
$X(1)$	0.903D 01	−0.680D 01
$X(2)$	0.194D 02	−0.211D 01
$X(3)$	0.143D 02	0.823D 01
$X(4)$	0.775D 01	−0.688D 01
$X(5)$	0.959D 01	−0.527D 00
$X(6)$	0.804D 01	0.523D 01
$X(7)$	0.714D−01	0.800D 00
$X(8)$	0.175D 00	0.753D 00
$X(9)$	0.427D 00	0.124D 01
eigenvalue	878.519	196.096

Table 10.17

Factor	Variance explained[a]	Cumulative proportion of total variance[b]
1	878.519228	0.609237
2	196.095890	0.745225
3	128.643106	0.834437
4	103.430205	0.906164
5	81.261078	0.962517
6	37.848834	0.988764
7	6.976383	0.993602
8	5.706471	0.997560
9	3.518806	1.000000

[a] The eigenvalue for each factor.
[b] Defined as the sum of the diagonal elements of the correlation (covariance) matrix.

The entries in Table 10.16 are for $\lambda_i^{1/2}\mathbf{a}_i$, $i=1,2$, where λ_1 and λ_2 are the eigenvalues for the eigenvectors $\mathbf{a}_1$ and $\mathbf{a}_2$. To obtain the normalized eigenvectors, one divides the columns by the square root of the eigenvalues. To obtain the correlations between the original variables $X(1),\ldots,X(9)$ and the components (Factor 1, Factor 2), one divides the rows of Table 10.17 by the standard deviations of the original variables. The other eigenvalues and the cumulative proportion of the total variance are given in Table 10.17.

If the principal component analysis is performed on the correlation matrix then the /FACTOR card can be omitted in the program cards. The resulting output is shown in Tables 10.18–22.

Table 10.18 gives the correlation matrix for the nine original variables and Table 10.19 gives the multiple correlation between each variable and all the other variables. The higher the values of the multiple R^2, the greater is the degree of multicollinearity of the original variables. Table 10.21 gives the eigenvalues of the correlation matrix and the cumulative proportion of the total variance. The number of factors chosen is usually the number of

Table 10.18 Correlation Matrix

	$X_1^{(1)}$	$X_2^{(2)}$	$X_3^{(3)}$	$X_4^{(4)}$	$X_5^{(5)}$	$X_6^{(6)}$	$X_7^{(7)}$	$X_8^{(8)}$	$X_9^{(9)}$
$X(1)$	1.000								
$X(2)$	0.657	1.000							
$X(3)$	0.411	0.689	1.000						
$X(4)$	0.574	0.509	0.275	1.000					
$X(5)$	0.340	0.755	0.536	0.693	1.000				
$X(6)$	0.213	0.548	0.722	0.295	0.725	1.000			
$X(7)$	−0.090	−0.017	0.039	0.055	0.059	0.030	1.000		
$X(8)$	−0.049	0.008	0.037	0.038	0.068	0.028	0.851	1.000	
$X(9)$	−0.033	0.028	0.091	0.023	0.074	0.102	0.817	0.901	1.000

Table 10.19 Squared Multiple Correlations (SMC) of Each Variable with All Other Variables[a]

Variable	SMC
$X(1)$	0.77408
$X(2)$	0.89054
$X(3)$	0.73293
$X(4)$	0.81457
$X(5)$	0.93041
$X(6)$	0.82152
$X(7)$	0.74879
$X(8)$	0.85929
$X(9)$	0.83182

[a]Condition number, 0.249664D 03.

Table 10.20 Communalities Obtained from Three Factors After One Iteration

Variable	Communality[a]
$X(1)$	0.7551
$X(2)$	0.8086
$X(3)$	0.7662
$X(4)$	0.7707
$X(5)$	0.7763
$X(6)$	0.8792
$X(7)$	0.8754
$X(8)$	0.9321
$X(9)$	0.9071

[a]The squared multiple correlation (covariance) of a variable with the factors.

Table 10.21

Factor	Variance explained[a]	Cumulative proportion of total variance[b]
1	3.703307	0.411479
2	2.713413	0.712969
3	1.053881	0.830067
4	0.704347	0.908328
5	0.300933	0.941764
6	0.233031	0.967657
7	0.168980	0.986432
8	0.091034	0.996547
9	0.031075	1.000000

[a]The eigenvalue for each factor.
[b]Defined as the sum of the diagonal elements of the correlation (covariance) matrix.

Table 10.22 Unrotated Factor Loadings (Pattern) for Principal Components

Variable	Loadings		
	Factor 1	Factor 2	Factor 3
$X(1)$	0.644	−0.194	0.550
$X(2)$	0.887	−0.135	0.057
$X(3)$	0.781	−0.048	−0.393
$X(4)$	0.697	−0.073	0.529
$X(5)$	0.877	−0.045	−0.074
$X(6)$	0.760	−0.031	−0.549
$X(7)$	0.119	0.927	0.049
$X(8)$	0.135	0.954	0.068
$X(9)$	0.168	0.938	0.003
eigenvalue	3.703	2.713	1.054

Table 10.23 Normalized Eigenvectors

Variable	Component		
	I	II	III
1	0.335	−0.118	0.536
2	0.461	−0.082	0.056
3	0.406	−0.029	−0.383
4	0.362	−0.042	0.515
5	0.456	−0.027	−0.072
6	0.395	−0.019	−0.535
7	0.062	0.563	0.044
8	0.070	0.579	0.062
9	0.087	0.569	0.003
eigenvalue	3.70	2.71	1.05

eigenvalues larger than 1. For this case, three factors were chosen, accounting for 83% of the total variation. The squared multiple correlation R^2 between the old variables and the three new components or factors is presented in Table 10.20. The lowest R^2 value is 0.75. Hence most of the variation in the original variables is accounted for by the three principal components. The correlations between the original variables and the first three components are listed in Table 10.22, and the normalized eigenvectors can be obtained by dividing the columns of Table 10.22 by the square root of the corresponding eigenvalue. The resulting normalized eigenvectors are given in Table 10.23.

The first component is a measure of the average density of the photo in the shoulder and middle-tone portions. However, unlike the results of Example 10.2.1, the second factor is now the average density of the toe portion. Finally, the third component measures the tint, the difference in hard and soft colors. The major difference in the analyses of the covariance matrix and the correlation matrix is the introduction of this new component, based on the toe region, in the analysis of the correlation matrix; the toe region was virtually ignored in the analysis based on the covariance matrix. With three factors, the unexplained or residual variance in the correlation matrix case is much smaller in the toe region than for the covariance.

Problems

10.1 The calculated correlation matrix for Problem 4.3 on soil composition is shown in Table 10.24. Perform a principal component analysis on the above correlation matrix and interpret the results. Note that the standard deviations for these variables are $s_1 = 0.6719$, $s_2 = 0.0672$, $s_3 = 0.2198$, $s_4 = 81.161$, $s_5 = 3.256$, $s_6 = 1.3682$, $s_7 = 0.2239$, $s_8 = 3.2889$, and $s_9 = 3.987$.

Table 10.24

	pH	N	BD	P	Ca	Mg	K	Na	Conductivity
pH	1								
N	−0.20	1							
BD	0.04	−0.56	1						
P	−0.31	0.67	−0.37	1					
Ca	0.40	0.38	−0.44	0.15	1				
Mg	−0.12	0.08	−0.18	−0.13	0.29	1			
K	0.18	0.12	−0.16	0.17	0.14	0.25	1		
Na	0.44	−0.17	0.12	−0.28	0.36	0.47	0.12	1	
Conductivity	0.12	−0.07	0.27	−0.15	0.08	0.25	−0.26	0.70	1

Table 10.25[a]

	1	2	3	4	5	6	7	8	9	10	11	12	13	14	15	16	17	18	19	20	21
1	1.00	0.78	0.70	0.91	0.91	0.95	0.63	0.92	0.75	0.92	0.90	0.78	0.86	0.90	0.55	0.63	0.77	0.67	0.83	0.59	0.41
2		1.00	0.51	0.84	0.77	0.84	0.66	0.75	0.59	0.72	0.79	0.65	0.61	0.64	0.43	0.45	0.59	0.55	0.67	0.57	0.41
3			1.00	0.62	0.64	0.64	0.46	0.77	0.46	0.64	0.64	0.52	0.56	0.59	0.25	0.51	0.46	0.58	0.70	0.48	0.35
4				1.00	0.93	0.94	0.66	0.89	0.66	0.91	0.90	0.68	0.72	0.78	0.51	0.60	0.68	0.62	0.80	0.60	0.39
5					1.00	0.94	0.64	0.88	0.74	0.91	0.89	0.66	0.73	0.80	0.52	0.53	0.68	0.65	0.80	0.55	0.41
6						1.00	0.63	0.96	0.77	0.90	0.90	0.75	0.77	0.84	0.64	0.56	0.80	0.69	0.80	0.58	0.40
7							1.00	0.48	0.61	0.55	0.56	0.41	0.50	0.57	0.49	0.60	0.60	0.37	0.47	0.46	0.46
8								1.00	0.94	0.95	0.89	0.69	0.71	0.78	0.33	0.50	0.53	0.60	0.81	0.59	0.37
9									1.00	0.70	0.84	0.51	0.53	0.58	0.53	0.38	0.57	0.72	0.69	0.55	0.56
10										1.00	0.93	0.71	0.82	0.83	0.61	0.50	0.64	0.71	0.80	0.62	0.36
11											1.00	0.52	0.79	0.83	0.73	0.69	0.73	0.65	0.75	0.71	0.37
12												1.00	0.68	0.73	0.41	0.50	0.57	0.57	0.68	0.41	0.33
13													1.00	0.89	0.40	0.55	0.66	0.67	0.79	0.54	0.41
14														1.00	0.44	0.67	0.70	0.65	0.82	0.53	0.44
15															1.00	0.39	0.51	0.42	0.68	0.41	0.33
16																1.00	0.63	0.45	0.59	0.39	0.38
17																	1.00	0.52	0.60	0.55	0.55
18																		1.00	0.75	0.43	0.33
19																			1.00	0.50	0.37
20																				1.00	0.27
21																					1.00

[a]Reproduced with the permission of the editors of the *Journal of Zoology* (*London*).

10.2 Table 10.25 (Machin, 1974) gives the correlations among 21 external measurements of male sperm whales, where the variables are

1. total length, tip of snout to notch of flukes,
2. projection of snout beyond tip of lower jaw,
3. tip of snout to blow-hole.
4. tip of snout to angle of gape,
5. tip of snout to center of eye,
6. tip of snout to tip of flipper,
7. center of eye to center of ear,
8. length of severed head from condyle to tip,
9. greatest width of skull,
10. skull length, condyle to tip of premaxilla,
11. height of skull,
12. notch of flukes to posterior emargination of dorsal fin,
13. notch of flukes to center of anus,
14. notch of flukes to umbilicus,
15. axilla to tip of flipper,
16. anterior end of lower border to tip of flipper,
17. greatest width of flipper,
18. width of flukes at insertion,
19. tail flukes, tip to notch,
20. vertical height of dorsal fin, and
21. length of base of dorsal fin.

Perform a principal component analysis on the above correlation matrix. Interpret the main components. [A similar study was performed on humpback whales by Machin and Kitchenham (1971).]

10.3 Huntingford (1976) gives the correlation matrix (Table 10.26) between behavior scores for 27 three-spined sticklebacks, with variables

1. number of lunges,
2. number of bites
3. number of zig-zags,
4. number of bouts of nest activity,

Table 10.26

	1	2	3	4	5	6	7	8	9	10	11
1	1	0.702	−0.108	−0.140	0.035	−0.139	−0.259	0.239	−0.216	0.153	−0.230
2		1	−0.018	−0.113	0.383	−0.097	−0.215	0.119	−0.083	0.009	−0.141
3			1	0.390	0.031	0.376	−0.104	−0.162	0.111	0.095	0.070
4				1	−0.017	0.981	0.144	−0.299	0.050	−0.006	−0.065
5					1	−0.039	0.001	−0.040	−0.076	0.139	−0.093
6						1	0.141	−0.239	−0.054	0.054	−0.045
7							1	−0.165	0.130	−0.522	0.162
8								1	−0.138	0.244	−0.135
9									1	−0.600	0.985
10										1	−0.595
11											1

5. number of spine-raising incidents,
6. total duration of nest activity,
7. total duration of spine raising,
8,9. mean bout length of facing from less than 10 and more than 20 cm, respectively, and
10,11. proportion of time spent facing from less than 10 and more than 20 cm, respectively.

The original data are given in the Table 10.27. Note that variables 4, 6, 9, and 11 have very few nonzero values.

(a) Compute the eigenvalues and eigenvectors of the correlation matrix.
(b) Give the coefficients of the original variables for the main components.
(c) Give the correlations between the original variables and the new components.
(d) Give the multiple R^2 between the original variables and the main components.
(e) Ignoring variables 4, 6, 9, and 11, test each variable for normality. For those variables that do not appear normal, make transformations along the lines of Section 3.7. Perform a principal component analysis on the transformed data.

Table 10.27

Fish	1	2	3	4	5	6	7	8	9	10	11
1	108	59	1	0	28	0	46	40.9	0	1.00	0
2	211	80	14	1	28	0	2	42.0	0	1.00	0
3	81	17	0	0	16	0	1	42.7	0	1.00	0
4	64	11	10	0	20	0	36	306.0	0	1.00	0
5	140	134	0	0	17	0	62	146.5	0	0.95	0
6	121	76	17	0	25	0	138	101.0	0	1.00	0
7	87	50	6	1	28	1	8	21.0	0	1.00	0
8	120	85	7	3	31	9	122	22.7	0	0.93	0
9	143	139	16	0	22	0	8	72.0	2	0.96	0.006
10	173	175	3	0	46	0	83	73.0	0	0.99	0
11	137	55	2	0	7	0	199	7.7	0	0.85	0
12	118	30	2	1	11	0	1	28.1	0	1.00	0
13	177	115	4	0	14	0	42	305.0	0	1.00	0
14	277	152	3	1	11	0	19	48.8	0	1.00	0
15	165	102	3	0	16	0	0	77.0	0	1.00	0
16	121	67	26	1	5	5	0	23.0	0	1.00	0
17	157	74	5	0	18	0	0	302.0	0	1.00	0
18	83	30	3	0	15	0	177	48.8	0	0.98	0
19	41	3	0	0	9	0	134	30.1	0	1.00	0
20	147	25	5	0	14	0	84	16.2	0	1.00	0
21	67	35	21	5	24	14	108	8.7	0	0.99	0
22	147	70	7	0	11	0	0	152.5	0	1.00	0
23	125	67	11	0	23	0	0	36.4	0	1.00	0
24	132	55	6	0	14	0	4	151.5	0	1.00	0
25	71	31	9	2	14	1	114	25.0	8	0.87	0.074
26	204	57	4	0	17	0	125	303.0	0	1.00	0
27	117	54	15	15	14	50	85	11.9	0	1.00	0

Table 10.28

	1	2	3	4	5	6	7	8	9
Mean	3.28	2.40	−0.45	2.20	1.74	1.59	1.64	2.09	1.13
s.d.	0.27	0.99	1.26	0.91	1.19	1.11	1.02	1.39	0.97

10.4 For the data of Table 10.28 (Dudzinski and Arnold, 1973) the nine original variables were transformed by principal component analysis into nine new components:

1. total dry matter,
2. green dry matter,
3. percent edible green,
4. green leaf,
5. dry leaf,
6. green clover,
7. dry clover,
8. stem,
9. inert.

The first four components were then used as independent variables in a multiple regression analysis. The *correlation matrix* was

$$\begin{bmatrix} 1.00 & 0.11 & 0.04 & 0.11 & 0.42 & 0.11 & 0.22 & 0.34 & -0.51 \\ & 1.00 & 0.86 & 0.98 & -0.11 & 0.76 & -0.36 & -0.48 & 0.13 \\ & & 1.00 & 0.84 & -0.33 & 0.80 & -0.57 & -0.71 & -0.11 \\ & & & 1.00 & -0.13 & 0.64 & -0.39 & -0.48 & 0.12 \\ & & & & 1.00 & -0.17 & 0.21 & 0.39 & -0.06 \\ & & & & & 1.00 & -0.24 & -0.43 & 0.06 \\ & & & & & & 1.00 & 0.72 & 0.30 \\ & & & & & & & 1.00 & 0.19 \\ & & & & & & & & 1.00 \end{bmatrix}$$

(a) Give the eigenvalues, eigenvectors, and coefficients for the first four main components.

Table 10.29[a]

	1	2	3	4	5	6	7	8	9
1	1								
2	0.934	1							
3	0.927	0.941	1						
4	0.909	0.944	0.933	1					
5	0.524	0.487	0.543	0.499	1				
6	0.799	0.821	0.856	0.833	0.703	1			
7	0.854	0.865	0.886	0.889	0.719	0.923	1		
8	0.789	0.834	0.846	0.885	0.253	0.699	0.751	1	
9	0.835	0.863	0.862	0.850	0.462	0.752	0.793	0.745	1
10	0.845	0.878	0.863	0.881	0.567	0.836	0.913	0.787	0.805
11	−0.458	−0.496	−0.522	−0.488	−0.174	−0.317	−0.383	−0.497	−0.356
12	0.917	0.942	0.940	0.945	0.516	0.846	0.907	0.861	0.848
13	0.939	0.961	0.956	0.952	0.494	0.849	0.914	0.876	0.877
14	0.953	0.954	0.946	0.949	0.452	0.823	0.886	0.878	0.883
15	0.895	0.899	0.882	0.908	0.551	0.831	0.891	0.794	0.818
16	0.691	0.652	0.694	0.623	0.815	0.812	0.855	0.410	0.620
17	0.327	0.305	0.356	0.272	0.746	0.553	0.567	0.067	0.300
18	−0.676	−0.712	−0.667	−0.736	−0.233	−0.504	−0.502	−0.758	−0.666
19	0.702	0.729	0.746	0.777	0.285	0.499	0.592	0.793	0.671

[a]Reproduced with the permission of the editors of the *Australian Journal of Agricultural Research*.

(b) Give the correlations between these components and the original variables and interpret the components.

10.5 Jeffers (1967) studied 40 observations on adelges (winged aphids); 19 characteristics were measured for each aphid. Measurements were taken at different magnifications; hence, the error variances are not equal. Table 10.9 shows the correlation matrix for the following variables:

1. LENGTH (body length),
2. WIDTH (body width),
3. FORWING (fore-wing length),
4. HINWING (hind-wing length),
5. SPIRAC (number of spiracles),
6. ANTSEG1 (length of antennal segment I),
7. ANTSEG2 (length of antennal segment II),
8. ANTSEG3 (length of antennal segment III),
9. ANTSEG4 (length of antennal segment IV),
10. ANTSEG5 (length of antennal segment V),
11. ANTSPIN (number of antennal spines),
12. TARSUS3 (leg length, tarsus III),
13. TIBIA3 (leg length, tibia III),
14. FEMUR3 (leg length, femur III),
15. ROSTRUM (rostrum),
16. OVIPOS (ovipositor),
17. OVSPIN (number of ovipositor spines),
18. FOLD (anal fold), and
19. HOOKS (number of hind-wing hooks).

Perform a principal component analysis on this data.

Table 10.29 (*continued*)

10	11	12	13	14	15	16	17	18	19
1									
−0.371	1								
0.902	−0.465	1							
0.901	−0.447	0.981	1						
0.891	−0.439	0.971	0.991	1					
0.848	−0.405	0.908	0.920	0.921	1				
0.712	−0.198	0.725	0.714	0.676	0.720	1			
0.384	−0.032	0.396	0.360	0.298	0.378	0.781	1		
−0.629	0.492	−0.657	−0.655	−0.687	−0.633	−0.186	0.169	1	
0.668	−0.425	0.696	0.724	0.731	0.694	0.287	−0.026	−0.775	1

Chapter 11
Factor Analysis

11.1 Introduction

In Chapter 10 we discussed the problem of data reduction by principal component analysis. Another method of data reduction that assumes a specific underlying model is called *factor analysis*. For example, suppose we wish to judge the abilities of high school students entering university. We may give them a test of 50 questions. These 50 questions, however, may fall into a few categories, such as reading comprehension, mathematics, and the arts. These categories are called *factors*. If there are only three factors, the score on the ith question may be modeled in the form

$$y_i = \mu_i + \lambda_{i1} f_1 + \lambda_{i2} f_2 + \lambda_{i3} f_3 + \varepsilon_i, \qquad i = 1, \ldots, 50,$$

where $\varepsilon_i \sim N(0, \psi_i)$.

Observations are taken on several students, say N. These observations may be used to estimate the parameters λ_{ij}, $i = 1, \ldots, 50$, $j = 1, \ldots, 3$, and to test whether the model is indeed correct. If the model is not adequate, more factors can be included. In general, let $f_1, \ldots, f_k$ be the k factors and let $y_1, \ldots, y_p$ ($p \geq k$) be the p scores on the test. Then we write

$$y_i = \mu_i + \sum_{j=1}^{k} \lambda_{ij} f_j + \varepsilon_i, \tag{11.1.1}$$

where $\varepsilon_i \sim N(0, \psi_i)$ and $\varepsilon_1, \ldots, \varepsilon_p$ are independent. Note that (11.1.1) appears to be a multiple linear regression problem except that the values of $f_1, \ldots, f_k$ are unknown. The parameter λ_{ij}, relating the ith variable with the jth factor, is called a *factor loading*. In this chapter we discuss methods of estimating the factor loadings λ_{ij} and the variances $\psi_1, \ldots, \psi_p$. By inspecting the values of the factor loadings we may relate the questions to the underlying factors. If an underlying factor analytic model does exist, we should be able to give a meaningful interpretation to these underlying factors. Note that the statistical analysis of data does not ensure meaningful factors.

We now give examples of the use of factor analysis. Its use to estimate the parameters of regression models of unknown form is presented by Rao (1958), Tucker (1958), and Sheth (1969).

Example 11.1.1 Subjects were asked 70 questions regarding attitudes toward marriage (Hoegervorst, 1978). The response to each question was assigned a score of from 1 (traditional) to 5 (equalitarian). The questionnaire covered five distinct areas: desire for children, care of children, financial support and employment, homemaking skills, and authority patterns. Factor analysis of the data separates the questions on the questionnaire into five groups, corresponding to the five underlying common factors. The *factor scores* were then used in a multiple regression analysis to find which demographic factors, such as age or socioeconomic status affect a person's attitudes toward marriage.

Example 11.1.2 Thirty-seven panelists compared a red wine against a reference wine on a scale of from -4 to $+4$, with regard to 33 terms, such as tartness, bitterness, and sweetness (Wu et al., 1977). The data were then summarized in terms of eight common factors, such as pungency, overall quality, and smoothness.

11.2 Estimation of Parameters

For conciseness we now write the model (11.1.1) in matrix form. Let $\mathbf{y}'=(y_1,\dots,y_p)$, $\mathbf{f}'=(f_1,\dots,f_k)$, $\boldsymbol{\varepsilon}'=(\varepsilon_1,\dots,\varepsilon_p)$, $\boldsymbol{\psi}'=(\psi_1,\dots,\psi_p)$, and $\Lambda=(\lambda_{ij})$. Then (11.1.1) becomes

$$\mathbf{y}=\boldsymbol{\mu}+\Lambda\mathbf{f}+\boldsymbol{\varepsilon}, \tag{11.2.1}$$

where $\boldsymbol{\mu}=(\mu_1,\dots,\mu_p)'$ and $\boldsymbol{\varepsilon}\sim N_p(0,\psi)$, with $\psi=\mathrm{diag}(\psi_1,\dots,\psi_p)$. By taking the mean and variance of $\mathbf{y}$, we obtain

$$E(\mathbf{y})=\boldsymbol{\mu}, \tag{11.2.2}$$

$$\mathrm{cov}(\mathbf{y})=\Lambda\Lambda'+\psi\equiv\Sigma. \tag{11.2.3}$$

The sufficient estimators of $\boldsymbol{\mu}$ and Σ are the sample mean $\bar{y}$ and the sample covariance matrix S based on N observations. Note that the value of Σ is unchanged if Λ is postmultiplied by any $k\times k$ orthogonal matrix. Hence there is no unique choice of Λ. In Section 11.2 we give methods for making an initial choice of Λ and ψ. In Section 11.3 we use factor rotation to arrive at a final selection of Λ. For further reading, the reader is referred to Guttman (1954a, 1956), Harman (1967), and Lawley and Maxwell (1970).

11.2.1 Maximum Likelihood Estimates

The original derivations of the maximum likelihood estimates of Λ and ψ are given by Lawley (1940) and Lawley and Maxwell (1970). In the appendix to this chapter we derive the estimators as solutions of the equations

$$\operatorname{diag} S = \operatorname{diag}(\Lambda\Lambda' + \psi) \qquad \text{and} \qquad \Lambda = S(\Lambda\Lambda' + \psi)^{-1}\Lambda. \quad (11.2.4)$$

Solutions may be found using such packages as BMDP4M and are not repeated here. The reader is referred to Jöreskog (1967), Jöreskog and Lawley (1968), and Jöreskog and Von Thillo (1971). Detailed examples are given in Section 11.4. Note that the method of maximum likelihood estimation requires $N > p$. If $N \leq p$, then one can estimate the factor loadings by such methods as principal factor analysis (to be discussed in Section 11.2.2). It should be noted that adding the same constant to each variable has no effect on the covariance matrix. Similarly, if any variable is multiplied by a constant, then the factor loadings of the variable are multiplied by that constant. For simplicity we shall assume that all variables are standardized to have mean 0 and variance 1. We can now write (11.2.1)

$$\mathbf{y} = \Lambda \mathbf{f} + \boldsymbol{\varepsilon}, \qquad \varepsilon_i \sim N(0, \psi_i), \quad i = 1, \ldots, p, \qquad (11.2.5)$$

where $E(\mathbf{y}) = \mathbf{0}$, and $\operatorname{cov} \mathbf{y} = \Lambda\Lambda' + \psi = \Sigma = (\sigma_{ij})$, and we assume that $\sigma_{ii} = 1$, $i = 1, \ldots, p$. Hence (11.2.4) may be written

$$\sum_{j=1}^{k} \lambda_{ij}^2 + \psi_i = 1 \qquad \text{and} \qquad \Lambda = S(\Lambda\Lambda' + \psi)^{-1}\Lambda,$$

where S is the sample covariance matrix of the standardized variables (or the correlation matrix of the original variables). To avoid confusion we shall assume all variables in the rest of this chapter are standardized.

11.2.2 Principal Factor Analysis

The factor analytic model assumes that $\Sigma = \Lambda\Lambda' + \psi$. If we have standardized our variables, $\operatorname{diag}(\Sigma) = I$. Hence we have

$$\sum_{j=1}^{k} \lambda_{ij}^2 + \psi_i = 1,$$

or equivalently,

$$\psi_i = 1 - \sum_{j=1}^{k} \lambda_{ij}^2, \qquad i = 1, \ldots, p. \qquad (11.2.6)$$

If these residual variances are known, then we may obtain an estimate of Λ by performing a principal component analysis on $S - \psi$. The columns of Λ correspond to the first k eigenvectors of $S - \psi$. We can then estimate ψ_i, $i = 1, \ldots, p$, from (11.2.6). The process is then iterated until a final solution is reached. Note that one choice of initial solution is to set $\psi_i = 0$, $i = 1, \ldots, p$.

Other forms of factor analysis also exist. For details on image factor analysis, the reader is referred to Jöreskog (1969), Kaiser (1970), and Kaiser and Rice (1974).

Table 11.1 Correlations Between Measurements on Rats[a]

Variable	Measurement	1	2	3	4	5	6	7	8	9	10	11	12	13	14	15	16	17	Mean	s.d.
1	head length	1	0.55	0.33	0.51	0.51	0.51	0.60	0.47	0.34	0.30	0.77	0.71	0.05	0.28	0.50	0.76	0.60	50.7	1.5
2	head width		1	0.53	0.49	0.55	0.76	0.63	0.21	0.30	0.39	0.65	0.63	0.36	0.34	0.57	0.60	0.88	24.9	0.7
3	shoulder width			1	0.51	0.10	0.62	0.32	0.33	0.24	0.65	0.51	0.40	0.40	0.40	0.14	0.33	0.55	29.1	1.1
4	head and neck length				1	0.11	0.57	0.51	0.37	0.26	0.52	0.61	0.39	0.34	0.27	0.24	0.39	0.52	68.4	2.6
5	trunk length					1	0.53	0.65	0.22	0.39	0.04	0.59	0.63	0.01	0.34	0.60	0.63	0.68	142.6	6.7
6	hip width						1	0.63	0.25	0.30	0.36	0.69	0.59	0.37	0.31	0.51	0.57	0.87	35.1	1.2
7	tail length							1	0.39	0.39	0.32	0.78	0.67	0.17	0.43	0.64	0.74	0.73	198.6	7.9
8	ear diameter								1	0.14	0.33	0.58	0.40	0.09	0.18	0.20	0.39	0.25	18.9	0.6
9	elbow width									1	0.19	0.48	0.53	0.03	0.44	0.27	0.37	0.40	7.4	0.3
10	knee width										1	0.48	0.23	0.20	0.50	0.01	0.70	0.33	9.6	0.4
11	lower foreleg length											1	0.73	0.22	0.48	0.50	0.72	0.77	34.1	0.8
12	lower rear leg length												1	0.23	0.28	0.51	0.72	0.73	51.0	1.5
13	wrist width													1	0.03	0.04	0.15	0.29	5.2	0.1
14	ankle width														1	0.44	0.46	0.36	6.4	0.3
15	forefoot length															1	0.69	0.63	19.1	0.8
16	rear foot length																1	0.68	39.0	0.9
17	weight																	1	271.1	28.4

[a]Data from Watson and Broadhurst (1976).

11.2.3 Choosing the Number of Factors

The model (11.1.2) may be tested by means of the likelihood ratio test. That is, we test the hypothesis

$$H:\ \Sigma=\Lambda\Lambda'+\psi \qquad \text{vs} \qquad A:\ \Sigma\neq\Lambda\Lambda'+\psi.$$

The likelihood ratio tests rejects H if $\lambda=|S|^{N/2}/|\hat{\Sigma}|^{N/2}$ is small, where, if $\hat{\Lambda}$ and $\hat{\psi}$ are the solutions of (11.2.1), $\hat{\Sigma}=\hat{\Lambda}\hat{\Lambda}'+\hat{\psi}$. Asymptotic theory gives the distribution of $-2\ln\lambda$ as χ^2_f where $f=\frac{1}{2}[(p-k)^2-(p+k)]$. Bartlett (1951) suggested the approximation

$$-[(n-(2p+5)/6-2k/3]\ln(|S|/|\hat{\Sigma}|)=\chi^2_f, \qquad n=N-1. \tag{11.2.7}$$

Hence the hypothesis H is rejected if $\chi^2_f\geq\chi^2_{f,\alpha}$ where $\chi^2_{f,\alpha}$ is the χ^2_f critical value given in Table A.3. If the hypothesis is rejected, then the factor analytic model we have chosen is inadequate. The number of factors k is then increased by 1 and the model tested again, and the process is repeated until an insignificant χ^2_f value is obtained.

An alternative rule of thumb often used in statistical packages chooses k to be the number of eigenvalues of S greater than 1. This choice of k can be used as an initial guess for the number of factors. The model can then be tested using (11.2.6).

Table 11.2 Initial Factor Loadings

	Factor				
Variable	1	2	3	4	Communality
1	0.565	0.280	−0.083	0.583	0.7440
2	0.812	0.340	0.077	−0.062	0.7855
3	0.440	0.400	0.605	0.016	0.7197
4	0.470	0.270	0.421	0.292	0.5558
5	0.603	0.340	−0.464	0.103	0.7048
6	0.816	0.310	0.173	−0.029	0.7926
7	0.634	0.430	−0.153	0.313	0.7086
8	0.229	0.180	0.168	0.621	0.4992
9	0.266	0.440	−0.099	0.153	0.2972
10	0.172	0.500	0.595	0.181	0.6656
11	0.663	0.480	0.077	0.456	0.8838
12	0.693	0.280	−0.130	0.353	0.7002
13	0.304	0.030	0.352	−0.043	0.2191
14	0.000	1.000	−0.000	−0.000	1.0000
15	0.510	0.440	−0.424	0.051	0.6357
16	0.572	0.460	−0.246	0.371	0.7370
17	0.929	0.360	−0.002	−0.043	0.9942
variance	5.440	3.144	1.603	1.456	

Example 11.2.1 The data of Table 11.1 are the correlations between measurements on 50 male rats. The eigenvalues of the correlation matrix are 8.41, 1.98, 1.30, 1.10, 0.80, 0.64, 0.58, 0.46, 0.41, 0.28, 0.26, 0.18, 0.18, 0.16, 0.13, 0.09, and 0.05. As there are only four eigenvalues greater than 1, we choose initially four factors for our model. Using the maximum likelihood procedure we obtain the factor loadings in Table 11.2.

To test for the adequacy of the model, we calculate the determinants of S by multiplying the eigenvalues of S; and we evaluate $|\hat{\Sigma}|$ by taking the determinant of $\hat{\Lambda}\hat{\Lambda}'+\hat{\psi}$. The value of $\hat{\Lambda}$ is given in Table 11.2, and $\hat{\psi}_i$ can be obtained by subtracting the ith communality from 1. We find that $\hat{\psi}_1=0.256$, $\hat{\psi}_2=0.0145,\ldots,\hat{\psi}_{17}=0.0058$. We obtain $|S|=7.18\times10^{-7}$ and $|\hat{\Sigma}|=1.47\times10^{-6}$. Hence, with $k=4$, $p=17$, and $n=49$ we have

$$-[n-(2p+5)/6-2k/3]\ln|S|/|\hat{\Sigma}|$$

$$=-(49-6.5-2.67)\ln 0.488=28.7=\chi^2_{94}.$$

Because $\chi^2_{94,0.95}\approx117$ we do not reject the null hypothesis of H: $\Sigma=\Lambda\Lambda'+\psi$. Hence the model fits very well. We obtain similar results using the principal factor analysis estimates instead of the maximum likelihood estimates.

11.3 Factor Rotation

11.3.1 Introduction

Once the initial choice of the factor loadings Λ is made, the next step is to "rotate" the factors to obtain ones that are easily interpreted. The model we considered in (11.1.2) assumed the factors $f_1,\ldots,f_k$ to be independently distributed as $N(0,1)$. That is,

$$\mathbf{f}\sim N_p(\mathbf{0},I). \tag{11.3.1}$$

Hence if we premultiply out factors by a $p\times p$ orthogonal matrix Γ, we have

$$\mathbf{f}^*=\Gamma\mathbf{f}\sim N_p(\mathbf{0},I). \tag{11.3.2}$$

Hence the model in (11.1.2) can be written

$$\mathbf{y}=\Lambda\mathbf{f}+\boldsymbol{\varepsilon}=\Lambda\Gamma'\Gamma\mathbf{f}+\boldsymbol{\varepsilon}=\Lambda\Gamma'\mathbf{f}^*+\boldsymbol{\varepsilon}. \tag{11.3.3}$$

That is, postmultiplying the factor loadings by an orthogonal matrix is equivalent to rotating our factors. If we wish to relax our condition of independent factors, we may premultiply our factors $\mathbf{f}$ by an arbitrary matrix A, obtaining

$$\mathbf{f}^{**}=A\mathbf{f}.$$

Hence (11.3.2) becomes

$$\mathbf{y}=\Lambda\mathbf{f}+\boldsymbol{\varepsilon}=\Lambda A^{-1}A\mathbf{f}+\boldsymbol{\varepsilon}=\Lambda A^{-1}\mathbf{f}^{**}+\boldsymbol{\varepsilon}, \tag{11.3.4}$$

where $\mathbf{f}^{**}\sim N_p(\mathbf{0},AA')$.

Such a transformation of the factors is called *oblique*. The advantage of an oblique transformation is that it tends to give factors that are more easily interpreted. However, the resulting factors are not independent.

11.3.2 Varimax

The *varimax procedure* (see Maxwell and Lawley, 1971), the most common of the orthogonal rotation methods, chooses a rotation that maximizes the variation in the squares of the column entries of the estimated factor loading matrix. That is, let

$$d_j = \sum_{i=1}^{p} l_{ij}^2, \qquad j=1,\ldots,k. \tag{11.3.5}$$

Then the following expression is maximized:

$$\sum_{j=1}^{k} \sum_{i=1}^{p} \left(l_{ij}^2 - p^{-1} d_j\right)^2. \tag{11.3.6}$$

Such a procedure tries to give either large (in absolute value) or zero values in the columns of L. Hence the procedure tries to produce factors with either a strong association with the responses or no association at all.

11.3.3 Quartimax

Several methods exist for choosing the final factor loadings. If the variables are to be related to as few factors as possible, then the *quartimax method* is used (Neuhaus and Wrigley, 1954). This method minimizes the cross products of the factor loadings. That is, if $L=(l_{ij})$ are the estimated factor loadings, then the following expression is minimized:

$$\sum_{1\le s<t\le k} \sum_{i=1}^{p} (l_{is} l_{it})^2. \tag{11.3.7}$$

Now, since any orthogonal rotation on L leaves LL' unchanged, the diagonal terms of LL' are invariant under orthogonal transformations on L. Hence

$$\left(\sum_{j=1}^{k} \sum_{i=1}^{p} l_{ij}^2 \right)^2 = \text{const.} \tag{11.3.8}$$

However,

$$\left(\sum_{j=1}^{k} \sum_{i=1}^{p} l_{ij}^2 \right)^2 = \sum_{j=1}^{k} \sum_{i=1}^{p} l_{ij}^4 + 2 \sum_{1\le s<t\le k} \sum_{i=1}^{p} (l_{is} l_{it})^2. \tag{11.3.9}$$

Hence minimizing (11.3.5) is equivalent to maximizing the sum of the fourth powers of l_{ij}—or maximizing

$$\sum_{j=1}^{k} \sum_{i=1}^{p} l_{ij}^4, \tag{11.3.10}$$

from which the name "quartimax" derives.

11.3.4 Oblimin Rotations

The rotations of Sections 11.3.2 and 11.3.3 were obtained by minimizing the expression (see Harman, 1967).

$$G = \sum_{1 \le i < j \le k} \left(\sum_{s=1}^{p} l_{si}^2 l_{sj}^2 - \delta p^{-1} d_i d_j \right), \tag{11.3.11}$$

where $d_t = \sum_{s=1}^{p} l_{st}^2$ and δ determines the type of rotation. For orthogonal rotations $\delta = 0$ corresponds to the quartimax procedure and $\delta = 1$ corresponds to the varimax procedure. With $\delta = k/2$ the procedure is called *equamax*. Low values of δ tend to simplify the rows of L, and high values of δ tend to simplify the columns.

By relaxing the condition of independent factors we can use oblique transformations that reduce the value of G. For $\delta = 0$, $\frac{1}{2}$, and 1 the

Table 11.3 Factor Loadings for the Measurements of Table 11.1

		Factor			
Variable	Measurement	1	2	3	4
1	head length	0.555	0.164	0.638	0.052
2	head width	0.684	0.556	0.059	0.062
3	shoulder width	0.116	0.777	0.178	0.265
4	head and neck length	0.200	0.575	0.416	0.112
5	trunk length	0.825	−0.026	0.119	0.094
6	hip width	0.624	0.626	0.104	0.036
7	tail length	0.697	0.215	0.380	0.178
8	ear diameter	0.124	0.166	0.671	0.079
9	elbow width	0.389	0.092	0.193	0.317
10	knee width	−0.058	0.607	0.317	0.440
11	lower foreleg length	0.601	0.397	0.561	0.222
12	lower rear leg length	0.684	0.239	0.419	0.017
13	wrist width	0.046	0.461	0.043	−0.051
14	ankle width	0.298	0.132	0.069	0.943
15	forefoot length	0.762	−0.020	0.072	0.223
16	rear foot length	0.710	0.099	0.419	0.219
17	weight	0.825	0.554	0.079	0.037
	variance	5.239	2.876	2.056	1.472

procedures are called *direct quartimin*, *biquartimin*, and *covarimin*, respectively. These procedures suffer from a tendency for the estimates of the correlation matrix of the factors to be biased. See, for example, Carroll (1957), Warburton (1963), and Harman (1967).

Example 11.3.1 Consider the data of Example 11.2.1. The factor loadings were estimated using the method of maximum likelihood with a varimax rotation. The final factor loadings are given in Table 11.3. There were four eigenvalues of the correlation matrix greater than 1—8.41, 1.97, 1.30, and 1.10—and hence we use four factors for the model. The first rotated factor loads heavily on variables measuring length, the second factor entirely on width. The second factor may therefore be labeled width, but the other two factors are not so easily labeled. Perhaps factor 3 measures the shape of the head and factor 4 the thickness of the ankle. The highest correlated variables with the first two factors are then trunk length and shoulder width. Hence a reasonable index of the body build of a rat is given by

$$(\text{body build index}) = [(\text{trunk length})/(\text{shoulder width})] \times 100.$$

Watson and Broadhurst (1976) gave similar results using rear foot length instead of trunk length.

11.4 Factor Scores

Once we obtain an appropriate factor analytic model we can estimate the factor scores of each subject. If the responses consist of marks on 50 questions on an exam, for example, we may want to estimate scores on some underlying factors, such as reading comprehension and mathematics. This procedure is often used in questionnaire data to reduce the number of response variables from as many as 200 to a more reasonable 4 or 5 that can be used in subsequent analysis.

A method to estimate factor scores is given by Thomson (1951). If f_r is the score of the rth factor corresponding to a response of $\mathbf{x}$, then we estimate f_r by a linear combination of the x_i. That is, we estimate f_r by $\hat{f}_r = \mathbf{a}_r\mathbf{x}$. We choose $\mathbf{a}_r$ such that

$$E\left[(f_r - f_r)^2\right] \quad \text{is minimized.} \tag{11.4.1}$$

If we denote in vector form $\mathbf{f}' = (f_1, \ldots, f_k)$, the solution to (11.4.1) is estimated by

$$\hat{\mathbf{F}} = \hat{\Lambda}'\hat{\Sigma}\mathbf{x}. \tag{11.4.2}$$

Hence for each set of responses $\mathbf{x}$ we obtain from (11.4.2) the vector of factor scores $\mathbf{f}$. These scores can then be used in MANOVA, regression, and other techniques.

Table 11.4[a]

Species	1	2	3	4	5	6	7	8	9	10
1	1.00									
2	0.37	1.00								
3	0.34	0.85	1.00							
4	−0.01	0.46	0.52	1.00						
5	−0.08	−0.09	0.02	0.00	1.00					
6	0.09	0.24	0.31	0.57	−0.18	1.00				
7	0.11	−0.13	0.00	−0.32	0.38	−0.02	1.00			
8	−0.42	−0.79	−0.80	−0.55	−0.09	−0.56	−0.21	1.00		
9	0.43	0.45	0.61	0.28	−0.09	0.05	−0.15	−0.41	1.00	
10	−0.13	−0.20	−0.18	0.37	−0.01	0.58	0.01	−0.26	−0.25	1.00

[a]From Poole (1971).

11.5 Computational Procedures

Factor analysis can be performed using several statistical packages, such as SAS, SPSS, and BMDP4M. The following example was run on BMDP4M.

Example 11.5.1 The matrix in Table 11.4 gives the correlations among the frequencies of ten species of *Drosophila* over 28 months at Pines Woods (Colombia):

1. *melanogaster*,
2. *pseudoobscura*,
3. *bandeirantorum*,
4. "*tripunctata* 20,"
5. *hydei*,
6. *immigrans*,
7. *viracochi*,
8. *mesophragmatica*,
9. *brncici*
10. *gasici*.

The program cards were

```
/PROBLEM     TITLE IS 'PLANTS'
/INPUT       TYPE IS CORREL.
             VARIABLES ARE 10.
             SHAPE = LOWER.
             FORMAT IS '(10F4.2)'.
/VARIABLES   NAMES ARE A,B,C,D,E,F,G,H,I,J.
/FACTOR      FORM = CORR.
             METH = MLFA.
/ROTATE      METH = QRMAX.
/PRINT       CORR.
             PART.
```

The problem card gives the title of the analysis. Input in this case was a lower triangular matrix of the form

$$\begin{pmatrix} 1.00 & & \\ 0.37 & 1.00 & \\ 0.34 & 0.85 & 1.00 \\ \ddots & \ddots & \ddots \end{pmatrix}.$$

The format specifies the format of the last row of the correlation matrix. The factor paragraph specifies a maximum likelihood factor analysis of the correlation matrix and the rotate paragraph specifies a quartimax rotation. The resulting output is shown in Tables 11.5–11.9. The eigenvalues of the unaltered correlation matrix were 3.780230, 1.964693, 1.513416, 0.989423, 0.651892, 0.478176, 0.293043, 0.219706, 0.092865, and 0.016557. The three eigenvalues greater than 1 were chosen as the factors.

Table 11.5 Squared Multiple Correlations (SMC) of Each Variable with All Other Variables

Species	Variable	SMC
1	*A*	0.44064
2	*B*	0.94583
3	*C*	0.87682
4	*D*	0.66250
5	*E*	0.40678
6	*F*	0.68593
7	*G*	0.75541
8	*H*	0.97144
9	*I*	0.61394
10	*J*	0.86324

Table 11.6 Partial Correlations[a]

Variable	Species	*A* 1	*B* 2	*C* 3	*D* 4	*E* 5	*F* 6	*G* 7	*H* 8	*I* 9	*J* 10
A	1	1.000									
B	2	−0.209	1.000								
C	3	−0.275	−0.012	1.000							
D	4	−0.189	0.069	0.243	1.000						
E	5	−0.175	−0.419	−0.059	0.250	1.000					
F	6	−0.079	−0.445	0.006	0.281	−0.457	1.000				
G	7	−0.215	−0.729	−0.116	−0.271	−0.075	−0.268	1.000			
H	8	−0.400	−0.866	−0.412	0.025	−0.407	−0.464	−0.761	1.000		
I	9	0.182	−0.491	0.278	0.103	−0.252	−0.315	−0.415	−0.383	1.000	
J	10	−0.308	−0.742	−0.463	0.216	−0.313	−0.140	−0.589	−0.829	−0.360	1.000

[a] The partial correlations of each pair of variables "partialed" on all other variables (i.e., holding all other variables fixed).

Table 11.7 Iteration for Maximum Likelihood

Iteration	Maximum change in SQRT (uniqueness)	Likelihood criterion to be minimized	Step halvings
1	0.519483	2.025501	
2	0.358101	1.686476	1
3	0.394641	1.620525	1
4	0.216303	1.599884	1
5	0.251118	1.443878	2
6	0.429661	1.336105	0
7	0.149015	1.286191	0
8	0.103187	1.256671	0
9	0.065937	1.251604	0
10	0.019366	1.250654	0
11	0.000961	1.250594	0
12	0.000003	1.250594	0

Table 11.8 Communalities Obtained from Three Factors After 12 Iterations[a]

Species	Variable	Communality
1	*A*	0.2379
2	*B*	0.9046
3	*C*	0.8379
4	*D*	0.5468
5	*E*	0.1448
6	*F*	0.5303
7	*G*	1.0000
8	*H*	1.0000
9	*I*	0.3717
10	*J*	1.0000

[a]The communality of a variable is its squared multiple correlation (covariance) with the factors.

Table 11.9 Variance

Factor	Variance explained	Cumulative proportion of total variance[a]
1	3.533655	0.353366
2	1.663381	0.519704
3	1.376984	0.657402

[a]Total variance is defined as the sum of the diagonal elements of the correlation (covariance) matrix.

As can be seen in Table 11.8 the variables *A*, *E*, and *I* are not explained much by the first three factors; the addition of a fourth factor may improve our results. However, we give the analysis only in the three-factor case, shown in Tables 11.10–11.12. Table 11.12, the sorted rotated factor loadings, gives the easiest interpretation of the factors. Factor 1 loads heavily on

Table 11.10 Unrotated Factor Loadings (Pattern) for Maximum Likelihood Canonical Factors

		Factor		
Variable	Species	1	2	3
A	1	0.458	−0.114	0.125
B	2	0.914	−0.239	−0.105
C	3	0.895	−0.193	0.018
D	4	0.549	0.286	−0.404
E	5	0.024	0.067	0.374
F	6	0.447	0.554	−0.153
G	7	0.026	0.212	0.977
H	8	−0.952	−0.278	−0.129
I	9	0.528	−0.285	−0.106
J	10	0.014	0.979	−0.203
	VP[a]	3.534	1.663	1.377

[a]The variance explained by the factor, the sum of the squares of the elements of the column of the factor loading matrix corresponding to that factor.

Table 11.11 Rotated Factor Loadings (Pattern)

		Factor		
Variable	Species	1	2	3
A	1	0.473	−0.021	0.119
B	2	0.944	0.027	−0.111
C	3	0.914	0.028	0.019
D	4	0.454	0.497	−0.305
E	5	0.010	−0.015	0.380
F	6	0.290	0.668	−0.006
G	7	−0.020	−0.012	1.000
H	8	−0.851	−0.471	−0.233
I	9	0.583	−0.113	−0.140
J	10	−0.237	0.971	0.017
	VP[a]	3.361	1.875	1.338

[a]The VP for each factor is the sum of the squares of the elements of the column of the factor pattern matrix corresponding to that factor. When the rotation is orthogonal, the VP is the variance explained by the factor.

variables *B*, *C*, and *H*, but the factor loading for *H* is negative. Hence there must be some underlying factor producing an increase in the frequency of *pseudoobscura* and *bandeirantorum* but a decrease in the frequency of *mesophragmatica*. Factor 2 specifies that an increase in the frequency of *gasici* occurs with a decrease, again, in *mesophragmatica* and an increase in "*tripunctata* 20" and *immigrans*. Factor 3 can essentially be labeled vira-

Table 11.12 Sorted Rotated Factor Loadings (Pattern)[a]

Variable	Species	Factor 1	Factor 2	Factor 3
B	2	0.944	0.0	0.0
C	3	0.914	0.0	0.0
H	8	−0.851	−0.471	0.0
I	9	0.583	0.0	0.0
J	10	0.0	0.971	0.0
F	6	0.290	0.668	0.0
G	7	0.0	0.0	1.000
A	1	0.473	0.0	0.0
D	4	0.454	0.497	−0.305
E	5	0.0	0.0	0.380
	VP	3.361	1.875	1.338

[a]The factor loading matrix has been rearranged so that the columns appear in decreasing order of variance explained by factors. The rows have been rearranged so that for each successive factor loadings greater than 0.5000 appear first. Loadings less than 0.2500 have been replaced by zero.

cochi with a correlation of 1 between the species and the factor. An increase in the frequency of *viracochi* occurs with a decrease in "*tripunctata* 20" and an increase in *hydei*.

Other types of factor analysis can be specified in the FACTOR paragraph and other types of rotation in the ROTATE paragraph.

Example 11.5.2 The following 21 questions were asked on a human kinetics questionnaire.

1. My average, so far, in the university is
   (a) much lower
   (b) lower
   (c) the same
   (d) higher
   (e) much higher
   than my grade 12 average.
2. In comparison to my fellow students in high school I feel that I stood
   (a) in the top 10% of my class.
   (b) above average in my class.
   (c) average in my class.
   (d) below average in my class.
   (e) in the bottom 10% of my class.
3. Generally, in comparison to others in my classes at the university I feel I stand
   (a) in the bottom 10%.
   (b) below average.

   (c) average.
   (d) above average.
   (e) in the top 10%.
4. I find the course content at the university
   (a) much more difficult than high school.
   (b) more difficult than high school.
   (c) the same degree of difficulty as high school.
   (d) easier than high school.
   (e) much easier than high school.
5. Concerning my academic performance in high school my parents
   (a) were always interested.
   (b) were often interested.
   (c) were sometimes interested.
   (d) were rarely interested.
   (e) didn't care.
6. Concerning my academic performance in high school my parents
   (a) were pleased.
   (b) accepted it.
   (c) didn't care.
   (d) felt I could do better.
   (e) were very displeased.
7. Concerning my academic performance at the university, so far, my parents
   (a) are always interested.
   (b) are often interested.
   (c) are sometimes interested.
   (d) are rarely interested.
   (e) don't care.
8. Concerning my academic performance at the university, so far, my parents
   (a) are very displeased.
   (b) feel I could do better.
   (c) don't care.
   (d) accept it.
   (e) are pleased.
9. I
   (a) care very much
   (b) care to some degree
   (c) care slightly
   (d) do not care very much
   (e) do not care at all
   about how my parents feel concerning my academic performance at the university.

10. Concerning my academic performance at the university I
    (a) am very displeased.
    (b) feel I could be doing better.
    (c) don't care.
    (d) am satisfied.
    (e) am very pleased.
11. The statement or word best describing how I feel if my marks aren't as good as they could be is
    (a) great anxiety.
    (b) upset.
    (c) I will do something about it.
    (d) I can accept it.
    (e) I don't care.
12. If my academics are not going well,
    (a) it always bothers me considerably.
    (b) it often bothers me.
    (c) it rarely bothers me.
    (e) I do not let it bother me.
13. I
    (a) worry all the time
    (b) often worry
    (c) sometimes worry
    (d) rarely worry
    (e) never worry
    about my marks.
14. It
    (a) is very important
    (b) is slightly important
    (c) may be important
    (d) is not very important
    (e) is very unimportant
    to me that I stand at or near the top of my class.
15. I
    (a) always
    (b) often
    (c) sometimes
    (d) rarely
    (e) never
    feel uptight about my schoolwork.
16. When exams are coming up, I
    (a) always
    (b) often
    (c) sometimes

(d) rarely
(e) never
experience feelings of anxiety.

17. Just before or during exams, I
    (a) always
    (b) often
    (c) sometimes
    (d) rarely
    (e) never
    become irritable.
18. My mood is
    (a) always
    (b) often
    (c) sometimes
    (d) rarely
    (e) never
    affected by how well my academics are going.
19. When I have an important exam the following day, I
    (a) always
    (b) often
    (c) sometimes
    (d) rarely
    (e) never
    find it difficult to sleep the night before.
20. I
    (a) always
    (b) often
    (c) sometimes
    (d) rarely
    (e) never
    enjoy writing exams.
21. I
    (a) always
    (b) often
    (c) sometimes
    (d) rarely
    (e) never
    feel that I will not complete the program I am in at the university.

A Likert scale was used where response of (a) received a score of 5 and response (e) received a score of 1. The following file was created on a C.M.S. file system; the program was run on the SAS package:

```
Data HKINET;   INPUT CARDS;
X1 X2 X3 X4 X5 X6 X7 X8 X9 X10 X11 X12 X13 X14 X15 X16 X17 X18 X19 X20 X21;
3  3  3  4  3  5  4  1  3  4   3   3   3   4   3   3   1   1   1   3   3
4  4  3  3  5  5  5  1  4  2   3   4   3   2   3   3   3   2   4   2   1
4  5  1  3  5  5  5  1  4  4   3   3   2   4   2   2   1   3   1   3   2
4  3  4  4  3  5  4  4  5  4   3   4   4   2   3   4   4   4   4   1   2
5  2  4  4  3  4  4  4  3  4   4   3   4   1   3   4   4   4   4   3   4
4  3  4  4  3  5  4  2  3  4   2   4   4   2   3   4   2   4   1   1   3
2  4  3  4  4  4  4  2  4  1   4   4   2   2   3   4   2   3   3   2   2
4  4  3  4  3  4  4  2  4  2   2   3   3   3   3   4   2   3   1   2   1
3  3  3  5  4  4  5  3  3  4   2   1   1   3   3   4   1   4   1   1   1
4  3  3  4  4  5  4  4  3  4   3   4   4   2   4   4   3   3   4   2   3
5  3  3  4  5  1  3  5  5  4   3   3   3   2   3   1   3   4   1   1   3
4  3  3  3  5  2  5  4  4  4   3   4   4   2   4   4   4   4   2   2   3
4  2  3  4  5  5  5  2  5  5   2   3   3   2   2   1   3   3   1   1   2
4  3  3  4  4  5  5  1  4  4   3   3   3   5   3   3   4   3   1   3   1
2  4  2  4  4  5  4  1  3  2   3   1   2   2   2   3   5   4   1   3   3
4  4  3  4  5  2  4  4  4  4   3   3   3   2   3   3   2   3   2   2   2
5  4  4  4  4  4  4  2  3  4   3   3   3   4   3   4   3   4   3   1   5
2  3  2  3  2  4  4  1  3  2   3   3   2   4   2   2   2   2   3   1   1
4  3  3  4  3  4  4  2  4  2   3   3   3   2   2   3   4   3   1   2   2
3  2  3  4  4  2  3  2  3  4   3   2   2   1   2   2   2   3   1   1   3
4  5  2  5  5  5  5  1  4  4   2   4   4   4   4   4   5   2   2   1   3
4  3  3  4  5  2  5  4  5  4   3   5   4   3   4   4   4   3   2   3   1
5  3  3  4  5  5  5  2  5  4   3   5   3   4   3   1   2   3   1   1   3
2  5  3  4  5  5  5  1  5  4   3   4   3   4   3   4   3   4   3   2   2
5  5  2  4  5  5  5  2  5  5   4   5   5   5   5   5   5   5   1   3   3
4  3  3  4  5  5  5  2  4  4   3   4   3   1   3   2   2   3   3   2   1
5  5  3  3  5  5  5  4  4  4   3   1   2   4   2   2   2   2   1   1   2
5  4  3  4  5  5  5  4  5  4   4   4   4   2   3   4   3   4   1   2   3
4  5  2  4  3  5  4  4  4  4   3   1   2   4   2   2   4   3   2   2   2
5  5  3  5  4  5  4  1  5  4   3   4   4   4   4   3   3   4   4   4   3
4  3  3  4  4  4  4  4  4  4   2   3   3   2   3   3   3   3   1   1   1
4  4  3  4  4  5  4  4  3  4   3   5   4   3   4   3   1   4   3   2   2
4  4  3  4  4  4  4  2  1  4   3   3   4   1   3   4   1   3   1   1   1
4  3  2  3  3  5  4  1  5  4   3   3   3   2   3   3   2   2   2   1   3
5  5  4  4  2  4  4  2  4  5   3   3   3   1   3   4   4   4   2   2   1
4  5  2  4  5  5  2  1  5  1   3   4   3   2   3   3   2   3   2   1   1
4  5  2  4  5  5  4  1  5  2   4   5   4   4   5   5   4   4   1   4   3
4  4  2  4  5  5  4  1  5  1   3   3   3   4   3   2   2   3   1   3   1
4  3  3  3  4  5  4  2  4  2   3   3   3   2   1   2   3   3   1   2   3
4  5  2  4  5  5  5  1  5  1   2   4   3   4   3   3   1   5   1   3   1

PROC FACTOR    OUT = BOOK    METHOD = ML    ROTATE = VARIMAX
SCORE PLOT MAXITER = 100;
```

The maximum change in communalities due to rotation error is 0.05150206. There were seven eigenvalues of the correlation matrix larger than 1; hence seven factors were returned.

The input, consisting of 21 variables, specifies that a data file is created called HKINET. The procedure statement signifies that a factor analysis is to be performed using the maximum likelihood procedure with 100 iterations and a_0 varimax rotations. Plots were also asked for. The resulting partial output is shown in Table 11.13. The *orthogonal transformation matrix*

Table 11.13 Rotated Factor Pattern (Factor Loadings)

	Factor						
Question	1	2	3	4	5	6	7
1	0.16241	−0.00950	0.90840	−0.02962	0.08298	0.16198	0.07390
2	−0.07276	−0.08457	−0.00410	0.41423	0.14256	0.48475	0.32741
3	0.15608	0.10912	0.34584	−0.27114	−0.01494	−0.68214	0.09542
4	0.00411	−0.06626	0.05810	0.02779	−0.19635	−0.03861	0.55156
5	0.21409	0.21389	0.05517	−0.13268	−0.00928	0.63062	−0.02992
6	0.06075	0.12705	0.01845	0.89729	0.01085	0.00619	−0.05368
7	0.06818	0.91160	0.03896	0.14426	−0.03234	0.22042	0.00935
8	−0.04966	0.08564	0.39725	−0.57184	0.10530	−0.18635	0.05343
9	0.30445	0.00929	0.15356	0.05899	0.03190	0.52585	−0.08746
10	−0.02416	0.37682	0.45868	−0.21388	−0.00844	−0.18205	0.15021
11	0.18310	−0.09504	−0.00526	−0.03401	0.87636	0.05320	−0.06175
12	0.94806	0.04116	0.01994	0.04008	0.08803	0.17064	0.16394
13	0.67352	0.02706	0.36606	0.02354	0.19423	−0.01960	0.40221
14	−0.03532	0.21648	−0.06162	0.37292	0.00951	0.50723	0.14475
15	0.53327	0.06871	0.07816	−0.04164	0.18190	0.21790	0.65590
16	0.22099	0.12095	−0.11138	0.07667	0.29599	−0.20245	0.73801
17	0.01274	0.12319	0.15539	−0.03549	0.34160	−0.00252	0.28956
18	0.13806	0.02934	0.17941	−0.09635	0.16374	0.06859	0.40441
19	0.31890	0.00207	−0.04561	0.05506	0.23688	−0.34971	0.06768
20	0.09474	0.08996	−0.10268	0.21343	0.44091	0.33525	0.23435
21	0.04606	−0.06082	0.27232	−0.04649	0.36777	−0.14576	0.07141

is

$$\begin{pmatrix} -0.90146 & -0.11081 & 0.14698 & -0.06100 & 0.17881 & -0.19844 & -0.28004 \\ -0.10112 & 0.69905 & 0.27992 & 0.51215 & 0.15029 & 0.36319 & -0.07716 \\ 0.20963 & -0.42960 & 0.85864 & 0.16866 & -4.92E{-}05 & -0.02926 & -0.07019 \\ -0.19825 & -0.44709 & -0.28345 & 0.61617 & -0.43208 & 0.33736 & 0.01716 \\ 0.22681 & -0.27023 & -0.23884 & 0.05977 & 0.62962 & 0.30085 & -0.57270 \\ -0.05230 & 0.07775 & 0.14374 & -0.52610 & -0.46431 & 0.57855 & -0.37878 \\ 0.19932 & 0.18850 & -0.06875 & 0.21335 & -0.38307 & -0.53741 & -0.66253 \end{pmatrix}$$

Factor 1 is highly correlated with questions 12 and 13 and is a measure of worry. Factor 2 is highly correlated with question 7 and may be interpreted as parental interest. Factor 3 is highly correlated with question 1 and to a lesser degree with question 10. This factor may be interpreted as academic success. Note that academic satisfaction (question 10) is highly correlated with factors 2 and 3. This would indicate that academic satisfaction depends on grade average and parental interest.

Appendix

The likelihood function for the model in (11.1.4) may be written

$$L(\mu, \Lambda, \psi) = \text{const}\,|\Sigma|^{-N/2}\exp\left[-(N/2)(\bar{y}-\mu)'\Sigma^{-1}(\bar{y}-\mu)\right] \times\exp\left[-(N/2)\operatorname{tr} S\Sigma^{-1}\right], \quad (11.A.1)$$

where $\Sigma = \Lambda\Lambda' + \psi$. By taking derivatives with respect to $\boldsymbol{\mu}$, Λ, and ψ we obtain

$$\boldsymbol{\mu} = \bar{\mathbf{y}}, \tag{11.A.2}$$

$$\Lambda = S(\Lambda\Lambda' + \psi)^{-1}\Lambda = S\Sigma^{-1}\Lambda, \tag{11.A.3}$$

$$\operatorname{diag}\Sigma^{-1} = \operatorname{diag}\Sigma^{-1}S\Sigma^{-1}. \tag{11.A.4}$$

Although these equations cannot be solved explicitly for Λ and ψ, they can be reduced to a simpler form. We first use Theorem 1.5.1 to express Σ^{-1} as

$$\Sigma^{-1} = (\Lambda\Lambda' + \psi)^{-1} = \psi^{-1} - \psi^{-1}\Lambda(I + \Lambda'\psi^{-1}\Lambda)^{-1}\Lambda'\psi^{-1}. \tag{11.A.5}$$

Then substituting this expression into (11.A.3) and (11.A.4), we obtain

$$\Lambda = S\psi^{-1}\Lambda - S\psi^{-1}\Lambda(I + \Lambda'\psi^{-1}\Lambda)^{-1}\Lambda'\psi^{-1}\Lambda \tag{11.A.6}$$

and

$$\begin{aligned} &\operatorname{diag}\left[\psi^{-1} - \psi^{-1}\Lambda(I + \Lambda'\psi^{-1}\Lambda)^{-1}\Lambda'\psi^{-1}\right] \\ &\quad = \operatorname{diag}\left[\psi^{-1} - \psi^{-1}\Lambda(I + \Lambda'\psi^{-1}\Lambda)^{-1}\Lambda'\psi^{-1}\right] \\ &\quad\quad \times S\left[\psi^{-1} - \psi^{-1}\Lambda(I + \Lambda'\psi^{-1}\Lambda)^{-1}\Lambda'\psi^{-1}\right]. \end{aligned} \tag{11.A.7}$$

Using the properties that $\operatorname{diag}\psi = \psi$ and that for any two symmetric matrices A and B, $\operatorname{diag}(AB) = \operatorname{diag}(BA)$ and $\operatorname{diag}(A + B) = \operatorname{diag}A + \operatorname{diag}B$, (11.A.7) may be reduced to

$$\begin{aligned} &I - \psi^{-1}\operatorname{diag}\Lambda(I + \Lambda'\psi^{-1}\Lambda)^{-1}\Lambda' \\ &\quad = \psi^{-1}\operatorname{diag}S - 2\psi^{-1}\operatorname{diag}S\psi^{-1}\Lambda(I + \Lambda'\psi^{-1}\Lambda)^{-1}\Lambda' \\ &\quad\quad + \psi^{-1}\operatorname{diag}S\psi^{-1}\Lambda(I + \Lambda'\psi^{-1}\Lambda)^{-1}\Lambda'\psi^{-1}\Lambda(I + \lambda'\psi^{-1}\Lambda)\Lambda'. \end{aligned} \tag{11.A.8}$$

From (11.A.6) we have

$$S\psi^{-1}\Lambda(I + \Lambda'\psi^{-1}\Lambda)^{-1}\Lambda'\psi^{-1}\Lambda = S\psi^{-1}\Lambda - \Lambda. \tag{11.A.9}$$

Substituting this expression into the last term in (11.A.8) we obtain

$$\begin{aligned} &I - \psi^{-1}\operatorname{diag}\Lambda(I + \Lambda'\psi^{-1}\Lambda)^{-1}\Lambda' \\ &\quad = \psi^{-1}\operatorname{diag}S - \psi^{-1}\operatorname{diag}S\psi^{-1}(I + \lambda'\psi^{-1}\Lambda)^{-1}\Lambda' \\ &\quad\quad - \psi^{-1}\operatorname{diag}\Lambda(I + \Lambda'\psi^{-1}\Lambda)^{-1}\Lambda'. \end{aligned} \tag{11.A.10}$$

Equivalently, we have

$$I = \operatorname{diag}S\left[\psi^{-1} - \psi^{-1}\Lambda(I + \Lambda'\psi^{-1}\Lambda)^{-1}\Lambda'\psi^{-1}\right]. \tag{11.A.11}$$

From the expression for Σ^{-1} in (11.A.5) this relation simplifies to

$$I = \operatorname{diag}S\Sigma^{-1}. \tag{11.A.12}$$

Hence from (11.A.3) and (11.A.12) the two defining relations for Λ and ψ may be summarized as

$$\operatorname{diag}(S\Sigma^{-1} - I) = 0 \tag{11.A.13}$$

and

$$\Lambda = S\Sigma^{-1}\Lambda. \tag{11.A.14}$$

To simplify these expressions further we notice that $\psi = \Sigma - \Lambda\Lambda'$. Equation (11.A.13) may therefore be written

$$\begin{aligned} 0 &= \psi \operatorname{diag}(S\Sigma^{-1} - I) = \operatorname{diag}(S\Sigma^{-1} - I)\psi = \operatorname{diag}(S\Sigma^{-1} - I)(\Sigma - \Lambda\Lambda') \\ &= \operatorname{diag} S - \operatorname{diag}\Sigma + \operatorname{diag}\Lambda\Lambda' - \operatorname{diag} S\Sigma^{-1}\Lambda\Lambda'. \end{aligned} \tag{11.A.15}$$

But from (11.A.14), $S\Sigma^{-1}\Lambda = \Lambda$. Hence (11.A.15) becomes

$$\operatorname{diag} S = \operatorname{diag}\Sigma. \tag{11.A.16}$$

That is, the sample variances estimate the population variances. We must resort to computer programs to solve the final two equations.

Problems

11.1 Measurements were taken on the fruit of Jonathan apple trees under four experimental conditions: x_1, total nitrogen in parts per million; x_2, protein nitrogen in parts per million; x_3, phosphorus in parts per million; x_4, potassium in parts per million; x_5, calcium in parts per million; x_6, magnesium in parts per million; x_7, mean fruit weight in grams; and x_8, percentage of bitter pit incidence. Calculations yielded the following correlation matrix with 37 df (Ratkowski and Martin, (1974):

	x_1	x_2	x_3	x_4	x_5	x_6	x_7	x_8
x_1	1							
x_2	0.709	1						
x_3	0.803	0.795	1					
x_4	0.491	0.616	0.664	1				
x_5	−0.026	−0.375	−0.289	−0.548	1			
x_6	0.684	0.539	0.761	0.731	−0.099	1		
x_7	−0.024	0.163	0.202	0.603	−0.608	0.309	1	
x_8	−0.056	0.324	0.302	0.459	−0.583	0.247	0.562	1

(a) Use the maximum likelihood method to obtain initial factor loadings.
(b) Use the varimax and equamax rotations to achieve the final factor loadings.
(c) Interpret the factors. In particular, how is the incidence of bitter pit disorder related to the presence of the minerals?

11.2 The following table (Schall and Pianka, 1978) gives the correlations between densities of species of various vertebrate taxa for North America:

Variable	1	2	3	4	5	6	7
1	1	−0.002	0.81	0.69	0.34	−0.77	−0.65
2		1	0.11	0.50	0.38	−0.14	0.40
3			1	0.80	0.42	−0.85	−0.65
4				1	−0.18	0.95	0.70
5					1	−0.47	0.05
6						1	0.61
7							1

where the variables are
1. frogs,
2. lizards,
3. turtles,
4. snakes,
5. insectivorous birds,
6. seed-eating birds, and
7. mammals or marsupials.

Find the underlying factors affecting species abundancy.

11.3 The following data were analyzed in Wu et al. (1977). Panelists were asked to score red wine for a variety of descriptive variables. Variables measured were

1. tart	12. watery	23. musty (earthy, moldy)
2. dry	13. aromatic	24. musk-like
3. sweet	14. hearty	25. yeasty
4. bitter	15. delicate (soft, light)	26. burnt-smokey
5. salty	16. insipid (flat)	27. spicy
6. metallic	17. medicinal	28. vinegary
7. biting	18. sharp	29. sulfurous
8. astringent (puckery)	19. winey	30. fresh
9. smooth	20. fruity	31. mature
10. coarse	21. grapey	32. balanced (round)
11. syrupy	22. woody	33. desirable aftertaste

The correlation matrix of Table 11.14 resulted. Analyze the data explaining response variation with approximately eight factors and label the factors.

11.4 Table 11.15 shows the correlation matrix for measurements of heavy metal concentration the blood of *Salmo gairdneri* (rainbow trout), given by Singh and Ferns (1978). Perform a factor analysis on the correlation matrix to investigate the biological relationship among the chemicals.

11.5 Perform a factor analysis of the data on means and standard deviations (s.d.) of items on a student rating of instruction form in Tables 11.16 and 11.17 to

Table 11.14[a]

Descriptors	1	2	3	4	6	7	8	9	10	11	12	13	14
1	1	56	56	57	66	66	64	35	24	31	−2	−11	6
2		1	72	49	26	53	57	−34	29	−49	7	−10	4
3			1	−56	−26	−53	−53	39	−30	51	−15	19	3
4				1	42	68	61	−48	33	−24	1	−19	−5
6					1	41	31	−32	23	−15	5	−13	−13
7						1	76	−51	40	−28	−4	−20	0
8							1	−50	39	−31	−5	−15	6
9								1	−62	16	−3	28	21
10									1	−20	−1	−19	−7
11										1	−27	1	0
12											1	−14	−16
13												1	25
14													1
15													
16													
17													
18													
19													
20													
21													
23													
28													
30													
31													
32													
33													

	15	16	17	18	19	20	21	23	28	30	31	32	33
1	−26	3	14	56	−7	−29	−30	21	43	−16	−7	−11	−16
2	−12	1	15	49	−11	−43	−42	14	34	−17	10	−1	−10
3	17	−6	−14	−44	20	53	51	−20	−40	22	−3	12	21
4	−33	−14	32	49	−20	−36	−34	37	47	−20	−14	−25	−31
6	−18	18	47	24	−16	−13	−15	24	40	−10	−22	−28	−24
7	−38	7	32	66	−12	−34	−33	26	54	−25	−12	−24	−30
8	−34	2	25	61	−4	−33	−34	20	45	−26	−4	−18	−25
9	49	−12	−36	−37	23	30	29	−30	−40	26	28	50	56
10	−28	6	21	36	−10	−16	−18	15	28	−12	−12	−27	−36
11	−10	1	3	−33	15	42	43	3	−19	7	−11	−7	−6
12	13	26	6	−6	−13	−16	−13	6	7	−6	−5	−9	−4
13	17	−18	−21	−11	18	26	29	−17	−23	22	21	26	26
14	14	−10	−16	−5	25	8	9	4	−11	13	47	44	36
15	1	5	−25	−30	14	17	17	−28	−31	27	38	50	49
16		1	33	−2	−15	−2	−1	27	11	1	−9	−16	−14
17			1	21	−8	−5	−4	27	31	−17	−22	−34	−32
18				1	−7	−31	−32	5	41	−19	−7	−18	−20
19					1	44	42	−10	−14	10	27	28	24
20						1	86	−14	−26	27	3	12	20
21							1	−13	−27	31	2	13	20
23								1	30	−5	−15	−25	−30
28									1	−23	−25	−34	−34
30										1	4	19	19
31											1	67	51
32												1	72
33													1

[a]Reproduced with the permission of the editors of the *Journal of Food Science*. © 1977 by the Institute of Food Technologists.

Table 11.15[a]

	Cr	Mn	Fe	Co	Ni	Cu	Zn	Cd	Pb	Ca	Mg	Na	K
Cr	1												
Mn	0.034	1											
Fe	0.438	−0.096	1										
Co	0.121	0.127	0.057	1									
Ni	−0.208	0.365	0.033	−0.117	1								
Cu	0.203	0.429	0.294	0.150	−0.031	1							
Zn	0.210	0.304	0.429	0.208	0.053	0.352	1						
Cd	0.149	0.472	0.140	0.124	0.443	0.163	0.237	1					
Pb	−0.112	−0.145	0.026	−0.140	−0.002	−0.115	−0.107	−0.350	1				
Ca	0.160	0.265	0.317	0.260	0.495	0.126	0.215	0.430	−0.002	1			
Mg	0.247	0.443	0.386	0.332	0.313	0.297	0.352	0.414	0.077	0.785	1		
Na	0.033	0.131	0.381	0.174	0.396	0.030	0.286	0.460	0.038	0.741	0.619	1	
K	0.231	0.498	0.349	0.225	0.066	0.420	0.496	0.476	−0.173	0.531	0.663	0.646	1

[a]Reproduced with the permission of the editors of the *Journal of Fish Biology*.

Table 11.16[a]

Variable	Mean	s.d.
1. explains, demonstrates, and presents material clearly and understandably.	5.57	1.33
2. makes students feel free to ask questions, disagree, express their ideas.	5.95	1.21
3. is actively helpful when students have difficulty.	5.47	1.37
4. presents material in a well-organized fashion.	5.46	1.48
5. motivates me to do my best.	4.74	1.60
6. gives examinations adequately reflecting course objectives and assignments.	5.62	1.45
7. is fair and impartial in dealings with students.	6.03	1.24
8. inspires class confidence in his (her) knowledge of subject.	5.60	1.40
9. takes an active, personal interest in the class.	5.65	1.35
10. gives students feedback on how they are doing in the course.	5.26	1.65
11. uses a fair grading system based on sufficient evidence.	5.79	1.50
12. makes good use of examples and illustrations.	5.64	1.48
13. specifies the objectives of the course clearly.	5.66	1.35
14. uses class time well.	5.50	1.48
15. helps to broaden my interests.	5.20	1.66
16. uses teaching methods which maximized my learning.	5.04	1.57
17. My view of psychology was consistent with the instructor's view.	4.85	1.32
18. Text is clear in presentation of concepts.	5.22	1.51
19. Text contributes to the course.	5.29	1.52
20. Compared to other instructors, my overall rating of this instructor is [from 1 (poor) to 5 (outstanding)].	3.47	1.20

[a]Data from Abbott and Perkins (1978). Reproduced with the permission of the editors of *Educational and Psychological Measurement*.

Table 11.17 Correlations of Student Rating Items ($N = 341$)[a]

	1	2	3	4	5	6	7	8	9	10	11	12	13	14	15	16	17	18	19	20
1	1																			
2	0.51	1																		
3	0.56	0.66	1																	
4	0.78	0.38	0.52	1																
5	0.66	0.52	0.63	0.60	1															
6	0.61	0.46	0.52	0.53	0.56	1														
7	0.50	0.52	0.59	0.43	0.51	0.59	1													
8	0.68	0.52	0.54	0.67	0.62	0.47	0.56	1												
9	0.66	0.58	0.63	0.57	0.64	0.54	0.58	0.64	1											
10	0.46	0.42	0.47	0.40	0.45	0.54	0.52	0.41	0.48	1										
11	0.52	0.44	0.46	0.46	0.48	0.65	0.72	0.55	0.55	0.57	1									
12	0.69	0.43	0.50	0.63	0.58	0.49	0.50	0.62	0.56	0.42	0.52	1								
13	0.56	0.48	0.53	0.54	0.54	0.52	0.56	0.56	0.47	0.47	0.53	0.54	1							
14	0.74	0.44	0.51	0.69	0.63	0.47	0.50	0.63	0.57	0.43	0.47	0.59	0.49	1						
15	0.69	0.45	0.51	0.59	0.75	0.59	0.57	0.66	0.63	0.48	0.54	0.60	0.55	0.67	1					
16	0.71	0.46	0.56	0.66	0.74	0.63	0.56	0.65	0.63	0.53	0.55	0.60	0.55	0.63	0.78	1				
17	0.51	0.45	0.46	0.48	0.55	0.50	0.44	0.49	0.50	0.42	0.41	0.51	0.43	0.58	0.61	0.55	1			
18	0.40	0.34	0.33	0.32	0.38	0.44	0.35	0.33	0.38	0.38	0.41	0.34	0.36	0.33	0.42	0.41	0.46	1		
19	0.23	0.22	0.26	0.19	0.30	0.31	0.27	0.21	0.29	0.34	0.33	0.27	0.23	0.18	0.29	0.31	0.31	0.77	1	
20	0.63	0.39	0.50	0.64	0.66	0.47	0.44	0.60	0.56	0.38	0.46	0.57	0.44	0.61	0.64	0.67	0.44	0.32	0.27	1

[a]Data from Abbott and Perkins (1978). Reproduced with the permission of the editors of *Educational and Psychological Measurement*.

Table 11.18[a]

Variable interpretation	1	2	3	4	5	6	7	8	9	10	11	12	13
1 Verbal communication	1	0.69	0.54	0.37	0.46	0.39	0.37	0.40	0.53	0.37	0.62	0.52	0.29
2 Experiential evaluation		1	0.31	0.28	0.31	0.21	0.23	0.20	0.32	0.36	0.49	0.29	0.18
3 Induction (visual)			1	0.43	0.33	0.42	0.31	0.43	0.46	0.38	0.39	0.39	0.26
4 Auditory induction				1	0.30	0.50	0.52	0.41	0.41	0.37	0.37	0.31	0.27
5 Memory span					1	0.49	0.33	0.25	0.28	0.33	0.28	0.36	0.26
6 Temporal tracking						1	0.60	0.50	0.36	0.36	0.48	0.37	0.19
7 Sound pattern recognition							1	0.45	0.33	0.22	0.37	0.37	0.29
8 Relation perception								1	0.31	0.36	0.33	0.28	0.11
9 Spatial scanning									1	0.55	0.50	0.28	0.25
10 Flexibility of closure										1	0.38	0.30	0.21
11 Perceptual speed											1	0.40	0.27
12 Masking												1	0.44
13 Tempo													1

[a] From Stankov (1979). Reproduced with the permission of the editors of *Multivariate Behavioural Research*.

summarize the results of the questionnaire. (Maximum likelihood factor analysis with a varimax rotation is one possible combination of methods.)

11.6 Table 11.18 shows the correlations among 13 audiovisual variables. Perform a maximum likelihood factor analysis of the data with a varimax rotation to obtain the main factors.

Chapter 12
Inference on Covariance Matrices

12.1 Introduction

In many of the problems considered in the previous chapters we have assumed some hypotheses or structure on the covariance matrices. For example, in the usual univariate analysis of variance, the basic model assumes not only that the observations $\mathbf{x}_1, \ldots, \mathbf{x}_N$ are independently distributed as $N_p(\boldsymbol{\mu}, \Sigma))$ but also that $\Sigma = \sigma^2 I$ and that we wish to test the hypothesis that the components $\mu_1, \ldots, \mu_p$ of $\boldsymbol{\mu}$ are all equal. Even in the generalization of this to correlated variables, where we lose $n = N-1$ degrees of freedom in one-way classification and none in the two-way classification, we assume that Σ is of the intraclass correlation form

$$\Sigma = \sigma^2[(1-\rho)I + \rho\mathbf{e}\mathbf{e}'],$$

where $\mathbf{e}' = (1, 1, \ldots, 1)$. In testing the equality of two mean vectors in Chapter 3 we assumed that the two populations have the same common covariance matrix. Obviously, these assumptions should be verified and tested. In this chapter we propose likelihood ratio tests. The connected problem of confidence intervals is somewhat involved; the reader is referred to Srivastava and Khatri (1979). As has been pointed out earlier, the *likelihood ratio* tests are in most cases the ratio of the determinants of the two estimates of the covariances under the hypotheses and the alternatives.

12.2 A Test for $\Sigma = \Sigma_0$

In some situations we may have a large data set from past surveys, and Σ may be considered known. However, we may wish to verify it on the basis of new evidence, that is, on the basis of current observations. In this section we provide a likelihood ratio test.

12.2.1 Statement of the Problem

Let $\mathbf{x}_1,\ldots,\mathbf{x}_N$ be independently distributed as $N_p(\boldsymbol{\mu},\Sigma)$. We wish to test the hypothesis

$$H\colon \Sigma=\Sigma_0,\quad \text{where } \Sigma_0 \text{ is known and specified},\qquad \text{vs}\qquad A\colon \Sigma\neq\Sigma_0.$$

Note that no assumption on $\boldsymbol{\mu}$ is made; that is, $\boldsymbol{\mu}$ is unknown in this problem (cf. Section 12.8).

12.2.2 The Likelihood Ratio Test and Its Asymptotic Distribution

Let

$$nS=\sum_{i=1}^{N}(\mathbf{x}_i-\bar{\mathbf{x}})(\mathbf{x}_i-\bar{\mathbf{x}})', \tag{12.2.1}$$

where $\bar{\mathbf{x}}=N^{-1}\Sigma_{i=1}^{N}\mathbf{x}_i$ and $n=N-1$, where N is the sample size. Then the likelihood ratio test is based on the statistic

$$\lambda'=(ne/N)^{pN/2}|S|^{N/2}|\Sigma_0|^{-N/2}\left[\operatorname{etr}\left(-\tfrac{1}{2}n\Sigma_0^{-1}S\right)\right], \tag{12.2.2}$$

where etr(A) stands for the exponential of the trace of the matrix A. The test based on λ' is *not* unbiased (that is, the power of the test is not always greater than or equal to the significance level, a desirable feature of any test). However, a simple modification of the above test in which we replace N by n, the degrees of freedom associated with S, has been shown to be unbiased. Thus we shall base our test on

$$\lambda=e^{pn/2}|S|^{n/2}|\Sigma_0|^{-n/2}\left[\operatorname{etr}\left(-\tfrac{1}{2}n\Sigma_0^{-1}S\right)\right] \tag{12.2.3}$$

and call it the *modified likelihood ratio test*. For large N, $-2\ln\lambda$ is asymptotically distributed as χ_f^2 with $f=\frac{1}{2}p(p+1)$. Thus the hypothesis H will be rejected at the $\alpha 100\%$ significance level if

$$-2\ln\lambda\geq\chi_{f,\alpha}^2, \tag{12.2.4}$$

where $\chi_{f,\alpha}^2$ is the upper $\alpha 100\%$ point of the chi-square distribution with f df, $f=\frac{1}{2}p(p+1)$.

To obtain p-values, one uses the expression

$$\begin{aligned}P(-2\ln\lambda\geq z)=\;&P\left(\chi_f^2\geq z\right)+n^{-1}B_2\left[P\left(\chi_{f+2}^2\geq z\right)-P\left(\chi_f^2\geq z\right)\right]\\&+\tfrac{1}{6}n^{-2}\left[\left(3B_2^2-4B_3\right)P\left(\chi_4^2\geq z\right)-6B_2^2P\left(\chi_{f+2}^2\geq z\right)\right.\\&\left.+\left(3B_2^2+4B_3\right)P\left(\chi_f^2\geq z\right)\right],\end{aligned}$$

where

$$B_2=p(2p^2+3p-1)/24\qquad\text{and}\qquad B_3=-p(p-1)(p+1)(p+2)/32.$$

12.2.3 A Test for $\Sigma = I$

The modified likelihood ratio test for $H: \Sigma = I$ vs $A: \Sigma \neq I$ is given by (12.2.4) with Σ_0 replaced by I in (12.2.3).

12.2.4 Derivation of the Likelihood Ratio Test

Since under the hypothesis H Σ is known to be Σ_0, no estimate of Σ is needed, and the rule of thumb that the likelihood ratio test is the ratio of the determinants of the two estimates of Σ under the hypothesis and the alternative does not apply. From Chapter 2, the likelihood function is given by

$$L(\boldsymbol{\mu}, \Sigma) = c|\Sigma|^{-N/2}|V|^{(n-p-1)/2}\operatorname{etr}\{-\tfrac{1}{2}\Sigma^{-1}[V + N(\bar{\mathbf{x}} - \boldsymbol{\mu})(\bar{\mathbf{x}} - \boldsymbol{\mu})']\}, \tag{12.2.5}$$

where c is a constant and $V = nS$. The maximum likelihood estimate of $\boldsymbol{\mu}$ is $\bar{\mathbf{x}}$. Hence under the hypothesis H the maximum of the likelihood function is given by

$$c|\Sigma_0|^{-N/2}|V|^{(n-p-1)/2}\left[\operatorname{etr}\left(-\tfrac{1}{2}\Sigma_0^{-1}V\right)\right]. \tag{12.2.6}$$

Under the alternative A the maximum likelihood estimate of Σ is $N^{-1}V$ and that of $\boldsymbol{\mu}$ is $\bar{\mathbf{x}}$. Hence the maximum of the likelihood function is given by

$$c|N^{-1}V|^{N/2}|V|^{(n-p-1)/2}e^{-pN/2} \tag{12.2.7}$$

since $\operatorname{etr}(-\tfrac{1}{2}NI) = \exp(-\tfrac{1}{2}pN)$ and $V = nS$. Hence, taking the ratio of (12.2.6) and (12.2.7), we get λ' given in (12.2.2).

12.3 A Test for Sphericity

In the analysis of variance and many other situations we may want to test the hypothesis that $\Sigma = \sigma^2 I$, where σ^2 is unknown. In this section we give the likelihood ratio test.

12.3.1 Statement of the Problem

Let $\mathbf{x}_1, \ldots, \mathbf{x}_N$ be independently distributed as $N_p(\boldsymbol{\mu}, \Sigma)$. We wish to test the hypothesis

$$H: \Sigma = \sigma^2 I, \qquad \sigma^2 > 0 \text{ and unknown,}$$

vs

$$A: \Sigma \neq \sigma^2 I.$$

Note that in the above problem we are assuming that $\boldsymbol{\mu}$ is *not* known (cf. Section 12.9).

12.3.2 Likelihood Ratio Test and Its Asymptotic Distribution

The likelihood ratio test for the hypothesis H vs A is based on the statistic

$$\lambda = |S|/(p^{-1}\operatorname{tr} S)^p, \tag{12.3.1}$$

where S is the sample covariance matrix defined in (12.2.1). As $N \to \infty$, $-[n-(2p^2+p+2)/6p]\ln\lambda$ has asymptotically a chi-square distribution with $f = \frac{1}{2}p(p+1)-1$ df. Hence the hypothesis H is rejected if

$$-\left[n-(2p^2+p+2)/6p\right]\ln\lambda > \chi^2_{f,\alpha}.$$

To obtain p-values, one uses the expression

$$P(-m\ln\lambda \geq z) = P\left(\chi_f^2 \geq z\right) + am^{-2}\left[P\left(\chi_{f+4}^2 \geq z\right) - P\left(\chi_f^2 \geq z\right)\right],$$

where

$$m = n-(2p^2+p+2)/6p,$$
$$a = (p+1)(p-1)(p+2)(2p^3+6p^2+3p+2)/288p^2.$$

12.3.3 Derivation of the Likelihood Ratio Test

Since we are estimating the covariance matrix under the hypothesis as well as under the alternative, we take the ratio of the determinants of the estimates of Σ under H and under A. The maximum likelihood estimate of σ^2 under the hypothesis H is given by

$$\frac{1}{pN}\sum_{i=1}^{p} v_{ii} = (Np)^{-1}\operatorname{tr} V,$$

where $V = nS$. Hence the estimate of Σ under H is

$$\left[(Np)^{-1}\operatorname{tr} V\right]I,$$

and the determinant of this quantity is $[(Np)^{-1}\operatorname{tr} V]^p$ since $|cI_p| = c^p$. The maximum likelihood estimate of Σ under A is given by $N^{-1}V$, where $V = nS$. Hence the likelihood ratio test is based on the statistic

$$|N^{-1}V|/\left[(Np)^{-1}\operatorname{tr} V\right]^p = |V|/\left[p^{-1}\operatorname{tr} V\right]^p = |S|/\left[p^{-1}\operatorname{tr} S\right]^p.$$

12.3.4 Some Comments

Let $l_1 > l_2 > \cdots > l_p$ be the ordered characteristic roots of S. Then the likelihood ratio test λ given in (12.3.1) is given by

$$\left(\prod_{i=1}^{p} l_i\right) \Big/ \left(p^{-1}\sum_{i=1}^{p} l_i\right)^p$$

since $|S| = \prod_{i=1}^{p} l_i$ and $\operatorname{tr} S = l_1 + \cdots + l_p$. Note that this test is also for testing the equality of all the characteristic roots of Σ, a generalization of the problem, considered in Chapter 10, of principal component analysis.

Example 12.3.1 In Example 6.1.3 eight cows were given a drug and a response was measured over four time intervals. If the errors are independently distributed as $N(0,\sigma^2)$, then simple regression techniques are applicable. To check this assumption, one applies the test for sphericity. That is, one tests $H: \Sigma = \sigma^2 I$ vs $A: \Sigma \neq \sigma^2 I$, where Σ is the covariance matrix between time intervals. The sample covariance matrix based on the eight observations was found to be

$$S = \begin{pmatrix} 30.6 & 12.7 & 59.4 & 53.9 \\ 12.7 & 55.1 & 5.0 & -12.1 \\ 59.4 & 5.0 & 0.86 & 147.3 \\ 53.9 & -12.1 & 147.3 & 149.5 \end{pmatrix}.$$

Now

$$\lambda = |S|\left(p^{-1}\operatorname{tr} S\right)^{-p} = 1572278.37\left[\tfrac{1}{4}(421.2)\right]^{-4} = 0.0128.$$

Hence

$$-\ln\lambda = 4.359$$

and

$$\begin{aligned}-\left[n-(2p^2+p+2)/6p\right]\ln\lambda &= \left(7-\tfrac{38}{24}\right)4.359 = 23.61 \\ &\geq \chi^2_{9,0.05} = 16.9.\end{aligned}$$

Hence one concludes that the assumption of independent errors with the same variance is invalid.

The calculations in this problem and others in this chapter involve only elementary mathematical operations and the calculation of determinants. The evaluation of a determinant is available in many computer packages, usually under the heading of a linear algebra program. The p-value for this example is obtained by calculating

$$\begin{aligned}P(-m\ln\lambda \geq 23.61) &= P\left(\chi^2_9 \geq 23.61\right) \\ &\quad + am^{-2}\left[P\left(\chi^2_{13} \geq 23.61\right) - P\left(\chi^2_9 \geq 23.61\right)\right] \\ &= 0.015,\end{aligned}$$

where $a = 4.65$ and $m = 5.417$.

12.4 A Test for an Intraclass Correlation Model

We have pointed out earlier that for the analysis of variance we really do not need the sphericity criterion considered in the previous section. What we really need is the intraclass correlation model. In this section we obtain the likelihood ratio test and give its asymptotic distribution.

12.4.1 Statement of the Problem

Let $\mathbf{x}_1, \mathbf{x}_2, \ldots, \mathbf{x}_N$ be independently distributed as $N_p(\boldsymbol{\mu}, \Sigma)$. We wish to test the hypothesis

$$H: \Sigma = \sigma^2[(1-\rho)I + \rho\mathbf{e}\mathbf{e}'], \qquad \sigma^2, \rho \text{ unknown},$$

vs

$$A: \Sigma \neq \sigma^2[(1-\rho)I + \rho\mathbf{e}\mathbf{e}'],$$

where $\mathbf{e} = (1, 1, \ldots, 1)'$, a p-vector of 1s. Thus we wish to test the hypothesis that the variances of all the p characters are equal and they are equally correlated.

12.4.2 Likelihood Ratio Test and Its Asymptotic Distribution

The likelihood ratio test for the hypothesis H vs A was derived by Wilks (1946). It is based on the statistic

$$\lambda = |S| / (s^2)^p (1-r)^{p-1}[1 + (p-1)r], \tag{12.4.1}$$

where $S = (s_{ij})$, the sample covariance matrix defined in (12.2.1),

$$s^2 = p^{-1} \sum_{i=1}^{p} s_{ii}, \qquad \text{and} \qquad s^2 r = \frac{2}{p(p-1)} \sum_{i<j}^{p} \sum s_{ij}. \tag{12.4.2}$$

For large N Box (1949, 1950) has shown that

$$Q = -\left[N - 1 - p(p+1)^2(2p-3)/6(p-1)(p^2+p-4)\right] \ln \lambda \tag{12.4.3}$$

is asymptotically distributed as chi-square with $f = \frac{1}{2}p(p+1) - 2$ df. Hence the hypothesis H is rejected if

$$Q \geq \chi^2_{f,\alpha}.$$

12.4.3 Derivation of the Likelihood Ratio Test

The matrix $\sigma^2[(1-\rho)I + \rho\mathbf{e}\mathbf{e}']$ has only two distinct roots, $\sigma^2[1 + (p-1)\rho]$ and $\sigma^2(1-\rho)$, the second repeated $p-1$ times. Hence the determinant of the covariance matrix under the hypothesis H is given by

$$\begin{aligned} |\sigma^2[(1-\rho)I + \rho\mathbf{e}\mathbf{e}']| &= \sigma^2[1 + (p-1)\rho](\sigma^2)^{p-1}(1-\rho)^{p-1} \\ &= (\sigma^2)^p (1-\rho)^{p-1}[1 + (p-1)\rho]. \end{aligned}$$

The maximum likelihood estimate of σ^2 is given by $\hat{\sigma}^2 = (Np)^{-1}\Sigma_{i=1}^{p} v_{ii}$ and that of $\sigma^2\rho$ by $\hat{\sigma}^2\hat{\rho} = [2/p(p-1)N]\Sigma_{i<j} v_{ij}$, where $V = (v_{ij})$ and $V = nS$. Hence the determinant of the estimate of Σ under the hypothesis is

$$(\hat{\sigma}^2)^p (1-\hat{\rho})^{p-1}[1 + (p-1)\hat{\rho}].$$

The estimate of Σ under the alternative is given by $N^{-1}V=(n/N)S$. Hence the likelihood test is based on the statistic in (12.4.1).

Example 12.4.1 For the data set of Rao (1948) in Example 6.1.4 (cork borings in four tree facings) the covariance matrix was

$$S=\begin{pmatrix} 290.41 & 223.75 & 288.49 & 226.27 \\ 223.75 & 219.93 & 229.06 & 171.37 \\ 288.49 & 229.06 & 350.00 & 259.54 \\ 226.27 & 171.37 & 259.54 & 226.00 \end{pmatrix}$$

on 27 df, $N=28$. The hypothesis tested was that the mean thickness of the cork deposits was the same for all four directions N, S, E, and W. Standard ANOVA techniques are applicable if the correlations between directions are equal. To test this hypothesis we apply the test for an intraclass correlation matrix. Calculations yield

$$s^2=p^{-1}\sum_{i=1}^{p} s_{ii}=270.085,$$

$$r=\left[\frac{2}{p(p-1)}\sum_{1\le i<j\le p} s_{ij}\right]\Big/ s^2=\frac{233.072}{270.085}=0.863,$$

and

$$|S|=207{,}222{,}725.$$

Hence

$$\lambda=\frac{207{,}222{,}725}{(270.085)^4(1-0.863)^3(1+3\times 0.863)}=0.422,$$

$$Q=-\left[N-1-\frac{p(p+1)^2(2p-3)}{6(p-1)(p^2+p-4)}\right]\ln\lambda$$

$$=\left[28-1-\frac{4\times 25\times 5}{6\times 3\times 16}\right]0.864=22.20.$$

Because $\chi^2_{8,0.05}=15.5$ we reject H and claim the covariance matrix is not an intraclass correlation matrix.

12.5 A Test for Equicorrelation

In the previous section we gave the test for the hypothesis of the intraclass correlation model. Validating this model is essential in carrying out the analysis of variance and is useful in the principal component analysis, for if the covariance matrix is indeed of intraclass correlation form, then the principal component analysis based on covariance is considerably simplified. However, if we are interested in a principal component analysis based on the correlation matrix, we have somewhat less to ascertain to achieve the simplification.

12.5.1 *Statement of the Problem*

Let $\mathbf{x}_1,\ldots,\mathbf{x}_N$ be independently and identically distributed as $N_p(\boldsymbol{\mu},\Sigma)$. Writing

$$\Sigma=(\sigma_{ij})=D_\sigma P D_\sigma, \tag{12.5.1}$$

where $D_\sigma=\operatorname{diag}(\sigma_{11}^{1/2},\ldots,\sigma_{pp}^{1/2})$ and $P(\rho_{ij})$ with $\rho_{ii}=1$, $i=1,2,\ldots,p$, we are interested in testing the hypothesis

$$H:\rho_{ij}=\rho$$

vs

$$A:\rho_{ij}\neq\rho, \qquad \text{for at least one pair } i\neq j.$$

Let

$$nS=(s_{ij})=\sum_{i=1}^{N}(\mathbf{x}_i-\bar{\mathbf{x}})(\mathbf{x}_i-\bar{\mathbf{x}})', \qquad \bar{\mathbf{x}}=N^{-1}\sum_{i=1}^{N}\mathbf{x}_i,$$

and

$$S=D_S R D_S, \tag{12.5.2}$$

where $D_S=\operatorname{diag}(s_{11}^{1/2},\ldots,s_{pp}^{1/2})$ and $R=(r_{ij})$ with $r_{ii}=1$, $i=1,2,\ldots,p$.

12.5.2 *A Test*

It is difficult to obtain the likelihood ratio test. Lawley (1963) proposed the following heuristic test: The hypothesis H is rejected if

$$Q=\frac{n}{\hat{\lambda}^2}\left[\sum_{i<j}^{p}\sum^{p}(r_{ij}-\bar{r})^2-\bar{\mu}\sum_{k=1}^{p}(\bar{r}_k-\bar{r})^2\right]>\chi^2_{f,\alpha},$$

where

$$f=\tfrac{1}{2}(p+1)(p-2), \qquad n=N-1, \quad \hat{\lambda}=1-\bar{r},$$

$$\bar{r}=\frac{2}{p(p-1)}\sum_{i<j}^{p}\sum^{p}r_{ij},$$

$$\hat{\mu}=\frac{(p-1)^2(1-\hat{\lambda}^2)}{p-(p-2)\hat{\lambda}^2},$$

$$\bar{r}_k=\frac{1}{p-1}\sum_{\substack{i=1\\ i\neq k}}^{p}r_{ik}.$$

Example 12.5.1 Measurements on male ocean perch off the Columbia River were given by Bernard (1981) for x_1, the asymptotic length (mm), $-x_2$, the coefficient of growth, and $-x_3$, the time at which length is zero (yr). The correlation matrix based on 76 observations was given by

$$R = \begin{pmatrix} 1.00 & 0.97 & 0.92 \\ 0.97 & 1.00 & 0.98 \\ 0.92 & 0.98 & 1.00 \end{pmatrix}.$$

To test for equal correlations we calculate

$$p=3, \qquad n=75, \qquad \bar{r}=\tfrac{1}{3}(2.87)=0.957,$$

$$\bar{\lambda}=0.043, \quad \bar{\lambda}^2=0.00185, \qquad \hat{\mu}=(4\times 0.998)/(3-0.00185)=1.33,$$

$$\bar{r}_1=0.945, \qquad \bar{r}_2=0.975, \qquad \bar{r}_3=0.950.$$

Hence

$$\begin{aligned} Q &= (75/0.00185)(0.002-1.33\times 0.0005) \\ &= 54.12 > \chi^2_{2,0.05}=5.99 \end{aligned}$$

and we claim at the 5% level that the correlations are significantly different.

12.6 A Test for Zero Correlation

In this section we give a test for the hypothesis the covariance matrix is a *diagonal* matrix.

12.6.1 Statement of the Problem

Let $\mathbf{x}_1, \ldots, \mathbf{x}_N$ be independently distributed as $N_p(\boldsymbol{\mu}, \Sigma)$. We wish to test the hypothesis

$$H: \Sigma = \begin{pmatrix} \sigma_{11} & & 0 \\ & \ddots & \\ 0 & & \sigma_{pp} \end{pmatrix} \qquad \text{vs} \qquad A \neq H.$$

Note that if we write Σ in the form of (12.5.1), the above hypothesis is equivalent to testing that the *correlations* are all zero.

12.6.2 Likelihood Ratio Test and Its Asymptotic Distribution

The likelihood ratio test is based on the statistic

$$|R|, \qquad \text{where } R=(r_{ij}) \tag{12.6.1}$$

and r_{ij} is the sample correlations for $i \neq j$ with $r_{ii}=1$, as defined in (12.5.2). Bartlett (1954) has shown that, under H, $-[N-1-(2p+5)/6]\ln|R|$ is for large n asymptotically distributed as χ_f^2 with

$$f=\tfrac{1}{2}p(p-1).$$

It is left as an exercise for the reader to show that (12.6.1) is the likelihood ratio test.

12.7 A Test for Equality of Covariance

In the multivariate analysis of variance we assumed that the k populations have the same covariance matrix. In this section we provide a test for it.

12.7.1 Statement of the Problem

Let $\mathbf{x}_{ij}$ be independently normally distributed as $N_p(\boldsymbol{\mu}_i, \Sigma)$, $j=1,2,\ldots,N_i$, $i=1,2,\ldots,k$. We wish to test the hypothesis

$$H\colon \Sigma_1=\Sigma_2=\cdots=\Sigma_k \qquad \text{vs} \qquad A\neq H.$$

Let

$$V_i=\sum_{j=1}^{N_i}(\mathbf{x}_{ij}-\bar{\mathbf{x}}_i)(\mathbf{x}_{ij}-\bar{\mathbf{x}}_i)', \qquad \bar{\mathbf{x}}_i=N_i^{-1}\sum_{j=1}^{N_i}\mathbf{x}_{ij},$$

$$V=\sum_{i=1}^{k}V_i.$$

12.7.2 Likelihood Ratio Test and Its Asymptotic Distribution

The likelihood ratio test is based on the statistic

$$\lambda'=\left(\prod_{i=1}^{k}|V_i|^{N_i/2}/|V|^{N/2}\right)\left(N^{pN/2}/\prod_{i=1}^{k}N_i^{pN_i/2}\right), \qquad (12.7.1)$$

where $V_i=n_iS_i$ and $N=\sum_{i=1}^{k}N_i$.

However, this test is not unbiased unless N_i is changed to n_i, the degrees of freedom associated with V_i. The modified likelihood ratio test (suggested by Bartlett on an intuitive ground) is based on

$$\lambda=\left(\prod_{i=1}^{k}|V_i|^{n_i/2}n^{pn/2}/|V|^{n/2}\prod_{i=1}^{k}n_i^{pn_i/2}\right), \qquad n=\sum_{k=1}^{k}n_i.$$

For large n such that $\lim_{n\to\infty}(n_i/n)\equiv\lim(r_i)>0$,

$$-m\ln\lambda\sim\chi_f^2,$$

where

$$f=\tfrac{1}{2}(k-1)p(p+1), \qquad m=2n^{-1}(n-2\alpha),$$

$$\alpha=\left(\sum_{j=1}^{k} r_j^{-1}-1\right)(2p^2+3p-1)/12(p+1)(k-1), \qquad r_i=n_i/n.$$

For $k=2$ the modified likelihood becomes

$$\lambda=|V_1|^{n_1/2}|V_2|^{n_2/2}n^{pn/2}/|V_1+V_2|^{n/2}n_1^{pn_1/2}n_2^{pn_2/2}$$

and $-m\ln\lambda\sim\chi_f^2$, where

$$f=\tfrac{1}{2}p(p+1), \qquad m=2n^{-1}(n-2\alpha), \qquad n=n_1+n_2,$$

$$\alpha=(n^2-n_1n_2)(2p^2+3p-1)/12(p+1)n_1n_2.$$

To obtain p-values, we use the formula

$$p(-m\ln\lambda\geq z)=P(\chi_f^2\geq z)+m^{-2}\gamma_2\left[P(\chi_f^2\geq z)-P(\chi_f^2\geq z)\right]+O(m^{-3}),$$

where $m=n-2\alpha$, $r_i=n_i/n$, $n=\Sigma_{i=1}^k n_i$,

$$\alpha=\left(\sum_{j=1}^{k} r_j^{-1}-1\right)\frac{2p^2+3p-1}{12(p+1)(k-1)},$$

$$\gamma_2=p(p+1)\left[(p-1)(p+2)\left(\sum_{j=1}^{k} r_j^{-2}-1\right)-24(k-1)\alpha^2\right]\Big/48,$$

and

$$f=\tfrac{1}{2}p(p+1)(k-1).$$

Example 12.7.1 The following covariance matrices were used in a Hotelling's T^2-test for the equality of means:

$$S_1=\begin{pmatrix} 3.622 & 3.564 & 3.415 & 4.282 \\ 3.564 & 3.566 & 3.419 & 4.312 \\ 3.415 & 3.419 & 3.551 & 4.994 \\ 4.282 & 4.312 & 4.994 & 8.032 \end{pmatrix} \qquad \text{on} \quad n_1=5 \text{ df},$$

and

$$S_2=\begin{pmatrix} 3.403 & 3.277 & 3.543 & 4.555 \\ 3.277 & 3.171 & 3.375 & 4.379 \\ 3.543 & 3.375 & 4.006 & 5.346 \\ 4.555 & 4.379 & 5.346 & 7.770 \end{pmatrix} \qquad \text{on} \quad n_2=5 \text{ df}.$$

The assumption of equal covariances should be tested. This hypothesis is rejected if λ is small where (for two populations)

$$\lambda=|V_1|^{n_1/2}|V_2|^{n_2/2}(n_1+n_2)^{p(n_1+n_2)/2}/|V_1+V_2|^{(n_1+n_2)/2}n_1^{pn_1/2}n_2^{pn_2/2},$$

where $V_1 = n_1 S_1$ and $V_2 = n_2 S_2$. Calculations yield

$$\lambda = \frac{(3.736)^{2.5}(1.262)^{2.5}(10)^{20}}{(190.997)^5 5^{10} 5^{10}} = 1.992 \times 10^{-4},$$

$$\alpha = \frac{(n^2 - n_1 n_2)(2p^2 + 3p - 1)}{12(p+1)n_1 n_2} = \frac{75 \times 43}{12 \times 5 \times 25} = 2.15,$$

$$m = \tfrac{2}{10}(10 - 2 \times 2.15) = 1.14.$$

Hence $-m \ln \lambda = 1.14 \times 8.52 = 9.71 \leq \chi^2_{10, 0.05} = 18.3$. Therefore the null hypothesis is not rejected.

We have considered the problem of testing for the equality of k covariances. Sometimes, however, we expect only one of them to be different and the remaining $k-1$ covariance to be equal. We may then desire to know which, if any, is the "outlier." This is known as a *slippage problem*, and the reader is referred to Srivastava (1966) for a detailed discussion.

12.8 A Test for Independence

Let $\mathbf{x}_1, \ldots, \mathbf{x}_N$ be a random sample of size N on $\mathbf{x}$, where $\mathbf{x} \sim N_p(\boldsymbol{\mu}, \Sigma)$. Let $\mathbf{x}$ and Σ be partitioned as

$$\mathbf{x} = \begin{pmatrix} \boldsymbol{\mu}_1 \\ \vdots \\ \mathbf{u}_k \end{pmatrix} \quad \text{and} \quad \Sigma = \begin{pmatrix} \Sigma_{11} & \Sigma_{12} & \cdots & \Sigma_{1k} \\ \Sigma'_{12} & \Sigma_{22} & \cdots & \Sigma_{2k} \\ \vdots & \vdots & & \vdots \\ \Sigma'_{1k} & \Sigma'_{2k} & \cdots & \Sigma_{kk} \end{pmatrix},$$

respectively, where the $\mathbf{u}_i$s are p_i-vectors and the Σ_{ii}s are $p_i \times p_i$ matrices such that $\Sigma_{i=1}^{k} p_i = p$. We assume that Σ is a positive definite symmetric matrix. Let

$$\mathbf{y} = N^{1/2}\bar{\mathbf{x}} = \begin{pmatrix} \mathbf{y}_1 \\ \vdots \\ \mathbf{y}_k \end{pmatrix} \quad \text{and} \quad V = \begin{pmatrix} V_{11} & V_{12} & \cdots & V_{1k} \\ V'_{12} & V_{22} & \cdots & V_{2k} \end{pmatrix},$$

where the $\mathbf{y}_i$s are p_i-vectors and the V_{ii}s are $p_i \times p_i$ matrices. Suppose we wish to test the hypothesis

$$H: \Sigma_{ij} = 0_{ij}, \qquad i \neq j, \quad i, j = 1, 2, \ldots, k,$$

vs

$$A: \Sigma_{ij} \neq 0_{ij} \qquad \text{for at least one pair } (i, j), \quad i \neq j,$$

where 0_{ij} denotes zero matrices of order $p_i \times p_j$, $i \neq j$, $i, j = 1, 2, \ldots, k$. That is, we wish to test the hypothesis that the k subvectors $\mathbf{u}_1, \ldots, \mathbf{u}_k$ of $\mathbf{x}$ are independently distributed. This problem was considered in Chapter 9 for $k=2$. The likelihood ratio test rejects the hypothesis H if

$$\lambda = |V| / \prod_{i=1}^{k} |V_{ii}| < c_\alpha,$$

where c_α is chosen so that the size of the test is α. For large N, $-m\ln\lambda$ is asymptotically distributed as chi-square with f df, where

$$m = N - 2\alpha,$$
$$\alpha = \left(p^3 - \sum_{i=1}^{k} p_i^3\right) + 9\left(p^2 - \sum_{i=1}^{k} p_i^2\right) / 6\left(p^2 - \sum_{i=1}^{k} p_i^2\right),$$
$$f = \tfrac{1}{2}\left(p^2 - \sum_{i=1}^{k} p_i^2\right).$$

For p-values we use the formula

$$p(-m\ln\lambda \geq z) = p\left(\chi_f^2 \geq z\right) + m^{-2}\gamma_2 \times \left[p\left(\chi_{f+4}^2 \geq z\right) - p\left(\chi_f^2 \geq z\right)\right] + O(m^{-3}),$$

where

$$m = n - 2\alpha,$$
$$\alpha = \left[\left(p^3 - \sum_{i=1}^{k} p_i^3\right) + 9\left(p^2 - \sum_{i=1}^{k} p_i^2\right)\right] \Big/ \left[6\left(p^2 - \sum_{i=1}^{k} p_i^2\right)\right],$$
$$\gamma_2 = \tfrac{1}{48}\left(p^4 - \sum_{i=1}^{k} p_i^4\right) - \tfrac{5}{96}\left(p^2 - \sum_{i=1}^{k} p_i^2\right) - \tfrac{1}{72}\left(p^3 - \sum_{i=1}^{k} p_i^3\right)^2 \Big/ \left(p^2 - \sum_{i=1}^{k} p_i^2\right)$$

and

$$f = \tfrac{1}{2}\left(p^2 - \sum_{i=1}^{k} p_i^2\right).$$

Example 12.8.1 For the correlation matrix of Problem 5.1 we would like to test for independence between first-grade scores and second-grade scores. The test statistic is given by

$$\lambda = |V| / |V_{11}||V_{22}|,$$

where

$$V = \begin{pmatrix} V_{11} & V_{12} \\ V_{12}' & V_{22} \end{pmatrix}$$

is the sum of squares matrix. As the problem is invariant under scale transformations on each characteristic, the covariance matrix S can be used and also the correlation matrix R. That is,

$$\lambda=|V|/|V_{11}||V_{22}|=|S|/|S_{11}||S_{22}|=|R|/|R_{11}||R_{22}|,$$

where

$$S=\begin{pmatrix} S_{11} & S_{12} \\ S_{12}' & S_{22} \end{pmatrix} \quad \text{and} \quad R=\begin{pmatrix} R_{11} & R_{12} \\ R_{12}' & R_{22} \end{pmatrix}.$$

The data set in this case is presented in terms of the correlation matrix R. R_{11} is the upper 12×12 matrix and R_{22} is the lower 4×4 matrix. $|R|=0.0004$, $|R_{11}|=0.0070$, and $|R_{22}|=0.6000$. Hence

$$\lambda=\frac{0.0004}{0.007\times 0.600}=0.095,$$

$$\alpha=\frac{[16^3-(12^3+4^3)]+9[16^2-(12^2+4^2)]}{6[16^2-(12^2+4^2)]}=5.5,$$

$$f=\tfrac{1}{2}(16^2-12^2+4^2)=48,$$

$$m=(N-2\alpha)=219-11=208.$$

Because $-m\ln\lambda=208\times 2.53=489.6\geq\chi^2_{48,0.05}=65.2$, the null hypothesis is rejected, and we claim that first-grade scores are not independent of second-grade scores.

12.9 A Test for Covariance Matrix Equal to the Identity Matrix and Zero Mean Vector

12.9.1 Statement of the Problem

In some situations we may want to test the covariance matrix and the mean vector simultaneously. Let $\mathbf{x}_1,\ldots,\mathbf{x}_N$ be independently and identically distributed as $N_p(\boldsymbol{\mu},\Sigma)$. We wish to test the hypothesis

$$H:\Sigma=I,\quad \boldsymbol{\mu}=\mathbf{0}, \qquad \text{vs} \qquad A:\Sigma\neq I,\quad \boldsymbol{\mu}\neq\mathbf{0}.$$

12.9.2 Likelihood Ratio Test and Its Asymptotic Distribution

The likelihood ratio test for the hypothesis H vs A is based on the statistic

$$\lambda=(e/N)^{pN/2}|V|^{N/2}\operatorname{etr}[-\tfrac{1}{2}(V+N\bar{\mathbf{x}}\bar{\mathbf{x}}')], \qquad (12.9.1)$$

where

$$V=n^{-1}S=\sum_{\alpha=1}^{N}(\mathbf{x}_\alpha-\bar{\mathbf{x}})(\mathbf{x}_\alpha-\bar{\mathbf{x}}), \qquad \text{and} \qquad \bar{\mathbf{x}}=N^{-1}\sum_{\alpha=1}^{N}\mathbf{x}_\alpha.$$

For large N, $-2\ln\lambda$ is asymptotically distributed as chi-squared with $f=\frac{1}{2}p(p+1)$ degrees of freedom.

It is left as an exercise for the reader to show that (12.9.1) is the likelihood ratio test.

12.10 A Test for Covariance Matrix Equal to $\sigma^2 I$ and Zero Mean Vector

12.10.1 Statement of the Problem

In this section we consider another problem involving simultaneously testing of the covariance and mean vector. Let $\mathbf{x}_1,\dots,\mathbf{x}_N$ be independently and identically distributed as $N_p(\boldsymbol{\mu},\Sigma)$. We wish to test the hypothesis

$$H:\Sigma=\sigma^2 I,\quad \boldsymbol{\mu}=\mathbf{0},\quad \sigma^2>0,\qquad \text{vs}\qquad A:\Sigma\neq\sigma^2 I,\quad \boldsymbol{\mu}\neq\mathbf{0}.$$

12.10.2 Likelihood Ratio Test and Its Asymptotic Distribution

The likelihood ratio test for the hypothesis H vs the alternative A is based on the statistic

$$\lambda=\left[p^p|V|/(\operatorname{tr}V+N\bar{\mathbf{x}}'\bar{\mathbf{x}})^p\right]^{N/2}. \tag{12.10.1}$$

As $N\to\infty$, $-[N-(2p^2+p+2)/6p]\ln\lambda$ has an asymptotically chi-square distribution with $f=\frac{1}{2}p(p+1)-1$ df. For the exact null and nonnull distributions, see Khatri and Srivastava (1973).

It is left as an exercise for the reader to show that (12.10.1) is the likelihood ratio test.

Problems

12.1 Let $\mathbf{x}_1,\dots,\mathbf{x}_N$ be independently distributed as $N_p(\boldsymbol{\mu},\Sigma)$. Suppose we wish to test the hypothesis

$$H:\Sigma=\Sigma_0,\qquad \boldsymbol{\mu}=\boldsymbol{\mu}_0,\quad \Sigma_0 \text{ and } \boldsymbol{\mu}_0 \text{ specified},$$

vs

$$A:\Sigma\neq\Sigma_0,\qquad \boldsymbol{\mu}\neq\boldsymbol{\mu}_0.$$

Obtain the likelihood ratio test.

12.2 Let $\mathbf{x}_1,\dots,\mathbf{x}_N$ be independently distributed as $N_p(\boldsymbol{\mu},\Sigma)$. Suppose we wish to test the hypothesis

$$H:\Sigma=\sigma^2 I,\qquad \boldsymbol{\mu}=\boldsymbol{\mu}_0,\quad \sigma^2>0 \text{ and unknown},$$

vs

$$A:\Sigma\neq\sigma^2 I,\qquad \boldsymbol{\mu}\neq\boldsymbol{\mu}_0.$$

Obtain the likelihood ratio test.

12.3 Show that the test given in Section 12.6 is the likelihood ratio test.

12.4 Show that the test given in Section 12.7 is the likelihood ratio test.

12.5 For a data set given by Abruzzi (1950) and analyzed by Nagarsenkar (1976), six variables were measured—the times for an operator to perform the following procedures: x_1, pick up and position a garment; x_2, press and repress a short dart; x_3, reposition a garment on ironing board; x_4, press three quarters of the length of a long dart; x_5, press the balance of a long dart; and x_6, hang a garment on the rack. The sample mean was

$$\bar{x} = \begin{pmatrix} 9.47 \\ 25.56 \\ 13.25 \\ 31.44 \\ 27.29 \\ 8.80 \end{pmatrix},$$

and the sample covariance and correlation matrices were

$$S = \begin{pmatrix} 2.57 & 0.85 & 1.58 & 1.79 & 1.33 & 0.42 \\ 0.85 & 37.00 & 3.34 & 13.47 & 7.59 & 0.52 \\ 1.56 & 3.34 & 8.44 & 5.77 & 2.00 & 0.50 \\ 1.79 & 13.47 & 5.77 & 34.01 & 10.50 & 1.77 \\ 1.33 & 7.59 & 2.00 & 10.50 & 23.01 & 3.43 \\ 0.42 & 0.52 & 0.50 & 1.77 & 3.43 & 4.59 \end{pmatrix},$$

$$R = \begin{pmatrix} 1.000 & 0.088 & 0.334 & 0.191 & 0.173 & 0.123 \\ 0.088 & 1.000 & 0.186 & 0.384 & 0.262 & 0.040 \\ 0.334 & 0.186 & 1.000 & 0.343 & 0.144 & 0.080 \\ 0.191 & 0.384 & 0.343 & 1.000 & 0.375 & 0.142 \\ 0.173 & 0.262 & 0.144 & 0.375 & 1.000 & 0.334 \\ 0.123 & 0.040 & 0.080 & 0.142 & 0.334 & 1.000 \end{pmatrix}.$$

test the hypothesis that the procedures are independent at $\alpha = 0.05$.

12.6 The within covariance matrix of Example 7.6.1 is

$$\begin{pmatrix} 0.001797916667 & 0.01003958333 & 0.007760416667 & 0.00474375 \\ 0.01003958333 & 0.4280270833 & 0.23261875 & 0.1198958333 \\ 0.007760416667 & 0.23261875 & 0.15111875 & 0.08336666667 \\ 0.00474375 & 0.1198958333 & 0.08336666667 & 0.04789375 \end{pmatrix}.$$

and the correlation matrix is

$$\begin{pmatrix} 1 & 0.3619055847 & 0.4708050193 & 0.5112079424 \\ 0.3619055847 & 1 & 0.9146394229 & 0.8373924564 \\ 0.4708050193 & 0.9146394229 & 1 & 0.9799272601 \\ 0.5112079424 & 0.8373924564 & 0.9799272601 & 1 \end{pmatrix}.$$

Test the assumption of an interclass correlation matrix.

12.7 Danford et al. (1960) give an example of a covariance matrix on 41 df; the variable was time (in days) after radiation (see Problem 7.2). The variances and covariances were

$$\begin{bmatrix} 4685 & 4522 & 5096 & 5202 & 5035 & 4746 & 5026 & 5090 & 4847 & 4933 \\ & 4799 & 5287 & 5408 & 5250 & 5026 & 5323 & 5280 & 5238 & 5228 \\ & & 6151 & 6192 & 6055 & 5773 & 6120 & 6092 & 6027 & 6132 \\ & & & 6525 & 6231 & 5923 & 6285 & 6204 & 6137 & 6252 \\ & & & & 6171 & 5850 & 6185 & 6150 & 6105 & 6175 \\ & & & & & 5745 & 5941 & 6000 & 5950 & 5959 \\ & & & & & & 6479 & 6332 & 6371 & 6382 \\ & & & & & & & 6523 & 6398 & 6390 \\ & & & & & & & & 6798 & 6566 \\ & & & & & & & & & 6753 \end{bmatrix}$$

(a) Can an interclass correlation model be assumed? Test at $\alpha = 0.05$.
(b) Are the scores for each time period independent? Test at $\alpha = 0.05$.
(c) Do the results of **a** and **b** seem reasonable?

12.8 The following data, from Hosseini (1978) were used in Chapter 3 for a T^2-test assuming equal covariances: The covariance matrices are

$$\begin{pmatrix} 0.260 & 0.181 \\ 0.181 & 0.203 \end{pmatrix}, \qquad N_1 = 252,$$

$$\begin{pmatrix} 0.303 & 0.206 \\ 0.206 & 0.194 \end{pmatrix}, \qquad N_2 = 154.$$

Test the assumption of equality of covariance matrices at $\alpha = 0.05$.

12.9 The following covariance matrix of six maternal attitudes, given by Greenhouse and Geisser (1959), was pooled from five groups with 125 df:

$$S = \begin{pmatrix} 3.100 & 0.101 & -0.279 & -0.083 & -0.009 & 1.557 \\ 0.101 & 5.780 & 1.013 & -0.114 & -1.014 & 0.039 \\ -0.279 & 1.013 & 5.560 & 1.039 & 1.366 & -0.169 \\ -0.083 & -0.114 & 1.039 & 5.600 & 3.080 & 0.258 \\ -0.009 & -1.014 & 1.366 & 3.080 & 6.820 & 0.272 \\ 1.557 & 0.039 & -0.169 & 0.258 & 0.222 & 5.170 \end{pmatrix}.$$

Test the hypothesis that this is an interclass correlation matrix. (That is, could a univariate split-plot analysis be performed on the data?)

12.10 For the data of Example 7.6.1 the correlation matrix is

$$R = \begin{pmatrix} 1 & 0.362 & 0.471 & 0.511 \\ 0.362 & 1 & 0.915 & 0.838 \\ 0.471 & 0.915 & 1 & 0.980 \\ 0.511 & 0.838 & 0.980 & 1 \end{pmatrix}$$

on 12 df. Test for an intraclass correlation model.

12.11 In Example 8.5.1 the following covariance matrices were assumed equal in a discriminant analysis. Test this assumption at $\alpha = 0.05$.

$$S_1 = \begin{pmatrix} 165.84 & 87.96 & 24.83 \\ 87.96 & 59.82 & 15.55 \\ 24.83 & 15.55 & 5.61 \end{pmatrix} \qquad \text{on} \quad 9 \text{ df},$$

$$S_2 = \begin{pmatrix} 296.62 & 119.71 & 43.47 \\ 119.71 & 63.51 & 15.76 \\ 43.47 & 15.76 & 9.21 \end{pmatrix} \quad \text{on} \quad 9 \text{ df},$$

$$S_3 = \begin{pmatrix} 135.51 & 74.73 & 30.16 \\ 74.73 & 66.4 & 22.9 \\ 30.16 & 22.9 & 11.3 \end{pmatrix} \quad \text{on} \quad 9 \text{ df}.$$

12.12 The following covariance matrixes are used in a Hotelling's T^2-procedure in Problem 3.9a.

$$S_1 = \begin{pmatrix} 294.74 & -0.60 & -32.57 \\ -0.60 & 0.0013 & 0.073 \\ -32.57 & 0.073 & 4.23 \end{pmatrix}$$

on 75 df and

$$S_2 = \begin{pmatrix} 1596.18 & -1.19 & -91.05 \\ -1.19 & 0.0009 & 0.071 \\ -91.05 & 0.071 & 5.76 \end{pmatrix}$$

on 75 df. Test for equality of covariance matrices at $\alpha = 0.05$ (The covariance matrices were not assumed equal in Problem 3.9a.)

12.13 The following matrices were assumed to be equal for a T^2-analysis in Problem 3.9b.

$$S_1 = \begin{pmatrix} 294.74 & -0.60 & -32.57 \\ -0.60 & 0.0013 & 0.073 \\ -32.57 & 0.073 & 4.23 \end{pmatrix}$$

on 75 df and

$$S_3 = \begin{pmatrix} 182.67 & -0.42 & -22.00 \\ -0.42 & 0.0010 & 0.056 \\ -22.00 & 0.056 & 3.14 \end{pmatrix}$$

on 75 df. Test for equality of covariances at $\alpha = 0.05$.

12.14 For the female carp measurement of Problem 3.9 consider observations on x_1, the asymptotic length (mm), x_2, -1 times the coefficients of growth, and x_3, -1 times the time at which length is zero (yr). The correlations between these variables are given by

$$R = \begin{pmatrix} 1 & 0.9929 & 0.9496 \\ 0.9929 & 1 & 0.9861 \\ 0.9496 & 0.9861 & 1 \end{pmatrix}$$

on 75 df. Test at $\alpha = 0.05$ that the correlations are equal using the methods of Section 12.5.2.

Table A.1 Cumulative Normal Probability Distribution: $F(Z) = P(Z \leq z)$

z	0.00	0.01	0.02	0.03	0.04
0.0	0.50000	0.50399	0.50798	0.51197	0.51595
0.1	0.53983	0.54380	0.54776	0.55172	0.55567
0.2	0.57926	0.58317	0.58706	0.59095	0.59483
0.3	0.61791	0.62172	0.62552	0.62930	0.63307
0.4	0.65542	0.65910	0.66276	0.66640	0.67003
0.5	0.69146	0.69497	0.69847	0.70194	0.70540
0.6	0.72575	0.72907	0.73237	0.73585	0.73891
0.7	0.75804	0.76115	0.76424	0.76730	0.77035
0.8	0.78814	0.79103	0.79389	0.79673	0.79955
0.9	0.81594	0.81859	0.82121	0.82381	0.82639
1.0	0.84134	0.84375	0.84614	0.84849	0.85083
1.1	0.86433	0.86650	0.86864	0.87076	0.87286
1.2	0.88493	0.88686	0.88877	0.89065	0.89251
1.3	0.90320	0.90490	0.90658	0.90824	0.90988
1.4	0.91924	0.92073	0.92220	0.92364	0.92507
1.5	0.93319	0.93448	0.93574	0.93699	0.93822
1.6	0.94520	0.94630	0.94738	0.94845	0.94950
1.7	0.95543	0.94537	0.95728	0.95818	0.95907
1.8	0.96407	0.96485	0.96562	0.96638	0.96712
1.9	0.97128	0.97193	0.97257	0.97320	0.97381
2.0	0.97725	0.97778	0.97831	0.97882	0.97932
2.1	0.98214	0.98257	0.98300	0.98341	0.98382
2.2	0.98610	0.98645	0.98679	0.98713	0.98745
2.3	0.98928	0.98956	0.98983	0.99010	0.99036
2.4	0.99180	0.99202	0.99224	0.99245	0.99266
2.5	0.99379	0.99396	0.99413	0.99430	0.99446
2.6	0.99534	0.99547	0.99560	0.99573	0.99585
2.7	0.99653	0.99664	0.99674	0.99683	0.99693
2.8	0.99744	0.99752	0.99760	0.99767	0.99774
2.9	0.99813	0.99819	0.99825	0.99831	0.99836
3.0	0.99865	0.99869	0.99874	0.99878	0.99882
3.1	0.99903	0.99906	0.99910	0.99913	0.99916
3.2	0.99931	0.99934	0.99936	0.99938	0.99940
3.3	0.99952	0.99953	0.99955	0.99957	0.99958
3.4	0.99966	0.99968	0.99969	0.99970	0.99971
3.5	0.99977	0.99978	0.99978	0.99979	0.99980
3.6	0.99984	0.99985	0.99985	0.99986	0.99986
3.7	0.99989	0.99990	0.99990	0.99990	0.99991
3.8	0.99993	0.99993	0.99993	0.99994	0.99994
3.9	0.99995	0.99995	0.99996	0.99996	0.99996
4.0	0.99997	0.99997	0.99997	0.99997	0.99997

Table A.1 (*continued*)

0.05	0.06	0.07	0.08	0.09
0.51994	0.52392	0.52790	0.53188	0.53586
0.55962	0.56356	0.56749	0.57142	0.57535
0.59871	0.60257	0.60642	0.61026	0.61409
0.63683	0.64058	0.64431	0.64803	0.65173
0.67364	0.67724	0.68082	0.68439	0.68793
0.70884	0.71226	0.71566	0.71904	0.72240
0.74215	0.74537	0.74857	0.75175	0.75490
0.77337	0.77637	0.77935	0.78230	0.78524
0.80234	0.80511	0.80785	0.81057	0.81327
0.82894	0.83147	0.83398	0.83646	0.83891
0.85314	0.85543	0.85769	0.85993	0.86214
0.87493	0.87698	0.87900	0.88100	0.88298
0.89435	0.89617	0.89796	0.89973	0.90147
0.91149	0.91309	0.91466	0.91621	0.91774
0.92647	0.92785	0.92922	0.93056	0.93189
0.93943	0.94062	0.94179	0.94295	0.94408
0.95053	0.95154	0.95254	0.95352	0.95449
0.95994	0.96080	0.96164	0.96246	0.96327
0.96784	0.96856	0.96926	0.96995	0.97062
0.97441	0.97500	0.97558	0.97615	0.97670
0.97982	0.98030	0.98077	0.98124	0.98169
0.98422	0.98461	0.98500	0.98537	0.98574
0.98778	0.98809	0.98840	0.98870	0.98899
0.99061	0.99086	0.99111	0.99134	0.99158
0.99286	0.99305	0.99324	0.99343	0.99361
0.99461	0.99477	0.99492	0.99506	0.99520
0.99598	0.99609	0.99621	0.99632	0.99643
0.99702	0.99711	0.99720	0.99728	0.99736
0.99781	0.99788	0.99795	0.99801	0.99807
0.99841	0.99846	0.99851	0.99856	0.99861
0.99886	0.99889	0.99893	0.99896	0.99900
0.99918	0.99921	0.99924	0.99926	0.99929
0.99942	0.99944	0.99946	0.99948	0.99950
0.99960	0.99961	0.99962	0.99964	0.99965
0.99972	0.99973	0.99974	0.99975	0.99976
0.99981	0.99981	0.99982	0.99983	0.99983
0.99987	0.99987	0.99988	0.99988	0.99989
0.99991	0.99992	0.99992	0.99992	0.99992
0.99994	0.99994	0.99995	0.99995	0.99995
0.99996	0.99996	0.99996	0.99997	0.99997
0.99997	0.99998	0.99998	0.99998	0.99998

Table A.2 Critical Values of the t-Distribution[a]

	α					
df	0.200	0.125	0.050	0.025	0.003	0.0005
1	1.3764	2.4142	6.3138	12.7062	63.6567	636.6193
2	1.0607	1.6036	2.9200	4.3026	9.9248	31.5990
3	0.9785	1.4227	2.3534	3.1825	5.8409	12.9240
4	0.9410	1.3444	2.1318	2.7765	4.6041	8.6103
5	0.9195	1.3010	2.0151	2.5706	4.0322	6.8688
6	0.9057	1.2734	1.9432	2.4469	3.7074	5.9588
7	0.8960	1.2543	1.8946	2.3646	3.4994	5.4078
8	0.8889	1.2404	1.8596	2.3060	3.3554	5.0413
9	0.8834	1.2297	1.8332	2.2621	3.2498	4.7809
10	0.8791	1.2213	1.8125	2.2282	3.1693	4.5869
11	0.8755	1.2145	1.7959	2.2010	3.1059	4.4370
12	0.8726	1.2089	1.7823	2.1788	3.0545	4.3178
13	0.8702	1.2042	1.7709	2.1603	3.0122	4.2208
14	0.8681	1.2002	1.7613	2.1447	2.9769	4.1404
15	0.8662	1.1967	1.7530	2.1315	2.9467	4.0727

[a] This table gives the value of t_α such that $P(t > t_\alpha) = \alpha$ for the associated degrees of freedom.

Table A.3 Critical Values of the Chi-Square Distribution[a]

	α			
df	0.01	0.05	0.10	0.30
1	0.0002	0.0039	0.0158	0.1485
2	0.0201	0.1026	0.2107	0.7133
3	0.1148	0.3518	0.5844	1.4237
4	0.2971	0.7107	1.0636	2.1947
5	0.5543	1.1455	1.6103	2.9999
6	0.8721	1.6354	2.2041	3.8276
7	1.2390	2.1673	2.8331	4.6713
8	1.6465	2.7326	3.4895	5.5274
9	2.0879	3.3251	4.1682	6.3933
10	2.5582	3.9403	4.8652	7.2672
11	3.0535	4.5748	5.5778	8.1479
12	3.5706	5.2260	6.3038	9.0343
13	4.1069	5.8919	7.0415	9.9257
14	4.6604	6.5706	7.7895	10.8215
15	5.2293	7.2609	8.5468	11.7212
16	5.8122	7.9616	9.3122	12.6243
17	6.4078	8.6718	10.0852	13.5307
18	7.0149	9.3905	10.8649	14.4399
19	7.6327	10.1170	11.6509	15.3517
20	8.2604	10.8508	12.4426	16.2659
21	8.8972	11.5913	13.2396	17.1823
22	9.5425	12.3380	14.0415	18.1007
23	10.1957	13.0905	14.8480	19.0211
24	10.8564	13.8484	15.6587	19.9432
25	11.5240	14.6114	16.4734	20.8670
26	12.1981	15.3792	17.2919	21.7924
27	12.8785	16.1514	18.1139	22.7192
28	13.5647	16.9279	18.9392	23.6475
29	14.2565	17.7084	19.7677	24.5770
30	14.9535	18.4927	20.5992	25.5078
40	22.1643	26.5093	29.0505	34.8719
50	29.7067	34.7643	37.6886	44.3133
100	70.0649	77.9295	82.3581	92.1289

[a] The table gives the value of χ^2_α such that $P(\chi^2 \leq \chi^2_\alpha) = \alpha$.

Table A.2 (*continued*)

	α					
df	0.200	0.125	0.050	0.025	0.003	0.0005
16	0.8647	1.1937	1.7458	2.1199	2.9207	4.0150
17	0.8633	1.1911	1.7396	2.1098	2.8983	3.9652
18	0.8620	1.1888	1.7340	2.1009	2.8785	3.9216
19	0.8610	1.1867	1.7291	2.0930	2.8609	3.8835
20	0.8600	1.1848	1.7247	2.0860	2.8453	3.8495
21	0.8591	1.1831	1.7207	2.0796	2.8313	3.8193
22	0.8583	1.1816	1.7171	2.0739	2.8187	3.7921
23	0.8575	1.1802	1.7138	2.0686	2.8074	3.7677
24	0.8569	1.1789	1.7108	2.0639	2.7970	3.7454
25	0.8562	1.1778	1.7081	2.0595	2.7875	3.7251
26	0.8557	1.1767	1.7056	2.0555	2.7787	3.7066
27	0.8551	1.1757	1.7032	2.0518	2.7707	3.6896
28	0.8546	1.1748	1.7011	2.0484	2.7633	3.6739
29	0.8542	1.1739	1.6991	2.0452	2.7564	2.6594
30	0.8538	1.1731	1.6972	2.0422	2.7500	3.6460
40	0.8507	1.1673	1.6838	2.0211	2.7045	3.5510
50	0.8489	1.1639	1.6759	2.0086	2.6778	3.4961

Table A.3 (*continued*)

		α		
0.50	0.70	0.90	0.95	0.99
0.4549	1.0742	2.7055	3.8415	6.6349
1.3863	2.4079	4.6052	5.9915	9.2103
2.3660	3.6649	6.2514	7.8147	11.3449
3.3567	4.8784	7.7794	9.4877	13.2767
4.3515	6.0644	9.2364	11.0705	15.0863
5.3481	7.2311	10.6446	12.5916	16.8119
6.3458	8.3834	12.0170	14.0671	18.4753
7.3441	9.5245	13.3616	15.5073	20.0902
8.3428	10.6564	14.6837	16.9190	21.6660
9.3418	11.7807	15.9872	18.3070	23.2093
10.3410	12.8987	17.2750	19.6751	24.7250
11.3403	14.0111	18.5493	21.0261	26.2170
12.3398	15.1187	19.8119	22.3620	27.6883
13.3393	16.2221	21.0641	23.6848	29.1412
14.3389	17.3217	22.3071	24.9958	30.5779
15.3385	18.4179	23.5418	26.2962	31.9999
16.3382	19.5110	24.7690	27.5871	33.4087
17.3379	20.6014	25.9894	28.8693	34.8053
18.3377	21.6891	27.2036	30.1435	36.1909
19.3374	22.7745	28.4120	31.4104	37.5662
20.3372	23.8578	29.6151	32.6706	38.9322
21.3370	24.9390	30.8133	33.9244	40.2894
22.3369	26.0184	32.0069	35.1725	41.6384
23.3367	27.0960	33.1962	36.4150	42.9798
24.3366	28.1719	34.3816	37.6525	44.3141
25.3365	29.2463	35.5632	38.8851	45.6417
26.3363	30.3193	36.7412	40.1133	46.9629
27.3362	31.3909	37.9159	41.3371	48.2782
28.3361	32.4612	39.0875	42.5570	49.5879
29.3360	33.5302	40.2560	43.7730	50.8922
39.3353	44.1649	51.8051	55.7585	63.6907
49.3349	54.7228	63.1671	67.5048	76.1539
99.3341	106.9058	118.4980	124.3421	135.8067

Table A.4 Percentage points of the F-distribution[a]

	df, ν_1								
df, ν_2	1	2	3	4	5	6	7	8	9
					$\alpha = 0.05$				
1	161.4	199.5	215.7	224.6	230.2	234.0	236.8	238.9	240.5
2	18.51	19.00	19.16	19.25	19.30	19.33	19.35	19.37	19.38
3	10.13	9.55	9.28	9.12	9.01	8.94	8.89	8.85	8.81
4	7.71	6.94	6.59	6.39	6.26	6.16	6.09	6.04	6.00
5	6.61	5.79	5.41	5.19	5.05	4.95	4.88	4.82	4.77
6	5.99	5.14	4.76	4.53	4.39	4.28	4.21	4.15	4.10
7	5.59	4.74	4.35	4.12	3.97	3.87	3.79	3.73	3.68
8	5.32	4.46	4.07	3.84	3.69	3.58	3.50	3.44	3.39
9	5.12	4.26	3.86	3.63	3.48	3.37	3.29	3.23	3.18
10	4.96	4.10	3.71	3.48	3.33	3.22	3.14	3.07	3.02
11	4.84	3.98	3.59	3.36	3.20	3.09	3.01	2.95	2.90
12	4.75	3.89	3.49	3.26	3.11	3.00	2.91	2.85	2.80
13	4.67	3.81	3.41	3.18	3.03	2.92	2.83	2.77	2.71
14	4.60	3.74	3.34	3.11	2.96	2.85	2.76	2.70	2.65
15	4.54	3.68	3.29	3.06	2.90	2.79	2.71	2.64	2.59
16	4.49	3.63	3.24	3.01	2.85	2.74	2.66	2.59	2.54
17	4.45	3.59	3.20	2.96	2.81	2.70	2.61	2.55	2.49
18	4.41	3.55	3.16	2.93	2.77	2.66	2.58	2.51	2.46
19	4.38	3.52	3.13	2.90	2.74	2.63	2.54	2.48	2.42
20	4.35	3.49	3.10	2.87	2.71	2.60	2.51	2.45	2.39
21	4.32	3.47	3.07	2.84	2.68	2.57	2.49	2.42	2.37
22	4.30	3.44	3.05	2.82	2.66	2.55	2.46	2.40	2.34
23	4.28	3.42	3.03	2.80	2.64	2.53	2.44	2.37	2.32
24	4.26	3.40	3.01	2.78	2.62	2.51	2.42	2.36	2.30
25	4.24	3.39	2.99	2.76	2.60	2.49	2.40	2.34	2.28
26	4.23	3.37	2.98	2.74	2.59	2.47	2.39	2.32	2.27
27	4.21	3.35	2.96	2.73	2.57	2.46	2.37	2.31	2.25
28	4.20	3.34	2.95	2.71	2.56	2.45	2.36	2.29	2.24
29	4.18	3.33	2.93	2.70	2.55	2.43	2.35	2.28	2.22
30	4.17	3.32	2.92	2.69	2.53	2.42	2.33	2.27	2.21
40	4.08	3.23	2.84	2.61	2.45	2.34	2.25	2.18	2.12
60	4.00	3.15	2.76	2.53	2.37	2.25	2.17	2.10	2.04
120	3.92	3.07	2.68	2.45	2.29	2.17	2.09	2.02	1.96
∞	3.84	3.00	2.60	2.37	2.21	2.10	2.01	1.94	1.88

Table A.4 (*continued*)

df, ν_1									
10	12	15	20	24	30	40	60	120	∞
				$\alpha = 0.05$					
6056	6106	6157	6209	6235	6261	6287	6313	6339	6366
99.40	99.42	99.43	99.45	99.46	99.47	99.47	99.48	99.49	99.50
27.23	27.05	26.87	26.69	26.60	26.50	26.41	26.32	26.22	26.13
14.55	14.37	14.20	14.02	13.93	13.84	13.75	13.65	13.56	13.46
10.05	9.89	9.72	9.55	9.47	9.38	9.29	9.20	9.11	9.02
7.87	7.72	7.56	7.40	7.31	7.23	7.14	7.06	6.97	6.88
6.62	6.47	6.31	6.16	6.07	5.99	5.91	5.82	5.74	5.65
5.81	5.67	5.52	5.36	5.28	5.20	5.12	5.03	4.95	4.86
5.26	5.11	4.96	4.81	4.73	4.65	4.57	4.48	4.40	4.31
4.85	4.71	4.56	4.41	4.33	4.25	4.17	4.08	4.00	3.91
4.54	4.40	4.25	4.10	4.02	3.94	3.86	3.78	3.69	3.60
4.30	4.16	4.01	3.86	3.78	3.70	3.62	3.54	3.45	3.36
4.10	3.96	3.82	3.66	3.59	3.51	3.43	3.34	3.25	3.17
3.94	3.80	3.66	3.51	3.43	3.35	3.27	3.18	3.09	3.00
3.80	3.67	3.52	3.37	3.29	3.21	3.13	3.05	2.96	2.87
3.69	3.55	3.41	3.26	3.18	3.10	3.02	2.93	2.84	2.75
3.59	3.46	3.31	3.16	3.08	3.00	2.92	2.83	2.75	2.65
3.51	3.37	3.23	3.08	3.00	2.92	2.84	2.75	2.66	2.57
3.43	3.30	3.15	3.00	2.92	2.84	2.76	2.67	2.58	2.49
3.37	3.23	3.09	2.94	2.86	2.78	2.69	2.61	2.52	2.42
3.31	3.17	3.03	2.88	2.80	2.72	2.64	2.55	2.46	2.36
3.26	3.12	2.98	2.83	2.75	2.67	2.58	2.50	2.40	2.31
3.21	3.07	2.93	2.78	2.70	2.62	2.54	2.45	2.35	2.26
3.17	3.03	2.89	2.74	2.66	2.58	2.49	2.40	2.31	2.21
3.13	2.99	2.85	2.70	2.62	2.54	2.45	2.36	2.27	2.17
3.09	2.96	2.81	2.66	2.58	2.50	2.42	2.33	2.23	2.13
3.06	2.93	2.78	2.63	2.55	2.47	2.38	2.29	2.20	2.10
3.03	2.90	2.75	2.60	2.52	2.44	2.35	2.26	2.17	2.06
3.00	2.87	2.73	2.57	2.49	2.41	2.33	2.23	2.14	2.03
2.98	2.84	2.70	2.55	2.47	2.39	2.30	2.21	2.11	2.01
2.80	2.66	2.52	2.37	2.29	2.20	2.11	2.02	1.92	1.80
2.63	2.50	2.35	2.20	2.12	2.03	1.94	1.84	1.73	1.60
2.47	2.34	2.19	2.03	1.95	1.86	1.76	1.66	1.53	1.38
2.32	2.18	2.04	1.88	1.79	1.70	1.59	1.47	1.32	1.00

(*continued*)

Table A.4 Percentage points of the *F*-distribution[a] (*continued*)

	df, ν_1								
df, ν_2	1	2	3	4	5	6	7	8	9
					$\alpha = 0.01$				
1	4052	4999.5	5403	5625	5764	5859	5928	5982	6022
2	98.50	9.00	99.17	99.25	99.30	99.33	99.36	99.37	99.39
3	34.12	30.82	29.46	28.71	28.24	27.91	27.67	27.49	27.35
4	21.20	18.00	16.69	15.98	15.52	15.21	14.98	14.80	14.66
5	16.26	13.27	12.06	11.39	10.97	10.67	10.46	10.29	10.16
6	13.75	10.92	9.78	9.15	8.75	8.47	8.26	8.10	7.98
7	12.25	9.55	8.45	7.85	7.46	7.19	6.99	6.84	6.72
8	11.26	8.65	7.59	7.01	6.63	6.37	6.18	6.03	5.91
9	10.56	8.02	6.99	6.42	6.06	5.80	5.61	5.47	5.35
10	10.04	7.56	6.55	5.99	5.64	5.39	5.20	5.06	4.94
11	9.65	7.21	6.22	5.67	5.32	5.07	4.89	4.74	4.63
12	9.33	6.93	5.95	5.41	5.06	4.82	4.64	4.50	4.39
13	9.07	6.70	5.74	5.21	4.86	4.62	4.44	4.30	4.19
14	8.86	6.51	5.56	5.04	4.69	4.46	4.28	4.14	4.03
15	8.68	6.36	5.42	4.89	4.56	4.32	4.14	4.00	3.89
16	8.53	6.23	5.29	4.77	4.44	4.20	4.03	3.89	3.78
17	8.40	6.11	5.18	4.67	4.34	4.10	3.93	3.79	3.68
18	8.29	6.01	5.09	4.58	4.25	4.01	3.84	3.71	3.60
19	8.18	5.93	5.01	4.50	4.17	3.94	3.77	3.63	3.52
20	8.10	5.85	4.94	4.43	4.10	3.87	3.70	3.56	3.46
21	8.02	5.78	4.87	4.37	4.04	3.81	3.64	3.51	3.40
22	7.95	5.72	4.82	4.31	3.99	3.76	3.59	3.45	3.35
23	7.88	5.66	4.76	4.26	3.94	3.71	3.54	3.41	3.30
24	7.82	5.61	4.72	4.22	3.90	3.67	3.50	3.36	3.26
25	7.77	5.57	4.68	4.18	3.85	3.63	3.46	3.32	3.22
26	7.72	5.53	4.64	4.14	3.82	3.59	3.42	3.29	3.18
27	7.68	5.49	4.60	4.11	3.78	3.56	3.39	3.26	3.15
28	7.64	5.45	4.57	4.07	3.75	3.53	3.36	3.23	3.12
29	7.60	5.42	4.54	4.04	3.73	3.50	3.33	3.20	3.09
30	7.56	5.39	4.51	4.02	3.70	3.47	3.30	3.17	3.07
40	7.31	5.18	4.31	3.83	3.51	3.29	3.12	3.99	2.89
60	7.08	4.98	4.13	3.65	3.34	3.12	2.95	2.82	2.72
120	6.85	4.79	3.95	3.48	3.17	2.96	2.79	2.66	2.56
∞	6.63	4.61	3.78	3.32	3.02	2.80	2.64	2.51	2.41

[a]Reproduced with the permission of the editors of *Biometrika*.

Table A.4 (*continued*)

df, ν_1									
10	12	15	20	24	30	40	60	120	∞
				$\alpha = 0.01$					
241.9	243.9	245.9	248.0	249.1	250.1	251.1	252.2	253.3	254.3
19.40	19.41	19.43	19.45	19.45	19.46	19.47	19.48	19.49	19.50
8.79	8.74	8.70	8.66	8.64	8.62	8.59	8.57	8.55	8.53
5.96	5.91	5.86	5.80	5.77	5.75	5.72	5.69	5.66	5.63
4.74	4.68	4.62	4.56	4.53	4.50	4.46	4.43	4.40	4.36
4.06	4.00	3.94	3.87	3.84	3.81	3.77	3.74	3.70	3.67
3.64	3.57	3.51	3.44	3.41	3.38	3.34	3.30	3.27	3.23
3.35	3.28	3.22	3.15	3.12	3.08	3.04	3.01	2.97	2.93
3.14	3.07	3.01	2.94	2.90	2.86	2.83	2.79	2.75	2.71
2.98	2.91	2.85	2.77	2.74	2.70	2.66	2.62	2.58	2.54
2.85	2.79	2.72	2.65	2.61	2.57	2.53	2.49	2.45	2.40
2.75	2.69	2.62	2.54	2.51	2.47	2.43	2.38	2.34	2.30
2.67	2.60	2.53	2.46	2.42	2.38	2.34	2.30	2.25	2.21
2.60	2.53	2.46	2.39	2.35	2.31	2.27	2.22	2.18	2.13
2.54	2.48	2.40	2.33	2.29	2.25	2.20	2.16	2.11	2.07
2.49	2.42	2.35	2.28	2.24	2.19	2.15	2.11	2.06	2.01
2.45	2.38	2.31	2.23	2.19	2.15	2.10	2.06	2.01	1.96
2.41	2.34	2.27	2.19	2.15	2.11	2.06	2.02	1.97	1.92
2.38	2.31	2.23	2.16	2.11	2.07	2.03	1.98	1.93	1.88
2.35	2.28	2.20	2.12	2.08	2.04	1.99	1.95	1.90	1.84
2.32	2.25	2.18	2.10	2.05	2.01	1.96	1.92	1.87	1.81
2.30	2.23	2.15	2.07	2.03	1.98	1.94	1.89	1.84	1.78
2.27	2.20	2.13	2.05	2.01	1.96	1.91	1.86	1.81	1.76
2.25	2.18	2.11	2.03	1.98	1.94	1.89	1.84	1.79	1.73
2.24	2.16	2.09	2.01	1.96	1.92	1.87	1.82	1.77	1.71
2.22	2.15	2.07	1.99	1.95	1.90	1.85	1.80	1.75	1.69
2.20	2.13	2.06	1.97	1.93	1.88	1.84	1.79	1.73	1.67
2.19	2.12	2.04	1.96	1.91	1.87	1.82	1.77	1.71	1.65
2.18	2.10	2.03	1.94	1.90	1.85	1.81	1.75	1.70	1.64
2.16	2.09	2.01	1.93	1.89	1.84	1.79	1.74	1.68	1.62
2.08	2.00	1.92	1.84	1.79	1.74	1.69	1.64	1.58	1.51
1.99	1.92	1.84	1.75	1.70	1.65	1.59	1.53	1.47	1.39
1.91	1.83	1.75	1.66	1.61	1.55	1.50	1.43	1.35	1.25
1.83	1.75	1.67	1.57	1.52	1.46	1.39	1.32	1.22	1.00

Table A.5 x_α Such That $\{ch_1(HE^{-1}) \geq x_\alpha\} = \alpha^a$

$p_0 = 2, \alpha = 0.10$	m_0					
n	1	2	3	4	5	6
3	0.800000	0.948683	0.965489	0.974004	0.979148	0.982543
5	0.683772	0.794990	0.846342	0.876651	0.890815	0.911247
7	0.535841	0.662760	0.730698	0.774635	0.805787	0.829165
9	0.437659	0.563194	0.636631	0.680542	0.724427	0.753460
11	0.369043	0.487960	0.561760	0.614610	0.655074	0.687340
13	0.313708	0.429771	0.501663	0.554775	0.596457	0.630379
15	0.280314	0.383658	0.452669	0.504939	0.546742	0.581321
17	0.250106	0.346372	0.412182	0.462977	0.504265	0.538877
19	0.225736	0.315498	0.378172	0.427252	0.467644	0.501919
21	0.205072	0.289663	0.349254	0.346520	0.435860	0.469521
25	0.174596	0.248815	0.302785	0.346454	0.383424	0.415533
31	0.142304	0.205285	0.252260	0.291069	0.324527	0.354060
35	0.126074	0.183815	0.226958	0.262965	0.294286	0.322157
41	0.108749	0.158870	0.197243	0.229647	0.258129	0.283713
61	0.073881	0.109348	0.137244	0.161348	0.182977	0.202779
101	0.045007	0.067333	0.085286	0.101100	0.115542	0.128983

$p_0 = 3, \alpha = 0.10$	m_0					
n	1	2	3	4	5	6
4	0.932170	0.965489	0.976859	0.982593	0.980050	0.988362
6	0.758637	0.846342	0.885733	0.908085	0.923837	0.934626
8	0.621840	0.730698	0.787054	0.822877	0.848000	0.860709
10	0.522997	0.636631	0.700812	0.744045	0.775687	0.800045
12	0.450036	0.561760	0.628695	0.675685	0.711183	0.739211
14	0.394446	0.501663	0.568094	0.017231	0.654801	0.685063
16	0.350850	0.452699	0.518459	0.567235	0.605726	0.637234
18	0.315813	0.412182	0.475996	0.524248	0.562923	0.595007
20	0.287073	0.378172	0.439734	0.487017	0.325417	0.557625
26	0.225361	0.302785	0.357382	0.400747	0.436976	0.468120
30	0.197059	0.267134	0.317500	0.358123	0.392513	0.422422
36	0.165791	0.226958	0.271843	0.308666	0.368179	0.368179
40	0.149920	0.206248	0.248016	0.252577	0.312494	0.339038
60	0.1013628	0.141553	0.172307	0.198435	0.221587	0.242572
100	0.061497	0.086932	0.106897	0.124223	0.139892	0.154348

Table A.5 (*continued*)

m_0						
7	8	9	10	12	15	20
0.985061	0.986916	0.988362	0.989519	0.991258	0.993001	0.994746
0.922106	0.930581	0.937383	0.942966	0.951588	0.960528	0.969811
0.847414	0.862081	0.874141	0.884240	0.900213	0.917288	0.935609
0.776762	0.795926	0.811989	0.825663	0.847728	0.871939	0.898687
0.713803	0.735966	0.754836	0.771117	0.797833	0.827810	0.861791
0.658681	0.682737	0.703482	0.721585	0.751712	0.786175	0.826136
0.610578	0.635749	0.657689	0.077017	0.709574	0.747444	0.792245
0.568501	0.594249	0.616896	0.637008	0.671242	0.711649	0.760308
0.531526	0.557480	0.580486	0.601060	0.636400	0.678651	0.730348
0.498855	0.524763	0.547881	0.568682	0.604697	0.648249	0.702303
0.443893	0.469244	0.492115	0.512901	0.549374	0.594336	0.651531
0.380528	0.404506	0.426406	0.440537	0.482405	0.527623	0.586853
0.347320	0.370271	0.391366	0.410871	0.445902	0.490592	0.550063
0.307014	0.328439	0.348279	0.366754	0.400259	0.443627	0.502482
0.221135	0.238296	0.254438	0.209694	0.297938	0.335665	0.389149
0.141637	0.153643	0.165096	0.176069	0.196773	0.225267	0.267431

m_0						
7	8	9	10	12	15	20
0.990016	0.991258	0.992226	0.993001	0.994164	0.995328	0.996494
0.942714	0.949009	0.954050	0.958179	0.964545	0.971128	0.977947
0.881230	0.892850	0.902370	0.910319	0.922849	0.936186	0.950433
0.819459	0.835339	0.848590	0.859828	0.877883	0.897585	0.919220
0.762022	0.781009	0.797094	0.810913	0.833474	0.858625	0.886933
0.710109	0.731258	0.749397	0.765153	0.791230	0.820845	0.854911
0.663671	0.686263	0.705843	0.723009	0.751754	0.784935	0.823847
0.622237	0.645741	0.666294	0.684456	0.715183	0.751158	0.794817
0.585228	0.609259	0.630433	0.649273	0.681433	0.719565	0.765777
0.495398	0.519614	0.541331	0.560967	0.595200	0.637037	0.689716
0.448895	0.472619	0.494081	0.513642	0.548112	0.590899	0.645875
0.393150	0.415774	0.436449	0.455469	0.489413	0.532341	0.588875
0.362962	0.384830	0.404791	0.423309	0.456581	0.499087	0.555831
0.261862	0.279766	0.296500	0.312226	0.341124	0.379322	0.432759
0.167864	0.180613	0.192713	0.204251	0.225897	0.255439	0.298099

(*continued*)

Table A.5 x_α Such That $\{\mathrm{ch}_1(HE^{-1}) \geq x_\alpha\} = \alpha^a$ (*continued*)

$p_0 = 4, \alpha = 0.10$			m_0			
n	1	2	3	4	5	6
5	0.948683	0.974004	0.982593	0.986916	0.989519	0.991258
7	0.804200	0.876651	0.908685	0.927215	0.939394	0.948041
9	0.679539	0.774635	0.822877	0.853224	0.874366	0.890038
11	0.583890	0.680542	0.744045	0.781974	0.809513	0.830591
13	0.510310	0.614610	0.675685	0.718013	0.749752	0.774578
15	0.452565	0.554775	0.617231	0.661860	0.696091	0.723476
17	0.400245	0.504939	0.567235	0.612827	0.648471	0.677443
19	0.365362	0.462977	0.524245	0.569955	0.606254	0.636148
21	0.336848	0.427252	0.487017	0.532316	0.568759	0.599104
25	0.287498	0.369831	0.425999	0.409623	0.505440	0.535794
31	0.235569	0.307476	0.358123	0.398469	0.432316	0.461545
35	0.210211	0.376310	0.323581	0.361702	0.394021	0.422194
41	0.180961	0.239779	0.282577	0.317578	0.347618	0.374092
61	0.123570	0.166298	0.198435	0.225447	0.249196	0.270589
101	0.075581	0.103025	0.124229	0.142462	0.158822	0.173839

$p_0 = 5, \alpha = 0.10$			m_0		
n	1	2	3	4	5
6	0.958732	0.979148	0.986050	0.989514	0.991607
8	0.835071	0.896815	0.923837	0.939394	0.949591
10	0.721417	0.805787	0.848000	0.874365	0.892652
12	0.630181	0.724427	0.775687	0.809513	0.833937
14	0.557683	0.655074	0.711183	0.749752	0.778383
16	0.499377	0.596457	0.654801	0.696091	0.727551
18	0.451727	0.546742	0.605726	0.648471	0.681660
20	0.412171	0.504265	0.562923	0.606254	0.640423
26	0.325975	0.408004	0.463081	0.505440	0.539975
30	0.285928	0.361588	0.413582	0.454306	0.488019
36	0.241348	0.308675	0.356116	0.394021	0.425940
40	0.218593	0.281107	0.325813	0.361844	0.392448
60	0.148479	0.194303	0.228240	0.256476	0.281108
100	0.080416	0.119960	0.142521	0.161759	0.178917

Table A.5 (*continued*)

m_0						
7	8	9	10	12	15	20
0.992502	0.993437	0.994164	0.994746	0.995620	0.996494	0.997369
0.954509	0.959536	0.963556	0.966846	0.971912	0.977143	0.982554
0.902160	0.911833	0.919742	0.926333	0.936703	0.977713	0.959442
0.847319	0.860953	0.872298	0.881898	0.897278	0.914002	0.932299
0.794716	0.811412	0.825509	0.837588	0.857242	0.879060	0.903502
0.746019	0.764972	0.781170	0.795197	0.818326	0.844468	0.874380
0.701617	0.722180	0.739934	0.755448	0.781325	0.811038	0.845682
0.661375	0.683047	0.701923	0.718546	0.746552	0.779162	0.817830
0.624960	0.647361	0.667020	0.684450	0.714075	0.748999	0.791046
0.562113	0.585131	0.605629	0.624010	0.655712	0.693875	0.741044
0.487267	0.510206	0.530871	0.549637	0.582554	0.623160	0.674947
0.447196	0.469666	0.490052	0.490052	0.508686	0.541660	0.636319
0.397821	0.419341	0.439032	0.457172	0.489618	0.530810	0.585385
0.290154	0.308233	0.325068	0.340835	0.369685	0.407590	0.460230
0.187815	0.200947	0.213367	0.225177	0.247244	0.277198	0.320766

(*continued*)

Table A.5 x_α Such That $\{ch_1(HE^{-1}) \geq x_\alpha\} = \alpha^a$ (*continued*)

$p_0 = 2, \alpha = 0.05$				m_0		
n	1	2	3	4	5	6
3	0.950000	0.97479	0.983048	0.987259	0.989794	0.991488
5	0.776393	0.857729	0.894267	0.915546	0.929586	0.939578
7	0.631597	0.736978	0.791858	0.826821	0.851366	0.869655
9	0.527129	0.038346	0.701738	0.744615	0.776067	0.800314
11	0.450720	0.560312	0.626689	0.673534	0.709032	0.737118
13	0.393038	0.498145	0.564642	0.613082	0.650717	0.681106
15	0.343816	0.447851	0.513022	0.561674	0.600224	0.631870
17	0.312344	0.406494	0.469638	0.517705	0.556405	0.588604
19	0.283129	0.371965	0.432779	0.479813	0.518183	0.550470
21	0.258866	0.342744	0.401137	0.446897	0.516709	0.516709
25	0.220922	0.296057	0.349732	0.392671	0.428715	0.459809
31	0.181036	0.245671	0.293095	0.331865	0.365025	0.344108
35	0.161566	0.220588	0.264454	0.300698	0.331985	0.359655
41	0.139108	0.191257	0.230590	0.263488	0.292195	0.317832
61	0.095034	0.132448	0.161485	0.186364	0.208544	0.228746
101	0.058155	0.081970	0.100891	0.117434	0.132457	0.146374

$p_0 = 3, \alpha = 0.05$				m_0		
n	1	2	3	4	5	6
4	0.966383	0.983048	0.988666	0.991488	0.993184	0.994317
6	0.831750	0.894267	0.921841	0.937761	0.948211	0.955623
8	0.704013	0.791858	0.836506	0.864592	0.884150	0.898640
10	0.603932	0.701738	0.756023	0.792203	0.818485	0.838603
12	0.526623	0.626689	0.685665	0.726640	0.757361	0.781475
14	0.465976	0.564642	0.625386	0.668932	0.702389	0.729180
16	0.417435	0.513022	0.573853	0.618539	0.653548	0.682039
18	0.377834	0.469638	0.529603	0.574524	0.610280	0.639771
20	0.344972	0.432779	0.491351	0.535938	0.571800	0.601895
26	0.273208	0.349732	0.403009	0.445032	0.479837	0.509641
30	0.239958	0.309846	0.359544	0.399326	0.432801	0.401772
36	0.202773	0.264454	0.309275	0.345788	0.404340	0.404340
40	0.183765	0.240878	0.282833	0.317317	0.347006	0.373230
60	0.161096	0.166480	0.197921	0.224479	0.247904	0.269052
100	0.076276	0.102822	0.123519	0.141400	0.157494	0.172299

Table A.5 (*continued*)

m_0						
7	8	9	10	12	15	20
0.992699	0.993609	0.994317	0.994884	0.995735	0.996586	0.997439
0.947065	0.952890	0.957554	0.961274	0.967260	0.973345	0.979645
0.883854	0.895218	0.904529	0.912304	0.924560	0.937605	0.951540
0.819658	0.835490	0.848709	0.859924	0.877949	0.897625	0.919243
0.760010	0.779085	0.795258	0.809162	0.831878	0.857224	0.885775
0.706304	0.727611	0.745906	0.761812	0.788162	0.818124	0.852633
0.658479	0.681255	0.701020	0.718366	0.747449	0.781071	0.820564
0.615997	0.639683	0.660425	0.678776	0.709867	0.740332	0.789927
0.578209	0.602403	0.623756	0.642780	0.675304	0.713942	0.760868
0.544491	0.568905	0.590598	0.610042	0.643520	0.683817	0.733418
0.487116	0.511409	0.533317	0.552998	0.587517	0.629802	0.683185
0.420036	0.443416	0.464686	0.484169	0.518727	0.562026	0.618319
0.384509	0.407078	0.427741	0.446782	0.480831	0.524005	0.581036
0.341065	0.362338	0.381965	0.400182	0.433079	0.475410	0.532446
0.247391	0.264756	0.281037	0.296379	0.324673	0.362265	0.415205
0.159426	0.171769	0.183510	0.194729	0.215829	0.244730	0.287247

m_0						
7	8	9	10	12	15	20
0.995127	0.995735	0.9962078	0.996586	0.997154	0.997723	0.998292
0.961163	0.965464	0.968904	0.971717	0.976046	0.980515	0.985133
0.909843	0.918779	0.926082	0.932166	0.941734	0.951880	0.962694
0.854567	0.807577	0.878402	0.887559	0.902226	0.918168	0.935601
0.801010	0.817208	0.830885	0.842604	0.801672	0.882836	0.906539
0.751248	0.769809	0.785675	0.799417	0.822079	0.847696	0.877006
0.705802	0.726082	0.743574	0.758863	0.784373	0.813673	0.847842
0.664684	0.686102	0.704766	0.721210	0.748927	0.781214	0.819516
0.627483	0.649670	0.669155	0.686439	0.715833	0.750507	0.792277
0.535630	0.558614	0.579159	0.597680	0.629851	0.668964	0.717910
0.487305	0.510103	0.530662	0.549346	0.582150	0.622662	0.674393
0.428749	0.450788	0.470865	0.489285	0.522041	0.563262	0.617216
0.396773	0.418155	0.437740	0.455799	0.488135	0.529244	0.583792
0.288427	0.306356	0.323069	0.338738	0.367442	0.405214	0.457761
0.186102	0.199088	0.211385	0.223087	0.244981	0.274747	0.314513

(*continued*)

Table A.5 x_α Such That $\{\mathrm{ch}_1(HE^{-1}) \geq x_\alpha\} = \alpha^a$ (*continued*)

$p_0 = 4, \alpha = 0.05$			m_0			
n	1	2	3	4	5	6
5	0.974879	0.987259	0.991488	0.993609	0.994884	0.995735
7	0.864650	0.915546	0.937761	0.950525	0.958878	0.964791
9	0.751395	0.826821	0.864592	0.888224	0.904506	0.916569
11	0.657408	0.744615	0.792203	0.823624	0.846308	0.863594
13	0.581803	0.673534	0.726640	0.763161	0.790351	0.811584
15	0.520703	0.613082	0.668932	0.708542	0.738748	0.762802
17	0.470679	0.561674	0.618539	0.659934	0.691979	0.717967
19	0.429136	0.517705	0.574524	0.616622	0.049869	0.677125
21	0.394163	0.479813	0.535938	0.578195	0.612009	0.640041
26	0.338681	0.418081	0.471820	0.513301	0.547188	0.575783
31	0.279396	0.349964	0.399326	0.438429	0.471081	0.499167
35	0.250124	0.315523	0.361994	0.399274	0.430740	0.458066
41	0.216106	0.274833	0.317317	0.351894	0.381446	0.407397
61	0.148596	0.191991	0.224479	0.251678	0.275508	0.296908
101	0.091398	0.119649	0.141400	0.160043	0.176723	0.191994

$p_0 = 5, \alpha = 0.05$			m_0		
n	1	2	3	4	5
6	0.979692	0.989794	0.993184	0.994884	0.995905
8	0.886622	0.929580	0.948211	0.958878	0.965847
10	0.785230	0.851366	0.884150	0.904506	0.918565
12	0.097399	0.776067	0.818485	0.846308	0.866307
14	0.624472	0.709032	0.757361	0.790351	0.814788
16	0.564102	0.650717	0.702389	0.738748	0.766325
18	0.513741	0.600224	0.653548	0.691979	0.721685
20	0.471285	0.556405	0.610280	0.649869	0.680950
26	0.376883	0.455026	0.507226	0.647180	0.579636
30	0.332202	0.405207	0.455150	0.494099	0.526222
36	0.281888	0.347734	0.393954	0.430740	0.461610
40	0.255981	0.317584	0.361384	0.396604	0.426422
60	0.175239	0.221230	0.255211	0.283405	0.307933
100	0.107359	0.137477	0.160423	0.179954	0.197336

Table A.5 (*continued*)

m_0						
7	8	9	10	12	15	20
0.996343	0.996799	0.997154	0.997439	0.997865	0.998292	0.998718
0.969205	0.972629	0.975436	0.977599	0.981037	0.984582	0.988242
0.925871	0.933276	0.939318	0.944346	0.952239	0.960599	0.969481
0.877264	0.888374	0.897598	0.905386	0.917834	0.931327	0.946039
0.828715	0.842875	0.854799	0.846993	0.881536	0.899835	0.920252
0.782527	0.799059	0.813149	0.825321	0.845334	0.867807	0.893534
0.734569	0.757884	0.773654	0.787400	0.810259	0.836401	0.866736
0.700039	0.719660	0.736702	0.751673	0.776816	0.805970	0.840370
0.603836	0.684384	0.702367	0.718271	0.745218	0.776849	0.814738
0.600437	0.022024	0.641151	0.658259	0.687674	0.722930	0.766277
0.523797	0.545697	0.565373	0.583198	0.614367	0.652651	0.701218
0.482235	0.503893	0.523492	0.541365	0.572896	0.612129	0.662755
0.430584	0.451555	0.470695	0.488288	0.519663	0.559330	0.611610
0.316425	0.334418	0.351134	0.366759	0.395273	0.432596	0.484175
0.206174	0.219470	0.232023	0.243938	0.266151	0.296204	0.389732

(*continued*)

Table A.5 x_α Such That $\{\mathrm{ch}_1(HE^{-1}) \geq x_\alpha\} = \alpha^a$ (*continued*)

$p_0 = 2, \alpha = 0.01$			m_0			
n	1	2	3	4	5	6
3	0.990000	0.994987	0.996655	0.997491	0.997992	0.998326
5	0.900000	0.937772	0.954219	0.963649	0.969812	0.974169
7	0.784556	0.849779	0.882588	0.903084	0.917285	0.927765
9	0.683772	0.763483	0.807396	0.836490	0.857523	0.873562
11	0.601892	0.687771	0.738126	0.772934	0.798929	0.819257
13	0.535841	0.623303	0.676968	0.715283	0.744612	0.765935
15	0.482052	0.568667	0.623676	0.663951	0.695402	0.720922
17	0.437658	0.522234	0.577367	0.618516	0.651217	0.678115
19	0.400515	0.482314	0.536866	0.578301	0.6116371	0.639375
21	0.369085	0.447870	0.501393	0.538836	0.576155	0.604339
25	0.318708	0.391524	0.442358	0.482365	0.515521	0.543823
31	0.264357	0.328965	0.375390	0.412777	0.444375	0.471817
35	0.237301	0.297166	0.340788	0.376317	0.406640	0.433204
41	0.205672	0.259444	0.299259	0.332116	0.360480	0.385583
61	0.142304	0.182108	0.212515	0.238267	0.261013	0.281567
101	0.087989	0.113969	0.134336	0.151967	0.167849	0.182464

$p_0 = 3, \alpha = 0.01$			m_0			
n	1	2	3	4	5	6
4	0.993322	0.996655	0.997769	0.998326	0.998661	0.998884
6	0.926040	0.954219	0.966395	0.973352	0.977889	0.981091
8	0.830214	0.882588	0.908603	0.924745	0.935882	0.944076
10	0.740031	0.807396	0.843444	0.867949	0.855204	0.898306
12	0.662810	0.738126	0.781562	0.811299	0.833350	0.850512
14	0.598092	0.676968	0.724544	0.758161	0.783705	0.803481
16	0.543850	0.623676	0.673508	0.709605	0.737577	0.260142
18	0.498065	0.577367	0.628151	0.665721	0.695307	0.719497
20	0.459064	0.536866	0.587850	0.626211	0.656805	0.682105
26	0.371069	0.442358	0.491393	0.529580	0.560977	0.587611
30	0.328606	0.395401	0.442259	0.479359	0.510294	0.536857
36	0.280375	0.340788	0.384158	0.419134	0.474559	0.474559
40	0.255325	0.311943	0.353068	0.386546	0.415135	0.440209
60	0.176316	0.218872	0.250893	0.277728	0.301233	0.322327
100	0.108798	0.136850	0.158583	0.177235	0.193929	0.209247

Table A.5 (*continued*)

m_0						
7	8	9	10	12	15	20
0.998505	0.998744	0.998884	0.998995	0.999163	0.999330	0.999498
0.977418	0.979936	0.981946	0.983589	0.986113	0.988714	0.991397
0.935843	0.942271	0.947513	0.951875	0.958716	0.965957	0.973645
0.886248	0.895662	0.905121	0.912350	0.923900	0.936417	0.950057
0.835674	0.849251	0.860690	0.870472	0.886350	0.903915	0.923514
0.787247	0.803378	0.817137	0.829031	0.848598	0.870643	0.895767
0.742177	0.760225	0.775782	0.789354	0.811946	0.837809	0.867851
0.700848	0.720227	0.737364	0.752011	0.777021	0.806061	0.840376
0.663033	0.683413	0.701320	0.717178	0.744085	0.775725	0.813693
0.628534	0.049629	0.668245	0.684833	0.713203	0.746941	0.787997
0.568459	0.590208	0.609634	0.627092	0.657381	0.0694101	0.739885
0.496165	0.517793	0.537424	0.555303	0.586784	0.625826	0.675956
0.456871	0.478211	0.497628	0.515422	0.547011	0.586667	0.638405
0.408158	0.428689	0.447518	0.464901	0.496075	0.535799	0.588675
0.300408	0.317851	0.334119	0.349375	0.377335	0.414153	0.465423
0.196091	0.208911	0.221050	0.232602	0.254209	0.283577	0.326354

m_0						
7	8	9	10	12	15	20
0.999043	0.999163	0.999256	0.999330	0.999442	0.999553	0.999665
0.983477	0.985324	0.986798	0.988002	0.989851	0.991755	0.993717
0.950378	0.955384	0.959461	0.962848	0.968156	0.973764	0.979707
0.908636	0.917019	0.923962	0.929798	0.936125	0.949205	0.960159
0.864343	0.875699	0.885263	0.893424	0.906635	0.921201	0.937394
0.820562	0.834423	0.846211	0.856375	0.872616	0.891752	0.912976
0.778847	0.794669	0.808262	0.820089	0.839707	0.862067	0.887908
0.739781	0.757110	0.772130	0.785301	0.807367	0.832865	0.862828
0.703529	0.721991	0.738115	0.752350	0.776411	0.804559	0.838137
0.610675	0.630947	0.648971	0.665142	0.693058	0.726712	0.768393
0.560111	0.580752	0.599268	0.616017	0.645245	0.681035	0.726263
0.497427	0.517957	0.536567	0.553563	0.583609	0.621106	0.669678
0.462582	0.482789	0.501208	0.518117	0.548218	0.586174	0.636023
0.341548	0.359249	0.375679	0.391022	0.418981	0.455500	0.505818
0.223388	0.236675	0.249212	0.261103	0.283249	0.313163	0.356389

(*continued*)

Table A.5 x_α Such That $\{ch_1(HE^{-1}) \geq x_\alpha\} = \alpha^a$ (*continued*)

$p_0 = 4, \alpha = 0.01$			m_0			
n	1	2	3	4	5	6
5	0.994987	0.997491	0.998326	0.998744	0.998995	0.999163
7	0.941097	0.963649	0.973352	0.978885	0.982488	0.985029
9	0.859132	0.903084	0.924745	0.938132	0.947343	0.954108
11	0.777927	0.836490	0.867949	0.888497	0.903244	0.914356
13	0.705686	0.7729939	0.811299	0.837391	0.856654	0.871595
15	0.643364	0.715283	0.758161	0.788245	0.811045	0.828979
17	0.590164	0.663951	0.209605	0.742434	0.767735	0.788063
19	0.544033	0.618516	0.665721	0.700345	0.727461	0.749537
21	0.504352	0.578301	0.626211	0.661934	0.690287	0.713630
25	0.439543	0.510816	0.558574	0.595109	0.624724	0.649549
31	0.367890	0.433707	0.479359	0.515234	0.544977	0.570401
35	0.331633	0.393725	0.437513	0.472377	0.501606	0.526836
41	0.288790	0.345669	0.386546	0.419587	0.447649	0.472153
61	0.201590	0.245214	0.277728	0.304800	0.328390	0.349485
101	0.125524	0.154773	0.177235	0.196409	0.213494	0.229074

$p_0 = 5, \alpha = 0.01$			m_0		
n	1	2	3	5	6
6	0.995988	0.997992	0.998661	0.99895	0.999196
8	0.950986	0.969812	0.977889	0.982488	0.985480
10	0.879347	0.917285	0.935882	0.947343	0.955215
12	0.805632	0.857523	0.855204	0.903244	0.916076
14	0.738094	0.798929	0.833350	0.856654	0.873801
16	0.678284	0.744612	0.783705	0.811045	0.831547
18	0.626415	0.695402	0.737577	0.767735	0.790889
20	0.581010	0.651217	0.695307	0.727461	0.752539
26	0.475439	0.544268	0.589970	0.624724	0.652772
30	0.423397	0.489512	0.534512	0.569392	0.597992
36	0.363294	0.424969	0.467624	0.501606	0.529969
40	0.331749	0.390022	0.431314	0.464350	0.492177
60	0.230942	0.276302	0.309769	0.337432	0.361402
100	0.143393	0.174072	0.197453	0.217305	0.234920

[a]For number of characteristics p_0, df for hypothesis m_0, and df error n.

Table A.5 (*continued*)

m_0						
7	8	9	10	12	15	20
0.999282	0.999372	0.999442	0.999498	0.999581	0.999665	0.999749
0.986921	0.988386	0.984554	0.990508	0.991972	0.993479	0.995032
0.959303	0.963426	0.966780	0.969566	0.973927	0.978530	0.983402
0.923125	0.930222	0.936094	0.941038	0.948912	0.957408	0.966624
0.883583	0.893446	0.901719	0.908769	0.920161	0.932695	0.946593
0.843644	0.855874	0.866254	0.575189	0.889813	0.906178	0.924691
0.804861	0.819031	0.831178	0.541725	0.859182	0.879015	0.901858
0.767985	0.783701	0.797384	0.809180	0.829050	0.851934	0.878721
0.733327	0.750249	0.764991	0.777978	0.780123	0.825385	0.855696
0.770828	0.689366	0.705716	0.720282	0.745196	0.774842	0.810975
0.592625	0.612192	0.629741	0.645576	0.673123	0.706712	0.748941
0.549033	0.568830	0.586667	0.602870	0.631311	0.666444	0.711372
0.493936	0.513548	0.531373	0.547698	0.576665	0.613030	0.660533
0.368635	0.386218	0.402495	0.417659	0.445209	0.481035	0.530134
0.243491	0.256963	0.269644	0.281646	0.303938	0.333930	0.377052

Appendix II
APL Programs

1. COV: This program estimates the covariance matrix for a single multivariate population.

2. HOTELLING: This program computes the Hotelling's T^2 and corresponding F-value for testing the equality of means of two normal populations.

3. SHIFT 1 estimates the point in time for a shift in mean.

4. SHIFT 2 gives the value of the test statistics for testing for a shift in mean.

5. MONOTONE 1 tests that the means of a bivariate monotone sample are zero.

6. MONOTONE 2 tests that the means of a bivariate monotone sample are equal.

7. MONOTONE 3 tests that the means of two bivariate monotone samples are equal.

8. INCOMPLETE estimates the mean and covariance matrix of an incomplete data set. In addition, the program gives the test statistic for testing that the mean is a specified vector.

9. MEAN: This program computes the mean of a multivariate normal sample. The format is MEAN Y.

10. CO: This program computes the covariance matrix of multivariate response Y with design matrix A. The format is CO Y.

11. RESCAL: This program rescales a matrix B $(p \times k, p > k)$ such that $B'B = I$ by Gram–Schmidt orthogonalization. The format is RESCAL B.

12. EST: This program estimates the parameter ξ of a growth curve model of the form $E(Y) = B\xi A$. The program prompts for Y, B, and A.

13. GCURVE: This program estimates the parameter of a growth curve model of the form $E(Y) = B\xi A$, as in program 12. In addition, it tests for the fit of the model in time and gives the values of V and R used for confidence intervals (see Chapter 6). This program is used in conjunction with the program LINES and RESCAL. Format is GCURVE. The program prompts for A and Y. The matrices B and C (for $CE(Y) = 0$) may be entered or just the degree of polynomial to be fitted and the time periods.

14. TEST: This program tests $C\beta D = 0$ for the model $E(Y) = \beta X$. The program prompts for Y, X, C, and D. In addition, parameter estimates are given.

15. PROFILE: This program computes the three test statistics for a profile analysis. The program prompts for the number of groups and the observation matrix for each group.

16. LINES: This program is a spacing program used in GCURVE.

17. TR: Calculates sum of squares treatment for the model $E(Y)=\beta A$. The format is TR Y.

18. CI: Calculates the values of R and V used in confidence intervals for growth curve analysis. This program is a shorter form of GCURVE. The program prompts for all information.

19. DET: Used with DETERP this program calculates the determinant of a square matrix, say, X. The format is DET X.

```
    ∇ S←COV Z;ZZ
[1]   ZZ←Z-⍉(⍴⍉Z)⍴MEAN(Z)
[2]   S←(ZZ+.×⍉ZZ)÷(¯1↑(⍴Z))-1
    ∇
```

```
    ∇ HOTELLING;N1;N2;P;T;F
[1]   'THIS PROGRAM COMPUTES THE HOTELLING''S STATISTIC AND
                                               CORRESPONDING'
[2]   'F-VALUE FOR TESTING THE EQUALITY OF MEANS OF TWO NORMAL
                                                   POPULATIONS.'
[3]   'ENTER THE (P×N1) MATRIX FOR THE FIRST POPULATION.' ◇ DATA1←⎕
[4]   'ENTER THE (P×N2) MATRIX FOR THE SECOND POPULATION. ' ◇ DATA2←⎕
[5]   'CONTINUE?'
[6]   →('N'∊⍞)/0
[7]   S1←COV(DATA1) ◇ S2←COV(DATA2)
[8]   S←((¯1+(⍴DATA1)[2])×S1)+((¯1+(⍴DATA2)[2])×S2)
[9]   S←S÷¯2+(⍴DATA1)[2]+(⍴DATA2)[2]
[10]  M1←MEAN(DATA1) ◇ M2←MEAN(DATA2)
[11]  'THE MEAN OF THE FIRST GROUP IS         ' ◇ M1
[12]  'THE MEAN OF THE SECOND GROUP IS        ' ◇ M2
[13]  'THE COVARINCE MATRIX OF THE FIRST GROUP IS     ' ◇ S1
[14]  'THE COVARIANCE MATRIX OF THE SECOND GROUP IS       ' ◇ S2
[15]  'THE POOLED COVARIANCE MATRIX IS' ◇ S
[16]  N1←(⍴DATA1)[2] ◇ N2←(⍴DATA2)[2] ◇ P←(⍴DATA1)[1]
[17]  T←(M1-M2)+.×(⌹S)+.×(M1-M2)×N1×N2÷N1+N2
[18]  F←(N1+N2-P+1)×T÷P×N1+N2-2
[19]  'THE HOTELLING T SQUARE VALUE IS     ';T
[20]  'THE CORRESPONDING F VALUE IS     ';F
    ∇
```

```
    ∇ SHIFT1;N;P;ALLT;R;MXR;MXNR;V;YR;SR
[1]   'THIS PROGRAM ESTIMATES THE POINT IN TIME FOR A SHIFT IN MEAN'
[2]   'FOR A MULTIVARIATE SET OF DATA.'
[3]   'ENTER THE DATA MATRIX (PXN).' ◇ DATA←⎕
[4]   N←(⍴DATA)[2] ◇ P←(⍴DATA)[1]
```

```
[5]   ALLT←⍳0
[6]   R←0
[7]  L1:R←R+1
[8]   MXR←MEAN(DATA[;⍳R])
[9]   MXNR←MEAN(DATA[;R+⍳N-R])
[10]  V←(N-1)×COV(DATA)
[11]  YR←(MXR-MXNR)×((N-R)×R÷N)*0.5
[12]  SR←YR+.×(⌹V)+.×YR
[13]  T←SR÷1-SR
[14]  ALLT←ALLT,T
[15]  →(R<N-1)/L1
[16]  'THE CALCULATED T VALUES ARE  ';ALLT
    ∇
```

```
    ∇ SHIFT2;N;P;A1;U1;V
[1]   'THIS PROGRAM TESTS FOR A SHIFT IN MEAN FOR  A MULTIVARIATE
                                                   SET OF DATA.'
[2]   'ENTER THE (P×N) DATA MATRIX.  ' ◇ DATA←⎕
[3]   N←(⍴DATA)[2] ◇ P←(⍴DATA)[1]
[4]   A1←(⍳N)-0.5×(N+1)
[5]   U1←(DATA+.×A1)÷(+/A1*2)*0.5
[6]   V←(N-1)×COV(DATA)
[7]   Q←U1+.×(⌹V)+.×U1
[8]   'THE CALCULATED TEST STATISTIC Q IS    ' ◇ Q
    ∇
```

```
    ∇ MONOTONE1;N1;N2;X12;BETA;BETAH;S21;S;L1;L2;RHO;FINAL;LRT
[1]   'THIS PROGRAM TESTS THAT THE MEANS OF A BIVARIATE MONOTONE'
[2]   'SAMPLE ARE EQUAL TO ZERO.'
[3]   'ENTER ALL THE OBSERVATIONS OF THE FIRST CHARACTERISTIC.' ◇ X1←⎕
[4]   'ENTER ALL THE OBSERVATIONS OF THE SECOND CHARACTERISTIC.' ◇ X2←⎕
[5]   N1←⍴X1 ◇ N2←⍴X2 ◇ X12←N2↑X1
[6]   SIGMA←S←(COV((2,N2)⍴X12,X2))×(N2-1)÷N2
[7]   MN←2⍴0
[8]   SIGMA[1;1]←(COV X1)×(N1-1)÷N1
[9]   MN[1]←MEAN(X1)
[10]  BETA←S[1;2]÷S[1;1]
[11]  S21←S[2;2]-S[1;2]×S[1;2]÷S[1;1]
[12]  MN[2]←(MEAN X2)-BETA×(MEAN X12)-MN[1]
[13]  SIGMA[2;1]←SIGMA[1;2]←BETA×SIGMA[1;1]
[14]  SIGMA[2;2]←S21+SIGMA[1;1]×BETA*2
[15]  LINES 2 ◇ 'THE ESTIMATE OF THE MEAN IS  ';MN
[16]  LINES 2 ◇ 'THE ESTIMATE OF THE COVARIANCE MATRIX IS  ' ◇ SIGMA
[17]  BETAH←(X12+.×X2)÷X12+.×X12
[18]  L1←(N1×SIGMA[1;1]÷+/X1*2)*N1÷2
[19]  L2←(N2×S21÷(+/X2*2)-(BETAH*2)×(+/X12*2))*N2÷2
[20]  LRT←L1×L2
[21]  RHO←1-0.25×(3÷N1)+5÷N2
[22]  FINAL←¯2×RHO×⍟LRT
[23]  LINES 2 ◇ 'THE CHI SQUARE STATISTIC IS  ';FINAL;' ON 2 DEGREES OF
                                                             FREEDOM.'
[24]  LINES 2 ◇ 'THE LIKELIHOOD RATIO TEST STATISTIC IS  ';LRT
    ∇
```

```
    ∇ MONOTONE2
[1]   'THIS PROGRAM TESTS THAT THE MEANS OF A BIVARIATE MONOTONE SAMPLE'
[2]   'ARE EQUAL.'
[3]   'ENTER ALL THE OBSERVATIONS OF THE FIRST CHARACTERISTIC.' ⋄ X1←⎕
[4]   'ENTER ALL THE OBSERVATIONS OF THE SECOND CHARACTERISTIC.' ⋄ X2←⎕
[5]   N1←⍴X1 ⋄ N2←⍴X2 ⋄ X12←N2↑X1
[6]   Z←X2-(MEAN X1)+(X12-MEAN X12)×(N2÷N1)*0.5
[7]   STAT←|(MEAN Z)÷((COV Z)÷N2)*0.5
[8]   'THE TEST STATISTIC IS    ';STAT;' ON ';N2-1;' DEGREES OF FREEDOM.'
    ∇
```

```
    ∇ MONOTONE3
[1]   'INPUT FIRST VECTOR OF FIRST GROUP.' ⋄ X1←⎕
[2]   'INPUT FIRST VECTOR OF SECOND GROUP.' ⋄ Y1←⎕
[3]   'INPUT SECOND VECTOR OF FIRST GROUP.' ⋄ X2←⎕
[4]   'INPUT SECOND VECTOR OF SECOND GROUP.' ⋄ Y2←⎕
[5]   M1←⍴X1 ⋄ N1←⍴Y1 ⋄ M2←⍴X2 ⋄ N2←⍴Y2 ⋄ T1←N1+M1 ⋄ T2←M2+N2
[6]   S1H←(COV X1,Y1)×(T1-1)÷T1
[7]   S1E←(((COV X1)×M1-1)+(COV Y1)×N1-1)÷T1
[8]   V12←(T2,2)⍴1 ⋄ W12←(T2,3)⍴0 ⋄ I2←(⍳T2)∘.=⍳T2
[9]   W12[;3]←V12[;2]←(M2↑X1),N2↑Y1 ⋄ Z22←X2,Y2
[10]  W12[;1]←(M2⍴1),N2⍴0 ⋄ W12[;2]←(M2⍴0),N2⍴1
[11]  S2H←Z22+.×(I2-V12+.×(⌹(⍉V12)+.×V12)+.×⍉V12)+.×Z22
[12]  S2E←Z22+.×(I2-W12+.×(⌹(⍉W12)+.×W12)+.×⍉W12)+.×Z22
[13]  LAMBDA←((S1E÷S1H)*T1÷2)×(S2E÷S2H)*T2÷2
[14]  RHO←1-0.25×((5÷T1)+7÷T2) ⋄ CHI←-2×RHO×⍟LAMBDA
[15]  H1←0.5×T1×1-RHO ⋄ H2←0.5×T2×1-RHO
[16]  W2←(1÷6×RHO*2)×(12×H1×(H1-1)÷T1*2)+12×H2×(H2-1)÷T2*2
[17]  'RHO= ';RHO ⋄ 'THE VALUE OF W2 IN THE ASYMPTOTIC EXPANSION IS ';W2
[18]  'THE CHI SQUARE TEST STATISTIC IS ';CHI
    ∇
```

```
    ∇ INCOMPLETE
[1]   'INPUT MATRIX OF COMPLETE OBSERVATIONS.' ⋄ X1←⎕
[2]   'INPUT EXTRA OBSERVATIONS ON THE FIRST VARIABLE ONLY.' ⋄ X2←⎕
[3]   'INPUT EXTRA OBSERVATIONS ON THE SECOND VARIABLE ONLY.' ⋄ X3←⎕
[4]   N1←(⍴X1)[2] ⋄ N2←⍴X2 ⋄ N3←⍴X3 ⋄ I←0
[5]   Z1←MEAN X1 ⋄ Z2←MEAN X2 ⋄ Z3←MEAN X3 ⋄ S←COV X1
[6]   B←T1←T2←T← 2 2 ⍴0 ⋄ T[1;1]←S[1;1]*0.5 ⋄ T[2;1]←S[2;1]÷T[1;1]
[7]   T[2;2]←(S[2;2]-T[2;1]*2)*0.5
[8]  LOOP1:A←N1×⌹S ⋄ A[1;1]←A[1;1]+N2÷S[1;1]
[9]   A[2;2]←A[2;2]+N3÷S[2;2]
[10]  MN←N1×(⌹S)+.×Z1 ⋄ MN[1]←MN[1]+N2×Z2÷S[1;1]
[11]  MN[2]←MN[2]+N3×Z3÷S[2;2] ⋄ MU←(⌹A)+.×MN
[12]  U1←X1-MU∘.×N1⍴1 ⋄ U2←X2-MU[1] ⋄ U3←X3-MU[2]
[13]  B[1;1]←(U2+.×U2)÷S[1;1]*2 ⋄ B[2;2]←(U3+.×U3)÷S[2;2]*2
[14]  BB←(S+.×B+.×S)+U1+.×⍉U1
[15]  AA←(⍉T)+.×A+.×T ⋄ T1[1;1]←AA[1;1]*0.5 ⋄ T1[2;1]←AA[2;1]÷T1[1;1]
[16]  T1[2;2]←(AA[2;2]-T1[2;1]*2)*0.5
[17]  T2[1;1]←BB[1;1]*0.5 ⋄ T2[2;1]←BB[2;1]÷T2[1;1]
[18]  T2[2;2]←(BB[2;2]-T2[2;1]*2)*0.5
[19]  T←(⌹T1)+.×T2 ⋄ I←I+1 ⋄ S←T+.×⍉T
[20]  →(I≤25)/LOOP1
[21]  'THE ESTIMATE OF THE MEAN IS ';MU
[22]  'THE ESTIMATE OF THE COVARIANCE MATRIX IS ';S
[23]  L1←(N1×⍟DET S)+(N2×⍟S[1;1])+(N3×⍟S[2;2])+B[1;1]+B[2;2]
```

```
[24]  L1←L1+(+/+/((⍳2)∘.=⍳2)×(⌹S)+.×U1+.×⍉U1)
[25]  'TO TEST THAT THE MEAN IS EQUAL TO A SPECIFIED VECTOR ENTER'
[26]  'THE VECTOR.' ⋄ MU←⎕
[27]  U1←X1-MU∘.×N1⍴1 ⋄ U2←X2-MU[1] ⋄ U3←X3-MU[2]
[28]  S←(U1+.×⍉U1)÷N1 ⋄ I←0
[29]  T[1;1]←S[1;1]*0.5 ⋄ T[2;1]←S[2;1]÷T[1;1] ⋄ T[2;2]←(S[2;2]
                                                   -T[2;1]*2)*0.5
[30] LOOP2:A←N1×⌹S ⋄ A[1;1]←A[1;1]+N2÷S[1;1]
[31]  A[2;2]←A[2;2]+N3÷S[2;2]
[32]  B[1;1]←(U2+.×U2)÷S[1;1]*2 ⋄ B[2;2]←(U3+.×U3)÷S[2;2]*2
[33]  BB←(S+.×B+.×S)+U1+.×⍉U1
[34]  AA←(⍉T)+.×A+.×T ⋄ T1[1;1]←AA[1;1]*0.5 ⋄ T1[2;1]←AA[2;1]÷T1[1;1]
[35]  T1[2;2]←(AA[2;2]-T1[2;1]*2)*0.5
[36]  T2[1;1]←BB[1;1]*0.5 ⋄ T2[2;1]←BB[2;1]÷T2[1;1]
[37]  T2[2;2]←(BB[2;2]-T2[2;1]*2)*0.5
[38]  T←(⌹T1)+.×T2 ⋄ I←I+1 ⋄ S←T+.×⍉T
[39]  →(I≤50)/LOOP2
[40]  L2←(N1×⍟DET S)+(N2×⍟S[1;1])+(N3×⍟S[2;2])+B[1;1]+B[2;2]
[41]  L2←L2+(+/+/((⍳2)∘.=⍳2)×(⌹S)+.×U1+.×⍉U1)
[42]  LRT←L2-L1 ⋄ 'THE CHI SQUARE STATISTIC IS ';LRT;' ON 2 D.F.'
    ∇


    ∇ MMM←MEAN DATA
[1]   MMM←+/DATA÷¯1↑(⍴DATA)
    ∇


    ∇ S←A CO Z
[1]   S←((Z+.×⍉Z)-Z+.×(⍉A)+.×(⌹A+.×⍉A)+.×A+.×⍉Z)÷(⍴Z)[2]-(⍴A)[1]
    ∇


    ∇ H1←RESCAL H;K;P;I
[1]   I←0 ⋄ H1←H ⋄ K←(⍴H)[2] ⋄ P←(⍴H)[1]
[2]  START:I←I+1
[3]   H1[;I]←H1[;I]÷(+/H1[;I]*2)*0.5
[4]   →(I=K)/0
[5]   H1[;I+⍳K-I]←H1[;I+⍳K-I]-H1[;I]∘.×H1[;I]+.×H1[;I+⍳K-I]
[6]   →START
    ∇


    ∇ E←EST
[1]   'ENTER THE DESIGN MATRIX A.' ⋄ A←⎕
[2]   'ENTER THE MATRIX B.' ⋄ B←⎕
[3]   'ENTER THE OBSERVATION MATRIX Y.' ⋄ Y←⎕
[4]   S←A CO Y
[5]   E←(⌹((⍉B)+.×(⌹S)+.×B))+.×(⍉B)+.×(⌹S)+.×Y+.×(⍉A)+.×⌹A+.×⍉A
[6]   'THE SAMPLE COVARIANCE MATRIX IS ' ⋄ LINES 2 ⋄ S
[7]   'THE ESTIMATE OF THE PARAMETER MATRIX IS' ⋄ LINES 2 ⋄ E
    ∇


    ∇ GCURVE
[1]   'THIS PROGRAM GIVES THE ESTIMATES OF THE PARAMETERS OF A '
[2]   'GROWTH CURVE MODEL OF THE FORM Y=B BETA A + EPSILON .'
```

```
[3]   'ENTER THE DESIGN MATRIX A.' ◇ A←⎕ ◇ Q←(⍴A)[1]
[4]   'ENTER THE OBSERVATION MATRIX Y.' ◇ Y←⎕ ◇ P←(⍴Y)[1] ◇ N←(⍴Y)[2]
[5]   'DO YOU WISH TO ENTER THE MATRIX B (YES) OR TO ENTER THE TIME'
[6]   'POINTS AND THE DEGREE OF THE POLYNOMIAL IN TIME (NO).'
[7]   →('Y'∊⍞)/START1 ◇ 'ENTER THE ';P;' TIME POINTS' ◇ TIME←⎕
[8]   'ENTER THE DEGREE OF THE POLYNOMIAL IN TIME TO BE FITTED.' ◇ K←⎕
[9]   BB←TIME∘.*¯1+⍳P ◇ B←BB[;⍳K+1] ◇ C←⍉(RESCAL BB)[;(K+1)+⍳P-K+1]
[10]  →SKIP
[11] START1:'ENTER THE MATRIX B.' ◇ B←⎕
[12]  'TO TEST FOR THE FIT OF THE MODEL ENTER THE MATRIX C.' ◇
                                                       C←⍉RESCAL⍉⎕
[13] SKIP:S←A CO Y ◇ V←(N-Q)×S ◇ W←C+.×V+.×(⍉C)
[14]  U←DET W+.×⌹C+.×Y+.×(⍉Y)+.×⍉C ◇ LINES 2
[15]  'THE U STATISTIC FOR TESTING THE FIT OF THE MODEL IS ';U
[16]  CHI←-(N-0.5×P+Q-K)×⍟U ◇ LINES 2
[17]  ' THE CHI SQUARE APPROXIMATION IS  ';CHI;' ON ';Q×P-K+1;' D.F.'
[18]  V1←⌹(⍉B)+.×(⌹V)+.×B ◇ A1←⌹A+.×⍉A ◇ Y1←Y+.×(⍉A)+.×A1
[19]  E←V1+.×(⍉B)+.×(⌹V)+.×Y1 ◇ LINES 2
[20]  'THE ESTIMATE OF THE PARAMETER MATRIX BETA IS ' ◇ LINES 2
[21]  'F12.4' ⎕FMT(E) ◇ LINES 3
[22]  'IN ORDER TO PRODUCE CONFIDENCE INTERVALS THE FOLLOWING'
[23]  'MATRICES ARE CALCULATED.' ◇ LINES 5
[24]  'THE MATRIX V1 IS ' ◇ LINES 2 ◇ V1 ◇ LINES 5
[25]  R←A1+(⍉Y1)+.×((⌹V)-(⌹V)+.×B+.×V1+.×(⍉B)+.×⌹V)+.×Y1
[26]  'THE MATRIX R IS ' ◇ LINES 2 ◇ R
    ∇
```

```
    ∇ TEST
[1]   'THIS PROGRAM GIVES ESTIMATES FOR THE PARAMETERS IN A '
[2]   'MULTIVARIATE REGRESSION OF THE FORM Y=BETA X + EPSILON.'
[3]   'IT ALSO GIVES THE U TEST STATISTIC FOR TESTING THE HYPOTHESIS'
[4]   ' C BETA D = 0. '
[5]   'ENTER THE (P×N) OBSERVATION MATRIX Y.' ◇ Y←⎕
[6]   ' ENTER THE DESIGN MATRIX X .' ◇ X←⎕
[7]   'ENTER THE MATRIX C.' ◇ C←⎕
[8]   'ENTER THE MATRIX D.   ' ◇ D←⎕
[9]   N←(⍴Y)[2] ◇ P←(⍴Y)[1] ◇ RR←(⍴C)[1] ◇ M←(⍴X)[1] ◇ T←(⍴D)[2]
[10]  S←X CO Y ◇ V←S×N-M ◇ DD←⌹(((⍳P)∘.=⍳P)×S)*0.5
[11]  R←DD+.×S+.×DD ◇ LINES 3
[12]  LINES 1 ◇ 'THE COVARIANCE MATRIX IS ' ◇ LINES 2 ◇ S
[13]  LINES 1 ◇ 'THE CORRELATION MATRIX IS ' ◇ LINES 2 ◇ R
[14]  X1←⌹X+.×⍉X ◇ BETA←Y+.×(⍉X)+.×X1
[15]  LINES 1 ◇ 'THE ESTIMATE OF BETA IS ' ◇ LINES 2 ◇ BETA
[16]  W1←C+.×BETA+.×D ◇ W←W1+.×X1+.×⍉W1 ◇ V1←C+.×V+.×⍉C
[17]  LINES 1 ◇ 'THE SUM OF SQURES MATRIX DUE TO THE HYPOTHESIS IS '
[18]  LINES 2 ◇ V1 ◇
[19]  LINES 1 ◇ 'THE SUM OF SQUARES MATRIX DUE TO ERROR IS '
[20]  LINES 2 ◇ W
[21]  →((1↑⍴V1)>1)/L1
[22]  F←W×(N-M)÷V1×RR
[23]  'THE F STATISTIC HAS A VALUE OF ';F;' ON ';RR;' AND ';N-M;'
                                                    D.F.' ◇ →0
[24] L1:U←DET V1+.×(⌹V1+W)
[25]  LINES 2 ◇ 'THE U STATISTIC IS ';U
[26]  CHI←-(N-M+0.5×RR-T-1)×⍟U ◇
[27]  LINES 2 ◇ 'THE CHI SQUARE APPROXIMATION IS ';CHI ◇
    ∇
```

```
    ∇ PROFILE
[1]   'THIS PROGRAM COMPUTES THREE TEST STATISTICS FOR A PROFILE
                                                        ANALYSIS.'
[2]   'ENTER THE NUMBER OF GROUPS ' ◇ K←⎕
[3]   'ENTER THE P×N1 MATRIX OF THE FIRST GROUP. ' ◇ Y←⎕
[4]   P←(ρY)[1] ◇ N←Kρ(ρY)[2] ◇ MN←(P,K)ρ0
[5]   MN[;1]←MEAN Y ◇ V←(COV Y)×N[1]-1
[6]   I←1
[7]  L1:I←I+1
[8]   'ENTER THE P×N';I;' MATRIX OF GROUP ';I;'.'
[9]   Y←⎕
[10]  N[I]←(ρY)[2] ◇ MN[;I]←MEAN Y ◇ V←V+(N[I]-1)×COV Y
[11]  →(I<K)/L1
[12]  I←0
[13] L2:I←I+1
[14]  LINES 2 ◇ 'THE MEAN OF GROUP ';I;' IS ' ◇ LINES 1 ◇ MN[;I]
[15]  →(I<K)/L2
[16]  N1←+/N ◇ S←V÷N1-K
[17]  LINES 2 ◇ 'THE POOLED COVARIANCE MATRIX IS ' ◇ LINES 2 ◇ S
[18]  C2←((P-1),P)ρ(1 ¯1 ,(P-1)ρ0)
[19]  C1←⍉((K-1),K)ρ(1 ¯1 ,(K-1)ρ0)
[20]  MNT←+/(MN×(Pρ1)∘.×N)÷N1
[21]  SS1←MN-MNT∘.×Kρ1 ◇ SSTR←SS1+.×⍉SS1
[22]  U←DET C2+.×V+.×(⍉C2)+.×⌹C2+.×(V+SSTR)+.×⍉C2
[23]  LINES 2 ◇ 'THE VALUE OF THE U STATISTIC FOR PARALLELISM IS ';U
[24]  U1←DET V+.×⌹V+SSTR ◇ U2←U1÷U
[25]  F←(N1+1-K+P)×(1-U2)÷U2×K-1
[26]  LINES 2 ◇ 'THE VALUE OF THE F TEST FOR THE LEVEL HYPOTHESIS IS ';F
[27]  'ON ';K-1;' AND ';N1+1-K+P;' D.F.'
[28]  U3←MNT+.×(⍉C2)+.×(⌹(C2+.×V+.×⍉C2))+.×C2+.×MNT
[29]  F←U3×N1×(N1+2-P+K)÷P-1
[30]  LINES 2 ◇ 'THE VALUE OF THE F STATISTIC FOR THE CONDITION
                                              HYPOTHESIS IS ';F
[31]  'ON ';P-1;' AND ';N1+2-P+K;' D.F.'
[32]  GAMMA←(+⌿(⌹V)+.×MN[;⍳K-1]-MN[;K]∘.×(K-1)ρ1)÷+/+/⌹V
[33]  LINES 2 ◇ ' THE VALUES OF THE DIFFERENCES IN LEVELS FOR THE
                                                    I TH -';K;'TH'
[34]  ' GROUPS FOR I= 1 TO ';K-1;'  ARE ' ◇ LINES 2
[35]  LINES 2 ◇ GAMMA
    ∇


    ∇ LINES K
[1]   (K,1)ρ' '
    ∇


    ∇ SSTR←TR Y
[1]   'ENTER THE DESIGN MATRIX A. ' ◇ A←⎕
[2]   SSTR←Y+.×(⍉A)+.×(⌹(A+.×⍉A))+.×A+.×⍉Y
    ∇


    ∇ CI
[1]   'THIS PROGRAM GIVES THE MATRICES NEEDED FOR CALCULATING'
[2]   'CONFIDENCE INTERVALS IN A GROWTH CURVE MODEL.'
[3]   'ENTER THE OBSERVATION MATRIX Y (PXN). ' ◇ Y←⎕
```

```
[4]   'ENTER THE MATRIX A.' ⋄ A←⎕
[5]   'ENTER THE MATRIX B.' ⋄ B←⎕
[6]   S←A CO Y ⋄ V←S×(ρY)[2]-(ρA)[1]
[7]   A1←⌹A+.×⍉A ⋄ Y1←Y+.×(⍉A)+.×A1 ⋄ V1←⌹(⍉B)+.×(⌹V)+.×B
[8]   'THE VALUE OF THE MATRIX V1 IS ' ⋄ LINES 2 ⋄ V1
[9]   R←A1+(⍉Y1)+.×((⌹V)-(⌹V)+.×B+.×V1+.×(⍉B)+.×(⌹V))+.×Y1
[10]  'THE VALUE OF R IS ' ⋄ LINES 2 ⋄ R
    ∇
```

```
    ∇ Z←DET M
[1]   →((×/(ρM),1)=1)/END
[2]   →((1==/ρM)⍲2=ρρM)/MSG
[3]   →0×ρρρZ←DETERP M
[4]  MSG:'***MUST START WITH A SQUARE MATRIX' ⋄ →0
[5]  END:Z←M
```

```
    ∇ Z←DETERP M;K;L
[1]   Z←1
[2]   →7×⍳0=L←⌈/|M[;1]
[3]   M[L,1;]←M[1,L←(|M[;1])⍳L;]
[4]   M← 1 1 ↓M-M[;1]∘.×M[1;]÷K←M[1;1]
[5]   →2×⍳2≤1↑(ρM),Z←Z×K×1-2×L≠1
[6]   →0,Z←Z×M[1;1]
[7]   Z←0
    ∇
```

Bibliography

Abbott, R. D., and Perkins, D. (1978). Development and construct validation of a set of student rating-of-instructor items, *Ed. Psych. Measurement* **38**, 1069–1075.

Abruzzi, A. (1950). Experimental procedures and criteria for estimating and evaluating industrial productivity, Ph.D. thesis, Columbia University.

Anderson, T. W. (1951). Classification by multivariate analysis, *Psychometrika* **16**, 31–50.

Anderson, T. W. (1957). Maximum likelihood estimates for a multivariate normal distribution when some observations are missing, *J. Amer. Statist. Assoc.* **52**, 200–203.

Anderson, T. W. (1963). Asymptotic theory for principal component analysis, *Ann. Math. Statist.* **34**, 122–148.

Andrews, D. F., Gnanadesekan, R., and Warner, J. L. (1971). Transformations of multivariate data, *Biometrics* **27**, 825–840.

Andrews, D. F., Gnanadesekan, R., and Warner, J. L. (1973). Methods for assessing multivariate normality, in *Multivariate Analysis* (P. R. Krishnaiah, ed.), Academic, New York, Vol. 3, pp. 95–116.

Anscombe, F. J. (1953). Sequential estimation, *J. Roy. Statist. Soc. Ser. B* **15**, 1–29.

Anscombe, F. J., and Tukey, J. W. (1963). The examination and analysis of residuals, *Technometrics* **5**, 141–160.

Barnard, M. M. (1935). The secular variations of skull characters in four series of Egyptian skulls, *Ann. Eugen.* **6**, 352–371.

Barnett, M. K., and Mead, F. C., Jr. (1956). A 2^4 factorial experiment in four blocks of eight: A study in radioactive decontamination, *Appl. Statist.* **5**, 122–131.

Bartlett, M. S. (1937). Properties of sufficiency and statistical tests, *Proc. Cambridge Phil. Soc. Ser. A* **160**, 268–282.

Bartlett, M. S. (1947a). Multivariate analysis, *J. Roy. Statist. Soc. Ser. B* **9**, 176–197.

Bartlett, M. S. (1947b). The use of transformations, *Biometrics* **3**, 39–52.

Bartlett, M. S. (1951). The effect of standardization on an approximation in factor analysis, *Biometrika* **38**, 337–344.

Bartlett, M. S. (1954). A note on the multiplying factors for various χ^2-approximations, *J. Roy. Statist. Soc. Ser. B* **16**, 296–298.

Batlis, N. C. (1978). Job involvement as a predictor of academic performance. *Ed. Psych. Measurement* **38**, 1177–1180.

Bennett, B. M. (1951). Note on a solution of the generalized Behrens–Fisher problem, *Ann. Inst. Statist. Math.* **2**, 87.

Bernard, D. R. (1981). Multivariate analysis as a means of comparing growth in fish, *Can. J. Fish Aquat. Sci.* **38**, 233–236.

Bhargava, R. P. (1962). Multivariates tests of hypothesis with incomplete data, Tech. Rep. No. 3, Stanford University.

Bhargava, R. P., and Srivastava, M. S. (1973). On Tukey's confidence intervals for the contrasts in the means of the intraclass correlation model, *J. Roy. Statist. Soc.* **35**, 147–152.

Bhattacharya, G. K., and Johnson, R. A. (1968). Approach to degeneracy and the efficiency of some multivariate tests, *Ann. Math. Statist.* **39**, 1654–1660.

Bock, R. D. (1963). Multivariate analyses of repeated measurements, in *Problems in Measuring Change* (C. W. Harris, ed.), University of Wisconsin Press, Madison, Wis., pp. 85–103.

Bowker, A. H. and Sitgreaves, R. (1961). An asymptotic expansion for the distribution function of the *W*-classification statistic, in *Studies in Item Analysis and Prediction* (H. Solomon, ed.), Stanford University Press, Stanford, Cal., pp. 293–310.

Box, G. E. P. (1949). A general distribution theory for a class of likelihood criteria, *Biometrika* **36**, 317–346.

Box, G. E. P. (1950). Problems in the analysis of growth and wear curves, *Biometrics* **6**, 362–389.

Box, G. E. P., and Cox, D. R. (1964). An analysis of transformations, *J. Roy. Statist. Soc. Ser. B* **26**, 211–252.

Brian, M. V. (1978). *Production Ecology of Ants and Termites*, Cambridge University Press, Cambridge, England.

Broffit, J., and Williams, J. S. (1973). Minimum variance estimators for misclassification probabilities in discriminant analysis, *J. Multivariate Analysis* **3**, 311–327.

Carpenter, D., and Carpenter, S. (1978). The concurrent validity of the Loren Hammill test of written spelling in relation to the California achievement test, *Ed. Psych. Measurement* **38**, 1201–1205.

Carroll, J. B. (1957). Biquartimin criterion for rotation to oblique simple structure in factor analysis, *Science* **126**, 1114–1115.

Carter, E. M. (1982). Asymptotic expansions for the distribution of multivariate regression estimators with incomplete data, University of Guelph tech. rep.

Carter, E. M., and Hubert, J. J. (1981). A multivariate approach to bioassay in time, University of Guelph Statistical Series.

Carter, E. M., Khatri, C. G., and Srivastava, M. S. (1979). The effect of inequality of variances on the *t*-test, *Sankhya Ser. B* **41**, 216–225.

Chernoff, H., and Zacks, S. (1964). Estimating the current mean of a normal distribution which is subjected to changes in time, *Ann. Math. Statist.* **35**, 999–1018.

Chow, Y. S., and Robbins, H. (1965). On the asymptotic theory of fixed-width sequential confidence intervals for the mean, *Ann. Math. Statist.* **36**, 457–462.

Clifford, M. (1972). Effects of competition as a motivational technique in the classroom, *Amer. Ed. Res. J.* **9**, 123–137.

Cole, J. W. L., and Grizzle, J. E. (1966). Application of multivariate analysis of variance to repeated measurement experiments, *Biometrics* **22**, 810–828.

Cox, D. R., and Small, N. J. H. (1978). Testing multivariate normality, *Biometrika* **65**, 263–272.

Cramèr, H. (1937). *Random Variables and Probability Distributions*, Cambridge Tracts in Mathematics No. 36, Cambridge University Press, Cambridge, England.

D'Agostino, R. B. (1971). An omnibus test of normality for moderate and large sample sizes, *Biometrika* **58**, 341–348.

Danford, M. B., Hughes, H. M., and McNee, R. C. (1960). On the analysis of repeated measurements experiments *Biometrics* **16**, 547–565.

Dantzig, G. B. (1940). On the nonexistence of "Students" test hypothesis having power functions independent of σ, *Ann. Math. Statist.* **11**, 186–192.

Das Gupta, S. (1974). Probability inequalities and errors in classification, *Ann. Statist.* **2**, 751–762.

Das Gupta, S., Anderson, T. W., and Mudholkar, G. (1964). Monotonicity of power functions of tests of the multivariate linear hypothesis, *Ann. Math. Statist.* **35**, 200.

Davis, A. W. (1980). On the effects of moderate multivariate normality on Wilk's likelihood ratio criterion, *Biometrika* **67**, 419–427.

Dempster, A. P., Laird, N. M., and Rubin, D. B. (1977). Maximum likelihood from incomplete data via the EM algorithm, *J. Roy. Statist. Soc. Ser. B* **39**, 29–38.

Draper, N. R., and Hunter, W. G. (1969a). Transformations, *J. Roy. Statist. Soc. Ser. B* **26**, 211–252.

Draper, N. R., and Hunter, W. G. (1969b). Transformations; some examples revisited, *Technometrics* **11**, 23–40.

Dudzinski, M. L., and Arnold, W. R. (1973). Comparisons of diets of sheep and cattle grazing together on sown pastures on the southern tablelands of New South Wales by principal component analysis, *Austral. J. Agric. Res.* **24**, 899–912.

Elston, R. C., and Grizzle, J. E. (1962). Estimation of time response curves and their confidence bands, *Biometrics* **18**, 148–159.

Fisher, R. A. (1921). On the "probable error" of a coefficient of correlation deduced from a small sample, *Metron* **1**, 1–32.

Fisher, R. A. (1936). The use of multiple measurement in taxonomic problems, *Ann. Eugen.* **7**, 179–188.

Fisher, R. A., and Yates, F. (1948). *Statistical Tables*, Oliver and Boyd, Edinburgh.

Fletcher, R., and Powell, M. J. D. (1963). A rapidly convergent descent method for minimization, *Comput. J.* **6**, 163–168.

Fornell, C. (1979). External single-set component analysis of multiple criterion/multiple predictor variables, *Multiv. Behav. Res.* **14**, 323–338.

Fujikoshi, Y. (1977). Asymptotic expansions for the distributions of some multivariate tests, in *Multivariate Analysis* (P. R. Krishnaiah, ed.), North Holland, New York, Vol. 4, pp. 55–71.

Fujikoshi, Y. (1978). Asymptotic expansions of the distribution of the likelihood ratio statistic for the equality of the q smallest latent roots of a covariance matrix (unpublished).

Fujikoshi, Y. and Veitch, L. G. (1979). Estimation of dimensionality in canonical correlation analysis, *Biometrika* **66**, 345–351.

Gill, J. L., and Hafs, H. D. (1971). Analysis of repeated measurements on animals, *J. Animal Sci.* **33**, 331–335.

Giri, N. C. (1964). On the likelihood ratio test of a normal multivariate testing problem, *Ann. Math. Statist.* **35**, 181–190, 1388.

Girshick, M. A. (1939). On the sampling theory of roots of determinantal equations, *Ann. Math. Statist.* **10**, 203–224.

Gnanadesikan, R. (1977). *Methods for Statistical Analysis of Multivariate Observations*, Wiley, New York.

Graybill, F. A. (1969). *Introduction to Matrices with an Introduction to Statistics*, Wadsworth, Belmont, Cal.

Greenhouse, S. W., and Geisser, S. (1959). On methods in the analysis of profile data, *Psychometrika* **24**, 95–112.

Griffiths, G. B. (1904). Measurements of one hundred and thirty criminals, *Biometrika* **3**, 60–63.

Grizzle, J. E., and Allen, D. M. (1969). Analysis of growth and dose response curves, *Biometrics* **25**, 357–381.

Guttman, L. (1954a). Some necessary conditions for common-factor analysis, *Psychometrika* **18**, 277–296.

Guttman, L. (1954b). A new approach to factor analysis theradex, in *Mathematical Thinking in the Social Science* (P. F. Lazarsfield, ed.), Columbia University Press, New York, pp. 258–348.

Guttman, L. (1956). Best possible systematic estimates of communalities, *Psychometrika* **21**, 273–285.

Habbema, J. P. E., and Hermans J. (1977). Selection of variables in discriminant analysis by F-statistic and error rate, *Technometrics* **19**, 487–493.

Hariton, G. (1972). Multivariate mixture model, Ph.D. thesis, University of Toronto.

Harman, H. H. (1967). *Modern Factor Analysis*, 2nd ed., University of Chicago Press, Chicago, Ill.

Healy, M. J. R. (1968). Multivariate normal plotting, *Appl. Statist.* **17**, 157–161.

Heck, D. L. (1960). Charts of some upper percentage points of the distribution of the largest characteristic root, *Ann. Math. Statist.* **31**, 625–642.

Hermans, J., and Habbema, J. P. E. (1976). A stepwise discriminant program using density estimators, *Compstat Proceedings in Computer Statistics*, Physica Verlag, Vienna, pp. 107–110.

Hoegervorst, J. E. (1978). Marital role expectations of adolescent girls, M. Sc. thesis, University of Guelph.

Hopkins, J. W., and Clay, P. P. F. (1963). Some empirical distributions of bivariate T^2 and homoscedasticity, criterion M under unequal variance and leptokurtosis, *J. Amer. Statist. Assoc.* **58**, 1048–1053.

Horton, L. F., Russell, J. S., and Moore, A. W. (1968). Multivariate-covariance and canonical analyses: A method for selecting the most effective discriminators in a multivariate situation, *Biometrics* **24**, 845–858.

Hosseini, A. A. (1978). The predictive validity of the scholastic aptitude test of the national organization for education valuation of the Iranian ministry of sciences and higher education for a group of Iranian students, *Ed. Psych. Measurement* **38**, 1041–1047.

Hotelling, H. (1933). Analysis of a complex of statistical variables into principal components. *J. Ed. Psych.* **24**, 417–441, 498–520.

Hubert, J. J. (1980). *Bioassay*, Kendall Hunt, Dubuque, Iowa.

Huntingford, F. A. (1976). An investigation of the territorial behavior of the three-spined stickleback (*Gasterosteus aculeatus*) using principal components analysis, *Animal Behav.* **24**, 822–834.

Ito, K. (1969). On the effect of heteroscedasticity and non-normality upon some multivariate test procedures, in *Multivariate Analysis* (P. R. Krishnaiah, ed.), Academic, New York, Vol. 2, pp. 87–120.

Jackson, J. E., and Morris, R. H. (1957). An application of multivariate quality control to photographic processing, *J. Amer. Statist. Assoc.* **52**, 186–199.

James, G. S. (1954). Tests of linear hypotheses in univariate and multivariate analysis when the ratios of the population variances are unknown, *Biometrika* **41**, 19–43.

James, A. T. (1969). Tests of equality of latent roots of the covariance matrix, in *Multivariate Analysis* (P. R. Krishnaiah, ed.), Academic, New York, Vol. 2.

Jeffers, J. N. R. (1967). Two case studies in the application of principal component analysis, *Appl. Statist.* **16**, 225–236.

Jensen, D. R. (1972). Some simultaneous multivariate procedures using Hotelling's T^2 statistics, *Biometrics* **28**, 39–53.

John, J. A., and Draper, N. R. (1980). An alternative family of transformations, *Appl. Statist.* **29**, 190–197.

John, S. (1961). Errors in discrimination, *Ann. Math. Statist.* **32**, 1125–1144.

John. S. (1971). Some optimal multivariate tests, *Biometrika* **58**, 123–127.

Jöreskog, K. G. (1963). *Statistical Estimation in Factor Analysis*, Almquist and Wiksell, Stockholm.

Jöreskog, K. G. (1967). Some contributions to maximum likelihood factor analysis, *Psychometrika* **34**, 51–76.

Jöreskog, K. G. (1969). Efficient estimation in image factor analysis, *Psychometrika* **34**, 51–75.

Jöreskog, K. G., and Lawley, D. N. (1968). New methods in maximum likelihood factor analysis, *British J. Math. Statist. Psych.* **21**, 85–96.

Jöreskog, K. G., and Von Thillo, M. (1971). New rapid algorithms for factor analysis by unweighted least squares, generalized least squares and maximum likelihood, Res. Memo. 71-5, Educational Testing Service, Princeton, N.J.

Josefawitz, N. (1979). Placing assertion within a situation: An investigation of how the perception of the other person in the interaction influences assertive behaviour, Ph.D. thesis, University of Toronto.

Kaiser, H. F. (1970). A second generation "Little Jiffy," *Psychometrika* **35**, 401–415.

Kaiser, H. F., and Rice, T. (1974). Little Jiffy IV, *Ed. Psych. Measurement* **34**, 111–117.

Kempthorne, O. (1952). *Design of Experiments*, Wiley, New York.

Khatri, C. G. (1966). A note on a MANOVA model applied to problems in growth curve, *Ann. Inst. Statist. Math.* **18**, 75–86.

Khatri, C. G., and Srivastava, M. S. (1973). On the non-null distribution of likelihood ratio criteria for covariance matrices, *Ann. Inst. Statist. Math.* **25**, 345–354.

Khatri, C. G., and Srivastava, M. S. (1976). Asymptotic expansions of the non-null distributions of the likelihood ratio criteria for covariance matrices. II. Proc. Carleton University, Ottawa, 1974, *Metron* **36**, 55–71.

Kiefer, J., and Schwartz, R. (1965). Admissable Bayes character of $T^2 - R^2$ and other fully invariant tests for classical multivariate normal problems, *Ann. Math. Statist.* **36**, 747–770.

Kleinbaum, D. G. (1973). A generalization of the growth curve model which allows missing data, *J. Multivariate Anal.* **3**, 117–124.

Kovacic, C. R. (1965). Impact-absorbing qualities of football helmets, *Res. Quart.* **36**, 420–426.

Kowalski, C. J., and Guire, K. E. (1974). Longitudinal data analysis, *Growth* **38**, 131–169.

Lachenbruch, P. A., and Mickey, M. R. (1968). Estimation of error rates in discriminant analysis, *Technometrics* **10**, 1–10.

Lam, Y. (1980). An empirical study of the unidimensionality of Raven's progressive matrices, master's thesis, Ontario Institute of Studies in Education, University of Toronto.

Lawley, D. N. (1940). The estimation of factor loadings by the method of maximum likelihood, *Proc. Roy. Statist. Soc. Ser. A* **60**, 64–82.

Lawley, D. N. (1956). Tests of significance for the latent roots of covariance and correlation matrices, *Biometrika* **43**, 128–136.

Lawley, D. N. (1963). On testing a set of correlation coefficients for equality, *Ann. Math. Statist.* **34**, 149–151.

Lawley, D. N., and Maxwell, A. E. (1970). *Factor Analysis as a Statistical Method*, Butterworth, London.

Lee, Y. S. (1972). Some results on the distribution of Wilks' likelihood-ratio criterion, *Biometrika* **59**, 649–664.

Leung, C. Y. (1977). Discriminant analysis and testing problems based on a general regression model, Ph.D. thesis, University of Toronto.

Lin, P. E. (1971). Estimation procedures for difference of means with missing data, *J. Amer. Statist. Assoc.* **66**, 634–636.

Lin, P. E. (1973). Procedures for testing the difference of means with incomplete data, *J. Amer. Statist. Assoc.* **68**, 699–703.

Lohnes, P. (1972). Statistical descriptors of school classes, *Amer. Ed. Res. J.* **9**, 547–556.

Lubischew, A. A. (1962). On the use of discriminant functions in taxonomy, *Biometrics* **18**, 455–477.

Mcdonald, L. L. (1971). On the estimation of missing data in the multivariate linear model, *Biometrics* **27**, 535–543.

McDonald, L. L. (1972). A multivariate extension of Tukey's one degree of freedom for non-additivity, *J. Amer. Statist. Assoc.* **67**, 674–675.

Machin, D. (1974). A multivariate study of the external measurements of the sperm whale, *J. Zool. London* **172**, 267–288.

Machin, D., and Kitchenham, B. L. (1971). A multivariate study of the external measurements of the humpback whale (*Megaptera novaeangliae*), *J. Zool. London* **165**, 415–421.

Malkovich, J. R., and Afifi, A. A. (1973). On tests for multivariate normality, *J. Amer. Statist. Soc.* **68**, 176–179.

Mardia, K. V. (1970). Measures of multivariate skewness and kurtosis with applications, *Biometrika* **57**, 519–530.

Mardia, K. V. (1971). The effect of non-normality on some multivariate tests and robustness to non-normality in the linear model, *Biometrika* **58**, 105–121.

Mardia, K. V. (1974). Applications of some measures of multivariate skewness and kurtosis in testing normality and robustness studies, *Sankhya Ser. B* **36**, 115–128.

Mardia, K. V. (1975). Assessment of multinormality and the robustness of Hotelling's T^2 test, *Appl. Statist.* **24**, 163–171.

Mathai, A. M., and Katiya, R. S. (1979). Exact percentage points for testing independence, *Biometrika* **66**, 353–356.

Mathai, A. M., and Saxena, R. K. (1973). *Generalized Hypergeometric Functions with Applications in Statistics and Physical Sciences*, Lecture Notes Series 348, Springer-Verlag, New York.

Maxwell, A. E., and Lawley, D. N. (1971). *Factor Analysis as a Statistical Method*, Elsevier, New York.

Mehta, J. S., and Gurland, J. (1969). Testing equality of means in the presence of correlation, *Biometrika* **56**, 119–126.

Mehta, J. S., and Gurland, J. (1973). A test for equality of means in the presence of correlation and missing data, *Biometrika* **60**, 211–213.

Milliken, G. A., and Graybill, F. A. (1970). Extensions of the general linear hypothesis, *J. Amer. Statist. Assoc.* **65**, 797–808.

Morrison, D. F. (1970). The optimal spacing of repeated measurements, *Biometrics* **26**, 281–290.

Morrison, D. F. (1973). A test for equality of means of correlated variates with missing data on one response, *Biometrika* **60**, 101–105.

Nagarsenkar, B. N. (1976). The distribution of the determinant of correlation matrix useful in principal component analysis, *Commun. Statist. B — Simulation Comput.* **5**(1), 1–13.

Neuhaus, J. O., and Wrigley, C. (1954). The quartimax method, an analytical approach to orthogonal simple structure, British J. Statist. Psych. **7**, 81–91.

Okamoto, M. (1963). An asymptotic expansion for the linear discriminant function, *Ann. Math. Statist.* **34**, 1286–1301, **39**, 1358.

Olkin, I., and Shrikhande, S. S. (1953). On a modified T^2 problem, *Ann. Math. Statist.* **25**, 808A.

Olkin, I., and Sylvan, M. (1977). Correlation analysis when some variances and covariances are known, in *Multivariate Analysis* (P. R. Krishnaiah, ed.), North-Holland, New York, Vol. 4, pp. 175–191.

Pearson, K. (1901). On lines and planes of closest fit to systems of points in space, *Philos. Mag. Ser. 6* **2**, 559–572.

Pearson, E. S., and Hartley, H. O. (1948). *Biometrika Tables for Statisticians*, Cambridge University Press, Cambridge, England, Vol. 1.

Perlman, M. D. (1974). On the monotonicity of the power function of tests based on traces of multivariate beta matrices, *J. Multivariate Anal.* **4**, 22–30.

Pillai, K. C. (1960). *Statistical Tables for Tests of Multivariate Hypotheses*, Statistical Center, University of Philippines, Manilla.

Pillai, K. C., and Gupta, A. K. (1969). On the exact distribution of Wilks' criterion, *Biometrika* **56**, 109–118.

Pillai, K. C. S., and Jayachandran, K. (1967). Power comparisons of tests of two multivariate hypotheses based on four criteria, *Biometrika* **54**, 195–210.

Poole, R. W. (1971). The use of factor analysis in modelling natural communities of plants and animals, *Illinois Natural History Survey Bull. Biol. Notes* **72**, 1–14.

Potthoff, R. F., and Roy, S. N. (1964). A generalized multivariate analysis of variance model useful especially for growth curve problems, *Biometrika* **51**, 313.

Ranek, L., Keiding, N., and Jensen, S. T. (1975). A morphometric study of normal human liver cell nucleii, *Acta. Path. Microbiol. Scand. Ser. A* **83**, 467–76.

Rao, C. R. (1948). Tests of significance in multivariate analysis, *Biometrika* **35**, 58–79.

Rao, C. R. (1949). On some problems arising out of discrimination with multiple characters, *Sankhya* **9**, 343–364.

Rao, C. R. (1952). *Advanced Statistical Methods in Biometric Research*, Wiley, New York.

Rao, C. R. (1955). Estimation and tests of significance in factor analysis, *Psychometrika* **20**, 93–111.

Rao, C. R. (1958). Some statistical methods for comparison of growth curves, *Biometrics* **14**, 1–17.

Rao, C. R. (1959). Some problems involving linear hypothesis in multivariate analysis, *Biometrika* **46**, 49–58.

Rao, C. R. (1964). The use and interpretation of principal component analysis in applied research, *Sankhya Ser. A* **26**, 329–358.

Rao, C. R. (1965). The theory of least squares when the parameters are stochastic and its application to the analysis of growth curves, *Biometrika* **52**, 447–458.

Rao, C. R. (1966). Covariance adjustments and related problems in multivariate analysis, in *Multivariate Analysis*, Vol. 2 (P. R. Krishnaiah, ed.), Academic, New York, pp. 321–328.

Rao, C. R. (1967). Least squares theory using an estimated dispersion matrix and its application to measurement of signals, *Proc. 5th Berkeley Symp. Math. Statist. Prob.* **1**, 355–372.

Rao, C. R. (1973). *Linear Statistical Inference and Its Applications*, Wiley, New York.

Rao, C. R., and Mitra, S. K. (1971). *Generalized Inverses of Matrices and Its Applications*, Wiley, New York.

Rao, C. R., and Slater, P. (1949). Multivariate analysis applied to differences between neurotic groups, *British J. Statist. Psych.* **2**, 17–29.

Ratkowski, D. A., and Martin, D. (1974). The use of multivariate analysis in identifying relationships among disorder and mineral element content in apples, *Austral. J. Agric. Res.* **25**, 783–790.

Roy, S. N. (1957). *Some Aspects of Multivariate Analysis*, Wiley, New York.

Russell, J. S., Moore, A. W., and Coaldrake, J. E. (1967). Relationships between subtropical semiarid forest of *Acacia Narpophylla*, micro-relief and chemical properties of gilgai soil, *Austral. J. Botany* **15**, 481–498.

Schall, J. J., and Pianka, E. R. (1978). Geographical trends in numbers of species, *Science* **201**, 679–686.

Schatzoff, M. (1966). Exact distribution of Wilks' likelihood ratio criterion, *Biometrika* **53**, 347–358.

Scheffé, H. (1943). On solution of the Behrens–Fisher problem based on the *t*-distribution, *Ann. Math. Statist.* **14**, 35–44.

Scheffé, H. (1953). A method for judging all contrasts in the analysis of variance. *Biometrika* **40**, 87–104.

Scheffé, H. (1959). *The Analysis of Variance*, Wiley, New York.

Sema, T., and Kano, N. (1969). Glycine in the spinal cord of cats with local tetanus rigidity, *Science* **164**, 571–572.

Sen, A. K., and Srivastava, M. S. (1973). On multivariate tests for detecting change in mean, *Sankhya Ser. A* **35**, 173–185.

Sen, A. K., and Srivastava, M. S. (1975a). On tests for detecting change in mean, *Ann. Statist.* **3**, 98–108.

Sen, A. K., and Srivastava, M. S. (1975b). On tests for detecting change in mean when variance is unknown, *Ann. Inst. Statist. Math. Tokyo* **27**, 479–486.

Sen, A. K., and Srivastava, M. S. (1975c). Some one-sided tests for change in level, *Technometrics* **17**, 61–64.

Sen, A., and Srivastava, M. S. (1982). On the power of some tests for change in the multivariate mean (unpublished).

Shapiro, S. S., and Wilk, M. B. (1965). An analysis of variance test for normality (complete samples), *Biometrika* **52**, 591–611.

Sheth, J. N. (1969). Using factor analysis to estimate parameters, *J. Amer. Statist. Assoc.* **64**, 808–822.

Singh, S. M., and Ferns, P. N. (1978). Accumulation of heavy metals in rainbow trout, *Salmo gairneri* (Richardson), maintained on a diet containing activated sewage sludge, *J. Fish Biol.* **13**, 277–286.

Sitgreaves, R. (1961). Some results on the distribution of the *W*-classification statistics, in *Studies in Item Analysis and Prediction* (H. Solomon, ed.), Stanford University Press, Stanford, Cal., pp. 241–251.

Small, N. J. H. (1978). Plotting squared radii, *Biometrika* **65**, 657–658.

Snedecor, G. W., and Cochran, W. G. (1980). *Statistical Methods*, 7th ed., Iowa State University Press, Ames.

Srivastava, J. N. (1964). On the monotonicity property of the three main tests for multivariate analysis of variance, *J. Roy. Statist. Soc. Ser. B* **26**, 77–81.

Srivastava, J. N., and Zaatar, N. K. (1972). On the mazimum likelihood classification rule for incomplete multivariate samples and its admissibility, *J. Multivariate Anal.* **2**. 115–126.

Srivastava, M. S. (1965). Some tests for the intraclass comparison model, *Ann. Math. Statist.* **36**, 1802–1806.

Srivastava, M. S. (1966). On a multivariate slippage problem I, *Ann. Inst. Statist. Math.* **18**, 299–305.

Srivastava, M. S. (1967). On fixed width confidence bounds for regression parameters and mean vectors, *J. Roy. Statist. Soc. Ser. B.* **29**, 132–140.

Srivastava, M. S. (1973a). Evaluation of misclassification errors, *Canad. J. Statist.* **1**, 35–50.

Srivastava, M. S. (1973b). A sequential approach to classification, *J. Multivariate Anal.* **3**, 173–183.

Srivastava, M. S. (1980a). Multivariate data with missing observations, Tech. Rep. No. 2, University of Toronto.

Srivastava, M. S. (1980b). Profile analysis of several groups (unpublished).

Srivastava, M. S. (1981). On tests for detecting change in the multivariate mean, in *Statistical Distributions in Scientific Work* (C. Taillie, G. P. Patil, and B. Baldessari, eds.), University of Toronto, Toronto, Vol. 5, p. 181.

Srivastava, M. S. (1982). Assessing multivariate normality using principal components (unpublished).

Srivastava, M. S., and Awan, H. M. (1982). On the distribution of Hotelling's T^2 and the distribution of linear and quadratic forms in sampling from a mixture of two multivariate normal populations, *Commun. Statist. Theory Methods* (to appear).

Srivastava, M. S., and Bhargava, R. P. (1979). On fixed width confidence region for the mean, *Metron* **37**, 163–174.

Srivastava, M. S., and Khatri, C. G. (1979). *An Introduction to Multivariate Statistics*, North Holland, New York.

Stankov, L. (1979). Hierarchical factoring based on image analysis and orthoblique rotations, *Multiv. Behav. Res.* **14**, 339–353.

Stein, C. (1945). A two-sample test for a linear hypothesis whose power is independent of the variance, *Ann. Math. Statist.* **16**, 243–258.

Thomson, G. H. (1951). *The Factorial Analysis of Human Ability*, University of London Press, London.

Thurstone, L. L. (1947). *Multiple Factor Analysis*, University of Chicago Press, Chicago, Ill.

Tiku, M. L. (1975). *Selected Tables in Mathematical Statistics*, Vol. 2, Inst. Math. Statist., Amer. Math. Soc., Providence, R.I.

Tucker, L. R. (1958). Determination of parameters of a functional relation by factor analysis, *Psychometrika* **23**, 19–23.

Tukey, J. (1949). One degree of freedom for nonadditivity, *Biometrics* **5**, 232–242.

Van Ness, J. W., and Simpson, C. (1976). On the effects of dimension in discriminant analysis, *Technometrics* **18**, 175–187.

Vasicek, O. (1976). A test for normality based on sample entropy, *J. Roy. Statist. Soc. Ser. B* **38**, 54–59.

Warburton, F. W. (1963). Analytical methods of factor rotation, *British J. Statist. Psych.* **16**(2), 165–174.

Watson, R. H. J., and Broadhurst, P. L. (1976). A factor analysis of body build in the rat, *Amer. J. Physical Anthrop.* **44**, 513–520.

Welch, B. L. (1937). The significance of the difference between two means when the population variances are unequal, *Biometrika* **29**, 350.

Welch, B. L. (1939). Note on discriminant function, *Biometrika* **31**, 218–220.

Welch, B. L. (1947). The generalization of "Students" problem when several different population variances are involved, *Biometrika* **34**, 28–35.

Wilding, G., Blumberg, B. S., and Vesell, E. S. (1977). Reduced warfarin binding of albumin variants, *Science* **195**, 991–994.

Wilks, S. S. (1946). Sample criteria for testing equality of means, equality of variances, and equality of covariances in a normal multivariate distribution, *Ann. Math. Statist.* **17**, 257–281.

Woodard, D. E. (1931). Healing time of fractures of the jaw in relation to delay before reduction, infection, syphillis, and blood calcium and phosphorous content, *J. Amer. Dental Assoc.* **18**, 419–442.

Woolson, R. F. (1980). Growth curve analysis of complete and incomplete longitudinal data, *Commun. Statist. Theory Methods* **A9**, 1491–1513.

Wu, L. S., Bargmann, R. E., and Powers, J. J. (1977). Factor analysis applied to wine descriptors, *J. Food Sci.* **42**, 944–952.

Yao, Y. (1965). An approximate degrees of freedom solution to the multivariate Behrens–Fisher problem, *Biometrika* **52**, 139–147.

Zerbe, G. O. (1979). Randomization analysis of the completely randomized design extended to growth and response curves, *J. Amer. Statist. Assoc.* **74**, 215–221.

Author Index

Subject Index